Differential Algebras in Topology

Research Notes in Mathematics

Volume 3

Differential Algebras in Topology

David Anick
Department of Mathematics
Massachusetts Institute of Technology
Cambridge, Massachusetts

CRC Press
Taylor & Francis Group
Boca Raton London New York

CRC Press is an imprint of the
Taylor & Francis Group, an **informa** business
AN A K PETERS BOOK

First published 1993 by A K Peters, Ltd.

Published 2018 by CRC Press
Taylor & Francis Group
6000 Broken Sound Parkway NW, Suite 300
Boca Raton, FL 33487-2742

First issued in paperback 2019

ISBN 13: 978-0-367-45003-8 (pbk)
ISBN 13: 978-1-56881-001-0 (hbk)

Visit the Taylor & Francis Web site at
http://www.taylorandfrancis.com

and the CRC Press Web site at
http://www.crcpress.com

Library of Congress Cataloging-in-Publication Data

Anick, David Jay.
Differential algebras in topology / David Anick.
p. cm.
Includes bibliographic references.
ISBN 1-56881-001-6
1. Algebraic topology. 2. Complexes. I. Title.
QA612.A76 1993
514'.2—dc20 93-12604
CIP

To Flora and George

Who always encouraged me

In my love of math

AUTHOR'S FOREWORD

For the gifts and the call of God are irrevocable.

Romans 11:29

I offer the ideas in this book as my parting gift to the mathematical community. Responding to what I think is an irrevocable call, I am leaving to become a medical doctor. As I write, my medical training is set to commence in less than a week. I want to share my reflections on the process of creating the book, and to acknowledge the assistance of others.

The ideas in this monograph have been brewing inside me on an intuitive level for several years. I didn't want them simply to be lost when I stopped doing mathematics. Motivated by the desire to leave with a minimum of loose ends, I undertook the project of writing it up.

When I started, I had no clue that what I envisioned as a sixty- to eighty-page paper would blossom into a three hundred page book! The turning point came at the end of the fall semester of 1990, when I finally grasped that the classical techniques I had been counting upon — slight generalizations of those of Cohen–Moore–Neisendorfer — might be inadequate for the task. Gradually my focus shifted. Instead of seeing the problem (of constructing D_k) as principally one of clever combinatorics and well-coordinated equations, I began to think in terms of larger questions and in terms of the development of wholly new methods. The result is what you see before you.

Every chapter of the book was written or substantially rewritten during the four-month period of May through August of 1991. This haste shows, I know, in the book's many rough edges — chapters of greatly unequal length, an inconsistent labeling system for results and formulas, large quantities of notation, minimal provision of background, and so on. For these weaknesses I ask the reader's forbearance and forgiveness. Nearly all of the results, including most of the concepts and definitions, were unknown to me just five months ago. I had to concentrate on working out the details, and the larger picture correspondingly suffered.

There is nothing like an absolute deadline to motivate someone. During this summer of writing I have been constantly aware that it was "now or never" for developing these ideas. I recalled frequently the image of Évariste Galois during the final hours of his life, frantically writing down as much as he could of his mathematical ideas so that the ideas would not die too. I have never before worked so intensely, nor with so single a focus.

In spite of the exhaustion and the sacrifices, the process has been satisfying. For a few months I have given myself over entirely to the pursuit of new knowledge, immersing myself in a totally abstract yet utterly beautiful portion of All That Is. I will never regret this experience, and I am grateful for the opportunity to have had it.

It has been said that in a field of wildflowers, an entire book could be written about any one of them. As I have dwelled in the world of the D_k's, I have simultaneously been aware that even this fascinating complex world represents but one flower in the vast field of mathematical truth. Still, I hope that my readers will agree the story of the D_k's is a story that should be told, and that others will be able to appreciate the beauty and intricacy of this particular flower.

There are many who helped to make this book a reality. Financial and institutional support came from the Massachusetts Institute of Technology, the University of Rochester, the Sloan Foundation, and the National Science Foundation. Technical typing was done by Joan Robinson, Margaret Beucler, Robert Becker, and Judy Romvos.

For appreciation I wish to single out Brayton Gray and Steven Halperin, without whose staunch encouragement I might well have quit the entire project at any number of points. Those with whom my mathematical discussions directly influenced the content of the book include Fred Cohen, Brayton Gray, Mike Hopkins, Haynes Miller, Joe Neisendorfer, Bill Richter, and Paul Selick. To these individuals, as well as to the many others who helped me to feel valued and successful during my eleven years as a professional mathematician, I extend my profound gratitude.

TABLE OF CONTENTS

ABSTRACT

We construct an infinite family $\{D_k\}_{k\geq 0}$ of spaces that generalize the odd-dimensional Moore space $D_0 = P^{2n+1}(p^r), p \geq 5$. Extending some work of Cohen, Moore, and Neisendorfer, we explore the homotopy-theoretic properties of these spaces and of several closely related spaces. In the process, we develop a variety of algebraic and geometric tools and techniques that may have wide applicability in unstable p-primary homotopy theory.

PREFACE

Welcome, dear reader, to ***Differential Algebras in Topology***! We hope you will find this monograph to be interesting, enlightening, and perhaps even occasionally entertaining. The preface contains fundamentals such as the purpose and level, as well as a section on "how to read this book".

Purpose

The ostensible purpose of the book is to construct a new family $\{D_k\}_{k\geq 0}$ of finite complexes and to explore their properties. In the course of the construction we discover a variety of new algebraic and topological techniques. The book is ultimately as much about these new tools and techniques as it is about the $\{D_k\}$ themselves.

Let $m \geq 3$, $r \geq 1$, and let p be a prime ≥ 5. In [CMN123], Cohen, Moore, and Neisendorfer analyzed in detail the Moore space $P^m(p^r)$. Their analysis included the construction of several related spaces, a proof that $P^m(p^r)$ has a homotopy exponent, and a complete factorization of $\Omega P^m(p^r)$ into indecomposable factors. The splitting of $\Omega P^m(p^r)$ when m is even is reasonably straightforward, but the case of m odd leads to some quite complicated algebra and topology.

Our spaces $D_k = D_k^n(p^r)$ are generalizations of the odd-dimensional Moore space. The space D_0 is exactly $P^{2n+1}(p^r)$. In its broad strokes our analysis of the D_k's follows the same path that Cohen-Moore-Neisendorfer laid out, but in the particulars many significant differences arise.

The Moore space is obviously an object worth studying, but why study D_k? First, the fact that other spaces besides the Moore space can function in the Cohen-Moore-Neisendorfer program helps us to identify the essential features which permit that program to go through. Second, if the ends justify the means in mathematics, some of our conclusions like the special space T_∞ (see Theorem 1.5) may justify the journey through the $\{D_k\}$ needed to get there. Third, certain homotopy operations associated with D_k (see Theorem 9.1 (vi)), which generalize iterated relative Samelson products, seem likely to be useful in p-primary homotopy theory. Fourth, the study of $\{D_k\}$ has led to numerous valuable spinoffs, which we shall list shortly.

The author's original reason for creating D_k, however, was none of these. The space D_1 first arose in considering the following problem. Given a finite 1-connected CW complex

X and a large prime p ($p \geq \dim(X)$ should be large enough), how can one write $\Omega X_{(p)}$ as a product of indecomposable H-space factors? For many spaces X, $\Sigma\Omega X$ is p-locally a bouquet of spheres and Moore spaces, and then $\Omega X_{(p)}$ is a product of factors belonging to a certain known list [A5]. The smallest space Y for which $\Sigma\Omega Y$ is not a bouquet of this type at a prime $p \geq \dim(Y)$ is the four-cell space defined by the cofibration

$$P^{4n}(p) \xrightarrow{[\iota,\,\pi\beta^1(\iota)]} P^{2n+1}(p) \to Y \quad (n \geq 1\ ,\ p \geq 4n+1)\ .$$

The attaching map is the mod p Whitehead product of the homotopy class of the identity, denoted ι, with the homotopy Bockstein of ι. An alternate description of Y is that Y is the $(4n+1)$–skeleton of $S^{2n+1}\{p\}$, $S^{2n+1}\{p\}$ being the homotopy–theoretic fiber of the degree p map on the $(2n+1)$–dimensional sphere.

The factor of ΩY that carries the first non-zero homotopy group, i.e. $\pi_{2n-1}(\Omega Y) = \mathbb{Z}_p$, is a space we denote by T_1. It is a retract of ΩD_1. A factorization of ΩY has the form

$$g:\ T_1 \times \Omega W \xrightarrow{\approx} \Omega Y\ ,$$

where W is a certain bouquet of Moore spaces of orders p and p^2. In order to construct the homotopy equivalence g, we construct maps from W to Y and from T_1 to ΩY. Mapping the Moore space summands of W of order p into Y involves iterated relative Samelson products only, but mapping the mod p^2 summands may require the homotopy operations associated with D_1 via a certain map $D_1 \xrightarrow{\alpha} Y$. The map from T_1 to ΩY is also mediated by D_1, since it is the composition $T_1 \longrightarrow \Omega D_1 \xrightarrow{\Omega\alpha} \Omega Y$.

Unfortunately, even this application of D_1 transcends what this book will cover. We will not even get to show that T_1 is a retract of ΩD_1. Nevertheless, we hope we have made the point. The spaces T_k associated with the D_k (see Chapter 14) occur frequently as retracts of ΩX for finite complexes X, so an understanding of the homotopy of T_k is essential for an understanding of the homotopy of X. For example, it can be proved in this way that the space Y above has a homotopy exponent, and that this exponent is either p^3 or p^4.

Spinoffs

Spinoffs are useful technologies or ideas discovered serendipitously during the pursuit of a seemingly unrelated goal. The goal of constructing the $\{D_k\}$ has had more than its share of spinoffs. We list the principal ones now.

Chapters 2–5 are intended as "service" chapters in which we analyze in detail certain specific differential graded Lie algebras (dgL's) that we later need. Even these chapters, the book's "narrowest," contain some potentially useful results. The differential graded algebra (dga) we call $\mathcal{U}M^k$ has been described as a non-commutative version of the Witt vectors, and the one we call $\mathcal{U}\overline{L}^k$ exhibits a pattern in its homology torsion that is at least aesthetically appealing. Chapter 5 contains some general results about the homology of dgL's having linear differentials.

Theorem 6.4 provides a generalization of sorts of the mod p^t Hurewicz theorem, expressed in terms of Adams-Hilton models. Since in this book we encounter over a dozen settings where we can apply the theorem, we suspect that the result may have wide applicability. We develop as well the theory of thin products of (any) three spaces, finding a Hilton-Milnor-like splitting of an associated fiber. We define and examine the ring of stable homotopy operations $\mathcal{O}_A$ under a space A, and we describe a functorial factorization of the Hurewicz homomorphism. Stable homotopy operations under A offer a plausible approach to the homotopy theory of finite complexes at large primes.

Our single most powerful spinoff is the concept of a coherent sequence of algebras (csa) or of Lie algebras (csL). Chapters 7 and 11, the two longest chapters in the book, are devoted entirely to this topic. A csL (resp. csa) is a sequence of dgL's (resp. dga's) that satisfies certain axioms. For $p \geq 5$ and for X 1-connected, $\{\pi_*(\Omega X; \mathbb{Z}_{p^t})\}_{t\geq 1}$ is a csL and $\{H_*(\Omega X; \mathbb{Z}_{p^t})\}_{t\geq 1}$ is a csa. Csa's and csL's are the ideal tool for handling objects that have p^t-torsion for many different t, whenever one needs more precision than one can get from the Bockstein spectral sequence.

Unifying Theme

This monograph covers a lot of ground. One theme that recurs throughout and ties it all together is the application of differential graded algebras or differential graded Lie algebras to unstable odd-primary homotopy theory. Specifically, we utilize constantly the Adams-Hilton model (a dga), the ring of stable operations $\mathcal{O}_A$ under A (another dga), various csL's, and various csa's.

Despite the general-sounding title, *Differential Algebras in Topology* does not attempt anything like a comprehensive survey of the many diverse differential algebras that occur in topology. After all, any term of any geometrically motivated multiplicative spectral sequence could be called a differential algebra in topology. We focus much more narrowly on those dga's and dgL's that are relevant to our generalization of the Cohen-Moore-Neisendorfer program in unstable p-primary homotopy theory, for $p \geq 5$.

Joseph Neisendorfer once succinctly summarized the mathematical contents of this book with the words, "Homology doesn't lie [about unstable homotopy]." For the spaces $\{D_k\}$ and their related spaces, homology calculations provide a far more accurate guide to the odd–primary homotopy than anyone could reasonably have expected. We may view the $\{D_k\}$ as prototypes of (unstable) p–torsion CW complexes, in the sense that their attaching maps exhibit certain features that would be useful to understand in general. By studying the $\{D_k\}$ we are implicitly exploring the extent to which "homology tells the truth" about homotopy for many finite spaces.

Level of Sophistication

This monograph is written at the research level, intended for practicing topologists or advanced graduate students. Some familiarity with dga's and dgL's would be helpful, though it is probably not absolutely essential. We include fly-by reviews of p-primary homotopy theory (Chapter 6) and of the Cohen-Moore-Neisendorfer program (Chapter 1), but the reader who has previously spent some time on these topics will have a definite advantage.

The level of sophistication is high. Many proofs are greatly abbreviated or omitted entirely as exercises. The reader is frequently expected to be able to perform on his/her own (or simply to accept) "routine" calculations involving the Kunneth formula, Poincaré series, or the Serre spectral sequence. Our guiding philosophy has been to illustrate a new technique (e.g. techniques involving the csL axioms) only once or twice, figuring that further elaboration during subsequent proofs would provide no new insights.

In order to keep down the book's total length and in order to keep the main points more clearly in focus, we frequently give not so much a rigorous proof but rather an outline or a report or summary of a proof. We provide what we found to be the key ideas or insights in the proof, leaving it up to the interested reader to reconstruct the details. The reader is going to have to think creatively and not just blithely follow along.

In a sense we have adopted the approach that guides scientific literature other than in mathematics: "here is what we did, and here is how the experiment turned out." The assumption is that other scientists who repeat the experiment will obtain the same results. Mathematicians who reject this approach and believe that some of our curtailed proofs should be written out in full are welcome to do so. The author would feel honored by this kind of attention being paid to his work, and the larger mathematical community may well concur that the solutions to certain of our "exercises" are worthy of publication as separate papers.

We frequently use the word "straightforward" when omitting material. The best operational definition of this word is "the first thing the author tried worked, and it worked in exactly the way he expected." Since luck and familiarity obviously played a role in this experience, others may not find all such proofs to be genuinely straightforward. We apologize for any instances where we have made poor judgments about what to exclude or include, and we again invite readers to fill in any true gaps.

How to Read this Book

How to read this book depends upon what you, the reader, hope to get out of it. It is written in a linear logic mathematically correct fashion, so that one could hypothetically start at the beginning and read it straight through until the end. We recommend, however, that you read the Preface and Introduction (Chapter 1), and then decide what you want.

If the reader's intent is to look up a result or to learn about a particular spinoff (e.g. csL's), you can go directly to the relevant part (e.g Chapter 7). Although in the writing we have given ourselves permission to refer at any moment to any previous material, in many cases such references can be skipped without loss of continuity (e.g. references to $\mathcal{U}M^k$ and even to $\pi_*(\Omega X;\mathbb{Z}_{p^t})$ in Chapter 7 can be skipped).

Suppose the reader wants to know what we did but you are not so concerned about the proofs, e.g., because you intend to use it to do something else. We recommend reading Chapter 1 and then paying attention to the definitions, notations, propositions, and theorems (i.e. skip lemmas and proofs). Definitely read anything that looks like prose (rather than mathematics) or is labeled "Discussion." The list of global notations is designed to help someone to be able to skim the book in that manner.

If the reader really wants to understand why the $\{D_k\}$ exist, we recommend an initial skimming as above to get an overview of where we are headed. Chapter 1 and the introductory paragraphs of each chapter or unit, along with the discussions (7.7, preceding 9.1, 11.A1, 11.E5), are intended to assist with this overview. Chapters 14 and 15 may also be somewhat understandable and would help to remind you of why you are interested.

Once you have a feel for the overall plan, we recommend reading in order: Chapter 2, 3 (omit the material on $\widetilde{M}^k$), 4, the first two units of 6 (omit thin products), the first two units of 7 (omit the examples), 8, 9, 10, the third unit of 13 (proof of 9.1(xii) only), 14, 15. The truly dedicated reader should then review Discussion 11.A1, and then read in sequence Chapter 5 and the other material that was skipped on the first pass.

Good luck!

SUMMARY OF GLOBAL NOTATIONS

For each notation, we give a brief description and/or the nearest formula or result number to the spot where the notation is first defined. "Shortly after Formula (6D)" is written as "> (6D)". If a notation is not on this list it is probably a local notation whose explanation can be found within the material discussing that particular result.

$(-)^a$	(6D)	adjoint of a map out of a suspension
$(-)^{\bar{a}}$	> (6D)	coadjoint of a map into a loop space
a_s	<3.2	For $0 \leq s \leq k$, one of the generators of M^k and of $\mathcal{U}M^k$. Corresponds to $x_{p^s} \in L^k$ and to $y_{p^s} \in \overline{L}^k$
$\hat{a}_k$	(10J)	
$\underline{a}_f(-)$	<6.3	The element $\mathcal{A}(f-)(v_q) \in \mathcal{A}(-)$
AH	<6.2	Adams-Hilton
ad	<(2K), (5B)	
ad_H	<6.11	
$\mathcal{A}(-)$	6.2	Adams-Hilton model of space or map
b	<3.2	A generator of M^k and of $\mathcal{U}M^k$
b^{p^s}	<4.7	For $0 \leq s \leq k$, either the p^s power of b in $\mathcal{U}M^k$, or the descendent of b^{p^s} that survives to $E^{r+s}(\mathcal{U}M^k)$
b_s	<5.9	Another notation for b^{p^s}
B^k	3.7	The true AH model for D_k (equals $\mathcal{U}\widetilde{M}^k$)
$\mathcal{B}_{[t]}$	>(11DI)	t^{th} properness obstruction group
BSS	>(2D)	Bockstein spectral sequence
c_s	4.3	A cycle and generator of $\mathcal{U}M^k$
C	< (8A)	Reduced cone (on a pointed space)
$C(n)$	(13AD)	Homotopy-theoretic fiber of the double double suspension (p-local)
$C_*(-)$	< (6M)	0th Eilenberg subcomplex of the singular chains on $-$
$C(\lambda^q, t)$	> (2K)	

$C(\mathcal{P}(\lambda))$	> (2K)	
CC	< (6T)	Cyclic cotensor construction
csa	7.1	coherent sequence of algebras
csL	7.1	coherent sequence of Lie algebras
csm	7.8	coherent sequence of modules
CMN	1.2	
d		Generic name for the differential in any chain complex or dga or dgL
$d_{[t]}$		Generic name for the differential on the tth stage of a csa or csL
D_k	9.1	The kth space in the family $\{D_k\}$ of generalizations of the odd-dimensional Moore space
D_{kk}	(12A)	
D_{kkk}	(12S)	
dga	(2C)	differential graded algebra
dgH	> (2C)	differential graded Hopf algebra
dgL	(2D)	differential graded Lie algebra
Dcsa	7.11	
$\bar{e}$		Generic name for inclusions among dgL's or dga's for a fixed k, e.g., $\bar{e} = L^k \longrightarrow \mathcal{U}M^k$
e^m		An m-cell in a CW complex
$e^{(i)}$	< 5.11	For $i = 1, 2, 3$, the ith inclusion of M^k into $M^k \amalg M^k$ or $M^k \amalg M^k \amalg M^k$
$\hat{e}^{(i)}$	12.3	For $i = 1, 2, 3$, the ith inclusion of D_k into $D_k \vee D_k$ or $D_k \vee D_k \vee D_k$
$e_j(-)$	(11BD)	
E_k	(9A)	homotopy-theoretic fiber of ρ_k
$E^t(\ )$	< 2.1	t^{th} term of homology BSS, mod p
$_\pi E^t(\ )$		t^{th} term of homotopy BSS, mod p
$\mathcal{E}$	11.D4	Universal csL on a chain complex
f_k	< 13.1	
f'_k	13.3	
f^k	(3B)	
F	< 8.2	
F_k	(9A)	Homotopy-theoretic fiber of $\tilde{\rho}\rho_k$
$\mathfrak{f}$	8.1	
$\mathcal{F}$	7.2(b)	
$\mathcal{F}^0$	11.B9	
fle	< (7BF)	formal linear expression
g		Frequently, a shorthand for g_{kl}
g_{kl}	5.12	Chain map from $\mathcal{U}M^k_+ \otimes \mathcal{U}M^l_+$ to N^{kl}
g^k	3.7	
$\mathfrak{g}$	8.1	
gcd	<4.2	Greatest common divisor of a set of integers
$\mathcal{G}(-,-,-)$	(6MM)	Homotopy-theoretic fiber of the map $\tilde{\lambda}$ from the wedge to the thin product
h	< 6.1	Integral or mod p^t Hurewicz homomorphism
h^k	3.8	
$[-,-]_H$	(5C)	Hopf commutator

$H_*(-)$		Homology of a space or chain complex; if coefficient group is unspecified, the default is $\mathbb{Z}_{(p)}$
$\overline{H}_*(-)$		Reduced homology (over $\mathbb{Z}_{(p)}$)
$H^*(-)$		Cohomology (coefficients in $\mathbb{Z}_{(p)}$)
H_M		Hilbert series of graded module M
HAK	(2F), <(5E)	Hopf algebra kernel
$\mathcal{H}$	8.6	
$\mathcal{H}_0$	8.6	
$\mathcal{H}_1$	8.6	
$\bar{\imath}$		Generic name for the inclusion of a dgL or dga at stage $k-1$ into the corresponding dgL or dga at stage k, e.g. $\bar{\imath}: \overline{L}^{k-1} \longrightarrow \overline{L}^k$
$\hat{\imath}$		Inclusion map from a space at stage $k-1$ to the corresponding space at stage k, e.g. $\hat{\imath}: D_{k-1} \longrightarrow D_k$, $\hat{\imath}: E_{k-1} \longrightarrow E_k$, etc.
I_s	(2N)	$\{jp^s \mid j \geq 2\}$
im$(-)$		Image of a homomorphism
$J(-,-,-)$	(6Z)	
J^k	(3N)	
J^k_s	4.6	
k		Non-negative integer; one of the four basic parameters for our family of spaces; the inductive variable in the grand inductive scheme
K_g	<8.2	homotopy-theoretic fiber of the map g
K^k	3.8(d)	
ker$(-)$		Kernel of a homomorphism
L^k	(2O)	
$\widetilde{L}^k$	3.7(d)	
$\overline{L}^k$	(2P)	
$\overline{L}^k_s$	<2.7	
$\mathcal{L}(-)$	< (2B)	Free Lie algebra on a positively graded $\mathbb{Z}_{(p)}$-module or chain complex
$\mathbb{L}\langle-\rangle$	>(2A)	Free Lie algebra generated by a graded set
M^k	< 3.2	
$\widetilde{M}^k$	3.7	
$\mathcal{M}_A$	(8L)	Category of map pairs under the space A
n		Positive integer; one of the four basic parameters for our family of spaces; represents a dimension or connectivity
N^k	3.8	
N^{kl}	(5R)	
N^{klm}	5.15	
$\mathcal{O}$	(5S)	
$\mathcal{O}_A[t]$	8.5	Ring of stable homotopy operations under A
$\mathcal{O}_A$	8.5	Another notation for $\mathcal{O}_A[t]$, when t is unambiguous
$\mathcal{O}_{(f)}$	< 8.10	The functor $\mathcal{O}_{(\ )}$ applied to a map f

p		A prime; one of the four basic parameters for our family of spaces. Where N^{kl} or N^{klm} occur, assume $p \geq 3$. For geometric results, assume $p \geq 5$.
p^t		t^{th} power of p. Also, the homomorphism or map given by multiplication by p^t
$P(-,-,-)$	> (6S)	
$P^m(p^t)$	<(6A)	Moore space of dimension m and order p^t
$\mathcal{P}(\lambda)$	> (2K)	An infinite sequence of formal power series in λ
$\underline{q}_{[t]}$	(7B)	
Q^t	(11DQ)	
$\mathbb{Q}$		Field of rational numbers
quism		A dga homomorphism that induces an isomorphism on homology (also: quasi-isomorphism)
r		Positive integer; one of the four basic parameters for our family of spaces; denotes a torsion order
R^k	< (2O)	
R^k_s	4.9	The sub-dga of $\mathcal{U}M^k$ generated by $\{b^{p^s}, c_s, \ldots, c_k\}$. For $t > k$ R^k_t is $\mathbb{Z}_{(p)}$.
$\mathcal{R}(-,-,-)$	6.8	Thin product of three spaces
s		Frequently, an index in the range $0 \leq s \leq k$
sM		Suspension of the graded module M
$s^{-1}M$		Desuspension of the graded module M
S^m		m-sphere
$S^m\{p^t\}$		Homotopy-theoretic fiber of $S^m \xrightarrow{p^t} S^m$
S^k_s	< 5.9	
t		Frequently, used to denote a torsion order in the range $1 \leq t \leq r+k$, or to index the stages of a csa or csL.
t		Used as a dummy variable in Hilbert series
T	<(2B)	Tensor algebra on a positively graded $\mathbb{Z}_{(p)}$-module or chain complex
T_k	14.2	
$u^{(m)}_{q-1}$	<6.2	$(q-1)$-dimensional generator of $\mathcal{A}(P^{q+1}(p^m))$
u_{q-1}	< 6.2	Same as $u^{(m)}_{q-1}$, when m is unambiguous
U	6.12	
U^k_0	< (14M)	
$\mathcal{U}$		Universal enveloping algebra of a Lie algebra (resp. dgL). It is a primitively generated Hopf algebra (resp. dgH).
$v^{(m)}_q$	< (6F)	q-dimensional generator of $\mathcal{A}(P^{q+1}(p^m))$
v_q		Same as $v^{(m)}_q$ when m is unambiguous; also, the generator of $\mathcal{A}(S^{q+1})$
$\hat{v}_q$	> (8J)	Canonical homotopy class
$\hat{v}$		Same as $\hat{v}_q$, when q is unambiguous
V_k	(13CD)	
$V^{(m)}$	(5M)	
V^{kl}	> (5R)	$\mathcal{U}M^k_+ \otimes \mathcal{U}M^l_+$ as a differential $\mathbb{Z}_{(p)}$-module

VG	> (7BI)	Virtual generator
VR	> (7BG)	Virtual relator
VS	> (7BJ)	Virtual syzygy
w_i	< 8.1	Canonical map
w_s^0	(7EC)	
w_s^1	(7EI)	
w_s^{12}	(7DK)	(See also (7EJ))
w_s^{21}	(7DM)	
W_k	13.19	
$\widehat{W}_k$	13.1	
$W^+(V^{(\geq m)})$	5.6	
$W^-(V^{(\geq m)})$	5.6	
$\widetilde{W}(V^{(\geq m)})$	(5P)	
$\widetilde{W}^{(t)}$	(11DD)	
$\mathcal{W}$	< 8.2	
$\mathcal{W}_s^t$	< 10.2	
x_j	(2O)	For $j \geq 1$, a generator of L^k and of $\mathcal{U}L^k$
x_s^1	(7EI)	
x_s^{12}	(7DK)	(See also (7EJ))
x_s^{21}	(7DM)	
$\widetilde{X}^{(t)}$	(11DK)	
y_j	> (2P)	For $j \geq 2$, a generator of $\overline{L}^k$ and of $\mathcal{U}\overline{L}^k$ that embeds in L^k as x_j
y_s^{12}	(7DL)	
y_s^{21}	(7DN)	
y_s^{13}	(7EG)	
$\tilde{y}_{p^k}$	9.1(xii)	
$\hat{y}^*_{p^k}$	(10F)	
$\hat{y}_{p^{k+1}}$	(9O)	
z_j	> (2P)	For $j \geq 2$, a generator of $\overline{L}^k$ and of $\mathcal{U}\overline{L}^k$; for $j \geq 3$, z_j embeds in L^k as $[x_1, x_{j-1}]$
z'_j	(2R)	For $j \geq 2$, a different and in many ways more useful choice of generator for $\overline{L}^k$ and $\mathcal{U}\overline{L}^k$.
z_s^{12}	(7DL)	
z_s^{21}	(7DN)	
z_s^{13}	(7EG)	
$\mathbb{Z}$		Ring or set of integers
$\mathbb{Z}_+$		Set of positive integers
$\mathbb{Z}_{(p)}$		Ring of integers localized at p
$\mathbb{Z}_{p^t}$		Ring or group of integers modulo p^t

$\alpha_k(-)$	(2L)	
$\alpha(-)$		Same as $\alpha_k(-)$, when k is unambiguous
α_t^k	12.6	t^{th} stage of a certain csL/csm homomorphism
β^t	< 2.1	t^{th} differential of homology or cohomology BSS
${}_\pi\beta^t$		t^{th} differential of homotopy BSS
β_t^k	(12E)	t^{th} stage of a certain csL/csm homomorphism

γ^k_t	(9J)	t^{th} stage of a certain csL/csm homomorphism
Γ	(11DI)	
$\Gamma(x)$	14.4	Divided power algebra
Γ^k_s	$<$ 5.9	
δ_k	(13CE)	Retract inclusion of V_k into ΩE_k
δ^t	7.2(b)	
${}_\pi\delta^t$	>(6E)	p^t-Bockstein differential on $\pi_*(-;\mathbb{Z}_{p^t})$
Δ_X		Diagonal map $X \longrightarrow X \times X$
Δ		Same as Δ_X; also, the diagonal on a Hopf algebra
$\epsilon(v,u)$	<4.1	
ε	<(2A)	The augmentation for a connected graded algebra
ε_k	(13BK)	
ζ^k	4.10	
ζ^s_t	7.1	One of the operations in a csL. On $\pi_*(-;\mathbb{Z}_{p^s})$, ζ^s_t denotes $(\rho^s_t)^{\#}$.
θ_m	$<$ 4.1	
θ	$<$ 4.1	Same as θ_m; must determine m from the context
θ_X	<6.2	A quism, $\theta_X : \mathcal{A}(X) \longrightarrow C_*(\Omega X)$
θ^k	9.1(iv)	A certain dgH quism
θ^t_s	7.1	One of the operations in a csL. On $\pi_*(-;\mathbb{Z}_{p^t})$, θ^t_s denotes $(\omega^t_s)^{\#}$.
κ_k	(9C)	
λ		Used as a dummy variable for formal power series
$\lambda(-,-,-)$	(6U)	
λ		Frequently, a shorthand for $\lambda(-,-,-)$
$\tilde{\lambda}(-,-,-)$	$>$ 6.8	
$\tilde{\lambda}$		Frequently, a shorthand for $\tilde{\lambda}(-,-,-)$
λ_k	(9A)	
$\Lambda(x)$		Exterior algebra on x
μ		Generic name for the multiplication on an H-space or on a Hopf algebra
μ_k	(9C)	(Except in Chapter 5)
ν_k	14.2	
ξ_j	(11BK)	
Ξ	(11DI)	
$\Xi(q,t)$	2.1	An abelian dgL
$\pi_q(-)$		q^{th} integral homotopy group
$\pi_q(-;\mathbb{Z}_{p^t})$	<(6A)	q^{th} mod p^t homotopy group
$\bar{\pi}_i$		Projection of a product (resp. wedge) onto the ith factor (resp . summand)
$\bar{\pi}_{21}$	(5G)	(See also the definition of "principal", preceding 6.10)
$\tilde{\pi}_j(X;\mathbb{Z}_{p^t})$	$<$ (8K)	Definition is $\pi_j(\widetilde{X};\mathbb{Z}_{p^t})$, for universal covering space $\widetilde{X}$ of X.
$\hat{\pi}_j$	12.5, 12.7	

$\tilde{\rho}$	(1L)	$\tilde{\rho}: S^{2n+1}\{p^r\} \longrightarrow S^{2n+1}$
ρ_k	(9A)	(Except in Chapter 3)
ρ^t	(6A)	
ρ_t^s	(6C)	
σ_j	(5I)	(Except in Chapter 4)
$\tilde{\sigma}_j$	(5K)	See also 11.E5 and (11EN)
Σ		Suspension functor, for spaces
Σ^2	(13AD)	
τ	(2E)	$\tau(x \otimes y) = (-1)^{\lvert x\rvert\lvert y\rvert} y \otimes x$
τ_j	(5H)	
$\tilde{\tau}_j$	(5K)	See also 11.E5 and (11EN) and (for$\tilde{\tau}_1$) 13.25
ϕ_k	(14A)	(Except in Chapter 5)
ϕ'_k	(14C)	
Φ	$>$ 8.8	
χ	(5A)	Anti-homomorphism for a Hopf algebra
$\hat{\chi}_k$	13.1	
χ_k	13.9	(Except in Chapter 8)
ψ_X	(6II)	
ψ_t^k	9.1(vi)	Likewise, for $\bar{\psi}_t^k$, Ψ_t^k, $\overline{\Psi}_t^k$
$\psi_t^{k_1k_2}$	12.1	Likewise, for $\bar{\psi}_t^{k_1k_2}$, $\Psi_t^{k_1k_2}$, $\overline{\Psi}_t^{k_1k_2}$
$\psi_t^{k_1k_2k_3}$	12.2	Likewise, for $\overline{\psi}_t^{k_1k_2k_3}$, $\Psi_t^{k_1k_2k_3}$, $\overline{\Psi}_t^{k_1k_2k_3}$
ω^s	(6A)	
ω_s^t	(6C)	
Ω		Loop space functor
$\underline{\Omega}$		Cobar construction
∇	5.13	Folding homomorphism or map
$\nabla^{(ij)}$	(7EH)	
∇_k	(13AI)	
$\nabla_k^{(ij)}$	(13AT)	
$\rtimes$		Half-smash: $X \rtimes Y = (X \times Y)/(x_0 \times Y)$
$\simeq$		Homotopic as maps; also, quism of dga's
$\approx$		Homotopy equivalence of spaces; also, isomorphism of modules
$\amalg$		Coproduct (= categorical sum) of associative algebras or of Lie algebras
$\langle - \rangle$		Two-sided ideal generated by $-$
$(-\vert-)$		Binomial coefficient, $(m\vert q) = m!/q!(m-q)!$
$\langle -\vert- \rangle_k$	(2M)	Altered binomial coefficient
$\langle -\vert- \rangle$	(2M)	Same as $\langle -\vert- \rangle_k$, when k is unambiguous
$[-,-]$	(2A), (6E)	Relative or absolute, Samelson or Whitehead product for spaces; (graded) commutator for algebras

CONVENTION: All spaces and homotopies are assumed to be pointed, and spaces are assumed to have the homotopy type of CW complexes.

CHAPTER 1

Introduction and Mathematical Overview

We offer in Chapter 1 an overview of what we plan to accomplish in this book. Since we are generalizing the work of Cohen-Moore-Neisendorfer [CMN123], we begin with a review of their program. We consider the ways in which our program diverges from theirs. Lastly we provide a chapter by chapter summary of the book, emphasizing the highlights but giving no proofs.

Review of the Cohen-Moore-Neisendorfer Theory

The single best-known theorem in all of [CMN123] is undoubtedly [CMN2, Corollary 1.3], which we now recall.

Theorem 1.1. [CMN2] *Let p be a prime ≥ 5 and let $n \geq 2$. Then p^n annihilates the p-torsion component of S^{2n+1}, i.e.,*

(1A) $$(p^n)\pi_{>2n+1}(S^{2n+1}_{(p)}) \;=\; 0\,.$$

(Formula (1A) is also true when $p = 3$ or when $n = 1$, but [CMN123] does not prove this; see instead [N4] and [S1].)

Theorem 1.1 is well-known because it is easy to state and to remember, because it states a profound and beautiful truth, and because of its obvious importance to unstable homotopy theory.

Nevertheless, Cohen-Moore-Neisendorfer did not start out by trying to prove Theorem 1.1. They were exploring the homotopy theory of Moore spaces, and Theorem 1.1 came as a serendipitous spinoff. Since this book is in a sense a generalization of the Cohen-Moore-Neisendorfer program, let us review their program now.

For $m \geq 2$ let $P^m(p^r)$ denote the m-dimensional Moore space of order p^r, i.e., $P^m(p^r)$ is the cofiber of the Brouwer degree p^r self-map on S^{m-1}.

In $\pi_*(P^{m+1}(p^r))$, $m \geq 2$, $p \geq 5$, $r \geq 1$, there are many elements of order p^r. To find a denumerable infinity of p^r-torsion elements, Cohen-Moore-Neisendorfer started with the generators $\hat{b} \in \pi_m(\Omega P^{m+1}(p^r); \mathbb{Z}_p)$ and $\hat{a} = {}_\pi\beta^r(\hat{b})$, where $\hat{b}$ is the adjoint of the homotopy class of the Moore space inclusion $\omega_1^r : P^{m+1}(p) \longrightarrow P^{m+1}(p^r)$, and ${}_\pi\beta^t$ denotes the tth

Bockstein differential in the mod p homotopy Bockstein spectral sequence (henceforth *BSS*). (See [N1] for a description of the homotopy BSS.) When $p \geq 5$ the Samelson product endows $\pi_*(\Omega P^{m+1}(p^r); \mathbb{Z}_p)$ with the structure of a graded Lie algebra over $\mathbb{Z}_p$, so there is a unique Lie algebra homomorphism

$$f : \mathbb{L}\langle b, a\rangle \longrightarrow \pi_*(\Omega P^{m+1}(p^r); \mathbb{Z}_p) \tag{1B}$$

determined by the two formulas $f(b) = \hat{b}$ and $f(a) = \hat{a}$. Here $\mathbb{L}\langle b, a\rangle$ denotes the free graded Lie $\mathbb{Z}_p$-algebra generated by the two elements b and a of dimensions m and $m-1$ respectively.

The Bockstein differential ${}_\pi\beta^r$ satisfies the product rule for Samelson products, namely

$$ {}_\pi\beta^r[\hat{x}, \hat{y}] = [{}_\pi\beta^r(\hat{x}), \hat{y}] + (-1)^{|\hat{x}|}[\hat{x}, {}_\pi\beta^r(\hat{y})] \tag{1C}$$

whenever ${}_\pi\beta^r(\hat{x})$ and ${}_\pi\beta^r(\hat{y})$ are both defined. Suppose we impose a differential d on $\mathbb{L}\langle b, a\rangle$ that satisfies $d(b) = a$ and $d(a) = 0$ and the product rule. Then the homomorphism f of (1B) has the property that all of $\mathrm{im}(f)$ survives to the rth term ${}_\pi E^r(\Omega P^{m+1}(p^r))$ of the homotopy BSS for $\Omega P^{m+1}(p^r)$, and f satisfies

$$ {}_\pi\beta^r f(x) = f d(x) \tag{1D}$$

for any $x \in \mathbb{L}\langle b, a\rangle$. (To prove (1D) use (1C) and induct on bracket length.)

Now suppose y and x are non-zero elements of $\mathbb{L}\langle b, a\rangle$ that satisfy $d(y) = x$. Then $f(y)$ and $f(x)$ survive to ${}_\pi E^r(\Omega P^{m+1}(p^r))$, where they satisfy ${}_\pi\beta^r(f(y)) = f(x)$. As a result, if $f(y)$ and $f(x)$ are non-zero, they describe a summand of $\mathbb{Z}_{p^r}$ in $\pi_{|x|}(\Omega P^{m+1}(p^r))$.

So why are $f(y)$ and $f(x)$ non-zero? Cohen-Moore-Neisendorfer used the Hurewicz homomorphism

$$h^t : {}_\pi E^t(\Omega P^{m+1}(p^r)) \longrightarrow E^t(\Omega P^{m+1}(p^r))$$

to show this, where $\{E^t(X), \beta^t\}_{t\geq 1}$ denotes the mod p homology BSS for a space X. The Hurewicz homomorphism has the properties that $(h^t)({}_\pi\beta^t) = \beta^t h^t$ and

$$h^t[\hat{x}, \hat{y}] = [h^t(\hat{x}), h^t(\hat{y})] ,$$

i.e., it commutes with Bockstein differentials and converts Samelson products to (graded) Pontrjagin ring commutators. As a result, when $E^r(\Omega P^{m+1}(p^r))$ is identified via the Bott-Samelson theorem with the free associative $\mathbb{Z}_p$-algebra $\mathbb{Z}_p\langle b, a\rangle$ on two generators of dimensions m and $m-1$, the composite

$$h^r f : \mathbb{L}\langle b, a\rangle \longrightarrow \mathbb{Z}_p\langle b, a\rangle$$

coincides with the embedding of $\mathbb{L}\langle b, a\rangle$ into its universal enveloping algebra. Since this embedding is one-to-one, so is f.

At this point, Cohen-Moore-Neisendorder made a crucial observation. While "most" of $\mathbb{L}\langle b, a\rangle$ is spanned by pairs of elements (y, x) that satisfy $d(y) = x$, such pairs do not cover all of $\mathbb{L}\langle b, a\rangle$. To say the same thing another way, $H_*(\mathbb{L}\langle b, a\rangle, d) \neq 0$. For simplicity we now concentrate on the case of an odd-dimensional Moore space, i.e. $m+1 = 2n+1$, $n \geq 1$. The elements $\tau_j(b) = \mathrm{ad}^{p^j-1}(b)(a)$, $j \geq 1$, are in the kernel but not in the image of d. The f-images $\{f(\tau_j(b))\}_{j\geq 1}$ must lie in the kernel of ${}_\pi\beta^r$, and it turns out that except when $n = r = j = 1$ they do not lie in the image of ${}_\pi\beta^r$. Consequently $f(\tau_j(b))$ survives to ${}_\pi E^{r+1}(\Omega P^{2n+1}(p^r))$ and represents an element of torsion order at least p^{r+1} in $\pi_{2np^j-2}(\Omega P^{2n+1}(p^r))$. However, these elements are not visible in the $(r+1)^{st}$ term of the homology BSS, because $h^r f(\tau_j(b))$ does lie in the image of β^r (it is $\beta^r(b^{p^j})$).

The original motivation behind the Cohen-Moore-Neisendorfer program was to understand this "hidden" higher torsion. To flush the "hidden" higher torsion out of hiding, they cleverly examined the fiber of the pinch map $\hat{\rho}: P^{2n+1}(p^r) \longrightarrow S^{2n+1}$ which we shall call F_0. I.e.,

$$F_0 \xrightarrow{\lambda_0} P^{2n+1}(p^r) \xrightarrow{\hat{\rho}} S^{2n+1}$$

is a fibration up to homotopy. Now $f(\tau_j(b))$ goes to zero under $(\Omega\hat{\rho})_\#$ (because $f(a) \in \ker(\Omega\hat{\rho})_\#$), so $f(\tau_j(b))$ factors through $\pi_*(\Omega F_0; \mathbb{Z}_p)$. The lift of $f(\tau_j(b))$ to $\pi_*(\Omega F_0; \mathbb{Z}_p)$ survives to ${}_\pi E^{r+1}(\Omega F_0)$ and is not killed by

(1E) $$h^{r+1} : {}_\pi E^{r+1}(\Omega F_0) \longrightarrow E^{r+1}(\Omega F_0).$$

Cohen-Moore-Neisendorfer then found a wedge W_0 of mod p^r Moore spaces together with a map

(1F) $$\chi_0' : W_0 \longrightarrow F_0,$$

where χ_0' and W_0 are maximal with respect to the property that $E^r(\Omega\chi_0')$ is an injective homomorphism. They also gave a decomposition of ΩF_0 up to homotopy type,

(1G) $$S^{2n-1} \times V_0 \times \Omega W_0 \xrightarrow{\approx} \Omega F_0,$$

where V_0 is the weak infinite product,

(1H) $$V_0 = \prod_{j=1}^{\infty} S^{2np^j-1}\{p^{r+1}\}.$$

To prove (1G) they showed that there is a fibration sequence up to homotopy,

$$\Omega W_0 \xrightarrow{\Omega\chi_0'} \Omega F_0 \longrightarrow S^{2n-1} \times V_0 \longrightarrow W_0 \xrightarrow{\chi_0'} F_0 ,$$

in which the third (i.e. second from right) arrow is null-homotopic.

The existence of χ_0' and the decomposition (1G) are proved in [CMN1], using some clever tricks involving differential graded Lie algebras (henceforth *dgL*'s). Specifically, the cobar construction on the coalgebra $H_*(F_0)$ is computed and shown to coincide with $\mathcal{U}(L^0, d)$ for a certain free dgL (L^0, d). Localizing at p, a maximal sub-dgL $(C^0, d) \subseteq (L^0, d)$ is found, where $\mathcal{U}(C^0, d)$ is acyclic with respect to $p^{-r}d$. The free dgL C^0 serves as a guide to the construction of W_0 and χ_0'. The geometric decomposition (1G) corresponds to an algebraic decomposition of $\mathcal{U}L^0$ as a tensor product of differential coalgebras.

In order to prove Theorem 1.1, Cohen-Moore-Neisendorfer discovered that they needed to be able to lift the maps $\chi_0' : W_0 \longrightarrow F_0$ and $\delta_0 : V_0 \longrightarrow \Omega F_0$ to the homotopy-theoretic fiber, denoted E_0, of a certain map $F_0 \xrightarrow{\mu_0} \Omega S^{2n+1}$. This space E_0 also participates in another fibration sequence involving $P^{2n+1}(p^r)$ and the homotopy-theoretic fiber $S^{2n+1}\{p^r\}$ of the Brouwer degree p^r map on S^{2n+1}. Cohen-Moore-Neisendorfer realized that they needed a concise way to summarize the relationships among all the spaces and maps in this and in similar settings. They invented a kind of diagram for this purpose that we call a "CMN diagram."

Definition 1.2. A *CMN diagram* is a finite or infinite matrix of spaces and maps

$$
\begin{array}{ccccccc}
 & & \vdots & & \vdots & & \vdots \\
 & & \downarrow & & \downarrow & & \downarrow \\
\cdots & \longrightarrow & A_{33} & \xrightarrow{b_{32}} & A_{32} & \xrightarrow{b_{31}} & A_{31} \\
 & & \downarrow{\scriptstyle a_{23}} & & \downarrow{\scriptstyle a_{22}} & & \downarrow{\scriptstyle a_{21}} \\
\cdots & \longrightarrow & A_{23} & \xrightarrow{b_{22}} & A_{22} & \xrightarrow{b_{21}} & A_{21} \\
 & & \downarrow{\scriptstyle a_{13}} & & \downarrow{\scriptstyle a_{12}} & & \downarrow{\scriptstyle a_{11}} \\
\cdots & \longrightarrow & A_{13} & \xrightarrow{b_{12}} & A_{12} & \xrightarrow{b_{11}} & A_{11}
\end{array}
\tag{1I}
$$

in which each row and each column is a fibration sequence up to homotopy, and each square commutes up to homotopy. The size may be arbitrary, but we will not call something a CMN diagram unless it is at least three by three.

Lemma 1.3. *In any CMN diagram, there are homotopy equivalences* $A_{i,j+3} \approx \Omega A_{ij} \approx A_{i+3,j}$*, and homotopies* $b_{i+3,j} \simeq \Omega b_{ij}$ *and* $a_{i,j+3} \simeq \Omega a_{ij}$.

Cohen-Moore-Neisendorfer proved the following in [CMN2].

Lemma 1.4. [CMN2] *Any homotopy-commutative square*

$$
\begin{array}{ccc}
A_{22} & \xrightarrow{b_{21}} & A_{21} \\
\downarrow{\scriptstyle a_{12}} & & \downarrow{\scriptstyle a_{11}} \\
A_{12} & \xrightarrow{b_{11}} & A_{11}
\end{array}
\tag{1J}
$$

can be embedded in a CMN diagram (1I). If (1J) commutes on the nose, or more generally if we specify the homotopy between $a_{11}b_{21}$ *and* $b_{11}a_{12}$*, then the homotopy types of the* A_{ij}*'s and the homotopy classes of all* a_{ij}*'s and* b_{ij}*'s in (1I) are uniquely determined.*

To organize the various fibrations involving F_0 and E_0, Cohen-Moore-Neisendorfer turned the cofibration $P^{2n+1}(p^r) \longrightarrow S^{2n+1} \xrightarrow{p^r} S^{2n+1}$ into a homotopy-commutative square (the lower right square of (1K)). They applied Lemma 1.4 to it to obtain the CMN diagram

(1K)
$$\begin{array}{ccccccc}
 & & E_0 & \xrightarrow{\lambda_0\eta_0} & P^{2n+1}(p^r) & \xrightarrow{\rho_0} & S^{2n+1}\{p^r\} \\
 & & {\scriptstyle \eta_0}\downarrow & & \approx\downarrow & & \downarrow{\scriptstyle \tilde{\rho}} \\
\Omega S^{2n+1} & \xrightarrow{\kappa_0} & F_0 & \xrightarrow{\lambda_0} & P^{2n+1}(p^r) & \xrightarrow{\hat{\rho}} & S^{2n+1} \\
{\scriptstyle \Omega p^r}\downarrow & & {\scriptstyle \mu_0}\downarrow & & \downarrow & & \downarrow{\scriptstyle p^r} \\
\Omega S^{2n+1} & \xrightarrow{\approx} & \Omega S^{2n+1} & \longrightarrow & * & \longrightarrow & S^{2n+1},
\end{array}$$

in which we have given names to some of the maps. The way we shall view this is that the CMN diagram (1K) serves to define the spaces E_0 and F_0, and the maps $\mu_0, \lambda_0, \eta_0, \kappa_0, \rho_0$ and $\tilde{\rho}$. In particular, $\tilde{\rho}$ denotes the fiber inclusion in the fibration sequence

(1L)
$$S^{2n+1}\{p^r\} \xrightarrow{\tilde{\rho}} S^{2n+1} \xrightarrow{p^r} S^{2n+1},$$

and

(1M)
$$\rho_0 : P^{2n+1}(p^r) \longrightarrow S^{2n+1}\{p^r\}$$

denotes a map that induces an isomorphism on π_{2n}. Because the upper right square of (1K) commutes up to homotopy, we will henceforth write $\tilde{\rho}\rho_0$ for the pinch map $\hat{\rho}$.

Let us review how the logic works in (1K), since it can be confusing as to what is assumed, what is deduced, and what is definition. We know from Lemma 1.4 that a diagram exists, but a priori the identities of only the spaces and maps in the lower right square are known to us. The homotopy-theoretic fiber of $* \longrightarrow X$ is always ΩX, so the space in position A_{13} (see (1I)) is ΩS^{2n+1}. The homotopy-theoretic fiber of $X \longrightarrow *$ is X, so we get $A_{14} = \Omega S^{2n+1}$ and $A_{32} = P^{2n+1}(p^r)$. The space A_{31} is the space that has long been denoted as $S^{2n+1}\{p^r\}$.

We defined F_0 to be the homotopy-theoretic fiber of $\hat{\rho}$, so for the space A_{23} we may use F_0. There exists some map a_{13} and we call it μ_0. We define E_0 to be the homotopy-theoretic fiber of μ_0. Define $\eta_0, \kappa_0, \lambda_0$ to be the maps a_{23}, b_{23}, b_{22} respectively. A conclusion is that the map b_{32} is homotopic to $\lambda_0\eta_0$. Because the upper right square of (1K) commutes up to homotopy, a computation using $H_{2n+1}(-;\mathbb{Z}_{p^r})$ shows that the map marked ρ_0 induces an isomorphism on H_{2n}, hence an isomorphism on π_{2n}. A final conclusion is that the top row,

$$E_0 \xrightarrow{\lambda_0\eta_0} P^{2n+1}(p^r) \xrightarrow{\rho_0} S^{2n+1}\{p^r\},$$

is a fibration sequence up to homotopy.

Cohen-Moore-Neisendorfer proved that the map χ_0' of (1F) fits into CMN diagram

$$
\text{(1N)}\qquad
\begin{array}{ccccccc}
\Omega^2 S^{2n+1} & \xrightarrow{\Omega\kappa_0} & \Omega F_0 & & & & \\
\Big\downarrow{\scriptstyle \Omega^2 p^r} & & \Big\downarrow & & & & \\
\Omega^2 S^{2n+1} & \xrightarrow[(\phi_0,\phi_0')]{} & S^{2n-1}\times V_0 & \longrightarrow & T_0 & \longrightarrow & \Omega S^{2n+1} \\
\Big\downarrow & & \Big\downarrow & & \Big\downarrow & & \Big\downarrow \\
* & \longrightarrow & W_0 & \xrightarrow{=} & W_0 & \longrightarrow & * \\
\Big\downarrow & & \Big\downarrow{\scriptstyle \chi_0'} & & \Big\downarrow{\scriptstyle \lambda_0\chi_0'} & & \Big\downarrow \\
\Omega S^{2n+1} & \xrightarrow{\kappa_0} & F_0 & \xrightarrow{\lambda_0} & P^{2n+1}(p^r) & \xrightarrow{\bar{\rho}_0\rho_0} & S^{2n+1},
\end{array}
$$

which serves to define the space T_0 and the maps $\phi_0 : \Omega^2 S^{2n+1} \longrightarrow S^{2n-1}$, $\phi_0' : \Omega^2 S^{2n+1} \longrightarrow V_0$.

We mentioned earlier that the map χ_0' lifts through $\eta_0 : E_0 \longrightarrow F_0$. Denote the lift by $\chi_0 : W_0 \longrightarrow E_0$. Letting $\Sigma^2 : S^{2n-1} \longrightarrow \Omega^2 S^{2n+1}$ denote the double suspension (i.e. the double adjoint of the identity on S^{2n+1}) and letting $C(n)$ denote the homotopy-theoretic fiber of Σ^2, Cohen-Moore-Neisendorfer proved that there is yet another CMN diagram,

$$
\text{(1O)}\qquad
\begin{array}{ccccc}
C(n)\times V_0 & \longrightarrow & S^{2n-1}\times V_0 & \xrightarrow{\Sigma^2\bar{\pi}_1} & \Omega^2 S^{2n+1} \\
\Big\downarrow & & \Big\downarrow & & \Big\downarrow \\
W_0 & \xrightarrow{=} & W_0 & \longrightarrow & * \\
\Big\downarrow{\scriptstyle \chi_0} & & \Big\downarrow{\scriptstyle \chi_0'} & & \Big\downarrow \\
E_0 & \xrightarrow{\eta_0} & F_0 & \xrightarrow{\mu_0} & \Omega S^{2n+1}.
\end{array}
$$

Here $\bar{\pi}_i$ denotes the projection map from a product (resp. wedge) onto the ith factor (resp. summand). In order to prove (1O) is a CMN diagram, they first showed that there is a homotopy equivalence

$$
\text{(1P)}\qquad C(n)\times V_0 \times \Omega W_0 \xrightarrow{\approx} \Omega E_0.
$$

Combining the upper right square of (1O) with the upper left square of (1N) enabled Cohen-Moore-Neisendorfer to prove the beautiful result,

$$
\text{(1Q)}\qquad \Sigma^2\phi_0 \simeq \Omega^2 p^r,
$$

which leads inexorably to Theorem 1.1 by putting $r = 1$.

Lastly, let us call attention to the third (from bottom) row of (1N), which enmeshes T_0 in the fibration

$$
\text{(1R)}\qquad S^{2n-1}\times V_0 \longrightarrow T_0 \longrightarrow \Omega S^{2n+1}.
$$

It is easy to see (using $H_{2n-1}(-)$ and (1Q)) that the generator of $H^{2n-1}(S^{2n-1})$ transgresses to p^r times the generator of $H^{2n}(\Omega S^{2n+1})$, in the cohomology Serre spectral sequence for (1R).

Generalizing the Cohen-Moore-Neisendorfer Program

The pattern discovered by Cohen-Moore-Neisendorfer for $P^{2n+1}(p^r)$ also holds, with only trivial modifications, for an infinite family $\{D_k\}$ of spaces, as well as for the colimit space D_∞. There is one such family for each parameter triple (n, p, r), where $n \geq 1$ is a connectivity parameter, $p \geq 5$ is a prime, and $r \geq 1$ measures a torsion order. The principal purpose of this book is to prove that assertion, by constructing the spaces $\{D_k\}$ and by carrying out the Cohen-Moore-Neisendorfer program for them. The Moore space $P^{2n+1}(p^r)$ is D_0, so what Cohen-Moore-Neisendorfer did is viewed by us as the 0th or initial step in a recursive construction and proof. The space D_k is obtained from D_{k-1} by attaching a $(2np^k)$-dimensional mod p Moore space, so a cell structure for D_k is

$$\text{(1S)}\quad D_k = P^{2n+1}(p^r) \cup (e^{2np} \underset{p}{\cup} e^{2np+1}) \cup (e^{2np^2} \underset{p}{\cup} e^{2np^2+1}) \cup \ldots \cup (e^{2np^k} \underset{p}{\cup} e^{2np^k+1}) .$$

Each of the results of Cohen-Moore-Neisendorfer listed above has a counterpart when $k > 0$ or when $k = \infty$. In fact, a reasonable summary of our results is obtained by everywhere replacing $P^{2n+1}(p^r)$ by D_k and the subscript of zero by a subscript of k, in the above description of the Cohen-Moore-Neisendorfer theory. Each D_k comes with a "pinch" map $\rho_k : D_k \longrightarrow S^{2n+1}\{p^r\}$, the loops on the fiber F_k of $\tilde{\rho}\rho_k$ can be factored like in (1G), the CMN diagram (1O) exists, and so on. Two subtleties are that the generalization W_k of W_0 is a bouquet of mod p^m Moore spaces where m can be anywhere in the range $r \leq m \leq r + k$, and that the correct generalization of V_0 (cf. (1H)) is the weak infinite product

$$\text{(1T)}\qquad V_k = \prod_{j=k+1}^{\infty} S^{2np^j-1}\{p^{r+k+1}\}.$$

When $k \longrightarrow \infty$, $V_k \longrightarrow (*)$ because V_k becomes more and more highly connected as $k \longrightarrow \infty$. Thus $V_k \approx (*)$. In place of (1R) we obtain a very special fibration.

Theorem 1.5. (To be proved as Proposition 15.1.) *Let $p \geq 5$, $n \geq 1$, $r \geq 1$. There is a fibration sequence up to homotopy,*

$$\text{(1U)}\qquad \Omega^2 S^{2n+1} \xrightarrow{\phi_\infty} S^{2n-1} \longrightarrow T_\infty \longrightarrow \Omega S^{2n+1} ,$$

having the following two properties. The generator of $H^{2n-1}(S^{2n-1})$ transgresses to p^r times the generator of $H^{2n}(\Omega S^{2n+1})$ in the Serre spectral sequence, and

$$\Sigma^2 \phi_\infty \simeq (\Omega^2 p^r) : \ \Omega^2 S^{2n+1} \longrightarrow \Omega^2 S^{2n+1}.$$

We think of (1U) as the "spinoff" of the $\{D_k\}_{k \geq 0}$ that is analogous to the spinoffs (1Q) and (1A) in the case $k = 0$. Theorem 1.5 is easy to state, understand, and remember, and it seems likely to have numerous applications in unstable homotopy theory at primes $p \geq 5$.

Although we have made it sound simple and straightforward, the process of generalizing the Cohen-Moore-Neisendorfer program turned out to be anything but simple. Almost everything that Cohen-Moore-Neisendorfer did not only became harder for $k > 0$, but even necessitated the invention of entirely new techniques. Let us run through the differences, explaining as we go why many of the "background" chapters in our book became necessary.

As one of the first steps in their program, Cohen-Moore-Neisendorfer had to understand $H_*(\Omega F_0;\mathbb{Z}_{(p)})$. When we say that the "hidden" p^{r+1}-torsion becomes visible in ΩF_0, this presupposes that we understand enough about the homology BSS for ΩF_0 that we can recognize p^{r+1}-torsion elements when they show up (e.g. in the image of (1E)). For this purpose, Cohen-Moore-Neisendorfer used a differential coalgebra equivalence

$$(H_*(F_0),0) \xrightarrow{\simeq} C_*(F_0)\ ; \tag{1V}$$

they applied the differential cobar construction $\underline{\Omega}$ to it and obtained the following sequence of chain algebra equivalences (or ***quasi-isomorphisms***) between differential graded algebras (***dga***'s):

$$\mathcal{U}L^0 \xrightarrow{\approx} \underline{\Omega}(H_*(F_0),0) \xrightarrow{\simeq} \underline{\Omega}C_*(F_0) \xrightarrow{\simeq} C_*(\Omega F_0).$$

The last quasi-isomorphism here is a theorem of Adams [Ad].

When $k>0$, by contrast, there is no simple coalgebra approximation to $C_*(D_k)$ like there is when $k=0$, so there is no clear-cut way to concoct an analog for (1V). In fact, it remains an open question, whether there exists a differential coalgebra equivalence (1V) if F_0 is replaced by F_k. Instead of working with small differential coalgebras we must work directly with free tensor algebra models for the dga $C_*(\Omega F_k)$. Such models are called Adams-Hilton models, since they were first constructed in [AH]. We denote an Adams-Hilton model for the space X by $\mathcal{A}(X)$.

In place of (L^0,d) we have the somewhat more complicated dgL (L^k,d). Whereas $H_*(\mathcal{U}L^0,d)$ has most of its torsion of order p^r, the remainder being a small amount of p^{r+1}-torsion ("small" means that its Hilbert series has subexponential growth), we compute that $H_*(\mathcal{U}L^k,d)$ has significant (i.e. exponential growth) p^m-torsion for $r\le m\le r+k$, with a small amount of p^{r+k+1}-torsion. We are required to develop techniques for handling dgL's, especially free dgL's, having homology torsion at many orders simultaneously. Like with L^0, it does turn out that there are isomorphisms and quasi-isomorphisms among dga's,

$$\mathcal{U}L^k \xrightarrow{\approx} \underline{\Omega}(H_*(F_k),0) \xrightarrow{\approx} \mathcal{A}(F_k) \xrightarrow{\simeq} C_*(\Omega F_k)\ .$$

A second complication cropped up as well, as we went from $k=0$ to $k>0$. The Adams-Hilton model of D_0, i.e. $\mathcal{A}(P^{2n+1}(p^r))$, is a free dga on two generators. This dga and its mod p homology are so simple that Cohen-Moore-Neisendorfer did not need to prove anything about it in order to use it implicitly. However, $\mathcal{A}(D_k)$ for $k>0$ has a lot of subtlety. This dga acts on L^k and also on the homotopy of certain spaces associated to D_k, so it is important to understand it. We end up analyzing a $\mathbb{Z}_{(p)}$-free dgL (M^k,d) together with a quasi-isomorphism

$$\mathcal{A}(D_k) \xrightarrow{\simeq} (\mathcal{U}M^k,d).$$

It turns out that $H_*(\mathcal{U}M^k,d)$ has p^m-torsion for $r\le m\le r+k$ only.

One major change from $k=0$ to $k>0$ is that the homotopy and homology BSS's are no longer adequate for keeping track of all the torsion of various orders, to the degree of precision that we need to keep track of them. The BSS is adequate to carry out the initial phases of the program, in the sense that the author was able to write detailed first drafts of Chapters 9 and 10 that used only the BSS. Because the construction of D_{k+1} from D_k is so delicate, however, we cannot afford to make arbitrary choices of homotopy extensions, and we rely upon Adams-Hilton models to direct us to the canonical choices. The BSS has a lot of indeterminacy built into it and does not dovetail well with Adams-Hilton models, so the BSS ultimately turned out to be unusable.

To replace the BSS, we develop the concept of a coherent sequence of algebras (csa) or of Lie algebras (csL) or of modules (csm) over a csa. A csL is a sequence of dgL's that satisfies certain axioms. A motivating example is the list $\{\pi_*(\Omega X; \mathbb{Z}_{p^t})\}_{t\geq 1}$, which we call the *homotopy csL* of X. When $k=0$, Cohen-Moore-Neisendorfer constructed a homomorphism of dgL's from $E^r(\bar{L}^0) = \bar{L}^0 \otimes \mathbb{Z}_p$ to ${}_\pi E^r(\Omega E_0)$. Our generalization is a homomorphism of csL's, from a certain csL associated with $\bar{L}^k$, to the csL $\{\pi_*(\Omega E_k; \mathbb{Z}_{p^t})\}_{t\geq 1}$. (Here $\bar{L}^k$ for $k \geq 0$ denotes a certain sub-dgL of L^k.)

Another big change from $k=0$ to $k>0$ is that relative Samelson products are no longer sufficient to construct all the homotopy classes we need. We require instead the more sophisticated concept of the ring of "stable homotopy operations under a space." This ring is a simultaneous generalization, of iterated relative Samelson products on one hand, and of ordinary stable homotopy operations on the other. One major property of D_k is that the $\mathbb{Z}_{p^t}$-algebra $H_*(\Omega D_k; \mathbb{Z}_{p^t})$ is a retract of the corresponding ring of stable homotopy operations under D_k, for each $t \geq 1$.

The construction of D_k from D_{k-1} has no counterpart in the Cohen-Moore-Neisendorfer theory. The cell structure (1S) comes from the fact that there is for each $k \geq 1$ a cofibration

(1W) $$P^{2np^k}(p) \xrightarrow{\tilde{\tilde{y}}_{p^k}} D_{k-1} \xrightarrow{i} D_k\ ,$$

so the problem of constructing D_k is really the problem of constructing the attaching map $\tilde{\tilde{y}}_{p^k} \in \pi_{2np^k}(D_{k-1}; \mathbb{Z}_p)$. Equivalently, it amounts to the construction of the adjoint class $(\tilde{\tilde{y}}_{p^k})^a \in \pi_{2np^k-1}(\Omega D_{k-1}; \mathbb{Z}_p)$.

When $k=1$, $(\tilde{\tilde{y}}_p)^a$ is precisely the homotopy class $f(\tau_1(b))$ that we looked at in connection with the occurrence of p^{r+1}-torsion in $\pi_*(\Omega D_0)$. (In fact, D_1 can be thought of as a $P^{2n+1}(p^r)$ that has been altered so that its homotopy does not have any p^{r+1}-torsion until dimension $4np-1$.) When $k=2$, however, we need to prove a great many properties of D_1 before we can construct $\tilde{\tilde{y}}_{p^2}$ and D_2. (Actually this is not quite true — we could construct as far as D_{r+1} without doing anything special, but the procedure soon "runs out of steam" unless the full array of properties of D_k is proved at each k.) We must embed the construction in a recursive framework, called the "grand inductive scheme," and we assume the existence and all properties of D_{k-1} in order to construct and prove the properties of D_k. The proof of the main theorem, namely the existence of the $\{D_k\}$, is a proof by mathematical induction that has dozens of parts and that takes over seventy pages!

Another aspect of the case $k>0$ that has no true counterpart at $k=0$ is that we must prove eight "Jacobi identities for D_k." These are equations that are used primarily to prove an equation in $\pi_{6np^k-2}(\Omega E_k; \mathbb{Z}_{p^{r+k}})$ called a "virtual syzygy." The proof of the virtual syzygy equation is the hardest thing in the book. All of Chapters 5, 11, and 12, and much of 6, 7, and 13, are directly or indirectly dedicated to it. How did Cohen-Moore-Neisendorfer prove the syzygy at $k=0$? At $k=0$ they sidestepped the issue by falling back upon the ordinary Jacobi identity for relative Samelson products in $\pi_*(\Omega F_0; \mathbb{Z}_p)$. They then exploited the $(2np-4)$-connectedness of $C(n)$ to get $(\Omega\eta_0)_\#$ to be monomorphic in dimensions $6n-2$, and they used $(\Omega\eta_0)_\#$ to transfer the information from $\pi_{6n-2}(\Omega F_0;\mathbb{Z}_p)$ to $\pi_{6n-2}(\Omega E_0;\mathbb{Z}_p)$. This approach fails as soon as $6np^k-2$ exceeds $2np-4$, i.e., it fails for all $k>0$.

Overview of Mathematical Contents

Let us spend a few paragraphs surveying the mathematical content of each chapter to come. We describe the highlights only. We are attempting to answer the question, "if a reader was going to understand or remember just one or two key ideas from each chapter, what ideas would we want those to be?"

Chapters 2–5 comprise a kind of "Part I" for the book. A subtitle could be "Properties of Certain Special DGL's." We describe in detail certain dgL's denoted $\bar{L}^k, L^k, M^k$, and N^{kl}, and we examine the BSS for the enveloping dga's $\mathcal{U}\bar{L}^k$, $\mathcal{U}L^k$, $\mathcal{U}M^k$. The highlights are Results 2.7, 2.8, 4.3, 4.8, and 5.12.

We study these particular dgL's in detail because an understanding of the torsion in their homology is our starting point for understanding the spaces D_k, F_k, E_k. Specifically, $\mathcal{U}L^k$ is an Adams-Hilton model for F_k, so by computing $H_*(\mathcal{U}L^k)$ we are actually computing $H_*(\Omega F_k)$. Similarly, $H_*(\mathcal{U}M^k)$ equals $H_*(\Omega D_k)$, and $H_*(\mathcal{U}N^{kl})$ equals $H_*(\Omega D_{kl})$, where D_{kl} is the homotopy-theoretic fiber of the inclusion $D_k \vee D_l \longrightarrow D_k \times D_l$.

Throughout this discussion, the parameters $n \geq 1$, $r \geq 1$, $k \geq 0$, and p will be present, but only k will be expressed in the notation. For each parameter triple (n,p,r) there are $\{\bar{L}^k\}_{k\geq 0}$, $\{L^k\}_{k\geq 0}$, and so on. For these purely algebraic results, p may be any prime, although when we get to N^{kl} we simplify the discussion by presuming p to be odd.

The dgL $\bar{L}^k$ is a free dgL (i.e. free as a Lie algebra over $\mathbb{Z}_{(p)}$). The generators of $\bar{L}^k$ as a free Lie algebra comprise a double infinite family $\{y_j, z_j | j \geq 2\}$ that form a nice, fractal-like pattern as far as the torsion of $H_*(\mathcal{U}\bar{L}^k)$ is concerned. The generator y_j (resp. z_j) occurs in dimension $2nj-1$ (resp. $2nj-2$). The fractal-like pattern comes into focus when we replace each z_j by a different choice of generator z'_j, where

$$z'_j \;=\; z_j \;+\; (\text{a certain quadratic decomposable})\,.$$

Let $I_s = \{jp^s | j \geq 2\} \subseteq \mathbb{Z}_+$. Thus I_s consists of all positive multiples of p^s, other than p^s itself. The differential d on $\bar{L}^k$ connects y_j and z'_j in the following manner. Letting Q denote the set of positive integral powers of p,

$$\text{(1X)} \qquad -d(y_j) \;=\; \begin{cases} p^r j z'_j, & \text{if } j \notin I_k \cup Q \\ p^{r+s-1} z'_j, & \text{if } j = p^s,\ 0 < s \leq k \\ p^r j z'_j + p^{r+k}\cdot(\text{a decomposable})\,, & \text{if } j \in I_k - Q \\ p^{r+k+1}\cdot(\text{something}), & \text{if } j = p^m,\ m \geq k+1. \end{cases}$$

The thing to notice about (1X) is that when $j \in I_s - I_{s+1}$, $0 \leq s \leq k$, the pair of generators y_j and z'_j contribute a copy of $\mathbb{Z}_{p^{r+s}}$ to $H_*(\mathcal{U}\bar{L}^k)$ (except for $j = p^{k+1}$, which contributes a $\mathbb{Z}_{p^{r+k+1}}$). Excluding temporarily those j divisible by p^{k+1}, the pattern is as follows. The p^r-torsion occurs for

$$j = 2,3,\ldots,p-1,p,p+1,\ldots,2p-1,2p+1,\ldots,3p-1,3p+1,\ldots;$$

the p^{r+1}-torsion occurs for

$$j = 2p, 3p, \dots, p^2, \dots (2p-1)p, (2p+1)p, \dots, (3p-1)p, (3p+1)p, \dots;$$

the p^{r+2}-torsion occurs for

$$j = 2p^2, 3p^2, \dots, p^3, \dots, (2p-1)p^2, (2p+1)p^2, \dots, (3p-1)p^2, (3p+1)p^2, \dots;$$

and so on. We call this pattern "fractal-like" because as the torsion order advances the same pattern recurs, but it becomes smaller and smaller (in the sense of the p-adic topology on $\mathbb{Z}_+$). In $\bar{L}^\infty$, the limiting case as $k \longrightarrow \infty$, we get precisely this fractal pattern for all torsion orders and for all $j \geq 2$, but when $k < \infty$ the pattern begins to change with the p^{r+k}-torsion, and it becomes something else entirely at the p^{r+k+1}-torsion.

The dga $\mathcal{U}M^k$ exhibits a similar pattern. Here is what is special to us about $\mathcal{U}M^k$. It is a differential graded Hopf algebra (dgH) that is $\mathbb{Z}_{(p)}$-free and primitively generated, yet the generator of its homology's first p^{r+s}-torsion is not primitive, for $0 < s \leq k$. This generator is a cycle named c_s, of dimension $2np^s - 1$, and it generates p^{r+s}-torsion because of the lovely formula,

$$d(b^{p^s}) = -p^{r+s}(c_s)\,, \quad 0 \leq s \leq k. \tag{1Y}$$

In the BSS term $E^{r+s}(\mathcal{U}M^k)$, b^{p^s} and c_s are the first non-zero positive-dimensional survivors, so they must be primitive in the Hopf algebra $E^{r+s}(\mathcal{U}M^k)$. But they are not primitive in $\mathcal{U}M^k$, for $0 < s \leq k$. Also, b^{p^s} is a pth power in $\mathcal{U}M^k$, but it is indecomposable in $E^{r+s}(\mathcal{U}M^k)$. These facts have a geometric analog. The homotopy operations corresponding to b^{p^s} and c_s are "secondary" homotopy operations in the sense that they are not describable in terms of ordinary iterated relative Samelson products. Another significant property of $\mathcal{U}M^k$ is that $p^{r+k}\bar{H}_*(\mathcal{U}M^k) = 0$, i.e. unlike $\bar{H}_*(\mathcal{U}\bar{L}^k)$ it has no p^t-torsion for $t > r+k$.

The dgL N^{kl} is defined as the kernel of the dgL surjection $M^k \amalg M^l \longrightarrow M^k \oplus M^l$, from the coproduct (or categorical sum in the category DGL) to the direct sum (or categorical product in DGL). What is special about it is that is isomorphic as a dgL with the free dgL on the chain complex $\mathcal{U}M^k_+ \otimes \mathcal{U}M^l_+$. Because we analyzed the torsion of $H_*(\mathcal{U}M^k_+)$ so thoroughly, we have a good handle on N^{kl} and its homology torsion.

Chapters 6–8 constitute Part II, whose title could be "Topological and Algebraic Preparations." We describe several new ideas and techniques apropos to unstable p-primary homotopy theory. All of them are extremely important to the proof and even to the statement of the main theorem on the existence and properties of the D_k's.

Chapter 6, entitled "Three Topological Tools," begins with a rapid review of the fundamentals of p-primary homotopy theory. One thing the reader should definitely take away from this review is our names $\omega^t, \rho^t, \omega^t_s, \rho^s_t$ for certain maps among Moore spaces and spheres. We shall use these notations constantly and without further explanation throughout the remaining chapters.

We next discuss Adams-Hilton models and a certain extension problem associated with them. The centerpiece of the discussion is Theorem 6.4, which gives sufficient conditions for the existence of a unique extension of a mod p^t homotopy class to a mod p^{t+1} class having a prescribed mod p^{t+1} Hurewicz image. There are two crucial aspects to this theorem. First, one of its hypotheses (the exactness of (6H) at $p^t\pi_*(\Omega X)$) generalizes ordinary connectedness. It looks at $p^t\pi_*(X)$ rather than $\pi_*(X)$, so an X whose p-local homotopy below the relevant dimensions consists only of p^s-torsion for $s \leq t$ can still be sufficiently "connected" for the theorem to apply. Second, the form of the conclusion is significant. The (unique) homotopy class constructed in Theorem 6.4 satisfies a certain set of three equations, which will be very important to our applications.

Chapter 6 finishes with a unit on thin products, which are a certain homotopy limit associated to a 3-tuple of spaces, and on the algebraic analog, called a cyclic cotensor product. The highlights are the definition of the thin product along with Propositions 6.24 and 6.25, which describe a fibration that is associated naturally with the thin product.

Chapter 7 introduces the concept of csa's, csm's, and csL's (coherent sequences of algebras/modules/Lie algebras), and proves a number of results about them. The vocabulary of csL's is essential since we use that language throughout the construction and proof of properties of the D_k's and their related spaces. The theme of several results is that csa's and csL's are very "coherent", so that simple conditions may be given for a morphism or structure defined on a small subset to determine a unique morphism or structure on the total object. The acme of the chapter is Theorem 7.15, which will be our principal tool for constructing and comparing csL/csm homomorphisms. The theorem gives sufficient conditions for the existence and uniqueness of a homomorphism preserving both a csL and a csm structure, by relating them to a presentation for a certain graded module over a certain graded $\mathbb{Z}_p$-algebra.

The concept that permits us to relate a csL to a module presentation is called a "virtual structure" for the csL. A virtual structure consists of virtual relators, virtual generators, and virtual syzygies, which parallel the non-virtual versions for a presentation of a module. Virtual generators are elements of the csL, while virtual syzygies are equations that hold between specific elements and virtual relators are "generic" equations akin to polynomial identities. Theorem 7.15 says that a csL having a known virtual structure can be mapped to another csL, if two conditions are met. First, the images of the virtual generators must satisfy two very simple and reasonable hypotheses. Second, each virtual syzygy and each virtual relator must also be true in the target csL. We have already alluded to the fact that proving a certain virtual syzygy equation in a certain homotopy csL is perhaps the greatest single challenge in the entire book.

The closing material of Chapter 7, on the examples called $\mathcal{F}(N^{kl})$ and $\mathcal{F}(N^{klm})$, will not be relevant until Chapters 11 and 12. The principal lesson in it is that a virtual structure need not be given explicitly; especially, the virtual syzygies need not be individually spelled out. Instead, it can sometimes be more useful to describe relations or "meta-syzygies" from which the virtual syzygies can be derived.

Chapter 8 discusses the ring $\mathcal{O}_A = \mathcal{O}_A[t]$ of stable homotopy operations under a space A. One of the most important things to understand – because it can so readily be misunderstood – is the precise object on which elements of $\mathcal{O}_A$ must functorially act. The object is the mod p^t homotopy of ΩK_g, where K_g denotes the homotopy-theoretic fiber of the second map in a composable pair of maps $A \xrightarrow{f} Y \xrightarrow{g} Z$. There exists an Hurewicz-like natural transformation $\mathcal{H}_0$ from $\mathcal{O}_A[t]$ to $H_*(\Omega A; \mathbb{Z}_{p^t})$, and we direct a lot of attention toward the question of when $\mathcal{H}_0$ has a section. A major concept is that there can exist "secondary" homotopy operations in $\mathcal{O}_A$ that are not expressible as iterated relative Samelson products. When $A = D_k$ and $r < t \leq r + k$, there are operations under D_k, denoted $\Psi_t^k(b^{p^{t-r}})$ and $\Psi_t^k(c_{t-r})$, that are secondary in this sense.

Part III of the book consists of Chapters 9-13. We state and prove Theorem 9.1, asserting the existence of the $\{D_k\}$ and giving a full dozen of their properties. While no aspect of Theorem 9.1 is unimportant, the cell structure (1S) of D_k, along with some notion of the origin of the attaching maps (1W), is the sine qua non of knowing what the theorem says. A comprehensive overview of Theorem 9.1, which we need not repeat here, immediately precedes the statement of the theorem.

Chapter 10 contains the proof of many of the properties of the D_k's. The overall plan of the proof is the single thing most worth remembering. Typically, our proofs involve three parameters in addition to the notationally suppressed n, p, r. Our results will be true for all $k \geq 0$ and for all torsion orders t satisfying $1 \leq t \leq r+k$. For a given k and torsion order t, the relevant subset is generated by elements indexed by s in the range $\max(0, t-r) \leq s \leq k$. So the parameter space consists of (k, t, s) satisfying $1 \leq t \leq r+s \leq r+k$.

To illustrate, we consider the proof of the existence of the stable homotopy operations under D_k denoted $\Psi_t^k(b^{p^s})$ and $\Psi_t^k(c_s)$, for $1 \leq t \leq r+s \leq r+k$. The proof has three parts that are easy, medium, and difficult. The entire construction of the D_k's is embedded in a "grand induction" on k. The first and easiest part consists of the case $t-r \leq s \leq k-1$. Within the inductive scheme on k, we have already proved the existence of $\Psi_t^{k-1}(b^{p^s})$ and $\Psi_t^{k-1}(c_s)$, so we can get these at k merely by using the inclusion $\hat{\imath}$ of D_{k-1} into D_k and the homomorphisms that $\hat{\imath}$ induces. The second part takes care of the case $t-r < s = k$. (We can no longer use $\hat{\imath}$ because $(k-1, t, k)$ is outside our parameter space.) Some creativity is needed at this stage because something genuinely new is being built, but typically it can be built out of materials already on hand. The third and hardest part occurs for $t = r+s = r+k$. We are constructing a p^{r+k}-torsion homotopy class where none has been found previously. We do this by choosing an extension of a mod p^{r+k-1} class, invoking Theorem 6.4 to make sure we make the right choice. For the existence of $\Psi_{r+k}^k(b^{p^k})$ and $\Psi_{r+k}^k(c_k)$, we are obliged to compute $p^{r+k-1}\pi_{\leq q+2np^k}(X)$ for $X = \Sigma\Omega D_k \wedge \Omega P^{q+1}(p^{r+k})$ and to verify one by one the numerous hypotheses of Theorem 6.4.

Chapter 11 diverges from the topological path we have been pursuing, to develop in greater depth the theory of csL's. We are preparing to tackle the virtual syzygy equation in $\pi_*(\Omega E_k; \mathbb{Z}_{p^{r+k}})$ that we have mentioned. We encounter a succession of snags as we attempt to prove the equation, and new algebraic techniques are needed in order to circumvent them.

Chapter 11 begins with a comprehensive discussion of what the snags are and how we plan to overcome them. A long technical process ensues. We replace one csL called $\mathcal{FL}(V, d)$ by a more managable csL called $\mathcal{EL}(V, d)$, while maintaining control over its virtual structure. The chapter ends with a look at what our analysis has incidentally taught us about the homotopy csL $\{\pi_*(\Omega X; \mathbb{Z}_{p^t})\}_{t\geq 1}$ for a space X. Probably the thing most worth knowing is again not any of the technical details but rather the heuristic or higher level plan, as outlined in the initial discussion.

Relying upon the algebra in Chapter 11, Chapter 12 builds two more csL/csm homomorphisms, from computable algebraically defined csL's to homotopy csL's. We are continuing to carry out the plan described during the opening discussion of the previous chapter. The technical details can overwhelm, but three concepts stand out as noteworthy. First is the idea, used in the proofs of both Theorem 12.4 and Theorem 12.6, of modifying the dimension-wise inductive framework of Theorem 7.15 so as to carry additional hypotheses along with it. Second is the perfect parallel between the algebraic and the geometric pictures. In particular, the notion of one component M^k of $M^k \amalg M^k$ acting on the other, which we exploited in our analysis of N^{kk}, has a precise topological counterpart in terms of how homotopy operations associated with one summand D_k of $D_k \vee D_k$ can act on homotopy classes associated with the other summand. Third is the method of proof for lifting the Jacobi identities from the algebra to the topology, using the variation 6.7 on Theorem 6.4.

Chapter 13 finishes the proof of Theorem 9.1. Using folding maps, we transfer information from the (geometric) Jacobi identities for D_k, first to $\pi_*(\Omega(\Sigma\Omega D_k \wedge \Omega D_k); \mathbf{Z}_{p^{r+k}})$ and ultimately to $\pi_*(\Omega E_k; \mathbf{Z}_{p^{r+k}})$. We obtain the long-awaited virtual syzygy equation, thus ending our detour around the snags discussed earlier. Via Theorem 7.15 we rapidly complete the construction and any outstanding properties of all csL/csm homomorphisms. In some sense this finishes the properties of D_k itself, but we need to construct a certain mod p^{r+k+1} homotopy class in order to help us prepare for $\tilde{\tilde{y}}_{p^{k+1}}$ (see (1W)) so that D_{k+1} will be constructible. In the process we obtain product decompositions for ΩF_k and for ΩE_k, analogous to (1G) and (1P). We tack onto Chapter 13 some discussion of the homotopy operation $\tilde{\tau}_1$; its properties mimic those of an algebraic operation $\tilde{\tau}_1$ that exists in some csL's.

Chapter 13 contains so many delightful tidbits that it is difficult to select most memorable moments. We call the reader's attention to the generalization of the Cohen-Moore-Neisendorfer method for decomposing ΩE_k, since we need the decomposition to provide the needed connectivity hypothesis in Theorem 6.4. We perform the decomposition twice, once with $t = r + k - 1$ in low dimensions in order to construct the first mod p^{r+k} homotopy class, and later with $t = r + k$ to construct a mod p^{r+k+1} class denoted $\hat{y}_{p^{k+1}}$. Notice as well the perfect dovetailing of the three formulas in the conclusion of Theorem 6.4 with the properties required of $\hat{y}_{p^{k+1}}$ in Theorem 9.1(xii). The properties of $\tilde{\tau}_1$ are an interesting aside, and their proofs provide further illustration of the power of Theorem 6.4.

Chapters 14 and 15 comprise Part IV, an introduction to the consequences and applications of the $\{D_k\}$. Chapter 14 defines and examines the space T_k. The space T_k generalizes the space $T^{2n+1}\{p^r\}$ defined in [CMN3]; what we call T_0 is the Cohen-Moore-Neisendorfer space $T^{2n+1}\{p^r\}$. The space T_k is the atomic H-space factor of ΩD_k that carries $\pi_{2n-1}(\Omega D_k) = \mathbf{Z}_{p^r}$, but we do not get to prove this in Chapter 14. We do give a sufficient condition for T_k to split off of ΩD_k as a retract. We also compute the cohomology BSS for T_k. A corollary of the computation is that the mod p Betti numbers for T_k grow subexponentially, but still faster than any polynomial.

In the final chapter, Chapter 15, we let $k \longrightarrow \infty$ and we examine the limit spaces $D_\infty, F_\infty, E_\infty$, and so on. Basically, everything we have said about L^k, $\mathcal{U}M^k$, N^{kk}, D_k, E_k, Ψ_t^k, etc. remains true when we replace k by ∞. An immediate corollary is Theorem 1.5. The existence of the limiting spaces, along with the fibration (1U), is the principal result from Chapter 15. In closing we put forth a number of conjectures and questions, hinting at the rich vein of open problems in homotopy theory that the construction of the $\{D_k\}$ has uncovered.

CHAPTER 2

The Differential Graded Lie Algebras L^k and $\bar{L}^k$

We begin Chapter 2 with a lengthy review of definitions and notations that will serve us throughout the monograph. After this review we define some "altered binomial coefficients" which are necessary for the construction of a special differential graded Lie algebra denoted (L^k, d). In Chapter 9 we will identify $(\mathcal{U}L^k, d)$ as the Adams-Hilton model for the space F_k, but for now we analyze it from a purely algebraic viewpoint. It will be useful to have the properties of (L^k, d) at our fingertips when the times comes to work with its geometric realization. The properties of L^k are inseparable from those of a Lie subalgebra denoted $\bar{L}^k$, so we take a hard look at $\bar{L}^k$ as well. In Chapters 10 and 13 $\bar{L}^k$ will serve as our guide to the homotopy of the space E_k.

Notations and Definitions. Let R be a principal ideal domain; all algebraic objects we consider will be R-modules. A *graded R-algebra* $A = \bigoplus_{m=-\infty}^{\infty} A_m$ is an associative R-algebra that satisfies $(A_s)(A_t) \subseteq A_{s+t}$. We call A *connected* if $A_s = 0$ for $s < 0$ and A_0 is a central subring isomorphic with R and containing a multiplicative identity denoted 1. When A is connected there is an *augmentation* $\varepsilon : A \longrightarrow R$ whose kernel, the *augmentation ideal*, is A_+.

For $x \in A$, we write $|x| = m$ and call x *homogeneous* if $x \in A_m$ for some m. The *homogeneity convention* says that any elements of A to which we actually assign names are assumed to be homogeneous, unless the contrary is stated. The *(graded) commutator* of x and y is given by

$$[x, y] = xy - (-1)^{|x||y|} yx. \tag{2A}$$

We call A *commutative* if and only if $[A, A] = 0$.

Letting S denote a positively graded set (i.e. a set together with a function $S \longrightarrow \mathbb{Z}_+$), we denote by

$R(S)$ or Span(S), the free R-module with basis S;
$R\langle S\rangle$, the free associative R-algebra generated by S;
$R[S]$, the symmetric R-algebra on S;
$\Lambda(S)$, the exterior R-algebra on S;
$\mathbb{L}\langle S\rangle$, the free Lie R-algebra generated by S;

$\mathbb{L}^{ab}[S]$, the free abelian Lie R-algebra generated by S.

We make the following observations. If $S = S_{\text{even}}$ then $R[S]$ is commutative, and if $S = S_{\text{odd}}$ then $\Lambda(S)$ is commutative. The graded R-algebra $R\langle S\rangle$ is connected. The Lie algebras $\mathbb{L}\langle S\rangle$ and $\mathbb{L}^{ab}[S]$ are ***graded Lie algebras***, i.e., they satisfy the Jacobi identities with signs (in some contexts they would be called "superalgebras" to allow for the presence of odd-dimensional elements). These two graded Lie algebras are also connected; we call the graded Lie algebra $L = \bigoplus_{m=-\infty}^{\infty} L_m$ ***connected*** if $L_m = 0$ for all $m \leq 0$.

Let V denote a positively graded free R-module (i.e., $V = \bigoplus_m V_m$ with $V_m = 0$ for $m \leq 0$, and each V_m is a free R-module). We denote by

TV or $T(V)$, the tensor algebra on V;

$\mathcal{L}V$ or $\mathcal{L}(V)$, the free Lie R–algebra generated by V.

Notice that $\mathcal{L}(R(S)) = \mathbb{L}\langle S\rangle$ and $T(R(S)) = R\langle S\rangle$. Denoting the universal enveloping algebra functor by $\mathcal{U}$, we also have isomorphisms

$$\text{(2B)} \qquad \mathcal{U}\mathcal{L}(V) \approx TV\ ,\ \mathcal{U}\mathbb{L}^{ab}[S] \approx R[S_{\text{even}}] \otimes \Lambda(S_{\text{odd}}).$$

In this monograph a ***Hopf algebra*** (A, μ, Δ) is always a connected graded R-algebra, with multiplication denoted $\mu : A \otimes A \longrightarrow A$, together with an algebra homomorphism $\Delta :\ A \longrightarrow A \otimes A$ called the ***diagonal***. The diagonal is assumed to be ***coassociative***, i.e. $(\Delta \otimes 1)\Delta = (1 \otimes \Delta)\Delta$, and to satisfy $\Delta(x) - 1 \otimes x - x \otimes 1 \in A_+ \otimes A_+$ for $x \in A_+$.

A ***chain complex*** is a graded R-module M together with a ***differential*** d, i.e., a linear endomorphism of degree -1 satisfying $d^2 = 0$. When we wish to draw attention to the presence of the ground ring we may also call a chain complex a ***differential R-module***. A ***differential graded R-algebra*** (henceforth ***dga***) is a graded R-algebra A together with a differential d that satisfies the product rule, i.e.,

$$\text{(2C)} \qquad d(xy) = d(x)y + (-1)^{|x|}xd(y)\ .$$

A ***differential graded Hopf algebra*** (henceforth ***dgH***) is a dga (A, d) which is simultaneously a Hopf algebra, on which the diagonal and the differential commute. (I.e., $d'\Delta = \Delta d$, where $d'(x \otimes y) = d(x) \otimes y + (-1)^{|x|}x \otimes d(y)$.) The differential on a ***differential graded Lie algebra*** (henceforth ***dgL***) satisfies

$$\text{(2D)} \qquad d([x, y]) = [d(x), y] + (-1)^{|x|}[x, d(y)]\ .$$

In this monograph, all dgL's and dga's are assumed to be connected, unless we state otherwise (the only exception will be the non-connected dga $\mathcal{O}_A[t]$ in Chapter 8). Then the functor $\mathcal{U}$ takes dgL's to dgH's. A dgL or dga is called ***free*** if the underlying algebra (i.e. forgetting the differential) is free as a Lie or associative algebra.

Let p denote any prime. Nearly everything we do in this book is p-local. The ground ring R will always be $\mathbb{Z}_{(p)}$ or $\mathbb{Z}_{p^t}$. Homology and cohomology are assumed to be taken with coefficients in $\mathbb{Z}_{(p)}$ unless a different coefficient group is named in the notation. The topological spaces we encounter will always be pointed and will nearly always be localized at p, but we will not do any topology until Chapter 6.

The *Bockstein spectral sequence* (henceforth *BSS*) of a differential $\mathbb{Z}_{(p)}$-module will be an essential tool in our study of dga's and dgL's. We recall briefly its salient features [N1]. Given a $\mathbb{Z}_{(p)}$-free differential graded module M, denote the mth term of its BSS by $E^m(M)$ and the mth differential by β^m. Then $E^0(M) = M \otimes \mathbb{Z}_p$, the E^1 term $E^1(M)$ equals $H_*(M; \mathbb{Z}_p)$, and $E^\infty(M)$ has a $\mathbb{Z}_p$-basis that bijects with a $\mathbb{Q}$-basis for $H_*(M; \mathbb{Q})$. The differential β^0 is the mod p reduction of the original differential on M. A pair of $\mathbb{Z}_p$-basis elements (y, x) in $E^0(M)$ that survive to $E^m(M)$ but then $\beta^m(y) = x$ correspond to a summand of $\mathbb{Z}_{p^m}$ in $H_*(M)$. In this way the BSS detects an sorts out the higher torsion in $H_*(M)$. Finally, if M is a dga (resp. dgL, dgH) over $\mathbb{Z}_{(p)}$, then each $(E^m(M), \beta^m)$ is a dga (resp. dgL, dgH) over $\mathbb{Z}_p$.

A useful example of a specific and well-known BSS computation is given next.

Lemma 2.1. *For $q \geq 2$, $t \geq 1$, define a dgL over $\mathbb{Z}_{(p)}$ by*

$$\Xi(q,t) = (\mathbb{L}^{ab}[u,v], e),$$

where $|u| = q-1$, $|v| = q$, $e(u) = 0$, and $e(v) = -p^t u$.

(a) *If q is odd then the BSS for $\mathcal{U}\Xi(q,t)$ has*

$$E^0 = E^1 = \ldots = E^t \text{ and } E^{t+1} = \mathbb{Z}_p \ (i.e., E_+^{t+1} = 0).$$

(b) *If q is even then the BSS for $\mathcal{U}\Xi(q,t)$ has*

$$E^m(\mathcal{U}\Xi(q,t), \beta^m) = \begin{cases} \mathcal{U}\Xi(q, t-m) \otimes \mathbb{Z}_p & \text{if } m \leq t \\ \mathcal{U}\Xi(qp^{m-t}, 0) \otimes \mathbb{Z}_p & \text{if } m \geq t. \end{cases}$$

Proof. Straightforward.

More definitions. In Chapters 2–5, all the dgH's we encounter will be *cocommutative*, i.e. $\tau\Delta = \Delta : A \longrightarrow A \otimes A$, where

$$\tau(x \otimes y) = (-1)^{|x||y|} y \otimes x. \tag{2E}$$

Any primitively generated Hopf algebra is cocommutative, and any Hopf algebra which is the universal enveloping algebra of a Lie algebra is primitively generated. If (A, d) is a cocommutative dgH over $\mathbb{Z}_{(p)}$, then each term $(E^m(A), \beta^m)$ of its BSS is also a cocommutative dgH.

When $f : A \longrightarrow B$ is a homomorphism between cocommutative Hopf R-algebras, the *Hopf algebra kernel* (henceforth *HAK*) of f is given by

$$\text{HAK}(f) = \{x \in A | (1 \otimes f)\Delta_A(x) \in A \otimes B_0\}, \tag{2F}$$

by [MM]. It is a sub-Hopf algebra of A. If in addition f is surjective and B is R-free, we call

$$R \longrightarrow \text{HAK}(f) \xrightarrow{\subseteq} A \xrightarrow{f} B \longrightarrow R \tag{2G}$$

a ***short exact sequence*** of Hopf algebras. If f is a homomorphism of dgH's then HAK(f) is a sub-dgH of A; in that case, if (2G) is a short exact sequence of Hopf algebras, we can also call it ***a short exact sequence of dgH's.*** By [MM], whenever (2G) is a short exact sequence of Hopf algebras (resp. dgH's), there is an isomorphism of coalgebras (resp. differential coalgebras),

$$\text{(2H)} \qquad A \approx \mathrm{HAK}(f) \otimes B \, .$$

Here is one common source of short exact sequences of Hopf algebras. Let

$$0 \longrightarrow M' \longrightarrow M \longrightarrow M'' \longrightarrow 0$$

be a short exact sequence of connected graded Lie R-algebras (resp. of dgL's), where M'' is free as an R-module. Then

$$R \longrightarrow \mathcal{U}M' \longrightarrow \mathcal{U}M \longrightarrow \mathcal{U}M'' \longrightarrow R$$

is a short exact sequence of Hopf R-algebras (resp. of dgH's).

Sometimes we want to know the BSS for the HAK of a dgH surjection. Under suitable hypotheses, this BSS has a particularly nice description.

Lemma 2.2. *Let $f : (A, d) \longrightarrow (A'', d'')$ be a homomorphism of cocommutative dgH's over $\mathbb{Z}_{(p)}$, and let (A', d') be its HAK. Suppose, for the associated BSS's, that $E^j(f)$ is surjective for $j \le m$. Then $E^j(A') = HAK(E^j(f))$ for $j \le m$. In particular, $E^j(A) = E^j(A') \otimes E^j(A'')$ as coalgebras, for $j \le m$.*

Proof. Suppose inductively that the conclusions hold for $j \le m-1$. Write B', B, B'', ϕ for $E^{m-1}(A'), E^{m-1}(A), E^{m-1}(A''), E^{m-1}(f)$. We are assuming that

$$\text{(2I)} \qquad \mathbb{Z}_p \longrightarrow B' \longrightarrow B \xrightarrow{\phi} B'' \longrightarrow \mathbb{Z}_p$$

is a short exact sequence of dgH's, i.e., ϕ is surjective with $B' = \mathrm{HAK}(\phi)$, and also that $H_*(\phi)$ is surjective.

Associated to (2I) is the filtration of B given by $F_s(B) = ((1 \otimes \phi)\Delta)^{-1}(B \otimes B''_{\le s})$; note that $F_0(B) = B'$. Associated to this filtration is a first quadrant homology spectral sequence having $E^2_{st} = H_s(B''; H_t(B'))$ and converging to $H_*(B)$ [MM, Lemma 1.4]. The differentials in this spectral sequence are compatible with the $H_*(B')$-module structure. Because $H_*(\phi)$ is onto, the differentials d^q_{s0} must all be zero, and it follows that all $d^q_{st} = 0$. Thus the spectral sequence collapses, and there is a filtration $\{F'_s\}$ of $H_*(B)$ such that $F'_s/F'_{s-1} \approx H_*(B') \otimes H_s(B'')$.

Now, consider the filtration $F_s(H_*B)$ arising from the short exact sequence

$$\mathbb{Z}_p \longrightarrow \mathrm{HAK}(H_*\phi) \longrightarrow H_*(B) \xrightarrow{H_*(\phi)} H_*(B'') \longrightarrow \mathbb{Z}_p$$

of Hopf algebras. Clearly $F'_s \subseteq F_s(H_*B)$ for each s. In particular, the canonical Hopf algebra homomorphism

$$H_*(B') \longrightarrow H_*(\mathrm{HAK}(\phi)) \xrightarrow{\eta} \mathrm{HAK}(H_*\phi)$$

injects. Notice [MM, Lemma 1.1] that

$$F_s(H_*B)/F_{s-1}(H_*B) \approx (\mathrm{HAK}(H_*\phi)) \otimes H_s(B'').$$

In fact, the diagram

$$\text{(2J)}\qquad \begin{array}{ccc} F'_s/F'_{s-1} & \xrightarrow{\approx} & H_*(B') \otimes H_s(B'') \\ \gamma_s \downarrow & & \downarrow \eta\otimes 1 \\ F_s(H_sB)/F_{s-1}(H_*B) & \xrightarrow{\approx} & \mathrm{HAK}(H_*\phi) \otimes H_s(B'') \end{array}$$

commutes. In particular, the inclusion-induced arrows denoted γ_s in (2J) are monomorphic for all s. Since $\bigcup_{s=0}^{\infty} F'_s = H_*B$, all the γ_s must also be onto. Deduce that $F'_s = F_s(H_*B)$ for all s, and in particular, η is an isomorphism.

More Notation. Let L be a graded Lie algebra, and let $y \in \mathcal{U}L$. For $x \in \mathcal{U}L$, let $\mathrm{Ad}(x)(y) = xy - (-1)^{|x||y|}yx$. Thus $\mathrm{Ad}(x)(y) = [x, y]$ if x and y both belong to L. For $x \in TL$, the graded tensor algebra on L, define $\mathrm{ad}(x)(y)$ to be $\mathrm{Ad}(x_1)\mathrm{Ad}(x_2)\dots\mathrm{Ad}(x_m)(y)$ if $x = x_1 \dots x_m$, $x_i \in L$. The Jacobi identities show that $\mathrm{ad} : TL \longrightarrow \mathrm{Hom}(\mathcal{U}L, \mathcal{U}L)$ factors through $\mathcal{U}L$, hence ad( ) may be viewed as a left action on $\mathcal{U}L$ on itself. If (L, d) is a dgL, then an easy calculation yields

$$\text{(2K)}\qquad d(\mathrm{ad}(x)(y)) = \mathrm{ad}(d(x))(y) + (-1)^{|x|}\mathrm{ad}(x)(d(y)).$$

The next lemma concerns dgL kernels. We will frequently encounter pairs of generators in a dgL which are "linked" in the sense that the boundary of one generator equals a power of p times the other generator. This setup occurs so often that we introduce a notation to handle it more efficiently.

For $q \geq 2$ and $t \geq 1$, let $C(\lambda^q, t)$ denote the dgL $(\mathbb{L}\langle u, v\rangle, d)$, where $|v| = q$ and $d(v) = p^t u$. Let $C(m\lambda^q, t)$ denote the coproduct (or "free product") of m copies of $C(\lambda^q, t)$. Given a formal power series $P(\lambda) = a_2\lambda^2 + a_3\lambda^3 + \dots$ with non-negative integral coefficients, let $C(P(\lambda), t) = \coprod_{q=2}^{\infty} C(a_q\lambda^q, t)$. Finally, given a sequence of such series $\mathcal{P}(\lambda) = (P_1(\lambda), P_2(\lambda), \dots)$, let $C(\mathcal{P}(\lambda)) = \coprod_{t=1}^{\infty} C(P_t(\lambda), t)$. The notation $P(\lambda)\mathcal{P}(\lambda)$ of course means $(P(\lambda)P_1(\lambda), P(\lambda)P_2(\lambda), \dots)$.

Lemma 2.3. *Suppose (M, d) is a locally finite $\mathbb{Z}_{(p)}$-free dgL such that $E^t(\mathcal{U}M) = E^0(\mathcal{U}M)$ in the BSS for some $t > 0$ (i.e., M is "minimal" and $H_*(\mathcal{U}M)$ has no summands of order smaller than p^t). Let $f' : (M', d') \longrightarrow (M, d)$ be a dgL surjection with kernel (K', d'), where M' is also $\mathbb{Z}_{(p)}$-free and f' is split as a differential graded $\mathbb{Z}_{(p)}$-module homomorphism. Suppose $\mathcal{P}(\lambda) = (P_1(\lambda), P_2(\lambda), \dots)$ satisfies $P_i(\lambda) = 0$ for $i > t$. Let $f : (M', d') \coprod C(\mathcal{P}(\lambda)) \longrightarrow (M, d)$ be the extension of f' in which all of $C(\mathcal{P}(\lambda))$ is sent to zero, and let $(K, e) = \ker(f)$. Then (K, e) is isomorphic as a dgL with $(K', d') \coprod C(\mathcal{U}M(\lambda)\mathcal{P}(\lambda))$. Here $\mathcal{U}M(\lambda)$ denotes the Hilbert series of the free graded $\mathbb{Z}_{(p)}$-module $\mathcal{U}M$.*

Proof. We may choose elements $\{\omega_j\}_{j\in J}$ in $\mathcal{U}M'$ whose f'-images form a $\mathbb{Z}_{(p)}$-basis for $\mathcal{U}M$ and whose d'-boundaries are divisible by p^t. Let the set I index the generator pairs for $C(\mathcal{P}(\lambda))$, i.e. $C(\mathcal{P}(\lambda)) = (\mathbb{L}\langle v_i, u_i | i \in I\rangle, d)$ where $d(v_i) = p^{r_i}u_i$ and $1 \leq r_i \leq t$. As

Lie algebras, $K = K' \coprod \mathbb{L}\langle \mathrm{ad}(\omega_j)(v_i), \mathrm{ad}(\omega_j)(u_i) | i \in I,\ j \in J\rangle$. Define the dgL (D, d) by setting $D = \mathbb{L}\langle u_{ij}, u_{ij} | i \in I,\ j \in J\rangle$ and $d(v_{ij}) = p^{r_i} u_{ij}$. Extend the inclusion $K' \longrightarrow K$ to $g:\ K' \coprod D \longrightarrow K$ by setting $g(v_{ij}) = \mathrm{ad}(\omega_j)(v_i)$ and $g(u_{ij}) = (-1)^{|\omega_j|}\mathrm{ad}(\omega_j)(u_i) + p^{t-r_i}\mathrm{ad}(d'(\omega_j))(v_i)$. Then g is a dgL homomorphism, and it is clear that D is isomorphic with $C(\mathcal{U}M(\lambda)\mathcal{P}(\lambda))$.

We are finally ready to begin discussing the specific dgL, denoted (L^k, d), whose study will occupy us for the remainder of the chapter. The coefficients which appear in the formula for the differential d on L^k are what we call "altered binomial coefficients". The prime p and the non-negative integer k should now be thought of as fixed.

Notation. For $m \geq 1$, $\alpha_k(m)$ denotes the largest integer not exceeding k for which $p^{\alpha_k(m)} \leq m$, i.e.,

(2L) $$\alpha_k(m) = \min(k, [\log_p(m)]).$$

When no ambiguity can result we suppress k (as well as p) from the notation and write $\alpha(m)$ for $\alpha_k(m)$. We write the binomial coefficient m choose i as

$$(m|i) = m!/(i!)(m-i)!\ .$$

For $m > i > 0$, put

(2M) $$\langle m|i\rangle_k = (m|i)p^{\alpha(i)+\alpha(m-i)-\alpha(m)}.$$

As with α_k, we will write $\langle m|i\rangle$ for $\langle m|i\rangle_k$ when the context allows for only one plausible k. Lastly, for $s \geq 0$ let I_s denote the set

(2N) $$I_s \ = \ \{jp^s | j \text{ is an integer } \geq 2\}$$

Lemma 2.4. *Let $m > i > 0$ be integers, and write $m = p^s q$, where $p \nmid q$.*

(a) *If $k = 0$, then $\langle m|i\rangle = (m|i)$.*
(b) *$\langle m|i\rangle$ is a positive integer, and $\langle m|i\rangle = \langle m|m-i\rangle$.*
(c) *If $q \neq 1$, then $\langle m|1\rangle = m$.*
(d) *If $q \neq 1$ and $s \leq k$ then $p^s \mid \langle m|i\rangle$.*
(e) *If $q \neq 1$ and $s > k$ then $p^k \mid \langle m|i\rangle$, but $p^{k+1} \nmid \langle m|p^s\rangle$.*
(f) *If $q = 1$ and $s \leq k$ then $\langle m|1\rangle = p^{s-1}$ and $p^{s-1} \mid \langle m|i\rangle$.*
(g) *If $q = 1$ and $s > k$ then $\langle m|1\rangle = m$ and $p^{k+1} \mid \langle m|i\rangle$ but $p^{k+2} \nmid \langle m|p^{s-1}\rangle$.*
(h) *If $m \notin I_{k+1}$ or $i \notin I_k$, then $\langle m|1\rangle$ divides $\langle m|i\rangle$ in $\mathbb{Z}_{(p)}$.*

Proof. Let p^a be the largest power of p dividing $(m|i)$. One characterization of a is that a equals the number of carry operations that occur when i is added to $m-i$ in base p. With the help of this fact, all parts are straightforward.

We define next a commutative ring R^k that is closely allied with L^k. (Remark: R^k is isomorphic with the cohomology ring $H^*(F_k; \mathbb{Z}_{(p)})$.) Two parameters in addition to k and p enter into the definition. They are denoted n, a measure of connectivity, and r, which is an exponent or torsion order. Like p, n and r will always be present whenever we discuss R^k or L^k, but also like p we will suppress them from the notation.

Define the ring R^k as follows. As a graded $\mathbb{Z}_{(p)}$-module, R^k equals Span$(1, \xi_1, \xi_2, \xi_3, \dots)$, where $|\xi_j| = 2nj$. Multiplication is determined by the rule

$$\xi_i \xi_j = p^r \langle i + j | i \rangle \xi_{i+j},$$

which is easily seen to render R^k a commutative connected graded algebra.

Define (L^k, d) to be the formal dgL associated to the ring R^k. Explicitly, $L^k = \mathbb{L}\langle x_1, x_2, x_3, \dots \rangle$, where $|x_m| = 2nm - 1$ and

$$d(x_m) = -(\frac{1}{2})p^r \sum_{i=1}^{m-1} \langle m | i \rangle [x_i, x_{m-i}]. \tag{20}$$

Note that $\mathcal{U}(L^k, d)$ is a dgH having $\mathcal{U}L^k = \mathbb{Z}_{(p)}\langle x_1, x_2, \dots \rangle$ and with differential given by (20).

One relationship between R^k and (L^k, d) is that $(\mathcal{U}L^k, d)$ equals $\underline{\Omega}(R^k)^*$, the cobar construction on the dual of R^k. (We view the dual $(R^k)^*$ of R^k as a differential coalgebra, with zero differential.) The "homological" grading on $(\mathcal{U}L^k, d)$ is the grading by bracket length; each generator x_j has homological degree one.

For convenience we call the homological grading the *$*$-grading* on $\mathcal{U}L^k$ and we write $|w|_*$ for the *$*$-degree* of a *$*$-homogeneous* element. We do not require cobars to see this; we may simply define a second grading on L^k by assigning each $|x_j|_* = 1$, and then we notice that d is homogeneous of $*$-degree $+1$ with respect to the new grading. Thus (L^k, d) (resp. $(\mathcal{U}L^k, d)$) is a bigraded dgL (resp. dga).

Because $(\mathcal{U}L^k, d) = \underline{\Omega}(R^k)^*$ we have

$$H_*(\mathcal{U}L^k, d) = H_*(\underline{\Omega}(R^k)^*) = \mathrm{Ext}^*_{R^k}(\mathbb{Z}_{(p)}, \mathbb{Z}_{(p)}).$$

Since $R^k \otimes \mathbb{Q}$ is isomorphic to the rational polynomial ring $\mathbb{Q}[\xi_1]$ on a single $(2n)$-dimensional generator, we see that

$$H_*(\mathcal{U}L^k, d) \otimes \mathbb{Q} = H_*(\mathcal{U}L^k \otimes \mathbb{Q}, d) = \mathrm{Ext}^*_{R^k \otimes \mathbb{Q}}(\mathbb{Q}, \mathbb{Q})$$

is an exterior algebra on a single $(2n-1)$-dimensional generator. We have shown that $E^\infty(\mathcal{U}L^k) = \Lambda(x_1) \otimes \mathbb{Z}_p$, i.e., 1 and x_1 (and their scalar multiples) are the only infinite cycles in the BSS for $\mathcal{U}L^k$.

In order to compute the full BSS for $\mathcal{U}L^k$, we first remove the infinite cycle x_1. Let $\bar{L}^k$ denote the dgL kernel of the surjection

$$(L^k, d) \longrightarrow (\mathbb{L}^{\mathrm{ab}}[x_1], 0),$$

i.e., we have a short exact sequence of dgL's,

$$0 \longrightarrow \bar{L}^k \xrightarrow{\bar{e}} L^k \longrightarrow \mathbb{L}^{\mathrm{ab}}[x_1] \longrightarrow 0.$$

Then $\bar{L}^k = \mathbb{L}\langle y_i, z_i | i \geq 2 \rangle$, where $|y_i| = 2ni - 1$, $|z_i| = 2ni - 2$, and the inclusion of $\bar{L}^k$ into L^k is given by

$$\bar{e}(y_i) = x_i, \ \bar{e}(z_2) = x_1^2, \text{ and } \bar{e}(z_i) = [x_1, x_{i-1}] \text{ for } i \geq 3.$$

On the y_i's the differential is given by

$$(2Q)\qquad \begin{aligned} d(y_2) &= -p^r\langle 2|1\rangle z_2; \\ d(y_m) &= -\frac{p^r}{2}\sum_{j=2}^{m-2}\langle m|j\rangle[y_j, y_{m-j}] - p^r\langle m|1\rangle z_m \text{ for } m \geq 3. \end{aligned}$$

We will now alter the set $\{y_i, z_i\}$ of generators. The following lemma, which empowers us to do this, is well-known and easily proved.

Lemma 2.5.. *Suppose $L' = \mathbb{L}\langle a_1, a_2, \dots\rangle$. Let $\{c_i\}$ be units in $\mathbb{Z}_{(p)}$ and let $b_i \in [L', L'] + (L'_{\text{odd}})^2$ satisfy $|b_i| = |a_i|$. Then the Lie algebra homomorphism $\mathbb{L}\langle a'_1, a'_2, \dots\rangle \longrightarrow L'$ sending a'_i to $c_i a_i - b_i$ is an isomorphism.*

Recall that I_k denotes $\{ip^k | i \geq 2\}$. By Lemma 2.1(h), the expressions $z'_2 = z_2$ and for $m \geq 3$

$$(2R)\qquad z'_m = z_m + \sum_{\substack{2\leq j\leq m-2 \\ j\notin I_k \text{ or } m-j\notin I_k}} (\frac{1}{2})\langle m|j\rangle\langle m|1\rangle^{-1}[y_j, y_{m-j}]$$

makes sense in L^k. Using Lemma 2.5 we have $L^k = \mathbb{L}\langle y_i, z'_i | i \geq 2\rangle$, where

$$(2S)\qquad d(y_2) = -p^r\langle 2|1\rangle z'_2;$$

$$(2T)\qquad d(y_m) = -p^r\langle m|1\rangle z'_m \text{ if } m \notin i_k \text{ or } m = 2p^k; \text{ and for } m \geq 3$$

$$(2U)\qquad d(y_{mp^k}) = -p^r\langle mp^k|1\rangle z'_{mp^k} - \frac{p^r}{2}\sum_{i=2}^{m-2}\langle mp^k|ip^k\rangle[y_{ip^k}, y_{(m-i)p^k}]\,.$$

There is a nice coproduct decomposition for $\bar{L}^k$. Let s be in the range $0 \leq s < k$. For $i \in I_s - I_{s+1}$, we see from Lemma 2.4 that p^{r+s} is the largest power of p that divides $p^r\langle i|1\rangle$. In view of (2S,T), we may embed $C(\lambda^{2ni-1}, r+s)$ into $(\bar{L}^k, d)$ by sending v to y_i and u to $-(\langle i|1\rangle p^{-s})z'_i$. Put

$$P^*_{r+s}(\lambda) = \sum_{i\in I_s - I_{s+1}} \lambda^{2ni-1} = \frac{\lambda^{4np^s-1}}{1-\lambda^{2np^s}} - \frac{\lambda^{4np^{s+1}-1}}{1-\lambda^{2np^{s+1}}}\,,$$

and let $\mathcal{P}^*(\lambda) = (0, \dots, 0, P^*_r(\lambda), P^*_{r+1}(\lambda), \dots P^*_{r+k-1}(\lambda), 0, \dots)$. We have given an embedding of $C(\mathcal{P}^*(\lambda))$ into $(\bar{L}^k, d)$ that "accounts for" all generators of $\bar{L}^k$ except for $\{y_j, z'_j | j \in I_k\}$.

To account for the generators indexed by $j \in I_k$ we use a trick. We may write $\bar{L}^k$ as $\bar{L}^k(n, p^r)$ in those rare instances when we do not wish to suppress n and p^r from the notation. Let $(\bar{L}^\#, d^\#)$ denote the dgL $\bar{L}^0(np^k, p^{r+k})$. Referring to (2Q) and 2.4(a), we may write $\bar{L}^\#$ as

$$(2V)\qquad \begin{aligned} \bar{L}^\# &= \mathbb{L}\langle y^\#_j ; z^\#_j\rangle_{j\geq 2}, \text{ with} \\ d^\#(y^\#) &= -(\frac{1}{2})p^{r+k}\sum_{j=2}^{m-2}\langle m|j\rangle[y^\#_j, y^\#_{m-j}] - p^{r+k} m z^\#_m\,. \end{aligned}$$

Lemma 2.6. *Let* $c = (p^k - 1)/(p-1)$ *and for* $m \geq 1$ *define*

$$\beta(m) = (m!)p^{cm}/(mp^k)!\ .$$

(a) $\beta(m)$ *is a unit in* $\mathbb{Z}_{(p)}$, *and for* $m > i > 0$

$$\beta(m)\langle mp^k|ip^k\rangle = \beta(i)\beta(m-i)(m|i)p^k.$$

(b) *The Lie algebra homomorphism* $g: \bar{L}^{\#} \longrightarrow \bar{L}^k$ *given by*

$$g(y_m^{\#}) = \beta(m)y_{mp^k},\ g(z_m^{\#}) = \beta(m)z'_{mp^k},$$

is a homomorphism of dgL's, i.e., $dg = gd^{\#}$.

(c) *There is an extension of* g *to an isomorphism of dgL's,*

$$f:\ C(\mathcal{P}^*(\lambda))\amalg \bar{L}^{\#} \longrightarrow \bar{L}^k.$$

Proof. (a) For each $t \geq 1$,

$$((t-1)p^k+1)\cdots(tp^k)/tp^c$$

is a unit in $\mathbb{Z}_{(p)}$. Multiplying these together for $1 \leq t \leq m$ gives that $\beta(m)^{-1}$ is a unit in $\mathbb{Z}_{(p)}$. By (2M),

$$\beta(m)\langle mp^k|ip^k\rangle = p^k(mp^k|ip^k)(m!)p^{cm}/(mp^k)!\ ,$$

which equals $p^k(m|i)\beta(i)\beta(m-i)$.

(b) Since $\bar{L}^{\#}$ is free as a Lie algebra g is a unique and well-defined Lie algebra homomorphism. It suffices to check on generators that g commutes with $d^{\#}$, i.e., it suffices to see that

$$gd^{\#}(y_m^{\#}) = \beta(m)(y_{mp^k}) \text{ and } gd^{\#}(z_m^{\#}) = \beta(m)d(z'_{mp^k}),\ m \geq 2.$$

Using part (a) and (2V) and (2U), we compute for $m \geq 2$ that

$$\begin{aligned} gd^{\#}(y_m^{\#}) &= -(\frac{1}{2})p^{r+k}\sum_{j=2}^{m-2}(m|j)[\beta(j)y_{jp^k}, \beta(m-j)y_{(m-i)p^k}] \\ &\quad - p^{r+k}m\beta(m)z'_{mp^k} \\ &= -(\frac{1}{2})p^r\beta(m)\sum_{j=2}^{m-2}\langle mp^k|jp^k\rangle[y_{jp^k}, y_{(m-j)p^k}] - p^r(mp^k)\beta(m)z'_{mp^k} \\ &= \beta(m)d(y_{mp^k})\ . \end{aligned}$$

Because $(d^{\#})^2 = 0$, we may apply $d^{\#}$ to both sides of (2V) and solve for $d^{\#}(z_m^{\#})$. We get

$$(m)d^{\#}(z_m^{\#}) = -(\frac{1}{2})\sum_{j=2}^{m-2}(m|j)([d^{\#}(y_j^{\#}), y_{m-j}^{\#}] - [y_j^{\#}, d^{\#}(y_{m-j}^{\#})]).$$

Apply g to both sides of this equation to get

$$(m)gd^{\#}(z_m^{\#}) = -(\frac{1}{2})p^{-k}\beta(m)\sum_{j=2}^{m-2}\langle mp^k|jp^k\rangle d[y_{jp^k}, y_{(m-j)p^k}].$$

Similarly, apply d to both sides of (2U) and solve for $d(z'_{mp^k})$. We find

$$(mp^k)d(z'_{mp^k}) = -(\frac{1}{2})\sum_{i=2}^{m-2}\langle mp^k|ip^k\rangle d[y_{ip^k}, y_{(m-i)p^k}].$$

So $gd^{\#}(z^{\#}_m) = \beta(m)d(z'_{mp^k})$, as desired.

(c) We defined the dgL homomorphism $C(\mathcal{P}^*(\lambda)) \longrightarrow \bar{L}^k$ above. Combining this with part (b), we get a dgL homomorphism f. This f is an isomorphism of Lie algebras by Lemma 2.5 (because each $\langle i|1\rangle p^{-s}$ for $i \in I_s - I_{s+1}$ and each $\beta(m)$ is a unit in $\mathbb{Z}_{(p)}$). Thus f is an isomorphism of dgL's.

Notation. For $0 \le s \le k$, define the Lie algebra $\bar{L}^k_s$ of $\bar{L}^k$ by

$$\bar{L}^k_s = \mathbb{L}\langle y_j, z'_j | j \in I_s\rangle.$$

Then $d(\bar{L}^k_s) \subseteq \bar{L}^k_s$, so $\bar{L}^k_s$ is a sub-dgL of $\bar{L}^k$. Notice that $\bar{L}^k_0 = \bar{L}^k$ and that Lemma 2.6(b) gives an isomorphism of $\bar{L}^{\#}$ with $\bar{L}^k_k$. Generalizing Lemma 2.6, there is a dgL isomorphism

$$\bar{L}^k_s \approx C(P^*_{r+s}(\lambda), r+s) \amalg \ldots \amalg C(P^*_{r+k-1}(\lambda), r+k-1) \amalg \bar{L}^{\#} .$$

Proposition 2.7.

(a) $E^0(\mathcal{U}\bar{L}^k) = \ldots = E^r(\mathcal{U}\bar{L}^k) = \mathcal{U}\bar{L}^k \otimes \mathbb{Z}_p$.
(b) For $0 \le s \le k$, $E^{r+s}(\mathcal{U}\bar{L}^k) = \mathcal{U}\bar{L}^k_s \otimes \mathbb{Z}_p$.
(c) $E^{r+k+1}(\mathcal{U}\bar{L}^k) = (\bigotimes_{j=1}^{\infty} \mathcal{U}\Xi(2np^{k+j} - 1, r+k+1)) \otimes \mathbb{Z}_p$.
(d) $E^t(\mathcal{U}\bar{L}^k) = \mathbb{Z}_p$ *for* $t > r+k+1$ *and for* $t = \infty$.
(e) *As coalgebras,* $E^t(\mathcal{U}L^k) \approx \Lambda(x_1) \otimes E^t(\mathcal{U}\bar{L}^k)$ *for all* t.

Proof. Using Lemma 2.6(c), we have

$$\mathcal{U}\bar{L}^k \approx \mathcal{U}C(P^*_r(\lambda), r) \amalg \ldots \amalg \mathcal{U}C(P^*_{r+k-1}(\lambda), r+k-1) \amalg \mathcal{U}\bar{L}^{\#} .$$

The term $\mathcal{U}C(P^*_{r+s}(\lambda), r+s)$ is β^{r+s}-acyclic, and β^t vanishes on $E^t(\mathcal{U}\bar{L}^{\#})$ for $t < r+k$. Part (a) is evident, and for (b),

$$E^{r+s}(\mathcal{U}\bar{L}^k) = E^{r+s}(\mathcal{U}C(P^*_{r+s}(\lambda), r+s)) \amalg \ldots \amalg E^{r+s}(\mathcal{U}C(P^*_{r+k-1}(\lambda), r+k-1)) \amalg E^{r+s}(\mathcal{U}\bar{L}^{\#}) = E^{r+s}(\mathcal{U}\bar{L}^k_s) = \mathcal{U}\bar{L}^k_s \otimes \mathbb{Z}_p.$$

Letting g be the homomorphism of Lemma 2.6(b), we see that $\mathcal{U}g$ induces an isomorphism on E^{r+k}, hence an isomorphism on E^t for all $t \ge r+k$. This means that $E^t(\mathcal{U}\bar{L}^k) = E^t(\mathcal{U}\bar{L}^{\#})$, for $t \ge r+k$. The BSS of $\mathcal{U}\bar{L}^{\#}$ was computed in [CMN1, Section 12]. The results are (c) and (d). For (e) apply $\mathcal{U}$ to (2P) and use Lemma 2.2; we have already seen that x_1 survives to $E^{\infty}(\mathcal{U}L^k)$, so the homomorphisms $E^t(\mathcal{U}L^k \longrightarrow \Lambda(x_1))$ are surjective for each t.

Let us take a closer look at the computation in [CMN1, Section 12], referred to in the above proof. In our notation it is actually a computation of the BSS for $\mathcal{U}L^0(m, p^0)$, but $E^q(\mathcal{U}L^0(m,p^0)) = E^{q+t}(\mathcal{U}L^0(m,p^t))$. In order to do the computation, Cohen-Moore-Neisendorfer find a sequence of sub-dgL's of $\bar{L}^\#$,

$$\bar{L}^\# = \bar{L}^\#_{(0)} \supseteq \bar{L}^\#_{(1)} \supseteq \bar{L}^\#_{(2)} \supseteq \dots ,$$

for which $\bar{L}^\#_{(\infty)} = \bigcap_{j=0}^\infty \bar{L}^\#_{(j)}$ equals $C(P^\#(\lambda), r+k)$ for a certain power series $P^\#(\lambda)$. The $\bar{L}^\#_{(j)}$'s fit into short exact sequences of dgL's,

$$\text{(2W)} \qquad 0 \longrightarrow \bar{L}^\#_{(j)} \longrightarrow \bar{L}^\#_{(j-1)} \longrightarrow \Xi(2np^{k+j}-1, r+k+1) \longrightarrow 0 .$$

Theorem 2.8. *There exists*

$$\mathcal{P}(\lambda) = (0, \dots, 0, P_r(\lambda), P_{r+1}(\lambda), \dots, P_{r+k}(\lambda), 0, \dots)$$

such that $C(\mathcal{P}(\lambda))$ embeds in $(\bar{L}^k, d)$. The embedding carries the summand $C(P_{r+s}(\lambda), r+s)$ into $\bar{L}^k_s$, for $0 \le s \le k$. As differential coalgebras,

$$\text{(2X)} \qquad \mathcal{U}(L^k, d) \approx \Big(\bigotimes_{j=1}^{\infty} \mathcal{U}\Xi(2np^{k+j}-1, r+k+1)\Big) \otimes \mathcal{U}C(\mathcal{P}(\lambda)) ;$$

$$\text{(2Y)} \qquad \mathcal{U}(L^k, d) \approx (\Lambda(x_1), 0) \otimes \mathcal{U}(\bar{L}^k, d) .$$

Proof. Because of (2P), (2Y) follows from (2X). By Lemma 2.6(c) we may put $\bar{L}^k_{(0)} = \bar{L}^k = C(\mathcal{P}_{(0)}(\lambda)) \amalg \bar{L}_{(0)}$, where $\mathcal{P}_{(0)}(\lambda) = \mathcal{P}^*(\lambda)$. We are going to apply Lemma 2.3. As the short exact sequence $0 \longrightarrow K' \longrightarrow M' \longrightarrow M \longrightarrow 0$ in Lemma 2.3 we use the short exact sequence

$$0 \longrightarrow \bar{L}^\#_{(1)} \longrightarrow \bar{L}^\#_{(0)} \longrightarrow \Xi(2np^{k+1}-1, r+k+1) \longrightarrow 0 ,$$

obtained by setting $j=1$ in (2W). As the "$C(\mathcal{P}(\lambda))$" in Lemma 2.3 we take $C(\mathcal{P}^*(\lambda))$. We obtain the short exact sequence

$$0 \longrightarrow \bar{L}^k_{(1)} \longrightarrow \bar{L}^k_{(0)} \longrightarrow \Xi(2np^{k+1}-1, r+k+1) \longrightarrow 0 ,$$

where $\bar{L}^k_{(1)}$ equals $\bar{L}^\#_{(1)} \amalg C(\mathcal{P}_{(1)}(\lambda))$ for a certain $\mathcal{P}_{(1)}(\lambda)$. One property that $\mathcal{P}_{(1)}(\lambda)$ inherits from $\mathcal{P}_{(0)}(\lambda)$ is that the tth formal power series entry in $\mathcal{P}_{(1)}(\lambda)$ is zero, except when $r \le t \le r+k-1$.

We next apply Lemma 2.3 again, with $0 \longrightarrow K' \longrightarrow M' \longrightarrow M \longrightarrow 0$ being the short sequence

$$0 \longrightarrow \bar{L}^\#_{(2)} \longrightarrow \bar{L}^\#_{(1)} \longrightarrow \Xi(2np^{k+2}-1, r+k+1) \longrightarrow 0$$

obtained by setting $j=2$ in (2W), and with the "$C(\mathcal{P}(\lambda))$" of 2.3 being $C(\mathcal{P}_{(1)}(\lambda))$. We get

$$0 \longrightarrow \bar{L}^k_{(2)} \longrightarrow \bar{L}^k_{(1)} \longrightarrow \Xi(2np^{k+2}-1, r+k+1) \longrightarrow 0 ,$$

where $\bar{L}^k_{(2)} = \bar{L}_{(2)} \amalg C(\mathcal{P}_{(2)}(\lambda))$ for some $\mathcal{P}_{(2)}(\lambda)$. Again, the tth entry of $\mathcal{P}_{(2)}(\lambda)$ is zero unless $r \le t < r+k$. Continuing in this manner, we obtain short exact sequences

$$\text{(2Z)} \qquad 0 \longrightarrow \bar{L}^k_{(j)} \longrightarrow \bar{L}^k_{(j-1)} \longrightarrow \Xi(2np^{k+j}-1, r+k+1) \longrightarrow 0 ,$$

where $\bar{L}^k_{(j)}$ has the form $\bar{L}^{\#}_{(j)} \amalg C(\mathcal{P}_{(j)}(\lambda))$ and $\mathcal{P}_{(j)}(\lambda)$ has zero entries outside the range $r \le t < r+k$. Put $\bar{L}^k_\infty = \bigcap_{j=0}^\infty \bar{L}^k_{(j)}$, which has the form $\bar{L}^{\#}_{(\infty)} \amalg C(\mathcal{P}_{(\infty)}(\lambda))$ for a certain $\mathcal{P}_{(\infty)}(\lambda)$. Let $\mathcal{P}(\lambda)$ be obtained by changing the $(r+k)^{\text{th}}$ entry of $\mathcal{P}_{(\infty)}(\lambda)$ from 0 to $P^{\#}(\lambda)$. Then $\bar{L}^k_{(\infty)} = C(\mathcal{P}(\lambda))$.

Now use (2Z) at $j = 1$ to see that, as differential coalgebras,

$$\mathcal{U}\bar{L}^k \;=\; \mathcal{U}\bar{L}^k_{(0)} \;\approx\; \mathcal{U}\Xi(2np^{k+1} - 1, r+k+1) \otimes \mathcal{U}\bar{L}^k_{(1)}\,.$$

By induction on m (for the inductive step use (2Z) at $j = m$),

$$\mathcal{U}\bar{L}^k \;\approx\; \Big(\bigotimes_{j=1}^{m} \mathcal{U}\Xi(2np^{k+j} - 1, r+k+1)\Big) \otimes \mathcal{U}\bar{L}^k_{(m)}\,.$$

Letting $m \longrightarrow \infty$ yields (2X).

Remark. The sub-dgL $\bar{L}^k$ of L^k inherits the $*$-grading from L^k. Specifically, $(\bar{L}^k, d)$ is bigraded, with $|y_j|_* = 1$ and $|z'_j|_* = |z_j|_* = 2$ for all $j \ge 2$, and $|d|_* = +1$. Likewise each $\bar{L}^k_s$ is bigraded. The $*$-grading will become significant in later chapters.

CHAPTER 3

The Differential Graded Lie Algebra M^k

Building upon the previous chapter, we analyze here another dgL, to be denoted (M^k, d), whose enveloping algebra will turn out to have the same quasi-isomorphism type as the Adams-Hilton model for the space D_k.

Again, it will be convenient to have the various algebraic properties already at our disposal when the time comes to correlate them with the topology. We focus in this section on $(\mathcal{U}M^k, d)$ as a dgH, postponing until Chapter 4 the explicit computation of its BSS.

We fix throughout this section a prime p, the exponent $r \geq 1$, and the connectivity parameter $n \geq 1$. We allow k to vary, however: indeed, we shall frequently see inductions with k as the inductive variable.

Lemma 3.1. *Write $L^k = \mathbb{L}\langle x_1, x_2, \dots\rangle$ as in Chapter 2, and write $L^{k+1} = \mathbb{L}\langle x'_1, x'_2, \dots\rangle$. There is a dgL monomorphism $\gamma^k : L^k \to L^{k+1}$ defined by $\gamma^k(x_j) = x_j$ if $j < p^{k+1}$ but $\gamma^k(x_j) = px'_j$ if $j \leq p^{k+1}$. Thus $L^{k+1} = (L^k \amalg \mathbb{L}\langle x'_{p^{k+1}}, x'_{p^{k+1}+1}, \dots\rangle)/I$, where I is the Lie ideal generated by all $px'_j - x_j$ for $j \geq p^{k+1}$.*

Proof. One needs only to verify that γ^k commutes with the differentials. Since γ^k may be written as $\gamma^k(x_j) = p^{\alpha_{k+1}(j)-\alpha_k(j)}x'_j$, we need only to check that

$$\frac{p^r}{2}\langle m|i\rangle_k p^{\alpha_{k+1}(i)-\alpha_k(i)} p^{\alpha_{k+1}(m-i)-\alpha_k(m-k)} = \frac{p^r}{2}\langle m|i\rangle_{k+1} p^{\alpha_{k+1}(m)-\alpha_k(m)} ,$$

since these are the respective coefficients of $[x'_i, x'_{m-i}]$ in $\gamma^k d(x_j)$ and in $d\gamma^k(x_j)$. This check is trivial.

We proceed to define M^k, but we hold off briefly on defining the differential on M^k. Put

$$M^k = \mathbb{L}\langle b, a_0, a_1, \dots, a_k\rangle / J ,$$

where $|b| = 2n$, $|a_s| = 2np^s - 1$ for $0 \leq s \leq k$, and J is the Lie ideal generated by all $\rho_s = \mathrm{ad}(b^{p^s(p-1)})(a_s) - pa_{s+1}$ for $0 \leq s < k$. Since M^{k+1} is obtained from M^k by adjoining the new generator a_{k+1} and dividing out by the new relation and $\mathrm{ad}(b^{p^k(p-1)})(a_k) - pa_{k+1}$, there is an obvious embedding of M^k into M^{k+1}. Denote this embedding by η^k. The mod p reduction of $\mathcal{U}M^k$ is obvious:

Lemma 3.2. *$\mathcal{U}M^k \otimes \mathbb{Z}_p = \mathbb{Z}_p\langle \overline{a}_0, \overline{b}, \overline{a}_1, \ldots, \overline{a}_k\rangle / \overline{J}$, where $\overline{J}$ is the ideal generated by all $\mathrm{ad}(\overline{b}^{p^s(p-1)})(\overline{a}_s)$, $0 \le s < k$. Also,*

$$\mathcal{U}M^{k+1} \otimes \mathbb{Z}_p = (\mathcal{U}M^k / \langle \overline{\rho}_k\rangle) \amalg \mathbb{Z}_p\langle \overline{a}_{k+1}\rangle, \tag{3A}$$

$\langle \overline{\rho}_k\rangle$ denoting the ideal generated by the single element $\overline{\rho}_k = \mathrm{ad}(\overline{b}^{p^k(p-1)})(\overline{a}_k)$.

Lemma 3.3. *For each $k \ge 0$, there is a short exact sequence of Lie $\mathbb{Z}_{(p)}$-algebras*

$$0 \to L^k \xrightarrow{f^k} M^k \to \mathbb{L}^{\mathrm{ab}}[b] \to 0 \tag{3B}$$

in which $f^k(x_j) = \mathrm{ad}(b^{j-p^s})(a_s)$, where $s = \alpha_k(j)$. Furthermore,

$$\eta^k f^k = f^{k+1}\gamma^k . \tag{3C}$$

Proof. When $k = 0$, see [CMN1, Prop. 9.2]. Suppose we have established that (3B) is short exact for some k and we want it for $k+1$. The surjection $M^k \to \mathbb{L}^{\mathrm{ab}}[b]$ obviously extends over $M^{k+1} = (M^k \amalg \mathbb{L}\langle a_{k+1}\rangle)/\langle \rho_k\rangle$, $\langle\rho_k\rangle$ denoting the Lie ideal generated by $\rho_k = \mathrm{ad}(b^{p^k(p-1)})(a_k) - pa_{k+1}$. Thus

$$\begin{aligned}\ker(M^{k+1} \to \mathbb{L}^{\mathrm{ab}}[b]) &= \left(L^k \amalg \mathbb{L}\langle \mathrm{ad}(b^j)(a_{k+1})\rangle_{j\ge 0}\right) / \langle \mathrm{ad}(b^j)(\rho_k)\rangle_{j \ge 0} \\ &= \left(L^k \amalg \mathbb{L}\langle x'_{p^{k+1}}, x'_{p^{k+1}+1}, \ldots\rangle\right) / \langle x_{p^{k+1}+j} - px'_{p^{k+1}+j}\rangle_{j\ge 0} \\ &= L^{k+1} .\end{aligned}$$

The reader can check (3C).

Now we insert the differential. Let $d(b) = -p^r a_0$, and let $d(a_s) = f^k d(x_{p^s})$ in M^k for $0 \le s \le k$.

Lemma 3.4. *(M^k, d) is a dgL, and (3B) is a short exact sequence of dgL's.*

Proof. For (M^k, d) to be a dgL, it suffices that J be a differential ideal (i.e., $d(J) \subseteq J$), and for (3B) to be exact, it suffices that f^k commute with the differentials. When $k = 0$ the first assertion is trivial and the second is essentially done in [CMN1]. Let δ denote the differential $\delta = p^{-r}d$ on L^0. Cohen-Moore-Neisendorfer show (see [CMN1, Prop. 9.2 and p. 151]) that the dgL kernel of $(M^0, \delta) \to (\mathbb{L}^{\mathrm{ab}}[b], 0)$, a dgL they denote as $L^{(0)}$, equals $\mathbb{L}\langle x_1, x_2, \ldots\rangle$, with the formula for the differential being given by [CMN1, Lemma 4.3]. Identifying their generator $x_i \in L^{(0)}$ with $-x_i$ in our L^0, we have an isomorphism of their $(L^{(0)}, d)$ with our (L^0, δ). The dgL kernel of $(M^0, \delta) \to (\mathbb{L}^{\mathrm{ab}}[b], 0)$ is (L^0, δ), and consequently the dgL kernel of $(M^0, d) \to (\mathbb{L}^{\mathrm{ab}}[b], 0)$ is our dgL (L^0, d).

Now suppose the lemma is true for some $k \ge 0$. Let $\{x'_j\}$ denote the generators of L^{k+1} and $\{x_j\}$ the generators of L^k. In M^{k+1} we have

$$\begin{aligned}d(\rho_k) &= d\left(\mathrm{ad}\left(b^{p^k(p-1)}\right)(a_k)\right) - d(pa_{k+1} = d\left(\eta^k f^k\left(x_{p^{k+1}}\right)\right) - pd(a_{k+1}) \\ &= \eta^k f^k d(x_{p^{k+1}}) - pf^{k+1} d(x'_{p^{k+1}}) = f^{k+1}\gamma^k d(x_{p^{k+1}}) - f^{k+1} d(px'_{p^{k+1}}) \\ &= f^{k+1} d(\gamma^k(x_{p^{k+1}}) - px'_{p^{k+1}}) = 0 ,\end{aligned}$$

so (M^{k+1}, d) is a dgL. Similarly, putting $c = p^{\alpha_{k+1}(j)-\alpha_k(j)}$,

$$\begin{aligned} cdf^{k+1}(x_j') &= df^{k+1}\gamma^k(x_j) = d\eta^k f^k(x_j) = \eta^k f^k d(x_j) \\ &= f^{k+1}\gamma^k d(x_j) = f^{k+1} d\gamma^k(x_j) = cf^{k+1}d(x_j), \end{aligned}$$

so f^{k+1} commutes with d, as desired.

Two immediate consequences of Lemma 3.4 are given next.

Corollary 3.5.

(a) *There is a short exact sequence of dgH's*

$$\mathbb{Z}_{(p)} \to (\mathcal{U}L^k, d) \to (\mathcal{U}M^k, d) \to (\mathbb{Z}_{(p)}[b], 0) \to \mathbb{Z}_{(p)} . \tag{3D}$$

(b) *$\mathcal{U}M^k$ is a twisted tensor product or semi-tensor product* [Sm] *of $\mathcal{U}L^k$ and $\mathbb{Z}_p[b]$. The action of $\mathbb{Z}_{(p)}[b]$ on $\mathcal{U}L^k$ which determines the twisting is given by*

$$b \cdot x_m = \begin{cases} px_{m+1} & \text{if } m+1 = p^s,\ 1 \le s \le k\,; \\ x_{m+1} & \text{otherwise.} \end{cases}$$

Lemma 3.6. *There is a short exact sequence of dgL's*

$$0 \to (\overline{L}^k, d) \xrightarrow{\overline{f}^k} (M^k, d) \xrightarrow{q} \Xi(2n, r) \to 0, \tag{3E}$$

where $\overline{f}^k$ denotes the restriction of f^k to $\overline{L}^k$.

Proof. Let q carry b and a_0 to v and to u, and let $q(a_s) = 0$ for $s \ge 1$. Then q is a dgL surjection whose kernel we denote by K. We get a commuting diagram of dgL homomorphisms.

$$\begin{array}{ccccccccc} & & & & 0 & & 0 & & \\ & & & & \downarrow & & \downarrow & & \\ 0 & \longrightarrow & \overline{L}^k & \xrightarrow{\overline{e}} & L^k & \longrightarrow & \mathbb{L}^{ab}[a] & \longrightarrow & 0 \\ & & \downarrow & & \downarrow{\scriptstyle f^k} & & \downarrow & & \\ 0 & \longrightarrow & K & \longrightarrow & M^k & \xrightarrow{q} & \Xi(2n, r) & \longrightarrow & 0, \\ & & & & \downarrow & & \downarrow & & \\ & & & & \mathbb{L}^{ab}[b] & \xrightarrow{=} & \mathbb{L}^{ab}[b] & & \\ & & & & \downarrow & & \downarrow & & \\ & & & & 0 & & 0 & & \end{array} \tag{3F}$$

in which the indicated rows and columns are short exact sequences. It follows easily that the induced homomorphism on kernels $\overline{L}^k \to K$ is an isomorphism.

Remark. Like (L^k, d) and $(\overline{L}^k, d)$, the dgL (M^k, d) is bigraded, and we call the second grading the $*$*–grade*. To define it, put $|b|_* = 0$ and $|a_s|_* = 1$ for $0 \le s \le k$. Each relator ρ_s for $0 \le s < k$ is $*$-homogeneous (of $*$-degree 1), so the $*$-grade is well-defined on (M^k, d). Furthermore, $|d|_* = +1$ and (3B) is a short exact sequence of bigraded dgL's. All the homomorphisms appearing in (3F) preserve $*$-degree too. Note that with respect to the $*$-grade, M^k is a non-connected Lie algebra.

Our remaining task for Chapter 3 is the construction of a "free model" for $(\mathcal{U}M^k, d)$, along with a certain closely related dgL. By a *free model* we mean a free dga (B^k, d), together with a *quasi-isomorphism* or *quism* $g : (B^k, d) \to (\mathcal{U}M^k, d)$, a quism being a dga homomorphism which induces an isomorphism on homology. To motivate this task, recall that we have been studying M^k because we want an Adams-Hilton model for a certain space. Since Adams-Hilton models are always free dga's, however, $\mathcal{U}M^k$ cannot qualify for this role. The free model (B^k, d) will serve us instead.

The dga (B^k, d) has the form $\mathcal{U}(\widetilde{M}^k, d)$ for a certain dgL $(\widetilde{M}^k, d)$, so we build $(\widetilde{M}^k, d)$ directly.

Proposition 3.7. *For each $k \ge 0$, there exists a dgL $(\widetilde{M}^k, d)$ and a surjective dgL homomorphism $g^k : \widetilde{M}^k \to M^k$ having the following properties.*

(a) *As a Lie algebra, $\widetilde{M}^k = \mathbb{L}\langle \tilde{u}_0, \tilde{v}_0, \tilde{u}_1, \tilde{v}_1, \ldots, \tilde{u}_k, \tilde{v}_k\rangle$, where $|\tilde{u}_s| = 2np^s - 1$ and $|\tilde{v}_s| = 2np^s$.*

(b) *When $k = 0$, $\widetilde{M}^0 \approx M^0$ and g^0 is the isomorphism given by $g^0(\tilde{v}_0) = b$, $g^0(\tilde{u}_0) = a_0$.*

(c) *For $k \ge 1$ the homomorphism $\overline{i} : \widetilde{M}^{k-1} \to \widetilde{M}^k$, given by $\overline{i}(\tilde{u}_s) = \tilde{u}_s$ and $\overline{i}(\tilde{v}_s) = \tilde{v}_s$ for $s \le k-1$, is a dgL inclusion, and this square commutes:*

$$\text{(3G)} \qquad \begin{array}{ccc} \widetilde{M}^{k-1} & \xrightarrow{g^{k-1}} & M^{k-1} \\ \overline{i}\big\downarrow & & \big\downarrow i \\ \widetilde{M}^k & \xrightarrow{g^k} & M^k. \end{array}$$

(d) *Letting $\widetilde{L}^k$ denote the dgL kernel of the composition*

$$\widetilde{M}^k \xrightarrow{g^k} M^k \to \mathbb{L}^{\mathrm{ab}}[b] \,,$$

we have $g^k(\widetilde{L}^k) \subseteq L^k$. Furthermore, L^k is a dgL retract of $\widetilde{L}^k$, i.e., there exists a dgL homomorphism $\nu^k : L^k \to \widetilde{L}^k$ for which $g^k\nu^k$ is the identity on L^k.

(e) *There is an isomorphism of dgL's,*

$$\widetilde{L}^k \approx L^k \amalg C(Q(\lambda), 0),$$

where $Q(\lambda) = (1 - \lambda^{2n})^{-1}(\Sigma_{s=1}^k \lambda^{2np^s})$.

(f) *The homomorphism $\mathcal{U}g^k : \mathcal{U}\widetilde{M}^k \to \mathcal{U}M^k$ and its restriction $\mathcal{U}g^k : \mathcal{U}\widetilde{L}^k \to \mathcal{U}L^k$ are quisms.*

Proof. Take (a) as the definition of $\widetilde{M}^k$ as a Lie algebra. Define $g^k : \widetilde{M}^k \to M^k$ by $g^k(\tilde{v}_0) = b$, $g^k(\tilde{v}_s) = 0$ for $0 < s \le k$, and $g^k(\tilde{u}_s) = a_s$ for $0 \le s \le k$. Define $\overline{i} : \widetilde{M}^{k-1} \to \widetilde{M}^k$ as in (c). Then g^k is surjective and (a)–(c) are true as statements about Lie algebras.

To define the differential on $\widetilde{M}^k$, we first construct two sequences of elements $\{\tilde{w}_m\}_{m\geq 1}$ and $\{\tilde{w}'_m\}_{m\geq 1}$ in $\widetilde{M}^k$ as follows. Let $\tilde{w}_1 = \tilde{w}'_1 = \tilde{u}_0$. Given $\tilde{w}_i$ and $\tilde{w}'_i$ for $i < m$, $m \geq 2$, define

$$\text{(3H)} \qquad \tilde{w}'_m = [\tilde{v}_0, \tilde{w}_{m-1}] + \sum_{s=1}^{\alpha(m-1)} (p^r)\langle m-1|p^s-1\rangle[\tilde{v}_s, \tilde{w}_{m-p^s}],$$

where α and $\langle\ |\ \rangle$ are given by (2L) and (2M). Define $\tilde{w}_m$ by

$$\tilde{w}_m = \begin{cases} \tilde{u}_s & \text{if } m = p^s,\ 1 \leq s \leq k \\ \tilde{w}'_m & \text{otherwise.} \end{cases}$$

Notice that

$$g^k(\tilde{w}_m) = \begin{cases} a_s & \text{if } m = p^s,\ 0 \leq s \leq k \\ [b, g^k(\tilde{w}_{m-1})] & \text{otherwise,} \end{cases}$$

so by induction on m we have

$$\text{(3I)} \qquad g^k(\tilde{w}_m) = f^k(x_m),$$

where $f^k : L \to M^k$ is the dgL inclusion appearing in (3B).

We can now define the differential on $\widetilde{M}^k$. Put $d(\tilde{u}_0) = 0$ and $d(\tilde{v}_0) = -p^r\tilde{u}_0$. Once $d(\tilde{u}_j)$ and $d(\tilde{v}_j)$ have been defined for $j < s$, where $1 \leq s \leq k$, observe that $d(\tilde{w}'_m)$ is determined (by (2D)) for all $m \leq p^s$. Recursively (with respect to s) we may therefore define

$$\text{(3J)} \qquad d(\tilde{u}_s) = p^{-1}d(\tilde{w}'_{p^s}), \quad d(\tilde{v}_s) = -p\tilde{u}_s + \tilde{w}'_{p^s}.$$

Then $d^2 = 0$ on all generators, hence d is a differential.

The key formula is

$$\text{(3K)} \qquad d(\tilde{w}_m) = -\frac{p^r}{2}\sum_{i=1}^{m-1}\langle m|i\rangle[\tilde{w}_i, \tilde{w}_{m-i}].$$

To prove (3K), suppose it is true for $i < m$. A long but straightforward computation using (3J) and (3H) yields

$$\text{(3L)} \qquad d(\tilde{w}'_m) = -\frac{p^r}{2}\sum_{i=1}^{m-1}\langle m|i\rangle p^{\alpha(m)-\alpha(m-1)}[\tilde{w}_i, \tilde{w}_{m-i}].$$

If m is not p^s for some s in the range $1 \leq s \leq k$, then (3L) is precisely (3K). If $m = p^s$ then

$$d(\tilde{w}_{p^s}) = d(\tilde{u}_s) = p^{-1}d(\tilde{w}'_{p^s}),$$

and by (3L) $p^{-1}d(\tilde{w}'_{p^s})$ equals the right–hand side of (3K). Formula (3K) also clears up any doubt that (3J) is well–defined (at issue was whether $p^{-1}d(w'_{p^s})$ lies in $\widetilde{M}^k$ or merely in $\widetilde{M}^k \otimes \mathbb{Q}$). Part (c) also follows, since $\bar{i}(\tilde{w}_m) = \tilde{w}_m$ for $m < p^k$, so the needed formula

$d\bar{\imath} = \bar{\imath}d$ holds on all generators of $\widetilde{M}^{k-1}$. The formula $g^k d = dg^k$ is also easily checked, using (3I).

For part (d), there is obviously an induced homomorphism between dgL kernels in the diagram

$$\begin{array}{ccccccccc} 0 & \longrightarrow & \widetilde{L}^k & \longrightarrow & \widetilde{M}^k & \longrightarrow & \mathbb{L}^{ab}[\tilde{v}_0] & \longrightarrow & 0 \\ & & \downarrow & & \downarrow{\scriptstyle g_k} & & \downarrow{\scriptstyle \approx} & & \\ 0 & \longrightarrow & L^k & \xrightarrow{f^k} & M^k & \longrightarrow & \mathbb{L}^{ab}[b] & \longrightarrow & 0. \end{array} \tag{3M}$$

Define ν^k by $\nu^k(x_j) = \tilde{w}_j$, which is a right inverse for $g^k|_{L^k}$ by (3I). To prove (e), write $C(Q(\lambda), 0)$ as

$$C(Q(\lambda), 0) = \mathbb{L}\langle v_{sj}, u_{sj} \mid 1 \le s \le k,\ j \ge 0\rangle \ ,$$

$d(v_{sj}) = u_{sj}$, $|v_{sj}| = 2n(p^s + j)$. Extend ν^k to

$$\nu^k : \ L^k \amalg C(Q(\lambda), 0) \to \widetilde{L}^k$$

by putting $\nu^k(v_{sj}) = \mathrm{ad}^j(\tilde{v}_0)(\tilde{v}_s)$, $\nu^k(u_{sj}) = d\nu^k(v_{sj})$. Then the extended ν^k is an isomorphism (Lemma 2.5 helps to show this). For (f), $\mathcal{U}\nu^k$ and $\mathcal{U}(g^k|_{L^k})$ are inverse quisms because $\mathcal{U}C(Q(\lambda), 0)$ is acyclic. To get $\mathcal{U}g^k$ to be a quism, apply $\mathcal{U}$ to (3M) and get a homomorphism between two short exact sequences of dgH's. The induced homomorphism between the associated Milnor–Moore spectral sequences (i.e. the spectral sequences of Lemma 2.2) is an isomorphism on the E^2 terms, hence an isomorphism on the E^∞ terms. It follows that $H_*(\mathcal{U}g^k)$ is an isomorphism.

Although we will not explicitly use it in this book, we consider one more dgL that is closely related to $\widetilde{M}^k$. We denote it as (N^k, d). As motivation, $\mathcal{U}(N^k, d)$ will turn out to be the Adams–Hilton model for a space G_k which plays an essential role in factoring ΩD_k.

Proposition 3.8. *For each $k \ge 0$ there exists a dgL N^k over $\mathbb{Z}_{(p)}$ and a dgL monomorphism $h^k : N^k \to \widetilde{M}^k$ having the following properties.*

(a) *As a Lie algebra, $N^k = \mathbb{L}\langle u_0, v_0, u_1, v_1, \ldots, u_k, v_k\rangle$, where $|u_s| = 2np^s - 1$ and $|v_s| = 2np^s$.*

(b) *When $k = 0$, $N^0 \approx M^0$ and h^0 is the isomorphism given by $h^0(v_0) = \tilde{v}_0$ and $h^0(u_0) = \tilde{u}_0$.*

(c) *For $k \ge 1$ the homomorphism $\bar{\imath} : N^{k-1} \to N^k$, given by $\bar{\imath}(u_s) = u_s$ and $\bar{\imath}(v_s) = v_s$ for $s \le k - 1$, is a dgL inclusion, and this square commutes:*

$$\begin{array}{ccc} N^{k-1} & \xrightarrow{h^{k-1}} & \widetilde{M}^{k-1} \\ {\scriptstyle \bar{\imath}}\downarrow & & \downarrow{\scriptstyle \bar{\imath}} \\ N^k & \xrightarrow{h^k} & \widetilde{M}^k. \end{array}$$

(d) *Letting K^k denote the dgL kernel of the composite*

$$N^k \xrightarrow{h^k} \widetilde{M}^k \xrightarrow{g^k} M^k \to \mathbb{L}^{ab}[b] \ ,$$

we have $h^k(K^k) \subseteq \widetilde{L}^k$. Furthermore, L^k is a dgL retract of K^k, and there exists a dgL homomorphism $\mu^k : L^k \to K^k$ for which $h^k\mu^k = \nu^k : L^k \to \widetilde{L}^k$.

(e) *The dgL kernel*

$$J^k = \ker(N^k \xrightarrow{g^k h^k} M^k) \tag{3N}$$

is a free dgL.

(f) *Let τ^k: $\mathcal{U}M^k \longrightarrow \mathcal{U}N^k$ denote any right inverse for $\mathcal{U}(g^kh^k)$ as a homomorphism of free $\mathbb{Z}_{(p)}$– modules, and let Z^k denote a $\mathbb{Z}_{(p)}$–basis for $\mathcal{U}M^k$. The set*

$$Y^k = \{\mathrm{ad}(\tau^k(z))(v_s), \ \mathrm{ad}(\tau^k(z))(-pu_s + w'_{p^s}) \mid 1 \le s \le k, \ z \in Z^k\}$$

is a set of generators for J^k as a free Lie algebra. I.e., ignoring the differential we have $J^k = \mathbb{L}\langle Y^k\rangle$.

Proof. We begin with the observation that

$$p^{s-1} \mid \langle q|p^s - 1\rangle_k \tag{3O}$$

whenever $1 \le s \le k$ and $q \ge p^s$; this can be proved by considering cases, as in the proofs for the parts of Lemma 2.4. Let N^k be the Lie algebra given by (a), and define $h^k(v_0) = \tilde{v}_0$, $h^k(u_s) = \tilde{u}_s$ for $0 \le s \le k$, $h^k(v_s) = p^{r+s-1}\tilde{v}_s$ for $1 \le s \le k$. Define two sequences $\{w_m\}$ and $\{w'_m\}$ in N^k by $w'_1 = w_1 = u_0$ and for $m \ge 2$

$$\begin{aligned} w'_m &= [v_0, w_{m-1}] + \sum_{s=1}^{\alpha(m-1)} p^{-(s-1)}\langle m-1|p^s - 1\rangle [v_s, w_{m-p^s}] , \\ w_m &= \begin{cases} u_s & \text{if } m = p^s, \ 1 \le s \le k \\ w'_m & \text{otherwise.} \end{cases} \end{aligned} \tag{3P}$$

Notice that w'_m makes sense as an element of N^k because of (3O). By induction on m using (3P) and (3H) we have $h^k(w_m) = \tilde{w}_m$ for all $m \ge 1$. Define the differential d on N^k to be what it must be to make h^k a dgL homomorphism, namely,

$$d(v_0) = -p^r u_0, \ d(u_s) = p^{-1}d(w'_{p^s}), \ d(v_s) = p^{r+s-1}(-pu_s + w'_{p^s}).$$

Because $h^kd = dh^k$ we have

$$d(w_m) = (h^k)^{-1}d(\tilde{w}_m) = -\frac{p^r}{2}\sum_{i=1}^{m-1}\langle m|i\rangle[w_i, w_{m-i}] , \tag{3Q}$$

so the homomorphism $\mu^k : L^k \to N^k$, defined by $\mu^k(x_m) = w_m$, is a homomorphism of dgL's. Clearly μ^k factors through K^k and satisfies $h^k\mu^k = \nu^k$.

For parts (e) and (f) we ignore the differential entirely. We have (cf. 3.7(e))

$$K^k = L^k \amalg \mathbb{L}\langle u_{sj}, v_{sj} \mid 1 \le s \le k , \ j \ge 0\rangle . \tag{3R}$$

We may write (3R) as $K^k = L^k \amalg \mathcal{L}(U^k)$, where $U^k = \mathrm{Span}(u_{sj}, v_{sj})$. Since J^k is also the Lie algebra kernel of the restriction g^kh^k: $K^k \longrightarrow L^k$, the homomorphism

$$\mathrm{ad}(\ )(\ ) : \mathcal{U}L^k \otimes U^k \to J^k$$

gives a bijection of $\mathcal{U}L^k \otimes U^k$ with the indecomposables of J^k. (The proof of this assertion is just like the proof of Lemma 2.3.) This homomorphism extends uniquely to an isomorphism,

$$\mathcal{L}(\mathcal{U}L^k \otimes U^k) \xrightarrow{\approx} J^k \, . \tag{3S}$$

For (f), let

$$S \;=\; \{v_s,\, -pu_s + w'_{p^s} \mid 1 \le s \le k\} \, . \tag{3T}$$

It is clear that $S \subseteq \ker(g^k h^k) = J^k$, hence all of Y^k belongs to J^k. It suffices to prove that the mod p reductions of the elements of Y^k freely generate $J^k \otimes \mathbf{Z}_p$ as a (non-differential) Lie $\mathbf{Z}_p$–algebra. Write $\overline{J}^k$ for $J^k \otimes \mathbf{Z}_p$ and $\overline{N}^k$ for $N^k \otimes \mathbf{Z}_p$. Let $\overline{S} \subseteq \overline{J}^k$ denote the set of mod p reductions of the elements of S.

The set $\overline{S}$ is inert or strongly free (the two terms "inert" and "strongly free" are interchangeable) in the free $\mathbf{Z}_p$–algebra $\mathcal{U}\overline{N}^k = \mathbf{Z}_p\langle \bar{u}_0, \bar{v}_0, \ldots, \bar{u}_k, \bar{v}_k\rangle$. To see this, put a total ordering on the generators of $\mathcal{U}N^k$ via $\bar{u}_0 > \cdots > \bar{u}_k > \bar{v}_0 > \cdots > \bar{v}_k$, and apply [A1, Lemma 3.2]. In [HL] Halperin and Lemaire introduced the notion of inert sets in a Lie algebra over a field. (They worked only over $\mathbf{Q}$ but the results we quote are valid over $\mathbf{Z}_p$ as well.) We have $\overline{S}$ is inert in $N^k \iff \overline{S}$ is inert in $\mathcal{U}N^k$, and (by [HL, Théoremè 3.3]) $\overline{S}$ is inert in $N^k \iff \overline{J}^k/[\overline{J}^k, \overline{J}^k]$ is freely generated as a $\mathcal{U}(\overline{N}^k/\overline{J}^k)$– module (via the adjoint action) by the set $\overline{S}$. We may conclude that $\{\mathrm{ad}(\bar{z})(\bar{x})\}$ is a $\mathbf{Z}_p$–basis for $\overline{J}^k/[\overline{J}^k, \overline{J}^k]$, as $\bar{z}$ runs through a $\mathbf{Z}_p$–basis for $\mathcal{U}(\overline{N}^k/\overline{J}^k)$ and $\bar{x}$ runs through $\overline{S}$. Since the quotient $\overline{N}^k/\overline{J}^k$ is $M^k \otimes \mathbf{Z}_p$, the set of mod p reductions of

$$\{\mathrm{ad}(\tau^k(z))(x) \mid z \in Z^k \, , \, x \in S\}$$

comprises a minimal set of indecomposable Lie $\mathbf{Z}_p$–algebra generators for $\overline{J}^k$, as desired.

Remark. Like L^k and M^k, the dgL's $\widetilde{M}^k$, N^k, etc. are bigraded, by dimension and $*$-degree. We define each v_s or $\tilde{v}_s$ to have $*$–degree zero, and we put $|u_s|_* = |\tilde{u}_s|_* = 1$.

CHAPTER 4

The Bockstein Spectral Sequence for $\mathcal{U}M^k$

We continue our analysis of the dgH $(\mathcal{U}M^k, d)$ by computing its homology. Because of the quism $B^k \to \mathcal{U}M^k$, we are simultaneously computing the homology of (B^k, d), i.e., the homology of the space ΩD_k. We first introduce a trick for converting facts about polynomials in one variable into relations among the elements of $\mathcal{U}M^k$. After gaining fluency in the use of this trick we establish several formulas that are valid in $\mathcal{U}M^k$, and we compute explicitly the BSS for $\mathcal{U}M^k$. The highlights of Chapter 4 are the innocent-looking formulas (4M), whose geometric counterparts will be pivotal, and the determination that $\mathcal{U}M^k$ has exponent p^{r+k}, i.e., $p^{r+k}\overline{H}_*(\mathcal{U}M^k, d) = 0$.

We know that $d(b) = -p^r a_0$ in $\mathcal{U}M^k$, and the goal of the next several pages is to compute $d(b^{p^s})$. We plan to construct cycles $c_0, \ldots, c_k$ for which $d(b^{p^s}) = -p^{r+s}c_s$. The geometric analogs of the algebraic operations $\mathrm{ad}(b^{p^s})$ and $\mathrm{ad}(c_s)$ will play major roles in our analysis of the homotopy theory of D_k.

Because we shall work with many linear combinations of elements of the form b^iub^j for various $u \in \mathcal{U}M^k$, the following notation will be very convenient.

Notation. Fix $m \geq 0$ and let $u \in \mathcal{U}M^k$. Let $(\mathcal{U}M^k)_t$ denote the t-dimensional component of $\mathcal{U}M^k$. Let $\theta_m(u)(\ \ )$ denote the homomorphism from (the abelian group of) polynomials of degree $\leq m$ in the indeterminate x to $(\mathcal{U}M^k)_{2nm+|u|}$ defined by $\theta_m(u)(x^i) = b^iub^{m-i}$. When there is no ambiguity about m or m can be determined from the context (e.g. from the requirement that all summands in an equation in $\mathcal{U}M^k$ must share a common dimension) we shall write θ for θ_m. This will be especially convenient for formulas that involve $\theta_{m_j}(u_j)$ for several different m_j's and u_j's but which share a common value for $2nm_j + |u_j|$. Define $\epsilon(u, v)$ by

$$\epsilon(v, u) = \sum_{i=0}^{p-1} v^i u v^{p-1-i} .$$

Lemma 4.1.

(a) $\mathrm{ad}(b)(\theta_m(u)(f)) = \theta_{m+1}(u)((x-1)f)$, $\deg(f) \leq m$.
(b) $\theta_m(u + \lambda u')(f) = \theta_m(u)(f) + \lambda\theta_m(u')(f)$, $\lambda \in \mathbf{Z}_{(p)}$.

(c) $\theta(\theta(u)(f))(g) = \theta(u)(fg)$
(d) $\mathrm{ad}(b^t)(\theta(u)(f)) = \theta(u)((x-1)^t f)$
(e) $\epsilon(b^t, u) = \theta(u)(1 + x^t + \cdots + x^{(p-1)t}) = \theta(u)\left(\frac{x^{p^t}-1}{x^t-1}\right)$.

Proof. Straightforward.

We digress next to accumulate a few facts about the relationships among the polynomials $x^{p^t} - 1$ and $(x-1)^{p^t}$. Let *gcd*( ) denote the greatest common divisor of a set of integers.

Lemma 4.2. *Fix p, and let $r_{q,t}(x)$ denote the remainder when $(x-1)^q$ is divided by $x^{p^t} - 1$. Write*

$$r_{q,t}(x) = \sum_{m=0}^{p^t-1} \omega_{m,t}(q) x^m .$$

(a) *$r_{q,0}(x) = 0$ for any $q \geq 1$. If $lp^t < q \leq (l+1)p^t$, then the coefficients $w_{m,t}(q)$ satisfy*

(4A) $$p^l \gcd(p^t, q) | \omega_{m,t}(q) \gcd(p^{t-1}, m), \text{ for } t \geq 1, q \geq 1, 0 \leq m < p^t .$$

(b) *For $0 \leq t < k$, $r_{p^k,t+1}(x) - r_{p^k,t}(x)$ is divisible by $p^{p^{k-t-1}}(x^{p^t} - 1)$ in $\mathbb{Z}[x]$, hence it is also divisible by $p^{k-t}(x^{p^t} - 1)$.*
(c) *For $0 \leq t < k$, if $p^{t-k}(x^{p^t} - 1)^{-1}(r_{p^k,t+1}(x) - r_{p^k,t}(x))$ is written as the polynomial $\sum_{j=1}^{p^t(p-1)-1} \lambda_j^{k,t} x^j$, then*

$$p^{k-1} \quad | \quad \lambda_j^{k,t} \gcd(p^t, j), \quad 0 \leq j < p^t(p-1) .$$

(d) *$\lambda_j^{k,k-1}$ is divisible by p except when j is a multiple of p^{k-1}, and*

$$\lambda^{k,k-1}_{mp^{k-1}} \equiv p^{-1}(-1 + (-1)^{p-1-m}(p-1|m)) \pmod p, \; k \geq 1, \; 0 \leq m \leq p-2 .$$

In particular, the mod p congruence class of $\lambda^{k,k-1}_{mp^{k-1}}$ is independent of k.

Proof. We begin with the observation that

(4B) $$p^t \quad | \quad (p^t|j)\gcd(p^{t-1}, j) \text{ and } p^{t+1} \not| \quad (p^t|j)\gcd(p^{t-1}, j), \text{ for } 0 < j < p^t .$$

Next, check that (4A) holds when $l = 0$, i.e., for $q \leq p^t$. If $0 < q < p^t$ then $\omega_{m,t}(q) = (q|m)(-1)^{q-m}$ because $r_{q,t}(x) = (x-1)^q$. The relation

$$\gcd(p^t, q) \quad | \quad (q|m)\gcd(p^{t-1}, m)$$

is easily established. The same proof works when $q = p^t$ and $m \geq 0$. The remaining case, namely $q = p^t$ and $m = 0$, is trivial.

We show next that the sequence $\{\omega_{m,t}(q)\}_{q \geq 1}$ satisfies a linear recursion of order $p^t - 1$. Let $\{w_j\}_{j \geq 0}$ be the sequence defined by $w_j = 1$ (resp. $w_j = 0$) if $j \equiv m$ (resp. $j \not\equiv m$) $(\bmod p^t)$. Since the sequence $\{w_j\}$ satisfies the obvious linear recursion $w_j = w_{j-p^t}$, whose characteristic polynomial $x^{p^t}-1$ has distinct roots $\{\xi_i\}$, we may write w_j as a linear combination of the jth powers of $\{\xi_1,\ldots,\xi_{p^t}\}$,

$$w_j=\sum_{i=1}^{p^t}\zeta_i(\xi_i)^j,\quad j\geq 0,$$

for certain complex constants ζ_i. Then

$$\omega_{m,t}(q)=\sum_{j=0}^{q}(-1)^{q-j}(q|j)w_j=\sum_{i=1}^{t}\zeta_i\sum_{j=0}^{q}(-1)^{q-j}(\xi_i)^j(q|j)=\sum_{i=1}^{p^t}\zeta_i(\xi_i-1)^q,$$

so $\{\omega_{m,t}(q)\}$ satisfies a linear recursion whose characteristic polynomial's roots are $\{\xi_i-1\}$. This polynomial is $(1+x)^{p^t}-1$, so we have

(4C) $$\omega_{m,t}(q)=-\sum_{j=1}^{p^t-1}(p^t|j)\omega_{m,t}(q-j)\quad\text{for } q\geq p^t.$$

Suppose inductively that (4A) holds below q and we wish to establish it at some $q>p^t$. Choose l so that $lp^t<q\leq(l+1)p^t$. We handle the cases $q<(l+1)p^t$ and $q=(l+1)p^t$ separately. Suppose $q<(l+1)p^t$, so $\gcd(p^t,q)|p^{t-1}$. For j in the range $1\leq j\leq p^t-1$ we have

$$\gcd(p^t,q)\gcd(p^t,j)\quad|\quad p^{t-1}\gcd(p^t,q,j)\quad|\quad p^{t-1}\gcd(p^t,q-j),$$

hence

$$(p)\gcd(p^t,q)\gcd(p^t,j)\quad|\quad p^t\gcd(p^t,q-j).$$

Multiply both sides of this by $(p^t|j)$ and simplify using (4B) to obtain

$$(p)\gcd(p^t,q)\quad|\quad(p^t|j)\gcd(p^t,q-j).$$

Inductively, since $q-j>(l-1)p^t$, we have by (4A)

$$p^{l-1}\gcd(p^t,q-j)\quad|\quad\omega_{m,t}(q-j)\gcd(p^{t-1},m).$$

Combining these last two relations gives

$$p^l\gcd(p^t,q)\quad|\quad(p^t|j)\omega_{m,t}(q-j)\gcd(p^{t-1},m)\quad\text{if } 1\leq j\leq p^t-1.$$

Using (4C), it follows by summing over j that

$$p^l\gcd(p^t,q)\quad|\quad\omega_{m,t}(q)\gcd(p^{t-1},m),$$

which is (4A) at q.

The case $q=(l+1)p^t$ is handled similarly. Now we know that $q-j>lp^t$ when $1\leq j\leq p^t-1$ and so

$$p^l\gcd(p^t,q-j)\quad|\quad\omega_{m,t}(q-j)\gcd(p^{t-1},m).$$

Since $p^t|q$, this may be rewritten as

$$p^l \gcd(j, p^t) \quad | \quad \omega_{m,t}(q-j) \gcd(p^{t-1}, m) .$$

Multiply through by $(p^t|j)$ to obtain

$$p^l p^t \quad | \quad \omega_{m,t}(q-j)(p^t|j) \gcd(p^{t-1}, m) \quad \text{if } 1 \le j \le p^t - 1 .$$

Sum over j to obtain

$$p^l p^t \quad | \quad \omega_{m,t}(q) \gcd(p^{t-1}, m) .$$

Since $p^t = \gcd(p^t, q)$, this is the desired conclusion.

(b) By replacing $\gcd(p^{t-1}, m)$ by p^{t-1} and q by p^k (so $l = p^{k-t} - 1$), we find that every $\omega_{m,t}(p^k)$ is divisible by $p^{p^{k-t}}$. The coefficient of x^m in $r_{p^k,t+1}(x) - r_{p^k,t}(x)$ equals $\omega_{m,t+1}(p^k) - \omega_{m,t}(p^k)$, which is divisible by $p^{p^{k-t-1}}$. The polynomial $r_{p^k,t+1}(x) - r_{p^k,t}(x)$ is divisible by $x^{p^t} - 1$ because $r_{p^k,t+1}(x)$ and $r_{p^k,t}(x)$ are both congruent to $(x-1)^{p^k}$ modulo $x^{p^t} - 1$.

(c) By part (b) the expression $p^{t-k}(x^{p^t} - 1)^{-1}(r_{p^k,t+1}(x) - r_{p^k,t}(x))$ is a polynomial with integer coefficients, and it has degree at most $p^{t+1} - p^t - 1$. The coefficient $\lambda_j^{k,t}$ is p^{t-k} times a certain sum of coefficients belonging to the set $\{\omega_{m,t}(p^k), \omega_{m,t+1}(p^k) | m \equiv j (\bmod p^t)\}$, so we deduce that

$$\gcd(p^{p^{k-t}-1} p^t,\ p^{p^{k-t-1}-1} p^{t+1}) p^{t-k} \quad | \quad \lambda_j^{k,t} \gcd(p^t, j) ,$$

and hence that

$$p^{p^{k-t-1}+2t-k} \quad | \quad \lambda_j^{k,t} \gcd(p^t, j) . \tag{4D}$$

Part (c) now follows from the trivial observation that $k - 1 \le p^{k-t-1} + 2t - k$ if $k > t > 0$, except for the case of $p = 2$ with t equal to $k-2$ or $k-3$. The proof when $t = 0$ works just the same way, with $\omega_{m,0}(p^k) = 0$. The remaining cases, $p = 2$ with $t = k-2$ or $t = k-3$, can be handled individually. We omit these arguments.

(d) Left as an exercise in working with binomial coefficients.

Proposition 4.3. *In $\mathcal{U}M^k$ there exist cycles $c_0, c_1, \ldots, c_k$ with the following properties:*

(a) *$c_0 = a_0$, and for $s > 0$*

$$pc_s = \epsilon(b^{p^{s-1}},\ c_{s-1}) . \tag{4E}$$

(b) *For $k \ge s > t \ge 0$ and $0 \le i < p^t(p-1)$, there are integers $\lambda_i^{s,t}$ such that*

$$a_s = c_s + \sum_{t=0}^{s-1} \sum_{i=0}^{p^t(p-1)-1} \lambda_i^{s,t} b^i c_t b^{p^s - p^t - i} \tag{4F}$$

and

$$p^{s-1} \quad | \quad \lambda_i^{s,t} \gcd(p^t, i) . \tag{4G}$$

(c) *For* $s \geq 1$,

$$a_s \equiv c_s + \sum_{i=0}^{p-2} \lambda_i^{1,0} (b^{p^{s-1}})^i c_{s-1} (b^{p^{s-1}})^{p-1-i} \quad (\mod p)\,.$$

(d) $d(b^{p^s}) = -p^{r+s} c_s$.

(e) *The following is an alternate definition of the dga* $(\mathcal{U}M^k, d)$. *In the free* $\mathbb{Z}_{(p)}$-*algebra on generators denoted* $\{b, c_0, c_1, \ldots, c_k\}$ *with* $|b| = 2n$ *and* $|c_t| = 2np^t - 1$, *let* $\sigma_t = pc_{t+1} - \epsilon(b^{p^t}, c_t)$ *for* $0 \leq t < k$. *Then* $\mathcal{U}M^k$ *is isomorphic with* $\mathbb{Z}_{(p)}\langle b, c_0, \ldots, c_k\rangle / \langle \sigma_0, \ldots, \sigma_{k-1}\rangle$, *with differential determined by* $d(b) = -p^r c_0$ *and* $d(c_t) = 0$ *for* $0 \leq t \leq k$.

Proof. Induct on k. When $k = 0$, we need only to set $c_0 = a_0$. Suppose the proposition is true for $\mathcal{U}M^{k-1}$. Because $\mathcal{U}\eta^{k-1}$ is an isomorphism of $\mathcal{U}M^{k-1}$ with $\mathcal{U}M^k$ in dimensions below $2np^k - 2$, there are elements $\{c_0, \ldots, c_{k-1}\}$ in $\mathcal{U}M^{k-1}$ whose $(\mathcal{U}\eta^{k-1})$-images satisfy (a) through (d) for $\mathcal{U}M^k$. We may assume, therefore, that $c_0, \ldots, c_{k-1}$ exist, and it remains only to construct c_k and to verify its properties. In other words, we may assume henceforth that $s = k$.

The integers $\{\lambda_i^{k,t}\}$ of (4F) will be precisely the coefficients $\{\lambda_i^{k,t}\}$ defined by Lemma 4.2(c). Our plan is to show that the equation

$$\text{(4H)} \qquad \operatorname{ad}(b^{p^k - p^{k-1}})(a_{k-1}) = \epsilon(b^{p^{k-1}}, c_{k-1}) + p \sum_{t=0}^{k-1} \sum_{i=0}^{p^t(p-1)-1} \lambda_i^{k,t} b^i c_t b^{p^k - p^t - i}$$

is true. Since the left-hand side of (4H) equals pa_k in $\mathcal{U}M^k$, this shows $\epsilon(b^{p^{k-1}}, c_{k-1})$ to be divisible by p in $\mathcal{U}M^k$.

We may therefore define c_k by

$$\text{(4I)} \qquad c_k = a_k - \sum_{t=0}^{k-1} \sum_{i=0}^{p^t(p-1)-1} \lambda_i^{k,t} b^i c_t b^{p^k - p^t - i}\,,$$

and then (4E) and (4F) will follow at once. Proposition 4.3(c) is obtained by combining (4I) with Lemma 4.2(d). Part 4.3(d) will also be an immediate consequence because (we know 4.3(d) is valid for $s < k$)

$$d(b^{p^k}) = d((b^{p^{k-1}})^p) = \epsilon(b^{p^{k-1}}, d(b^{p^{k-1}})) = -p^{r+k-1}\epsilon(b^{p^{k-1}}, c_{k-1}) = -p^{r+k} c_k\,.$$

For 4.3(e), note that (4F) shows that $\{b, c_0, \ldots, c_k\}$ generates $\mathcal{U}M^k$; the relators $\{\sigma_t\}$ obviously vanish in $\mathcal{U}M^k$, so $\mathcal{U}M^k$ is a quotient of $\mathbb{Z}_{(p)}\langle b, c_0, \ldots, c_k\rangle / \langle \sigma_0, \ldots, \sigma_{k-1}\rangle$. Both algebras are rationally free on $\{b, c_0\}$, and their Hilbert series when tensored with $\mathbb{Z}_p$ coincide too (since $\{\sigma_0, \ldots, \sigma_{k-1}\}$ also becomes a strongly free set over $\mathbb{Z}_p$), so this quotient homomorphism is an isomorphism. Thus all of Proposition 4.3 hinges on formula (4H).

Since $a_{k-1} = p^{-k+1} \operatorname{ad}(b^{p^{k-1}-1})(a_0)$, the left-hand side of (4H) may also be written as $p^{-k+1} \operatorname{ad}(b^{p^k - 1})(a_0)$. Using Lemma 4.1, (4H) is equivalent to

$$\text{(4J)} \qquad p^{-k+1}\theta(a_0)((x-1)^{p^k - 1}) = \theta(c_{k-1})\left(\frac{x^{p^k} - 1}{x^{p^{k-1}} - 1}\right) + p \sum_{t=0}^{k-1} \theta(c_t)(f_t)\,,$$

where $f_t(x)$ is the polynomial

$$f_t(x) = \sum_{i=0}^{p^t(p-1)-1} \lambda_i^{k,t} x^i .$$

Using (4E) we may replace pc_t by $\theta(c_{t-1})\left(\frac{x^{p^t}-1}{x^{p^{t-1}}-1}\right)$; applying this fact repeatedly gives

$$p^t c_t = \theta(c_0)\left(\frac{x^{p^t}-1}{x-1}\right) . \tag{4K}$$

Multiplying (4J) through by p^{k-1} and applying (4K) gives

$$\theta(a_0)((x-1)^{p^{k-1}}) = \theta(a_0)\left(\frac{x^{p^t}-1}{x-1}\right) + \sum_{k=0}^{t-1} p^{k-t} f_t(x)\left(\frac{x^{p^t}-1}{x-1}\right) .$$

It therefore suffices to verify the equation in $\mathbb{Z}[x]$

$$(x-1)^{p^k-1} = \left(\frac{x^{p^t}-1}{x-1}\right) + \sum_{k=0}^{t-1} p^{k-t} f_t(x)\left(\frac{x^{p^t}-1}{x-1}\right) ,$$

which is of course equivalent to

$$(x-1)^{p^k} = (x^{p^k}-1) + \sum_{k=0}^{t-1} p^{k-t} f_t(x)(x^{p^t}-1) .$$

By Lemma 4.2(c) we have $p^{k-t} f_t(x)(x^{p^t}-1) = r_{p^k,t+1}(x) - r_{p^k,t}(x)$. The sum telescopes (because $r_{p^k,0}(x) = 0$) to the expression

$$(x-1)^{p^k} = (x^{p^k}-1) + r_{p^k,k}(x) ,$$

which is obviously true. This completes the proof.

The next lemma is a further application of our "polynomial trick" θ. Denote the image of $x_i \in \mathcal{U}L^k$ in $\mathcal{U}M^k$ also by x_i. Thus $x_i = p^{-\alpha(i)}\,\mathrm{ad}(b^{i-1})(a_0)$, where $\alpha(\ \ )$ is the function given by (2L). As before, we let $I_s = \{jp^s | j \text{ is an integer } \geq 2\}$.

Lemma 4.4.

(a) *For* $0 \leq s \leq k$,

$$c_s = \sum_{i=1}^{p^s} (p^s|i) p^{\alpha(i)-s} x_i b^{p^s-i} .$$

(b) *For* $0 \leq s \leq k$, *if* $m \in I_s$ *we have in* $\mathcal{U}M^k$

$$c_s \equiv \sum_{i=1}^{p^s} (\langle m|i\rangle\langle m|1\rangle^{-1}) x_i b^{p^s-i} \quad (\mod p^s) .$$

(c) $x_2 = [b, c_0]$, *and for* $0 < s \le k$,

$$b^{p^s} c_s - c_s b^{p^s} - x_{2p^s} = \sum_{i=2}^{2p^s-1} \mu_i x_i b^{2p^s - i} .$$

Here the coefficients $\{\mu_i\}$ *satisfy*

$$(p^{s-t})(p^{\alpha(i)-\alpha(i-1)}) \quad | \quad \mu_i ,$$

where t *is the largest integer such that* $i \in I_t$.

(d) *For* $0 \le q < k$,

$$c_{q+1} - c_q b^{p^q(p-1)} \quad = \quad x_{p^{q+1}} \quad + \quad \sum_{i=2}^{p^{q+1}-1} \nu_i x_i b^{p^{q+1}-i} .$$

The coefficients $\{\nu_i\}$ *satisfy*

$$p^{q-t} \quad | \quad \nu_i ,$$

where t *is the largest integer such that* $i \in I_t$.

Proof.

(a) Multiply through by p^s; use the polynomial trick to replace $p^s c_s$ by $\left(\frac{x^{p^s}-1}{x-1}\right)$ and $x_i b^{p^s - i}$ by $p^{-\alpha(i)}(x-1)^{i-1}$. Lemma 4.4(a) is equivalent to the assertion that

$$x^{p^s} - 1 = \sum_{i=1}^{p^s} (p^s|i)(x-1)^i ,$$

which is merely the binomial theorem.

(b) Using part (a), 4.4(b) is equivalent to the assertion that

$$\text{(4L)} \qquad \langle m|i\rangle \langle m|1\rangle^{-1} \equiv (p^s|i) p^{\alpha(i)-s} \quad (\mod p^s)$$

for any $m \in I_s$. The left-hand side of (4L) is $(m|i)(m|1)^{-1} p^{s_0}$, where the exponent s_0 is

$$s_0 = \alpha(i) + \alpha(m-i) - \alpha(m) - \alpha(1) - \alpha(m-1) + \alpha(m) ,$$

which equals $\alpha(i)$ because $\alpha(m-1) = \alpha(m-i)$ when $m \in I_s$ and $1 \le i \le p^s$. Writing $m = jp^s, j \ge 2$, congruence (4L) reduces to showing

$$(jp^s|i) j^{-1} \equiv (p^s|i) \quad (\mod p^{2s-\alpha(i)}),\ 1 \le i \le p^s ,$$

which we leave to the reader to check.

(c) The equation that determines the μ_i's is equivalent to

$$p^{-s} x^{p^s} \left(\frac{x^{p^s}-1}{x-1}\right) - p^{-s}\left(\frac{x^{p^s}-1}{x-1}\right) - p^{-s}(x-1)^{2p^s-1} = \sum_{i=1}^{2p^s-1} \mu_i p^{-\alpha(i)} (x-1)^{i-1} ,$$

which we rewrite as

$$(x^{p^s}-1)^2 = (x-1)^{2p^s} + \sum_{i=2}^{2p^s-1} \mu_i p^{s-\alpha(i)}(x-1)^i .$$

Substituting $y = x - 1$, this becomes

$$(y^{p^s} + (p^s|p^s-1)y^{p^s-1} + \cdots + (p^s|1)y)^2 = ((1+y)^{p^s}-1)^2 = y^{2p^s} + \sum_{i=2}^{2p^s-1} \mu_i p^{s-\alpha(i)} y^i .$$

We leave it as an exercise to verify that the coefficient of y^i in $(y^{p^s} + \cdots + (p^s|1)y)^2$ is divisible by $p^{2s-t-\alpha(i-1)}$.

(d) Left as an exercise.

We have two descriptions of $(\mathcal{U}M^k, d)$. The description of Proposition 4.3(e) is convenient to work with, while the description of Chapter 3 shows us that $(\mathcal{U}M^k, d)$ is simultaneously an enveloping algebra. Combining these two views of $(\mathcal{U}M^k, d)$ yields some powerful results. In particular, the fact that $(\mathcal{U}M^k, d)$ is the enveloping algebra of a dgL means that the formula (2K) is valid on $(\mathcal{U}M^k, d)$. From this observation we deduce at once

Lemma 4.5. *Let* $0 \le s \le k$.

(a) *In the dgH* $\mathcal{U}(M^k \amalg C(\lambda^q, t))$ *(notation from Chapter 2), we have*

$$\text{(4M)} \qquad \begin{aligned} d(\mathrm{ad}(b^{p^s})(v)) &= p^t \,\mathrm{ad}(b^{p^s})(u) - p^{r+s}\,\mathrm{ad}(c_s)(v), \\ d(\mathrm{ad}(c_s)(v)) &= -p^t\,\mathrm{ad}(c_s)(u) . \end{aligned}$$

(b) $\overline{L}^k$ *and* L^k *are Lie ideals of* M^k, *hence they are closed under the operations* $\mathrm{ad}(b^{p^s})$ *and* $\mathrm{ad}(c_s)$. *If* $w_1 \in \overline{L}^k$ *and* $d(w_1) = p^{r+s}w_2$, *then*

$$\text{(4N)} \qquad \begin{aligned} d(\mathrm{ad}(b^{p^s})(w_1)) &= p^{r+s}(\mathrm{ad}(b^{p^s})(w_2) - \mathrm{ad}(c_s)(w_1)) , \\ d(\mathrm{ad}(c_s)(w_1)) &= -p^{r+s}\,\mathrm{ad}(c_s)(w_2) . \end{aligned}$$

Notation. Let Y_s^k be the $\mathbb{Z}_{(p)}$-submodule of $\overline{L}^k$ spanned by $\{[y_i, y_j] \mid i \in I_s,\ j \in I_s\}$. Let

$$J_s^k = p^s Y_0^k + \cdots + pY_{s-1}^k + Y_s^k .$$

We also denote by J_s^k the image of J_s^k in $\mathcal{U}\overline{L}^k$, in M^k, or in $\mathcal{U}M^k$.

Lemma 4.6. *Let* $0 \le s \le k$. *The following are true in* $\overline{L}^k$.

(a) J_s^k *is closed under* $\mathrm{ad}(b^{p^s})$.

(b) $\mathrm{ad}(b^{p^s})(z'_{2p^s}) \equiv z'_{3p^s} \pmod{J_s^k}$.

(c) *If* $m - p^s \in I_s$, *then* $z'_m \equiv \mathrm{ad}(c_s)(y_{m-p^s}) \pmod{J_s^k}$.

Proof.
(a) We have

$$\mathrm{ad}(b^{p^s})[y_i, y_j] = \sum_{l=0}^{p^s} (p^s|l)[y_{i+l}, y_{j+p^s-l}] .$$

It suffices to prove that the coefficient $(p^s|l)$ is divisible by p^{t-t_1}, whenever i and j belong to I_t while $i+l$ and $j+p^s-l$ belong to I_{t_1}. This assertion is vacuous if $t_1 \geq t$, and when $t_1 < t$ we have $\gcd(p^t, l) = p^{t_1}$, whence

$$p^{t-t_1} \quad | \quad p^{s-t_1} \quad | \quad (p^s|l)$$

by (4B).

(b) Expanding the definition (2R) of z'_{2p^s} gives us

$$(2p^{2s})z'_{2p^s} = \left(\frac{1}{2}\right) \sum_{j=1}^{2p^s-1} (2p^s|j)[\mathrm{ad}(b^{j-1})(a_0),\ \mathrm{ad}(b^{2p^s-1-j})(a_0)] .$$

Apply $\mathrm{ad}(b^{p^s})$ to both sides to obtain

$$\begin{aligned}
&(2p^{2s})\,\mathrm{ad}(b^{p^s})(z'_{2p^s}) \\
&= \left(\frac{1}{2}\right) \sum_{m=1}^{2p^s-1} ((3p^s|m) - (p^s|m) - (p^s|3p^s - m))[\mathrm{ad}(b^{m-1})(a_0), \mathrm{ad}(b^{3p^s-1-m})(a_0)] \\
&\equiv \sum_{m=1}^{p^s} ((3p^s|m) - (p^s|m))[\mathrm{ad}(b^{m-1})(a_0), \mathrm{ad}(b^{3p^s-1-m})(a_0)] \quad (\bmod p^{2s}J^k_s) .
\end{aligned}$$

On the other hand, expanding (2R) for z'_{3p^s} gives us

$$(3p^{2s})z'_{3p^s} = \left(\frac{1}{2}\right) \sum_{m=1}^{3p^s-1} (3p^s|m)[\mathrm{ad}(b^{m-1})(a_0), \mathrm{ad}(b^{3p^s-1-m})(a_0)] \quad (\bmod p^{2s}J^k_s) .$$

Lemma 4.6(b) can now be deduced from the fact that

$$\gcd(p,3)p^{2s-t-\alpha(m)}|(3p^s|m) - 3(p^s|m) \quad \text{for } 1 \leq m \leq p^s ,$$

where t is the largest integer for which $m \in I_t$. We omit the proof of this fact.

(c) Compute $\mathrm{ad}(c_s)(y_{m-p^s})$ using Lemma 4.4(b), and compare the result with formula (2R) for z'_m.

We now turn our attention to the BSS for $(\mathcal{U}M^k, d)$. We shall abuse notation when working with this BSS in the following manner. When w is an element of a $\mathbb{Z}_{(p)}$-free chain complex, and when the mod p reduction of w survives to the $E^t(\ )$ term of the BSS, we shall write w (rather than $\overline{w}$ or $w_{(t)}$ or some other notation) for the class of w in the subquotient $E^t(\ )$. The one confusion that might result, and which we hope to forestall by calling the reader's attention to it now, is that $E^{r+s}(\mathcal{U}M^k)$ will have an element denoted "b^{p^s}" even though $E^{r+s}(\mathcal{U}M^k)$ contains no "b" for $s > 0$. I.e., the element denoted $b^{p^s} \in E^{r+s}(\mathcal{U}M^k)$ is not a pth power of anything else, but we keep the notation to remind us of its origin as $(b)^{p^s}$ in $\mathcal{U}M^k$.

Lemma 4.7.

(a) b^{p^s} *and* c_s *survive to the* $r+s$ *term of the BSS for* $\mathcal{U}M^k$, *and* $\beta^{r+s}(b^{p^s}) = -c_s$.
(b) *In the BSS for* $\mathcal{U}(M^k \amalg C(\lambda^q, t))$, *if* $t \geq r+s$, $0 \leq s \leq k$, *we have*

$$\text{(4O)}\qquad \begin{aligned} \beta^{r+s}\,\mathrm{ad}(b^{p^s})(v) &= \mathrm{ad}(b^{p^s})\beta^{r+s}(v) - \mathrm{ad}(c_s)(v)\,, \\ \beta^{r+s}\,\mathrm{ad}(c^s)(v) &= -\,\mathrm{ad}(c^s)\beta^{r+s}(v)\,. \end{aligned}$$

(c) *In the BSS for* $\mathcal{U}M^k$, *if* w *survives to* $E^{r+s}(\mathcal{U}M^k)$,$0 \leq s \leq k$, *then*

$$\text{(4P)}\qquad \begin{aligned} \beta^{r+s}\,\mathrm{ad}(b^{p^s})(w) &= \mathrm{ad}(b^{p^s})\beta^{r+s}(w) - \mathrm{ad}(c_s)(w)\,, \\ \beta^{r+s}\,\mathrm{ad}(c^s)(w) &= -\,\mathrm{ad}(c^s)\beta^{r+s}(w)\,. \end{aligned}$$

Proof. (a) This follows immediately from 4.3(d), provided that c_s survives to the E^{r+s} term. Since c_s is a cycle in $\mathcal{U}M^k$, it fails to survive only if it is in the image of β^m for some $m < r+s$. To see that this is impossible, consider that c_s has $*$-degree $+1$, so it can only be the β^m-image of something of $*$-degree zero. But the only elements in $\mathcal{U}M^k$ of $*$-degree zero and of dimension $|c_s|+1 = 2np^s$ are the scalar multiples of b^{p^s}.

(b) These follow from (4M).

(c) The fact that $w \in \mathcal{U}M^k$ survives to $E^{r+s}(\mathcal{U}M^k)$ means that there is some $w' \in \mathcal{U}M^k$ whose image in $E^{r+s}(\mathcal{U}M^k)$ is the same as that of w, and $p^{r+s}|d(w')$. Use the dga homomorphism

$$\mathcal{U}(M^k \amalg C(\lambda^{|w|}, r+s)) \to \mathcal{U}M^k$$

that is the identity on $\mathcal{U}M^k$ and which carries the generator v to w', to derive the two equations of (4P) from the two equations of (4O).

Theorem 4.8. *Let* $(E^t = E^t(\mathcal{U}M^k), \beta^t)$ *denote the Bockstein spectral sequence for* $\mathcal{U}M^k$.
(a) $E^1 = E^r = \mathcal{U}M^k \otimes \mathbb{Z}_p$, and for $0 \leq s \leq k$

$$E^{r+s} = \mathbb{Z}_p\langle b^{p^s}, c_s, c_{s+1}, \ldots c_k\rangle / \langle \overline{\sigma}_s, \ldots, \overline{\sigma}_{k-1}\rangle,$$

as a $\mathbb{Z}_p$-algebra, where $\overline{\sigma}_t = \epsilon((b^{p^s})^{p^{t-s}}, c_t)$.
(b) E^{r+s} *is generated as a* $\mathbb{Z}_p$*-algebra* $\{b^{p^s}, c_s, a_{s+1}, \ldots, a_k\}$.
(c) $E^{r+k+1} = \mathbb{Z}_p$, *i.e.,* $(E^{r+k+1})_{>0} = 0$. *Thus* $p^{r+k}\overline{H}_*(\mathcal{U}M^k, d) = 0$.

Proof. When $k = 0$ the theorem is trivial, since the dga $(\mathbb{Z}_{(p)}\langle b, a_0\rangle, d(b) = -p^r a_0)$ is well-known to be β^r-acyclic. Our proof proceeds by induction on k. Our inductive hypothesis is that $k \geq 1$ and that the theorem holds for $\mathcal{U}M^{k-1}$ for any parameter values of n and r. Our proof will utilize a comparison between the BSS of the dga $\mathcal{U}M^{k-1}$ constructed for the parameter set $(np, p, r+1, k-1)$ and the BSS for the dga $\mathcal{U}M^k$ constructed for parameter set (n, p, r, k). The generators of this $\mathcal{U}M^{k-1}$ will be denoted $\{b^*, c_0^*, \ldots, c_{k-1}^*\}$ and the families known as $\{y_j\}$ and $\{z_j\}$ in $\mathcal{U}\overline{L}^{k-1}$ will be denoted as $\{y_j^*\}$ and $\{z_j^*\}$. This leaves the unstarred notations available for their usual meanings in $\mathcal{U}M^k$.

Consider the diagram of dga homomorphisms

$$\text{(4Q)}\qquad \begin{array}{ccccc} \mathcal{U}\overline{L}^{k-1} & \xrightarrow{\mathcal{U}\overline{f}^{k-1}} & \mathcal{U}M^{k-1} & \xrightarrow{\mathcal{U}q^{k-1}} & \mathcal{U}\Xi(2np, r+1) \\ \downarrow{\scriptstyle g_1} & & \downarrow{\scriptstyle g^k} & & \downarrow{\scriptstyle g_0} \\ \mathcal{U}\overline{L}^k & \xrightarrow{\mathcal{U}\overline{f}^k} & \mathcal{U}M^k & \xrightarrow{\mathcal{U}q^k} & \mathcal{U}\Xi(2n, r)\,, \end{array}$$

in which the rows are obtained by applying $\mathcal{U}$ to (3E). Here g_0 carries the generators v^* and u^* to v^p and $v^{p-1}u$, respectively. Define g^k by $g^k(b^*) = b^p$ and $g^k(c_s^*) = c_{s+1}$. Then g^k is a dga (not dgH) homomorphism. The right-hand square of (4Q) commutes, and each row is a short exact sequence of Hopf algebras. The dotted arrow g_1 cannot be filled in.

The homomorphism q^k takes b^{p^s} to v^{p^s} and c_s to uv^{p^s-1}, $0 \le s \le k$. Since b^{p^s} and c_s survive to E^{r+s}, and since v^{p^s} and uv^{p^s-1} generate $E^{r+s}(\mathcal{U}\Xi(2n,r))$ as an algebra, it follows that $E^{r+s}(q^k)$ is surjective. Lemma 2.2 applies, and $E^{r+s}(\mathcal{U}\overline{L}^k)$ is the HAK of $E^{r+s}(q^k)$, for $0 \le s \le k$. One consequence of this is that the lowest dimensional non-trivial elements of $E^{r+1}(\mathcal{U}M^k)$ can be ascertained to the preimages of v^p and uv^{p-1}; these elements must be b^p and c_1. Since $(E^{r+1}(\mathcal{U}M^k), \beta^{r+1})$ is a dgH over $\mathbb{Z}_p$, conclude that b^p and c_1 are primitive in it.

Let us dispense with part (b) now. In view of 4.3(c), in E^r put

$$e_s = \sum_{i=0}^{p-2} \lambda_i^{1,0}(b^{p^{s-1}})^i c_{s-1}(b^{p^{s-1}})^{p-1-i}$$

for $1 \le s \le k$. Since a_s survives to E^{r+s-1} (it has $*$-degree $+1$ and $d(a_s)$ is divisible by p^{r+s-1}), we have $a_t = c_t + e_t$ in E^{r+s-1} whenever $1 \le s \le t \le k$. Since e_t is decomposable in E^{r+s-1}, 4.8(b) follows from 4.8(a). We need not say anything more about 4.8(b).

Now we use (4Q) to establish 4.8(a). It is obvious that 4.8(a) is true when $s = 0$. Apply the functor $E^{r+1}(\quad)$ to the diagram (4Q):

$$\text{(4R)}\qquad \begin{array}{ccccc} E^{r+1}(\mathcal{U}\overline{L}^{k-1}) & \longrightarrow & E^{r+1}(\mathcal{U}M^{k-1}) & \longrightarrow & E^{r+1}(\mathcal{U}\Xi(2np, r+1)) \\ \downarrow{\scriptstyle \overline{g}_1} & & \downarrow{\scriptstyle E^{r+1}(g^k)} & & \approx\downarrow{\scriptstyle E^{r+1}(g_0)} \\ E^{r+1}(\mathcal{U}\overline{L}^k) & \longrightarrow & E^{r+1}(\mathcal{U}M^k) & \longrightarrow & E^{r+1}(\mathcal{U}\Xi(2n, r))\,. \end{array}$$

We assert that $E^{r+1}(g^k)$ is a homomorphism of Hopf algebras. To see this, it suffices to check that the primitive generators $\{b^*, a_0^*, \ldots, a_{k-1}^*\}$ are carried to primitives in $E^{r+1}(\mathcal{U}M^k)$. We have already seen that the images b^p and c_1 of b^* and a_0^* are primitive. Since $E^{r+1}(g^k)((b^*)^{p^{s-1}}) = (b^p)^{p^{s-1}}$ and $E^{r+1}(g^k)(c_{s-1}^*) = c_s$, 4.3(c) gives $E^{r+1}(g^k)(e_s^*) = e_{s+1}$ for $s \ge 1$. Deduce that $E^{r+1}(g^k)(a_s^*) = E^{r+1}(g^k)(c_s^* + e_s^*) = c_{s+1} + e_{s+1} = a_{s+1}$, which is primitive in $E^{r+1}(\mathcal{U}M^k)$.

It follows that $E^{r+1}(g^k)$ is a homomorphism of dgH's, whence the induced homomorphism $\overline{g}_1$ between HAK's in (4R) can be filled in.

We show next that $\overline{g}_1$ is an isomorphism. By Proposition 2.7, $E^{r+1}(\mathcal{U}\overline{L}^{k-1}) = \mathbb{Z}_p\langle y_i^*, z_i^* | i \ge 2\rangle$ while $E^{r+1}(\mathcal{U}\overline{L}^k) = \mathbb{Z}_p\langle y_{ip}, z_{ip}' | i \ge 2\rangle$. It suffices to show that

(4S) $\quad \overline{g}_1(y_i^*) = y_{ip}$, and

(4T) $\quad \overline{g}_1(z_i^*) = z_{ip}' +$ (decomposable Lie elements in $E^{r+1}(\mathcal{U}\overline{L}^k)$).

Lemma 4.4(c) gives $\overline{g}_1(y_2^*) = y_{2p}$, hence $\overline{g}_1(z_2^*) = (\frac{1}{2})\beta^{r+1}\overline{g}_1(y_2^*) = (\frac{1}{2})\beta^{r+1}(y_{2p}) = z'_{2p}$ (use β^{r+2} instead of $(\frac{1}{2})\beta^{r+1}$ if $p = 2$). Because $\overline{g}_1$ is an induced homomorphism between HAK's, it satisfies

$$\text{(4U)} \qquad \overline{g}_1(\mathrm{ad}(b^*)(w)) = \mathrm{ad}(b^p)(\overline{g}_1(w)),$$

$$\text{(4V)} \qquad \overline{g}_1(\mathrm{ad}(a_0^*)(w)) = \mathrm{ad}(c_1)(g_1(w))$$

for $w \in E^{r+1}(\mathcal{U}\overline{L}^{k-1})$. Induction on i using (4U) establishes (4S). The induction stops whenever $i = p^s$ for $s \leq k-1$, but the already-proved relation $E^{r+1}(g^k)(a_s^*) = a_{s+1}$, yielding $\overline{g}_1(a_s^*) = a_{s+1}$, gets it started again (because $y_{p^s}^* = a_s^*$). Formula (4V), together with the definition of z_i^* as $[y_{i-1}^*, a_0^*]$, gives (4T) upon applying Lemma 4.6(c). Having proved $\overline{g}_1$ to be an isomorphism, it follows that $E^{r+1}(g^k)$ is an isomorphism (because it respects the filtration $F_*(\ )$ of Lemma 2.2 and induces an isomorphism on associated graded algebras). Subsequent terms of the BSS's must also be mapped isomorphically by g^k [CE], i.e., $E^t(g^k)$ is an isomorphism for all $t \geq r+1$. The inductive hypothesis that 4.8(a,c) hold for $\mathcal{U}M^{k-1}$ now gives us 4.8(a,c) for $\mathcal{U}M^k$.

Corollary 4.9.. *For $0 \leq s \leq k$, let R_s^k denote the $\mathbb{Z}_{(p)}$-subalgebra of $\mathcal{U}M^k$ generated by $\{b^{p^s}, c_s, \ldots, c_k\}$.*

(a) $R_0^k = \mathcal{U}M^k$ *and* $R_k^k = \mathbb{Z}_{(p)}\langle b^{p^k}, c_k\rangle$.

(b) *A presentation for R_s^k is*

$$R_s^k = \mathbb{Z}_{(p)}\langle b^{p^s}, c_s, \ldots, c_k\rangle / \langle \sigma_s, \sigma_{s+1}, \ldots, \sigma_{k-1}\rangle .$$

(c) $d(R_s^k) \subseteq R_s^k$, *i.e., (R_s^k, d) is a sub-dga of $\mathcal{U}M^k$.*

(d) $E^{r+s}(R_s^k) \approx R_s^k \otimes \mathbb{Z}_p$, *and the inclusion $R_s^k \to \mathcal{U}M^k$ induces an isomorphism on $E^t(\ )$, for all $t \geq r+s$.*

Proof. Straightforward.

We finish Chapter 4 by re–examining the dgL's of Proposition 3.8 in terms of the BSS's for their universal enveloping algebras.

Proposition 4.10. *Recall from Proposition 3.8 the short exact sequence of dgL's,*

$$\text{(4W)} \qquad 0 \to J^k \to N^k \xrightarrow{g^k h^k} M^k \to 0 .$$

Let $\zeta^k = \mathcal{U}(g^k h^k) : \mathcal{U}N^k \to \mathcal{U}M^k$.

(a) *The induced homomorphism on BSS's, $E^t(\zeta^k) : E^t(\mathcal{U}N^k) \to E^t(\mathcal{U}M^k)$, is surjective for all t.*

(b) *As a homomorphism of differential $\mathbb{Z}_{(p)}$–modules, ζ^k has a right inverse.*

(c) *Let S be the set (3T), and view $\mathbb{Z}_{(p)}(S)$ as a differential submodule of J^k. Then $\mathcal{U}M^k \otimes \mathbb{Z}_{(p)}(S)$ is a chain complex, and there is an isomorphism of dgL's*

$$\text{(4X)} \qquad \mathcal{L}(\mathcal{U}M^k \otimes \mathbb{Z}_{(p)}(S)) \xrightarrow{\approx} J^k .$$

(d) $E^t(\mathcal{U}J^k) = \mathbb{Z}_p$ *for $t \geq r+k$, i.e., $p^{r+k-1}\overline{H}_*(\mathcal{U}J^k) = 0$.*

(e) $E^t(\zeta^k)$ *is an isomorphism for $t \geq r+k$.*

(f) $p^{r+k}\overline{H}_*(\mathcal{U}N^k) = 0$.

Proof. We prove all parts together, in an induction on k. When $k = 0$ (a)–(f) are trivial because g^k and h^k are isomorphisms and the set S is empty. Suppose $k \geq 1$ and (a)–(f) hold at $k-1$.

To prove (a), consider separately the cases $t > r+k$, $t < r+k$, and $t = r+k$. When $t > r+k$, part (a) is trivial because $E^t(\mathcal{U}M^k) = \mathbb{Z}_p$ by Theorem 4.8(c). When $t < r+k$, let $s = \max(0, t-r)$. Since $E^t(\zeta^k)$ is multiplicative, by Theorem 4.8(b) it suffices to show that the set $\{b^{p^s}, c_s, \ldots, a_k\}$ lies in $\mathrm{im}(E^t(\zeta^k))$. The inductive hypothesis tells us that $\{b^{p^s}, c_s, \ldots, a_{k-1}\} \subseteq \mathrm{im}(E^t(\zeta^{k-1}))$. Deduce that $\{b^{p^s}, c_s, \ldots, a_{k-1}\} \subseteq \mathrm{im}(E^t(\bar{\imath}\zeta^{k-1})) \subseteq \mathrm{im}(E^t(\zeta^k))$. To get a_k to belong to $\mathrm{im}(E^t(\zeta^k))$, note simply that $\zeta^k(u_k) = a_k$, and that u_k survives to $E^t(\mathcal{U}N^k)$ because $d(u_k)$ is divisible by p^{r+k-1}, by (3Q) and Lemma 2.4(f).

When $t = r+k$ we need only to show that c_k and b^{p^k} belong to $\mathrm{im}(E^{r+k}(\zeta^k))$. Let e_s denote the double summation in (4F), so that $e_k \in \mathcal{U}M^{k-1}$, and the cycle $c_k \in \mathcal{U}M^k$ equals $a_k - \bar{\imath}(e_k)$. In $\mathcal{U}M^{k-1}$ we have

$$\epsilon(b^{p^{k-1}}, c_{k-1}) \;=\; a'_k \;-\; pe_k$$

by (4H), where $a'_k = \mathrm{ad}^{p^k - p^{k-1}}(b)(a_{k-1})$. Let q_k denote $p^{-(r+k)}d(a'_k) \in \mathcal{U}M^{k-1}$. Since $\epsilon(b^{p^{k-1}}, c_{k-1}) = -p^{-(r+k-1)}d((b^{p^{k-1}})^p)$ is a cycle we have $d(e_k) = p^{r+k-1}q_k$. Because $\zeta^{k-1}(w'_{p^k}) = a'_k$, we also have $\zeta^{k-1}(p^{-(r+k)}d(w'_{p^k})) = q_k$.

By our inductive hypothesis that (f) holds at $k-1$, p^{r+k-1} times the cycle $p^{-(r+k)}d(w'_{p^k})$ must be a boundary in $\mathcal{U}N^{k-1}$. Let $e'' \in \mathcal{U}N^{k-1}$ satisfy $d(e'') = p^{-1}d(w'_{p^k})$. Now $\zeta^{k-1}(e'' - e_k)$ is a cycle in $\mathcal{U}M^{k-1}$, so (because (b) holds at $k-1$) it is the ζ^{k-1}–image of a cycle $e' \in \mathcal{U}N^{k-1}$. Thus $\zeta^{k-1}(e'' - e') = e_k$ and $d(e'' - e') = d(e_k) = p^{-1}d(w'_{p^k})$. The element $w'_{p^k} - p(e'' - e')$ is consequently a cycle in $\mathcal{U}N^{k-1}$, so there exists some $b'' \in \mathcal{U}N^{k-1}$ such that $d(b'') = p^{r+k-1}w'_{p^k} - p^{r+k}(e'' - e')$.

Now let us look at some elements of $\mathcal{U}N^k$. Let $c'_k = u_k - \bar{\imath}(e'' - e')$; it satisfies $\zeta^k(c'_k) = a_k - \bar{\imath}(e_k) = c_k$ while $d(c'_k) = 0$. This shows that c'_k survives to $E^{r+k}(\mathcal{U}N^k)$ and has $E^{r+k}(\zeta^k)(c'_k) = c_k$, i.e. $c_k \in \mathrm{im}(E^{r+k}(\zeta^k))$. Compute similarly that $d(v_k - \bar{\imath}(b'')) = -p^{r+k}c'_k$ in $\mathcal{U}N^k$. Deduce that $b^{p^k} - \zeta^k(v_k - \bar{\imath}(b'')) = b^{p^k} + \zeta^k\bar{\imath}(b'')$ is a cycle in $\mathcal{U}M^k$. Let b' be a cycle in $\mathcal{U}N^k$ satisfying $\zeta^k(b') = b^{p^k} + \zeta^k\bar{\imath}(b'')$. Then $\zeta^k(b' + v_k - \bar{\imath}(b'')) = b^{p^k}$ while $d(b' + v_k - \bar{\imath}(b'')) = 0 - p^{r+k}c'_k$ is divisible by p^{r+k}. It follows that $b' + v_k - \bar{\imath}(b'')$ survives to $E^{r+k}(\mathcal{U}N^k)$ and that its $E^{r+k}(\zeta^k)$–image is b^{p^k}.

Part (b) follows easily from (a) (see [D]). Denote a right inverse for ζ^k as chain complexes by τ^k, and use this differential–preserving τ^k as the τ^k in 3.8(f). Proposition 3.8(f) gives us an isomorphism of $\mathcal{L}(\mathcal{U}M^k \otimes \mathbb{Z}_{(p)}(S))$ with J^k. Because τ^k is a chain map, the isomorphism commutes with d (use (2K) to see this), i.e., it is an isomorphism of dgL's, as needed for (c). Because $p^{r+k-1}\overline{H}_*(\mathbb{Z}_{(p)}(S))$ vanishes we also have $p^{r+k-1}\overline{H}_*(\mathcal{U}M^k \otimes \mathbb{Z}_{(p)}(S)) = 0$ and hence $p^{r+k-1}\overline{H}_*(T(\mathcal{U}M^k \otimes \mathbb{Z}_{(p)}(S))) = 0$. This settles (d). Part (e) now follows from the Milnor–Moore spectal sequence (i.e. the spectral sequence of Lemma 2.2). Part (f) follows immediately from (e) and Theorem 4.8(c).

CHAPTER 5

The Differential Graded Lie Algebra N^{kl}

We examine in Chapter 5 a sub-dgL N^{kl} of $M^k \amalg M^l$. Topologically, N^{kl} corresponds to $\Sigma(\Omega D_k) \wedge (\Omega D_l)$. We begin with some general discussion of Hopf commutators and Hopf algebra kernels. We explore the obstructions to a free dgL having a "linear differential" when its enveloping algebra has one. Specializing to the case of N^{kl}, we construct an isomorphism $g_{kl} : \mathcal{L}(V^{kl}) \to N^{kl}$, where $V^{kl} = (\mathcal{U}M^k)_+ \otimes (\mathcal{U}M^l)_+$, that enables us to access explicitly the elements of N^{kl}. The main result of Chapter 5 is that g_{kl} is an isomorphism of dgL's, and hence that (N^{kl}, d) has a "linear differential." This fact is the algebraic analog of $\Sigma(\Omega D_k) \wedge (\Omega D_l)$ being a wedge of Moore spaces, which will be proved in Chapter 12.

The dgL N^{kl} is defined as the kernel of the surjection $M^k \amalg M^l \to M^k \oplus M^l$. We need to study it because it acts as our guide to the homotopy of the fiber of the map $D_k \vee D_l \to D_k \times D_l$. We need to understand the homotopy of this fiber because it will permit us to prove a certain crucial formula (the formula will be dubbed a "virtual syzygy") involving the homotopy of E_k, without which the attaching map to build D_{k+1} from D_k could not be constructed. In order to analyze N^{kl}, we begin with some general discussion of the Hopf algebra kernel (HAK) of $A \amalg B \to A \otimes B$.

Recall from [MM] that every connected graded Hopf algebra (A, μ, Δ) comes equipped with a canonical anti-homomorphism $\chi : A \to A$ such that the composition

$$(5A) \qquad A \xrightarrow{\Delta} A \otimes A \xrightarrow{1\otimes\chi} A \otimes A \xrightarrow{\mu} A$$

coincides with the augmentation $A \xrightarrow{\epsilon_0} A_0 \longrightarrow A$. By "anti-homomorphism" we mean that $\chi(xy) = (-1)^{|x||y|}\chi(y)\chi(x)$. When (A, d) is a dgH then χ is automatically a chain map. If $x \in A$ is primitive then $\chi(x) = -x$. If (A, μ, Δ) is co-commutative then χ is an involution, but in any case χ is an isomorphism of graded modules.

The *conjugation* operation, denoted $\mathrm{ad}()() : A \otimes A \to A$, is defined to be the composition

$$(5B) \qquad A \otimes A \xrightarrow{\Delta\otimes 1} A \otimes A \otimes A \xrightarrow{1\otimes\tau} A \otimes A \otimes A \xrightarrow{1\otimes 1\otimes\chi} A \otimes A \otimes A \xrightarrow{\mu(\mu\otimes 1)} A,$$

where as usual $\tau : A \otimes A \to A \otimes A$ is given by $\tau(a \otimes b) = (-1)^{|a||b|} b \otimes a$. If $A = \mathcal{U}L$ for a Lie algebra L then this definition of $\mathrm{ad}(x)(y)$ coincides with the definition of ad (not of

Ad) that we have been using. When A is a dgH we have

$$d(\mathrm{ad}(x)(y)) = \mathrm{ad}(d(x))(y) + (-1)^{|x|}\,\mathrm{ad}(x)(d(y))$$

because every factor in the composition (5B) commutes with d.

Define the ***Hopf commutator*** of two elements x and y in A, denoted $[x,y]_H$, to be the image of $x \otimes y$ under the composition

$$\text{(5C)}\qquad \begin{array}{l} A\otimes A \xrightarrow{\Delta\otimes\Delta} A\otimes A\otimes A\otimes A \xrightarrow{1\otimes\tau\otimes 1} A\otimes A\otimes A\otimes A \\ \xrightarrow{1\otimes 1\otimes\chi\otimes\chi} A\otimes A\otimes A\otimes A \xrightarrow{\mu\otimes\mu} A\otimes A \xrightarrow{\mu} A. \end{array}$$

We always have the antisymmetry formula,

$$\chi([x,y]_H) = (-1)^{|x||y|}[y,x]_H.$$

Another identity describes for us the diagonal of a Hopf commutator, namely, this diagram commutes:

$$\text{(5D)}\qquad \begin{array}{ccc} A\otimes A & \xrightarrow{\Delta\otimes\Delta} & A\otimes A\otimes A\otimes A \\ {\scriptstyle =}\downarrow & & \downarrow{\scriptstyle 1\otimes\tau\otimes 1} \\ A\otimes A & & A\otimes A\otimes A\otimes A \\ {\scriptstyle [\,,\,]_H}\downarrow & & \downarrow{\scriptstyle [\,,\,]_H\otimes[\,,\,]_H} \\ A & \xrightarrow{\Delta} & A\otimes A \end{array}$$

One important consequence of (5D) is that, when B and C are sub-Hopf algebras of A, the set of Hopf commutators

$$[B,C]_H = \mathrm{Span}\{[x,y]_H | x \in B,\ y \in C\}$$

is a subcoalgebra, hence the subalgebra of A generated by $[B,C]_H$ is another sub-Hopf algebra. If $[B,A]_H \subseteq B$ we call B a ***Hopf ideal.*** In general the ***Hopf algebra kernel*** (or ***HAK***) of a homomorphism $f : A \to B$ between Hopf algebras may be defined as the largest Hopf ideal of A that is contained in $f^{-1}(B_0)$.

Let us list a few more properties of Hopf commutators. if x and y are primitive then $[x,y]_H = [x,y]$. If y is primitive then $[x,y]_H = \mathrm{ad}(x)(y)$ and if x is primitive then $[x,y]_H = -(-1)^{|x||y|}\,\mathrm{ad}(y)(x)$. We also have

$$[x,[y,z]_H] = [[x,y],z]_H + (-1)^{|x||y|}[y,[x,z]_H]$$

for any primitive element x; and in the presence of a differential d making A into a dgH,

$$\text{(5E)}\qquad d([x,y]_H) = [d(x),y]_H + (-1)^{|x|}[x,d(y)]_H.$$

We will focus in this discussion on the HAK of the surjection $A \amalg B \to A\otimes B$ when A and B are connected graded Hopf algebras over $\mathbb{Z}_{(p)}$ that are free as $\mathbb{Z}_{(p)}$-modules. Denote this HAK by C, so that

$$C \to A \amalg B \to A \otimes B$$

is a short exact sequence of Hopf algebras. Consider

$$f_0 : A \otimes B \to A \amalg B$$

given by $f_0(x \otimes y) = [x, y]_H$, where the Hopf commutator makes sense because x and y are both being viewed as elements of $A \amalg B$. Then $f_0(a \otimes 1) = 0$ for $|a| > 0$ and $f_0(1 \otimes b) = 0$ for $|b| > 0$, and we may extend f_0 to an algebra homomorphism $f : T(A_+ \otimes B_+) \to A \amalg B$. According to (5D), im($f$) must be a sub-Hopf algebra of $A \amalg B$. (5A) and (5C) make it plain that im(f) $\subseteq$ ker($A \amalg B \to A \otimes B$), so the sub-Hopf algebra im(f) lies in the HAK of C. Thus we think of f as an algebra homomorphism,

(5F) $$f : T(A_+ \otimes B_+) \to C$$

Lemma 5.1.

(a) *The homomorphism (5F) is an isomorphism.*

(b) *Furthermore, if A and B are dgH's, then C is a sub-dgH of $A \amalg B$, and (5F) is a dgH isomorphism.*

Proof. Part (b) follows from (5E) and part (a), so we concentrate on (a). To begin with, the element $a_1 b_1 \cdots a_m b_m$ of $A \amalg B$, where $a_j \in A$ and $b_j \in B$, has *bidegree* $(|a_1| + \cdots + |a_m|, |b_1| + \cdots + |b_m|)$. An element of $A \amalg B$ is *bihomogeneous* if it is a sum of such terms sharing a uniform bidegree. Define the *rank* of the element $a_1 b_1 \cdots a_m b_m$ of $A \amalg B$ to be $\Sigma_{i<j} |b_i||a_j|$. The rank counts how often (with weights) a "b" precedes an "a". Define the *high term* of a non-zero element x of $A \amalg B$, denoted $HT(x)$ to be the non-zero component of maximum rank. Notice that $HT(x_1 x_2) = HT(x_1)HT(x_2)$, when x_1 and x_2 are bihomogeneous. In particular, $HT(f(a \otimes b)) = (-1)^{|a||b|} b\chi(a)$, whence

$$HT(f(a_1 \otimes b_1) f(a_2 \otimes b_2) \cdots f(a_m \otimes b_m)) = (\pm) b_1 \chi(a_1) b_2 \chi(a_2) \cdots b_m \chi(a_m).$$

The reader should now be able to finish the argument that f is monomorphic.

To show that (5F) is onto, first reduce to the case of finite type by a direct limit argument. When A and B have finite type their Hilbert series are defined; we denote these as H_A and H_B. From [MM] it is clear that

$$H_C H_{A \otimes B} \leq H_{A \amalg B},$$

i.e., $H_C H_A H_B \leq [H_A^{-1} + H_B^{-1} - 1]^{-1}$, leading to

$$H_C \leq [H_A + H_B - H_A H_B]^{-1}$$

At the same time, $H_{T(A_+ \otimes B_+)} = [1 - (H_A - 1)(H_B - 1)]^{-1} = [H_A + H_B - H_A H_B]^{-1}$. Since f is a monomorphism, this can only happen if $H_C = H_{T(A_+ \otimes B_+)}$. The same argument works if A and B are first tensored with $\mathbb{Z}_p$. Deduce that f is an isomorphism.

Remark/Notation. With part (b) we have implicitly introduced a new notation when we refer to the object $T(A_+ \otimes B_+)$ of (5F) as a dgH or dga. The tensor algebra on a $\mathbb{Z}_{(p)}$-module V gets a differential automatically whenever V has a differential. While this is obvious enough, it is worth mentioning that we are now viewing T as a functor from $\mathbb{Z}_{(p)}$-free positively graded chain complexes to dga's over $\mathbb{Z}_{(p)}$. Our notation is $T(V, d)$, which has a much more restrictive meaning than the notation (TV, d). For the latter object, $d(V)$ may lie anywhere in TV, whereas with $T(V, d)$ we must have $d(V) \subseteq V = T^1V$.

Likewise, the free Lie algebra $\mathcal{L}(V)$ becomes a free dgL, denoted $\mathcal{L}(V, d)$, if (V, d) is a $\mathbb{Z}_{(p)}$-free positively graded chain complex. Again, we caution the reader to distinguish carefully between $\mathcal{L}(V, d)$ and $(\mathcal{L}V, d)$.

We take it to be obvious that if $A = \mathcal{U}L$ and $B = \mathcal{U}M$ for Lie algebras L and M in the above discussion, then the HAK of C equals $\mathcal{U}N$, where N is defined to be the kernel in the short exact sequence of Lie algebras,

$$0 \to N \to L \amalg M \to L \oplus M \to 0.$$

This is "obvious" because $\mathcal{U}N$ is clearly contained in C, and a Hilbert series argument along with the Poincare–Birkhoff–Witt theorem enables us to compute that

$$H_{\mathcal{U}N} H_{\mathcal{U}(L \oplus M)} = H_{\mathcal{U}(L \amalg M)},$$

whence $H_{\mathcal{U}N} = H_C$. Since $C = T(A_+ \otimes B_+)$ is a tensor algebra, it follows from [Le] that N is a free Lie algebra on an isomorphic vector space, so we can write $N \approx \mathcal{L}(A_+ \otimes B_+)$. There are two important subtleties, however, in the Lie algebra case. First, there is no obvious natural homomorphism like (5F) for Lie algebras; in particular $f(\mathcal{L}(A_+ \otimes B_+))$ will not lie in N. Second, when our Lie algebras are dgL's, it is not clear in general whether (N, d) must equal $\mathcal{L}(A_+ \otimes B_+, d)$ as a dgL.

Following up on one idea introduced during the proof of Lemma 5.1, we have the next lemma. We think of (5G) as providing a means for detecting decomposability.

Lemma 5.2. *Let $\bar{\pi}_{21}$ denote the module projection of $\mathcal{U}L \amalg \mathcal{U}M$ onto its direct summand $(\mathcal{U}M)_+ \otimes (\mathcal{U}L)_+$. The composite*

$$\text{(5G)} \qquad h : (\mathcal{U}N)_+ \longrightarrow \mathcal{U}L \amalg \mathcal{U}M \xrightarrow{\bar{\pi}_{21}} (\mathcal{U}M)_+ \otimes (\mathcal{U}L)_+$$

has the following properties. Its kernel is precisely the ideal of decomposable elements; this statement remains true for the restriction $h|_N$ to the Lie algebra. It is a chain map, and $h(x) = HT(x)$ whenever $h(x) \neq 0$. Lastly, $hf_0 = (1 \otimes \lambda)\tau$, if $f_0 = (\mathcal{U}L)_+ \otimes (\mathcal{U}M)_+ \to (\mathcal{U}N)_+$ is the Hopf commutator homomorphism described earlier.

Proof. Left as an easy exercise.

Definition. A free dga (TV, d) which is isomorphic with $T(W, e)$ for some chain complex (W, e) is said to ***have a linear differential.*** Likewise, a free dgL $(\mathcal{L}(V), d)$ ***has a linear differential*** if and only if it is isomorphic with $\mathcal{L}(W, e)$ for some chain complex (W, e).

It is trivial that $\mathcal{U}(L, d)$ has a linear differential whenever (L, d) has a linear differential. Unfortunately, the converse is false. To see how the converse can fail, we introduce the following notation, which will become extremely important in Chapters 7 and 11.

Notation 5.3. Given an even-dimensional element w of a dgL (L, d), let

$$\text{(5H)} \qquad \tau_j(w) = \mathrm{ad}^{p^j - 1}(w)(d(w))$$

for $j \geq 1$. Let

(5I) $$\sigma_j(w) = \sum_{i=1}^{p-1}(2p)^{-1}(p^j|i)[\mathrm{ad}^{i-1}(w)(d(w)), \mathrm{ad}^{p^j-1-i}(w)(d(w))].$$

According to [CMN1, p. 134] we have

(5J) $$d(\tau_j(w)) = (p)\sigma_j(w).$$

When L is $\mathbb{Z}_{(p)}$-free and p^m divides $d(w)$, we let $\tilde{\tau}_j(w)$ denote $p^{-m}\tau_j(w)$ and we let $\tilde{\sigma}_j(w)$ denote $p^{-2m}\sigma_j(w)$. Then

(5K) $$d(\tilde{\tau}_j(w)) = (p^{m+1})\tilde{\sigma}_j(w)$$

in (L, d). Whenever we employ the notations $\tilde{\tau}_j$ and $\tilde{\sigma}_j$ the intended value of m will be clear from the context.

Because of (5K), the elements $\tilde{\tau}_j(w)$ and $\tilde{\sigma}_j(w)$ can be responsible for the appearance of p^{m+1}-torsion in $H_*(\mathcal{L}(V, e))$ even when $H_*(V, e)$ contains at most p^m-torsion. We will return to this point shortly, but first let us construct the counterexample to the above converse that we have promised the reader. Let $L = \mathbb{L}\langle a, b, x, y\rangle$ over $\mathbb{Z}_{(p)}$ with $|b|$ even, and put $d(a) = d(x) = 0$, $d(b) = p^m a$, and $d(y) = p^{m+1}x - p^m\tilde{\sigma}_j(b)$, for some $m \geq 1$ and $j \geq 1$. Then (L, d) does not have a linear differential. (This will become clear with Corollary 5.7.) However $\mathcal{U}(L, d)$ does have a linear differential because it is isomorphic with $T(V, d)$, where $V = \mathrm{Span}(a, b, x, y_1)$ with $y_1 = y + \tilde{e}_j(b)$. Here $\tilde{e}_j(b)$ denotes $p^{-m-1}e_j(b)$, where $e_j(b) = d(b^{p^j}) - \tau_j(b)$, i.e.,

$$\tilde{e}_j(b) = \sum_{i=0}^{p^j-1} p^{-1}(1 - (-1)^i(p^j - 1|i))b^{p^j-i-1}ab^i.$$

We can determine quite precisely the conditions under which a free dgL, whose enveloping algebra has a linear differential, has a linear differential too. Let (L, d) be a free dgL over $\mathbb{Z}_{(p)}$. Suppose (L, d) has a linear differential in dimensions $\leq q$ and we want to know whether this property extends to dimension $q + 1$. Write $L = \mathcal{L}(V)$, where $d(V_{\leq q}) \subseteq V_{<q}$, and let x by a typical $(q+1)$-dimensional basis element of V. Write $d(x)$ as $d(x) = d_1(x) + d_2(x)$, where $d_1(x) \in V_q$ and $d_2(x) \in [L, L]$. (Note: if $p = 2$, the notation $[L, L]$ is to be understood as signifying the Lie ideal $[L, L] + (L_{\mathrm{odd}})^2$.) Clearly, both $d_1(x)$ and $d_2(x)$ are cycles. The linearity of the differential can be extended to a generator of the form x + (something in $[L, L]$) if and only if $d_2(x) \in d([L, L])$. Since we are focusing on a setting where $\mathcal{U}(L, d)$ does have a linear differential, however, we deduce via similar reasoning that $d_2(x)$, when viewed as a cycle in $\mathcal{U}L$, is indeed the boundary of a decomposable, i.e., $d_2(x) \in d((\mathcal{U}L)^2_+)$. Thus the algebraic obstruction to extending the linearity of the differential over x can be viewed as the cycle $d_2(x)$ modulo $d([L, L])$. The obstruction group is the kernel of the homomorphism induced by the inclusion of L into $\mathcal{U}L$,

(5L) $$\ker(H_q(\mathcal{L}(V_{<q}, d)) \to H_q(T(V_{<q}, d))).$$

The linearity property of d extends over $x \in V_{q+1}$ if and only if $d_2(x)$, which represents an element of (5L), represents zero.

We compute (5L) after quoting some relevant facts from the literature.

Lemma 5.4. *Any $\mathbb{Z}_{(p)}$-free chain complex (V, d) of finite type can be written as a direct sum of minimal chain complex summands. Each minimal summand is either free of rank one, so it is $\mathbb{Z}_{(p)}(x)$ (i.e., $\mathrm{Span}_{\mathbb{Z}_{(p)}}(x)$) with $d(x) = 0$, or it is free of rank two, so it is $\mathbb{Z}_{(p)}(y, x)$ with $d(y) = p^m x$ for some $m \geq 0$.*

Proof. See [D].

According to Lemma 5.4, it is always possible to split a $\mathbb{Z}_{(p)}$-free finite type chain complex (V, d) into subcomplexes,

$$V = (V^{(0)} \oplus V^{(1)} \oplus V^{(2)} \oplus \cdots) \oplus V^{(\infty)}, \tag{5M}$$

where $V^{(\infty)}$ is the direct sum of all the minimal summands of the form $\mathbb{Z}_{(p)}(x)$ and $V^{(m)}$ is the direct sum of all the minimal summands of the form $\mathbb{Z}_{(p)}(y, x)$ with $d(y) = p^m x$. Clearly $V^{(m)}$ is acyclic with respect to the differential $p^{-m}d$.

For simplicity in stating our results, we now restrict our attention to odd p only.

Lemma 5.5. *Let (V, d) be an acyclic chain complex of finite type, i.e., $V = V^{(0)}$ in (5M). There are differential submodules of $\mathcal{L}(V)$, denoted $W^+(V)$ and $W^-(V)$, having the following properties.*

- (a) *There is a decomposition for $\mathcal{L}(V)$ as in (5M), $\mathcal{L}(V) = L^{(0)} \oplus L^{(1)}$, with $L^{(0)} = W^+(V) \oplus W^-(V)$.*
- (b) *$W^+(V)$ (resp. $W^-(V)$) is the direct sum of all minimal summands of $L^{(0)}$ of the form $\mathbb{Z}_{(p)}(y, x)$ with $|y|$ even (resp. $|y|$ odd).*
- (c) *If we write $W^+(V) = \bigoplus_{\alpha} \mathbb{Z}_{(p)}(y_\alpha, x_\alpha)$, with $|y_\alpha|$ even and $d(y_\alpha) = x_\alpha$, then $L^{(1)} = \bigoplus_{j=1}^{\infty} \bigoplus_{\alpha} \mathbb{Z}_{(p)}(\tau_j(y_\alpha), \sigma_j(y_\alpha))$.*
- (d) *$H_*(\mathcal{L}(V); \mathbb{Z}_p)$ is the $\mathbb{Z}_p$-vector space with basis $\{\tau_j(y_\alpha), \sigma_j(y_\alpha)\}_{\alpha; j \geq 1}$.*
- (e) *$H_*(\mathcal{L}(V))$ is the $\mathbb{Z}_p$-vector space with basis $\{\sigma_j(y_\alpha)\}_{\alpha; j \geq 1}$.*

Proof. See [CMN1, Section 4].

Lemma 5.6. *Let (V, d) be a $\mathbb{Z}_{(p)}$-free finite type chain complex, and decompose V as in (5M). Let $V^{(>m)}$ denote $(V^{(m+1)} \oplus V^{(m+2)} \oplus \cdots) \oplus V^{(\infty)}$ and let $V^{(\geq m)}$ denote $V^{(m)} \oplus V^{(>m)}$. There are differential submodules of $\mathcal{L}(V^{(\geq m)})$, denoted $W^+(V^{(\geq m)})$ and $W^-(V^{(\geq m)})$, having the following properties.*

- (a) *It is possible to decompose $\mathcal{L}(V^{(\geq m)})$ as in (5M),*

$$\mathcal{L}(V^{(\geq m)}) = (L^{(m,m)} \oplus L^{(m,m+1)} \oplus L^{(m,m+2)} \oplus \cdots) \oplus L^{(m,\infty)}, \tag{5N}$$

  so that $L^{(m,m)} = W^+(V^{(\geq m)}) \oplus W^-(V^{(\geq m)})$.
- (b) *$W^+(V^{(\geq m)})$ (resp. $W^-(V^{(\geq m)})$) is the direct sum of all minimal summands of $L^{(m,m)}$ of the form $\mathbb{Z}_{(p)}(y, x)$ with $|y|$ even (resp. $|y|$ odd).*
- (c) *If we write $W^+(V^{(\geq m)}) = \bigoplus_{\alpha} \mathbb{Z}_{(p)}(y_\alpha, x_\alpha)$ with $|y_\alpha|$ even and $d(y_\alpha) = p^m x_\alpha$, then*

$$L^{(m,>m)} = \mathcal{L}(V^{(>m)}) \oplus \widetilde{W}^m(V^{(\geq m)}), \tag{5O}$$

where

$$\widetilde{W}^m(V^{(\geq m)}) = \bigoplus_{j=1}^{\infty} \bigoplus_{\alpha} \mathbb{Z}_{(p)}(\tilde{\tau}_j(y_\alpha), \tilde{\sigma}_j(y_\alpha)). \tag{5P}$$

By (5K), $L^{(m,m+1)}$ can be chosen so that it contains $\tilde{W}^m(V^{(\geq m)})$ as a submodule.

(d) *The $(m+1)^{st}$ term of the BSS for $\mathcal{L}(V)$ is $L^{(m,>m)} \otimes \mathbb{Z}_{(p)}$.*

Proof. The kernel of $\mathcal{L}(V^{(\geq m)}, d) \to \mathcal{L}(V^{(>m)}, d)$ is free with a linear differential, by Lemma 2.4. Denote it as $\mathcal{L}(W^m, d)$, where (W^m, d) is acyclic with respect to the differential $d_{[m]} = p^{-m}d$. Apply Lemma 5.5 to $\mathcal{L}(W^m, d_{[m]})$.

Corollary 5.7.

(a) *When $q \not\equiv -2 \pmod{2p}$, the kernel (5L) is zero.*

(b) *Suppose $q \equiv -2 \pmod{2p}$. For each $m < q$, let $I^{(m)}$ denote an indexing set such that*

$$W^+(V^{(\geq m)}) = \bigoplus_{\alpha \in I^{(m)}} \mathbb{Z}_{(p)}(y_\alpha, x_\alpha)$$

as in Lemma 5.6(c). Let $S = \{(\beta, j, m) | \beta \in I^{(m)} \text{ and } p^j|y_\beta| = q+2, j \geq 1\}$. The kernel (5L) equals the $\mathbb{Z}_p$-vector space with basis

$$\{p^m \tilde{\sigma}_j(y_\beta) | (\beta, j, m) \in S\}.$$

Proof. Using (5N) and (5O), we have

$$\mathcal{L}(V) = L^{(0,0)} \oplus L^{(0,>0)} = L^{(0,0)} \oplus \mathcal{L}(V^{(\geq 1)}) \oplus \widetilde{W}(V),$$
$$\mathcal{L}(V^{(\geq 1)}) = L^{(1,1)} \oplus L^{(1,>1)} = L^{(1,1)} \oplus \mathcal{L}(V^{(\geq 2)}) \oplus \widetilde{W}(V^{(\geq 1)}),$$

and so on, so that

$$\mathcal{L}(V) = \mathcal{L}(V^{(\infty)}) \oplus (\bigoplus_{m=0}^{\infty} L^{(m,m)}) \oplus (\bigoplus_{m=0}^{\infty} \widetilde{W}^m(V^{(\geq m)})). \tag{5Q}$$

Since (5Q) is a decomposition as chain complexes, and since $L(V^{(\infty)})$ and each $L^{(m,m)}$ are retracts of $\mathcal{U}L(V)$ as chain complexes, all of the kernel (5L) lies in $\bigoplus_{m=0}^{\infty} \widetilde{W}^m(V^{(\geq m)})$. When each $\widetilde{W}^m(V^{(\geq m)})$ is written as the direct sum (5P), each summand contributes a single copy of $\mathbb{Z}_p$ to the kernel (5L). We obtain 5.7(b) by grouping together the summands which contribute their copy of $\mathbb{Z}_p$ in dimension q.

We have already alluded to the next point we wish to make, which concerns exponents for the homology groups of a dgL with linear differential. As is well-known and easily proved, $p^t H_{\leq q}(V, d) = 0$ implies $p^t \bar{H}_{\leq q}(T(V, d)) = 0$. Combining this fact with Corollary 5.7 yields the following result, whose proof is left as an exercise.

Lemma 5.8. *Let (V, d) be a $\mathbb{Z}_{(p)}$-free chain complex of finite type. Let $t \geq 0$, and suppose $p^t H_*(V, d) = 0$.*

(a) *$p^{t+1} H_* \mathcal{L}(V, d) = 0$ and $p^{t+1} H_* \mathcal{L}^{\geq 2}(V, d) = 0$. In the BSS for $\mathcal{L}(V, d)$, we have $E^{t+2}(\mathcal{L}(V, d)) = 0$.*

(b) *If $q \not\equiv -2 \pmod{2p}$ then $p^t H_q \mathcal{L}(V, d) = 0$ and $p^t H_q \mathcal{L}^{\geq 2}(V, d) = 0$.*

Definition and Properties of N^{kl}

Let $n \geq 1$, $r \geq 1$ as usual; assume $p \geq 3$, and let k, $l \geq 0$ be arbitrary. We now turn our attention to the specific dgL (N^{kl}, d), defined by the short exact sequence of dgL's

(5R) $$0 \longrightarrow N^{kl} \longrightarrow M^k \amalg M^l \longrightarrow M^k \oplus M^l \longrightarrow 0.$$

After giving some suitable notation for working with elements of N^{kl} and proving three preparatory lemmas, we prove Theorem 5.12, which shows N^{kl} to have a linear differential. We also glance briefly at the folding homomorphism and at a related dgL N^{klm}.

Let C^{kl} denote the HAK of the surjection $UM^k \amalg UM^l \to UM^k \otimes UM^l$. We observed earlier that C^{kl} is isomorphic as a dgH with UN^{kl}, and that $N^{kl} \approx \mathcal{L}(V^{kl})$ as Lie algebras, where $V^{kl} = (\mathcal{U}M^k)_+ \otimes (\mathcal{U}M^l)_+$. We have an example of the situation discussed above, i.e., $\mathcal{U}(N^{kl}, d) = (C^{kl}, d) = T(V^{kl}, d)$ as dga's by Lemma 5.1, and we aim to show that the dgL (N^{kl}, d) also has a linear differential. We will accomplish this by constructing a chain map $g = g_{kl} : V^{kl} \to N^{kl}$ which covers all the indecomposables of N^{kl}.

Let us put N^{kl} on hold for a short while and digress to establish a few more combinatorial properties of the algebra $\mathcal{U}M^k$.

Notation. In $\mathcal{U}M^k$, let b_s denote b^{p^s}, for $0 \leq s \leq k$. Thus $d(b_s) = -p^{r+s}c_s$. Let Γ_s^k denote the set $\{b_s, \cdots, b_k, c_s, \cdots, c_k\}$ for $0 \leq s \leq k$, and let S_s^k be the $\mathbb{Z}_{(p)}$-span of Γ_s^k. Then $S_0^k \supseteq S_1^k \supseteq \cdots \supseteq S_k^k$ is a descending chain of differential submodules of $\mathcal{U}M^k$.

Notice that $\lambda(b) = -b$ because b is primitive, where λ is the antiautomorphism discussed earlier, for $\mathcal{U}M^k$. Since p is odd, $\lambda(b_s) = (-b)^{p^s} = -b_s$. Differentiating this equation and dividing by $-p^{r+s}$ yields $\lambda(c_s) = -c_s$.

Lemma 5.9. *Let $\mu = \mu_k : \mathcal{U}M^k \otimes S_0^k \to \mathcal{U}M_+^k$ denote the multiplication homomorphism. There exists*

$$\phi = \phi_k : (\mathcal{U}M^k)_+ \to \mathcal{U}M^k \otimes S_0^k$$

having these properties: (i) $\mu\phi = 1$; (ii) $d\phi = \phi d$; and (iii) $\phi((R_s^k)_+) \subseteq R_s^k \otimes S_s^k$.

Proof. View $\mathcal{U}M^k$ as a quotient of the free algebra $\mathbb{Z}_{(p)}\langle\Gamma_0^k\rangle$. Impose a linear ordering on the set of generators by $b_0 > \cdots b_k > c_0 > \cdots > c_k$. Using the presentation of Proposition 4.3(e) for $\mathcal{U}M^k$ and the diamond lemma [Be], we find easily that

(5S) $$\mathcal{O} = \{b_s b_t, b_s c_t | 0 \leq t < s \leq k\} \cup \{b_s^p, b_s^{p-1}c_s | 0 \leq s < k\}$$

is a complete set of obstructions (the term comes from [A2]) for this ordered set of generators. By this we mean that if B denotes the set of all words on the alphabet $\Gamma_0^k = \{b_0, \cdots, c_k\}$ that contain no element of $\mathcal{O}$ as a subword, then B is a $\mathbb{Z}_{(p)}$-basis for $\mathcal{U}M^k$. Furthermore, any element of $\mathcal{O}$ equals a linear combination of lexicographically lower words, in $\mathcal{U}M^k$.

Define $\phi : B - \{1\} \to B \times \Gamma_0^k$ by splitting off the last letter, i.e., if $w = w_{i_1} \cdots w_{i_m}$ with $w_{i_j} \in \Gamma_0^k$ then $\phi(w) = (w_{i_1} \cdots w_{i_{m-1}}, w_{i_m})$. Viewing $B - \{1\}$ as a $\mathbb{Z}_{(p)}$-basis for $\mathcal{U}M_+^k$ and $B \times \Gamma_0^k$ as a $\mathbb{Z}_{(p)}$-basis for $\mathcal{U}M^k \otimes S_0^k$, we extend linearly to obtain $\phi : \mathcal{U}M_+^k \to \mathcal{U}M^k \otimes S_0^k$. Properties (i) and (iii) are trivial. For (ii), observe that if $w_{i_1} \cdots w_{i_m} \in B$ then for each j $w_{i_1} \cdots d(w_{i_j}) \cdots w_{i_m}$ is either zero (if $w_{i_j} = c_s$) or a multiple of an element of B (because when $w_{i_j} = b_s$, if $w_{i_1} \cdots c_s \cdots w_{i_m}$ contains an obstruction word then $w = w_{i_1} \cdots b_s \cdots w_{i_m}$ contains one also). It follows that d commutes with ϕ on $B - \{1\}$, hence on all of $\mathcal{U}M_+^k$.

Lemma 5.10. *Let $\langle\Gamma_s^k\rangle$ denote the two-sided ideal of $\mathcal{U}M^k$ generated by Γ_s^k, and let B be as in Lemma 5.9. Let C_s denote the set of all words on the alphabet Γ_0^k which contain at least one occurrence of b_t or c_t with $s \le t \le k$.*

(a) *$\langle\Gamma_s^k\rangle$ equals the $\mathbb{Z}_{(p)}$-span of $B \cap C_s$.*

(b) *$\langle\Gamma_s^k\rangle$ is a differential ideal (i.e., it is closed under d), and as a chain complex it is a retract of $\mathcal{U}M^k$.*

Proof. (a) As part of the theory of obstructions, each element of (5S) comes with a "straightening rule" that expresses it as a linear combination of elements of B. The $\mathbb{Z}_{(p)}$-span in $\mathcal{U}M^k$ of C_s is a two-sided ideal contained in $\langle\Gamma_s^k\rangle$ and containing Γ_s^k, so $\mathbb{Z}_{(p)}(C_s) = \langle\Gamma_s^k\rangle$ in $\mathcal{U}M^k$. Since $\Gamma_s^k \subseteq \mathbb{Z}_{(p)}(B \cap C_s) \subseteq \mathbb{Z}_{(p)}(C_s) = \langle\Gamma_s^k\rangle$, it suffices to show that $\mathbb{Z}_{(p)}(B\cap C_s)$ is a two-sided ideal. For this, it suffices to check that every obstruction that lies in C_s has a straightening rule that lies entirely in $\mathbb{Z}_{(p)}(B \cap C_s)$. This is easily checked.

(b) The two-sided ideal generated by a differential submodule in dga is always a differential ideal; in this instance, the differential submodule is S_s^k. To see that it is a chain complex retract, observe that

$$\mathcal{U}M^k = \mathbb{Z}_{(p)}(B) = \mathbb{Z}_{(p)}(B \cap C_s) \oplus \mathbb{Z}_{(p)}(B - C_s)$$

is a decomposition as chain complexes, since $d(B - C_s) \subseteq \mathbb{Z}_{(p)}(B - C_s)$.

Now we begin to work directly with N^{kl}.

Notation.

Let $e^{(1)} : M^k \to M^k \amalg M^l$ and $e^{(2)} = M^l \to M^k \amalg M^l$ denote the evident dgL inclusions. Write $b_s^{(i)}$ for each $e^{(i)}(b_s)$, $i = 1, 2$, and likewise for $c_s^{(i)}$. Let $\mathrm{ad}^{(i)}(x)$ denote $\mathrm{ad}(e^{(i)}(x))$, which can operate on either $M^k \amalg M^l$ or N^{kl}. Define $d^{(1)}$ to be the unique derivation on $M^k \amalg M^l$ for which $d^{(1)}e^{(1)} = e^{(1)}d$ and $d^{(1)}e^{(2)} = 0$, and likewise for $d^{(2)}$. The usual differential d for $M^k \amalg M^l$ is the "total differential," i.e., $d = d^{(1)} + d^{(2)}$. Notice also that $d^{(1)}$ and $d^{(2)}$ commute, i.e., $d^{(1)}d^{(2)} + d^{(2)}d^{(1)} = 0$.

Since M^k and M^l each have two gradings (dimension and *-degree), $M^k \amalg M^l$ has four gradings. We denote them by $|x|^{(1)}$, $|x|_*^{(1)}$, $|x|^{(2)}$, $|x|_*^{(2)}$. N^{kl} inherits all four gradings. When we refer to the dimension or *-degree of $x \in M^k \amalg M^l$, we intend the total dimension or total *-degree, i.e., $|x| = |x|^{(1)} + |x|^{(2)}$ and $|x|_* = |x|_*^{(1)} + |x|_*^{(2)}$. Total dimension and total *-degree are respected by the total differential d, but $|x|^{(1)}$ and $|x|_*^{(1)}$ are not. However, the sum of these two does define a third linearly independent grading that d respects. Define the *$\#^{(1)}$-grade* of $x \in M^k \amalg M^l$ by $|x|_\#^{(1)} = (2n)^{-1}(|x|^{(1)} + |x|_*^{(1)})$, and similarly put $|x|_\#^{(2)} = (2n)^{-1}(|x|^{(2)} + |x|_*^{(2)})$. We find that

$$\text{(5T)} \qquad |b_s^{(1)}|_\#^{(1)} = p^s = |c_s^{(1)}|_\#^{(1)} \quad , \quad |b_s^{(2)}|_\#^{(1)} = 0 = |c_s^{(2)}|_\#^{(1)},$$

and these same equations hold if (1) and (2) are everywhere switched. Notice that both the $\#^{(1)}$-grade and the $\#^{(2)}$-grade of any element of N^{kl} (resp. $M^k \amalg M^l$) must be positive (resp. non-negative). The *#-grade* of x is the pair $(|x|_\#^{(1)}, |x|_\#^{(2)})$, and the *trigrade* is $(|x|_\#^{(1)}, |x|_\#^{(2)}, |x|_*)$. The differential d is trihomogeneous with trigrade $(0, 0, +1)$. Dimension is a linear combination of these, specifically, $|x| = (2n)|x|_\#^{(1)} + (2n)|x|_\#^{(2)} - |x|_*$.

Lemma 5.11. *Let (V^{kl}, d) be the chain complex $UM_+^k \otimes UM_+^l$. Let $h : UN^{kl} \to UM_+^k \otimes UM_+^l$ be the homomorphism constructed by putting $L = M^k$ and $M = M^l$ in (5G).*

(a) *Let $g_0 : S_0^k \otimes S_0^l \to N^{kl}$ be a chain map satisfying $hg_0 = -\tau$, where as usual τ denotes $\tau(x \otimes y) = (-1)^{|x||y|} y \otimes x$. Then g_0 can be extended to a chain map*

$$g : V^{kl} \to N^{kl}$$

satisfying $hg = (1 \otimes \lambda)\tau$.

(b) *Suppose $g : V^{kl} \to N^{kl}$ is a chain map satisfying $hg = (1 \otimes \lambda)\tau$. Then g can be extended uniquely to a dgL isomorphism*

(5U) $$g_1 : \mathcal{L}(V^{kl}, d) \to (N^{kl}, d).$$

(c) *In particular, (N^{kl}, d) has a linear differential if any chain map g_0 exists satisfying $hg_0 = -\tau$.*

Proof. (a) We use the homomorphisms ϕ_k, ϕ_l of Lemma 5.9. Define g to be the composition

(5V) $$UM_+^k \otimes UM_+^l \xrightarrow{\phi_k \otimes \phi_l} \mathcal{U}M^k \otimes S_0^k \otimes \mathcal{U}M^l \otimes S_0^l \xrightarrow{1 \otimes \tau \otimes 1}$$
$$UM^k \otimes UM^l \otimes S_0^k \otimes S_0^l \xrightarrow{1 \otimes 1 \otimes g_0} UM^k \otimes UM^l \otimes N^{kl} \xrightarrow{\mathrm{ad}^{(1)}(\,)\,\mathrm{ad}^{(2)}(\,)(\,)} N^{kl}$$

The last arrow of (5V) sends $x \otimes y \otimes z$ to $\mathrm{ad}^{(1)}(x)(\mathrm{ad}^{(2)}(y)(z))$. When $g(x \otimes y)$ is included into $\mathcal{U}(M^k \amalg M^l)$ and written as a sum of components in this coproduct algebra, the only term of the form $e^{(2)}(w_1)e^{(1)}(w_2)$ is $(-1)^{|x||y|} e^{(2)}(y) e^{(1)}(\lambda(x))$.

(b) Clearly g_1 exists and satisfies $hg_1 = (1 \otimes \lambda)\tau$. As a homomorphism between free Lie algebras, g_1 induces a bijection on indecomposables, hence g_1 is an isomorphism of Lie algebras. Since g_1 is also a chain map, it is a dgL isomorphism.

(c) Obvious, from (a) and (b).

Theorem 5.12. *Suppose p is odd and let $k, l \geq 0$. There exists a chain map $g_{kl} : S_0^k \otimes S_0^l \to N^{kl}$ for which $hg_{kl} = -\tau$. This g_{kl} extends to a dgL isomorphism (5U), and (N^{kl}, d) has a linear differential.*

Proof. By Lemma 5.11, only the existence of $g_{kl} : S_0^k \otimes S_0^l \to N^{kl}$ needs to be proved. The proof is by induction on k and l. When $l = 0$ define $g_{0l}(x \otimes y)$ to be $-(-1)^{|x||y|} \mathrm{ad}^{(2)}(e^{(1)}(x))$. When $k = 0$ define $g_{k0}(x \otimes y)$ to be $\mathrm{ad}^{(1)}(x)(e^{(2)}(y))$. These definitions concur if $k = l = 0$.

Our inductive hypothesis will be that $k, l > 0$ and that $g_{k-1,l}$ and $g_{k,l-1}$ have been defined in a manner that is consistent on the intersection of their domains, which is

$$(S_0^{k-1} \otimes S_0^l) \cap (S_0^k \otimes S_0^{l-1}) = S_0^{k-1} \otimes S_0^{l-1}.$$

Combine $g_{k-1,l}$ and $g_{k,l-1}$ to form

(5W) $$g_0 : (S_0^{k-1} \otimes S_0^l) + (S_0^k \otimes S_0^{l-1}) \to N^{kl}$$

Although we suppressed k from the notation B in Lemma 5.9, we now redesignate that set by B^k, so we can distinguish B^k from B^l. Let $\bar{B}^k = B^k - \{1\}$, $\bar{B}^l = B^l - \{1\}$. Let

$$B^{kl} = \{(x,y) \in \bar{B}^k \times \bar{B}^l \mid \phi_k(x) \in B^k \times S_0^{k-1} \quad \text{or} \quad \phi_l(y) \in B^l \times S_0^{l-1}\},$$

and put $U^{kl} = \mathrm{Span}(x \otimes y | (x,y) \in B^{kl})$. Using (5W) as our g_0 in the composition (5V), we obtain a chain map

$$g : U^{kl} \to N^{kl}$$

satisfying $hg = (1 \otimes \lambda)\tau$ (cf. the construction in Lemma 5.11(a)). Extend g (uniquely) to a dgL monomorphism

$$g : \mathcal{L}(U^{kl}) \to N^{kl} \tag{5X}$$

(cf. 5.11(b)).

In view of (4F) we write $c_s = a_s - e_s$, where $e_s \in \mathcal{U}M^{s-1}$ is the double summation. Let $u_s = p^{-(r+s-1)}d(a_s) \in \mathcal{U}M^{s-1}$, so that $d(a_s) = d(e_s) = p^{r+s-1}u_s$. Since u_s is primitive, $\lambda(u_s) = -u_s$. Since a_s is primitive and we saw earlier that $\lambda(c_s) = -c_s$ when p is odd, we have $\lambda(e_s) = -e_s$. Write $u_s^{(i)}$ for $e^{(i)}(u_s)$ and $a_s^{(i)}$ for $e^{(i)}(a_s)$, $i = 1, 2$.

Recall that all we need to do is to find four elements in N^{kl} to serve as the g_{kl}-images of the $\mathbb{Z}_{(p)}$-basis $\Gamma_k^k \otimes \Gamma_l^l$ for $S_k^k \otimes S_l^l$. These four elements must have the right h-images, i.e., the equation $hg_{kl} = -\tau$ should be true, and they should have the right boundary relationships to make g_{kl} a chain map, e.g., $d(g_{kl}(b_k \otimes b_l))$ should equal $-p^{r+k}g_{kl}(c_k \otimes b_l) - p^{r+l}g_{kl}(b_k \otimes c_l)$, etc. We will construct several elements of N^{kl} and at the end pick out the ones we want. All elements we construct will be trihomogeneous with #-grade (p^k, p^l), and our choice of subscript (e.g., the "3" on w_3) will always reveal an element's $*$-degree.

To begin, note that (5X) is an isomorphism in dimensions below $q = 2n(p^k + p^l) - 2$ (because $U_{<q}^{kl} = V_{<q}^{kl}$) and an isomorphism on decomposables in dimensions $\leq q+1$ (because $(N^{kl})_{<2} = 0$). Furthermore, decomposables in $\mathcal{L}(U^{kl})$ in #-grade (p^k, p^l) must lie in $\mathcal{L}^{\geq 2}(V^{k-1,l-1})$.

The proof is going to divide into three cases according to whether $k > l$ or $k = l$ or $k < l$. When $k > l$ we need to perform a certain set of constructions, and when $k < l$ we need a different set. When $k = l$ we need both sets.

Suppose $k \geq l$. Let

$$w_3 = u_k \otimes c_l + g^{-1}(\mathrm{ad}^{(2)}(c_l)(u_k^{(1)})) \in \mathcal{L}(U^{kl}),$$

which is clearly a cycle. Because

$$hg(w_3) = (1 \otimes \lambda)\tau(u_k \otimes c_l) + h(\mathrm{ad}^{(2)}(c_l)(u_k^{(1)})) = -c_l \otimes u_k + c_l \otimes u_k = 0,$$

we know that w_3 is a decomposable cycle, hence it is a decomposable cycle in $\mathcal{L}(V^{k-1,l-1})$ in a dimension that is $\equiv -3(\mathrm{mod} 2p)$. By Lemma 5.8, $p^{r+l-1}w_3 = d(w_2)$ for some $w_2 \in \mathcal{L}^{\geq 2}(V^{k-1,l-1})$. Put

$$y_2 = g(e_k \otimes c_l) - \mathrm{ad}^{(2)}(c_l)(a_k^{(1)}) - p^{k-l}g(w_2). \tag{5Y}$$

We find $d(y_2)=0$, so y_2 is a cycle in N^{kl}, and

$$h(y_2)=(1\otimes\lambda)\tau(e_k\otimes c_l)-c_l\otimes a_k-(0)=c_l\otimes e_k-c_l\otimes a_k=-c_l\otimes c_k.$$

Next, consider

$$z_2=g(u_k\otimes b_l)+\mathrm{ad}^{(2)}(b_l)(u_k^{(1)})+pg(w_2). \tag{5Z}$$

We find $d(z_2)=0$ and $h(z_2)=-b_l\otimes u_k+b_l\otimes u_k=0$, so z_2 is a decomposable cycle in N^{kl} of dimension $2n(p^k+p^l)-2$. It follows that $z_2=g(v_2)$ for some unique decomposable cycle $v_2\in\mathcal{L}^{\geq 2}(V^{k-1,l-1})$. By an argument called "detection" that we postpone until later, the homology class represented by $p^{r+k-1}v_2$ in $\mathcal{L}(V^{k-1,l-1})$ is zero, hence there exists $w_1\in\mathcal{L}^{\geq 2}(v^{k-1,l-1})$ for which $d(w_1)=p^{r+k-1}v_2$. Put

$$y_1=g(e_k\otimes b_l)+\mathrm{ad}^{(2)}(b_l)(a_k^{(1)})-g(w_1) \tag{5AA}$$

and compute that

$$d(y_1)=p^{r+l}y_2$$

while

$$h(y_1)=-b_l\otimes e_k+b_l\otimes a_k-(0)=b_l\otimes c_k.$$

If $k>l$ we also put

$$y_0=\mathrm{ad}^{(1)}(b_{k-1}^{p-1})g(b_{k-1}\otimes b_l).$$

Then $h(y_0)=b_l\otimes b_{k-1}(b_{k-1}^{p-1})=b_l\otimes b_k$. Also, $d(y_0)$ is divisible in N^{kl} by p^{r+l} if $l\leq k-1$, so the cycle

$$z_1=p^{-(r+l)}d(y_0)$$

is defined. Compute that

$$h(z_1)=p^{-(r+l)}dh(y_0)=-(c_l\otimes b_k)-p^{k-l}(b_l\otimes c_k).$$

When $k\leq l$ we perform the same constructions with (1) and (2) switched, leaving all computations to the reader. Let w_3' be the decomposable cycle

$$w_3'=c_k\otimes u_l-g^{-1}(\mathrm{ad}^{(1)}(c_k)(u_l^{(2)}))\in\mathcal{L}^{\geq 2}(V^{k-1,l-1}),$$

and choose $w_2'\in\mathcal{L}^{\geq 2}(V^{k-1,l-1})$ such that $d(w_2')=p^{r+k-1}w_3'$. Put

$$y_2'=g(c_k\otimes e_l)-\mathrm{ad}^{(1)}(c_k)(a_l^{(2)})+p^{l-k}g(w_2')$$

Then $d(y_2')=0$ while $h(y_2')=-c_l\otimes c_k$. Let

$$z_2'=g(b_k\otimes u_l)-\mathrm{ad}^{(1)}(b_k)(u_l^{(2)})+pg(w_2'),$$

a decomposable cycle in N^{kl}. Let $v_2'=g^{-1}(z_2')$ and choose $w_1'\in\mathcal{L}^{\geq 2}(V^{k-1,l-1})$ such that $d(w_1')=p^{r+l-1}v_2'$. (If $k=l$ a "detection" argument, postponed to the end of the proof, is needed in order to be sure that such a w_1 exists.) Put

$$y_1'=g(b_k\otimes e_l)-\mathrm{ad}^{(1)}(b_k)(a_l^{(2)})-g(w_1'). \tag{5BB}$$

Then $h(y_1') = c_l \otimes b_k$ while $d(y_1') = -p^{r+k}y_2'$. If $k < l$ put

$$y_0' = \mathrm{ad}^{(2)}(b_{l-1}^{p-1})g(b_k \otimes b_{l-1}).$$

Then $h(y_0') = -(b_{l-1}^{p-1})(b_{l-1} \otimes b_k) = -b_l \otimes b_k$, and $d(y_0')$ is divisible by p^{r+k} in N^{kl}. Put

$$z_1' = p^{-(r+k)}d(y_0').$$

We have $d(z_1') = 0$ and $h(z_1') = b_l \otimes c_k + p^{l-k}c_l \otimes b_k$.

When $k = l$ we make a few more constructions. Consider $y_2'' = y_2 - y_2'$. Since $h(y_2'') = 0$, y_2'' is a decomposable cycle of dimension $4np^k - 2$. We can write $y_2'' = g(w_2'')$, where w_2'' is a decomposable cycle in $\mathcal{L}^{\geq 2}(V^{k-1,k-1})$. By Lemma 5.8 we may choose $w_1'' \in \mathcal{L}^{\geq 2}(V^{k-1,l-1})$ such that $d(w_1'') = p^{r+k}w_2''$. Put

$$y_1'' = y_1 + y_1' - g(w_1'').$$

Then $d(y_1'') = p^{r+k}y_2 - p^{r+k}y_2' - p^{r+k}(y_2 - y_2') = 0$ while

$$h(y_1'') = b_k \otimes c_k + c_k \otimes b_k.$$

Viewing y_1'' as an element of $\mathcal{U}N^{kk}$, we have $|y_1''| = 4np^k - 1$. Applying Corollary 5.7(a) to the homology class represented by $p^{r+k}y_1''$ we discover that there must exist $z_0 \in N^{kk}$ for which

$$d(z_0) = p^{r+k}y_1''.$$

Since we may choose z_0 to be homogeneous of *-degree zero and #-grade (p^k, p^l), the fact that $dh(z_0) = p^{r+k}h(y_1'')$ implies

$$h(z_0) = -b_k \otimes b_k.$$

We are finally ready to construct g_{kl}. Consider three cases. When $k > l$, define

$$\begin{aligned} g_{kl}(c_k \otimes c_l) &= -y_2; \\ g_{kl}(c_k \otimes b_l) &= -y_1; \\ g_{kl}(b_k \otimes c_l) &= z_1 + p^{k-l}y_1; \\ g_{kl}(b_k \otimes b_l) &= -y_0. \end{aligned}$$

When $k < l$, define

$$\begin{aligned} g_{kl}(c_k \otimes c_l) &= -y_2'; \\ g_{kl}(c_k \otimes b_l) &= -z_1' + p^{l-k}y_1'; \\ g_{kl}(b_k \otimes c_l) &= -y_1'; \\ g_{kl}(b_k \otimes b_l) &= y_0'. \end{aligned}$$

When $k = l$, define

$$\begin{aligned} g_{kk}(c_k \otimes c_k) &= -y_2; \\ g_{kk}(c_k \otimes b_k) &= -y_1; \\ g_{kk}(b_k \otimes c_k) &= -y_1'' + y_1; \\ g_{kk}(b_k \otimes b_k) &= z_0. \end{aligned}$$

The reader can check, in each of the three cases, that all the needed properties of the four elements are satisfied.

We need to tie up our loose end, which is the postponed argument that $p^{r+k-1}v_2$ and $p^{r+l-1}v_2'$ are boundaries. We do the case of v_2 only, since they are so similar. The reason we need a special argument for v_2 is that it is in a dimension $\equiv -2 \pmod{2p}$, so $p^{r+k-1}v_2$ is not automatically a boundary for dimensional reasons only. Consider the cases $k > l$ and $k = l$ separately. When $k > l$, then $r + k - 1 \geq r + l$, and Lemma 5.8 supplies a decomposable w_1 for which $d(w_1) = p^{r+k-1}v_2$. The difficult case is $k = l$, which we henceforth assume.

Our goal is to get $p^{r+k-1}v_2$ to be a boundary in $\mathcal{L}(V^{k-1,k-1})$. Let

$$m = |v_2| = 4np^k - 2$$

denote the dimension in which we are working. By applying Corollary 5.7(b) we see that $p^{r+k-1}H_m(\mathcal{L}(V^{k-1,k-1}))$ equals a single copy of $\mathbb{Z}_p$, with the homology class of the cycle

$$p^{r+k-1}\tilde{\sigma}_1((b_{k-1} \otimes b_{k-1})) \tag{5CC}$$

being a generator. Thus

$$p^{r+k-1}v_2 = (c)p^{r+k-1}\tilde{\sigma}((b_{k-1} \otimes b_{k-1})) \text{ in } H_m(\mathcal{L}(V^{k-1,k-1})), \tag{5DD}$$

for some $c \in \mathbb{Z}_p$. We want to show that $c = 0$. We shall construct a dgL homomorphism $\gamma_0 : (N^{kk}, d) \to (\widehat{N}^k, \hat{d})$ for which $p^{r+k-1}H_m(\gamma_0 g)$ is one-to-one, yet $\gamma_0 g(v_2)$ is a boundary. Consequently the cycle $p^{r+k-1}v_2$ is a boundary too.

Recall the dgL surjection of (3B), which we now denote as

$$\gamma : M^k \to \mathbb{L}^{\mathrm{ab}}[b].$$

Using γ we build the commuting diagram of short exact sequences of dgL's,

$$\begin{array}{ccccccccc} 0 & \longrightarrow & N^{kk} & \longrightarrow & M^k \amalg M^k & \longrightarrow & M^k \oplus M^k & \longrightarrow & 0 \\ & & \big\downarrow{\scriptstyle \gamma_0} & & \big\downarrow{\scriptstyle \gamma \amalg 1} & & \big\downarrow{\scriptstyle \gamma \oplus 1} & & \\ 0 & \longrightarrow & \widehat{N}^k & \longrightarrow & \mathbb{L}^{\mathrm{ab}}[b] \amalg M^k & \longrightarrow & \mathbb{L}^{\mathrm{ab}}[b] \oplus M^k & \longrightarrow & 0, \end{array} \tag{5EE}$$

which serves to define both the dgL $(\widehat{N}^k, \hat{d})$ and the dgL homomorphism γ_0.

The dgL $\widehat{N}^k$ is easier to analyze and to understand than N^{kk}. It is an example of the dgL kernel that was considered in general in Lemma 5.2, so

$$h : \widehat{N}^k \longrightarrow \mathcal{U}M_+^k \otimes \mathcal{U}(\mathbb{L}^{\mathrm{ab}}[b])_+ = \mathcal{U}M_+^k \otimes \mathbb{Z}_{(p)}[b]_+$$

is defined and is a chain map, for $\widehat{N}^k$. The construction in Lemma 5.2 is functorial, so

$$\begin{array}{ccc} N^{kk} & \xrightarrow{\gamma_0} & \widehat{N}^k \\ \big\downarrow{\scriptstyle h} & & \big\downarrow{\scriptstyle h} \\ \mathcal{U}M_+^k \otimes \mathcal{U}M_+^k & \xrightarrow{1 \otimes \mathcal{U}\gamma} & \mathcal{U}M_+^k \otimes \mathbb{Z}_{(p)}[b]_+ \end{array} \tag{5FF}$$

commutes. We may define

$$\hat{g}(b^j \otimes x) = \mathrm{ad}^{(1)}(b^{j-1})\,\mathrm{ad}^{(2)}(x)(b^{(1)}).$$

Then $\hat{g} : \mathbb{Z}_{(p)}[b] \otimes \mathcal{U}M_+^k \to \widehat{N}^k$ is a chain map satisfying

$$h\hat{g} = (1 \otimes \lambda)\tau,$$

so $\hat{g}$ extends uniquely to a dgL isomorphism

$$\hat{g} : \mathcal{L}(\mathbb{Z}[b]_+ \otimes \mathcal{U}M_+^k, d) \to \widehat{N}^k. \tag{5GG}$$

This isomorphism $\hat{g}$ shows that $\widehat{N}^k$ has a linear differential.

Let $U = b^{p^{k-1}} \otimes S_{k-1}^{k-1} = \mathrm{Span}(b^{p^{k-1}} \otimes b_{k-1}, b^{p^{k-1}} \otimes c_{k-1})$. Then U is a retract of $\mathbb{Z}_{(p)}[b]_+ \otimes \mathcal{U}M_+^k$ as chain complexes, so $\mathcal{L}(U, d)$ is a dgL retract of $\mathcal{L}(\mathbb{Z}_{(p)}[b]_+ \otimes \mathcal{U}M_+^k, d)$ as dgL's, hence also a dgL retract of $\widehat{N}^k$. Let γ_1 denote these retractions, where we leave it up to the context to clarify which one we intend. Let

$$q : \mathcal{L}(W) \to \mathcal{L}(W)/\mathcal{L}^{\geq 2}(W) \approx W$$

denote the retraction as chain complexes, for any $\mathbb{Z}_{(p)}$-free chain complex W. Let U' be the differential submodule of $\mathcal{U}M_+^k \otimes \mathcal{U}M_+^k$ given by

$$U' = \mathrm{Span}(b_{k-1} \otimes b_{k-1}, b_{k-1} \otimes c_{k-1} + c_{k-1} \otimes b_{k-1}).$$

Then this continuation of diagram (5FF) commutes:

$$\begin{array}{ccccccc}
\mathcal{L}(U') & \xrightarrow{g} & N^{kk} & \xrightarrow{\gamma_0} & \widehat{N}^k & \xrightarrow{\gamma_1} & \mathcal{L}(U) \\
{\scriptstyle -\tau q}\downarrow & & {\scriptstyle h}\downarrow & & \downarrow{\scriptstyle h} & & {\scriptstyle q}\downarrow \\
U' & \xrightarrow{\subseteq} & \mathcal{U}M_+^k \otimes \mathcal{U}M_+^k & \xrightarrow{1 \otimes \mathcal{U}\gamma} & \mathcal{U}M_+^k \otimes \mathbb{Z}_{(p)}[b]_+ & \xrightarrow{-\tau\gamma_1} & U.
\end{array} \tag{5HH}$$

Start with the element $b_{k-1} \otimes b_{k-1} \in \mathcal{L}(U')$ in (5HH), and go down and then to the right to get to U. The image is clarly $-b^{p^{k-1}} \otimes b_{k-1}$. Likewise $c_{k-1} \otimes b_{k-1} + b_{k-1} \otimes c_{k-1}$ goes to $-b^{p^{k-1}} \otimes c_{k-1}$. This implies that $\gamma_1\gamma_0 g$ is an isomorphism of dgL's, hence that $H_*(\gamma_0 g)$ is one-to-one. In particular, (5CC) is not in the kernel of $H_m(\gamma_0 g)$. Consequently if (5DD) belongs to $\ker(H_m(\gamma_0 g))$ then $c = 0$. We think of γ_0 as a "simplifying homomorphism" that provides a means to "detect" whether $c = 0$.

To finish the proof, compute that

$$\gamma_0 g(p^{r+k-1} v_2) = p^{r+k-1}\gamma_0(z_2) = 0 + 0 + p^{r+k}\gamma_0 g(w_2), \tag{5II}$$

which must be a boundary since $p^{r+k} H_*(\widehat{N}^k) = 0$ by Lemma 5.8. To do the computation (5II) we use the definition (5V) of g and we use the fact that γ_0 converts $u_k^{(1)}$ and every $\mathrm{ad}^{(1)}(c_s)$ to zero.

Remark. We gave short shrift to the prime 2 during the exposition of N^{kl}, since we will use the results only for $p \geq 5$. Theorem 5.12 is still true at p=2, but several minor modifications must be made to the proof.

Notation/Definition 5.13. Let $\nabla : M^k \amalg M^k \to M^k$ denote the folding homomorphism, i.e., $\nabla(e^{(1)}(x)) = \nabla(e^{(2)}(x)) = x$. Consider the diagram of short exact sequences of dgL's,

$$\begin{array}{ccccccccc} 0 & \longrightarrow & N^{kk} & \longrightarrow & M^k \amalg M^k & \longrightarrow & M^k \oplus M^k & \longrightarrow & 0 \\ & & \nabla\big\downarrow & & \nabla\big\downarrow & & \big\downarrow & & \\ 0 & \longrightarrow & \bar{L}^k & \longrightarrow & M^k & \longrightarrow & \mathbf{L}^{ab}[b, a_0] & \longrightarrow & 0, \end{array} \tag{5JJ}$$

in which the upper row is (5R) and the lower row is (3E). The kernel of $M^k \amalg M^k$ is the Lie ideal of $M^k \amalg M^k$ generated by all $[e^{(1)}(x), e^{(2)}(y)]$, so the ∇-image of this kernel consists entirely of decomposables, and all decomposables in M^k lie in the kernel of $M^k \to \mathbf{L}^{ab}[b, a_0]$. As a result the right-hand square of (5JJ) commutes. We denote the induced homomorphism on dgL kernels also by ∇.

Lemma 5.14. *Let* $p \geq 3$.

(a) ∇ *is compatible with the inclusion of* M^{k-1} *into* M^k, *i.e.,*

$$\begin{array}{ccc} N^{k-1,k-1} & \longrightarrow & N^{k,k} \\ \nabla\big\downarrow & & \big\downarrow\nabla \\ \bar{L}^{k-1} & \xrightarrow{\ i\ } & \bar{L}^k \end{array} \tag{5KK}$$

commutes.

(b) $\nabla(w) = 0$, *if* $|w|_* = 0$ *or if* $|w|_* = 1$ *and* w *is decomposable.*

(c) *For* $0 \leq s$, $t \leq k$, *let* $u = \max(s,t)$, $m = \min(s,t)$. *Then*

$$\begin{aligned} \nabla g(b_s \otimes b_t) &= 0 \\ \nabla g(b_s \otimes c_t) &= p^{u-t} y_{p^s+p^t} \\ \nabla g(c_s \otimes b_t) &= -p^{u-s} y_{p^s+p^t} \\ \nabla g(c_s \otimes c_t) &= (p^{s-m} + p^{t-m}) z'_{p^s+p^t} \end{aligned}$$

where y_j *and* z'_j *are the generators of* $\bar{L}^k$ *described in Chapter 2.*

Proof. Part (a) is trivial. For (b), note that $\bar{L}^k$ has no *-degree zero component, and consequently no decomposables in *-degree 1. For (c), the fact that $g(b_s \otimes b_t) \in \ker(\nabla)$ follows from (b).

To evaluate the other three expressions in (c), consider $s \geq t$ and $s < t$ separately. When $s \geq t$, evaluate $\nabla g(c_s \otimes b_t)$ by thinking of it as $-\nabla(y_1)$ when $s = k$ and $t = l$, where y_1 is given by (5AA). Referring to (5AA), when the term $g(e_s \otimes b_t)$ is expanded using (5V) and (4F), each summand has the form $\operatorname{ad}^{(1)}(-)\operatorname{ad}^{(2)}(-)g(b_j \otimes b_t)$ for some $j < s$. We saw in (b) that $\nabla g(b_j \otimes b_t) = 0$, so $\nabla g(e_s \otimes b_t) = 0$. The third term in (5AA), namely $g(w_1)$, is decomposable of *-degree one, so $\nabla g(w_1) = 0$. Since ∇ converts $\operatorname{ad}^{(i)}(x)$ to $\operatorname{ad}(x)$, (5AA) tells us

$$\begin{aligned} \nabla g(c_s \otimes b_t) &= -\nabla \operatorname{ad}^{(2)}(b_t)(a_s^{(1)}) = -\operatorname{ad}(b^{p^t})(a_s) \\ &= -\operatorname{ad}(b^{p^t})(y_{p^s}) = -p^{\alpha(p^s+p^t)-s} y_{p^s+p^t}, \end{aligned}$$

where $\alpha(\ )$ is the α of (2A). Since $p \geq 3$, $\alpha(p^s + p^t) = \max(s,t) = u$. To compute $\nabla g(b_s \otimes c_t)$, solve for it using $\nabla(dg(b_s \otimes b_t)) = 0$ and

$$dg(b_s \otimes b_t) = -p^{r+t}(p^{s-t}g(c_s \otimes b_t) + g(b_s \otimes c_t)).$$

To compute $\nabla g(c_s \otimes c_t)$, apply $p^{-(r+t)}d$ to both sides of the equation

$$\nabla g(c_s \otimes b_t) = -p^{u-s}y_{p^s+p^t}$$

and use (2I) to simplify.

The computation when $s < t$ follows exactly the same pattern, using (5BB) instead of (5AA) as its starting point.

We finish Chapter 5 by defining a dgL cousin of N^{kl}. As motivation, note that an alternate definition of N^{kl} is

$$N^{kl} = \text{Span}\{\text{homogeneous } x \in M^k \amalg M^l \mid x \text{ has positive bidimension}\}.$$

By "x has positive bidimension" we mean that both $|x|^{(1)}$ and $|x|^{(2)}$ are positive. The corresponding ideal of $M^k \amalg M^l \amalg M^m$ will be of interest to us.

Definition/Notation 5.15. For $i = 1,2,3$, let $|\,|^{(i)}$, $|\,|_*^{(i)}$, $|\,|_\#^{(i)}$ be the gradings on $M^k \amalg M^l \amalg M^m$ induced by the corresponding gradings on M^k, on M^l, and on M^m. Say that a trihomogeneous element x has ***positive tridimension*** if $|x|^{(i)}$ are all positive, $i = 1,2,3$. Define

$$N^{klm} = \text{Span}\{x \in M^k \amalg M^l \amalg M^m \mid x \text{ has positive tridimension}\}.$$

Let $e^{(12)} : M^k \amalg M^l \to M^k \amalg M^l \amalg M^m$ denote the evident dgL inclusion, and likewise for $e^{(13)}$ and $e^{(23)}$. The meanings of $e^{(i)}$, $\text{ad}^{(i)}$, and $d^{(i)}$ should be clear from their analogs for $M^k \amalg M^l$.

Lemma 5.16. *For any $k, l, m \geq 0$, N^{klm} is a differential Lie ideal of $M^k \amalg M^l \amalg M^m$. As a Lie ideal, N^{klm} is generated by*

$$\text{(5LL)}\qquad \text{ad}^{(1)}(\mathcal{U}M_+^k)e^{(23)}(N^{lm}) + \text{ad}^{(2)}(\mathcal{U}M_+^l)e^{(13)}(N^{km}) + \text{ad}^{(3)}(\mathcal{U}M_+^m)e^{(12)}(N^{kl}).$$

Proof. This is straightforward. For any $x \in N^{klm}$, write it as a sum of linearly independent iterated brackets, where each entry in each bracket belongs to M^k or to M^l or to M^m. Since $x \in N^{klm}$, each bracket contains at least one entry from each of M^k, M^l, and M^m.

We will see in Chapter 6 that N^{klm}, like N^{kl}, is free with a linear differential.

Remark 5.17. Although the construction of g_{kl} in Theorem 5.12 is constructive in the sense that a computer could hypothetically perform it, there are a lot of places where arbitrary choices are made. It is possible to give instead an explicit combinatorial definition for g_{kl}. The idea goes as follows. Letting w_{ij} denote the Hopf commutator $[(b^{(1)})^i, (b^{(2)})^j]_H$, we find primitive elements $\tilde{w}_{ij}$ which differ from w_{ij} by decomposables. There is an explicit and very beautiful formula for the $\tilde{w}_{ij}$'s. It is possible to use $\tilde{w}_{p^s,p^t}$ as $g(b_s \otimes b_t)$, and to let $g(c_s \otimes b_t)$ be $-p^{-(r+s)}d^{(1)}(\tilde{w}_{p^s,p^t}) + \tilde{y}_{st}$. The decomposable perturbation term $\tilde{y}_{st}$ is needed because $d^{(2)}(-p^{-(r+s)}d^{(1)}(\tilde{w}_{p^s,p^t}))$ is not divisible by p^{r+t}, but p^{r+t} does divide the $d^{(2)}$-image of the perturbed expression. Then we set $g(c_s \otimes c_t)$ equal to $p^{-(r+t)}d^{(2)}(\tilde{y}_{st} - p^{-(r+s)}d^{(1)}(\tilde{w}_{p^s,p^t}))$.

The author got this far with the explicit approach before deciding that the additional length and difficulty of those arguments outweighed their elegance and intrinsic appeal. Furthermore, more work was going to be needed in order to make the connection with, e.g., $\mathrm{ad}^{(1)}(b_k)(a_l^{(2)})$, a connection which falls naturally out of the approach adopted here. There is no way around the fact that N^{kl} is a very subtle and complicated dgL. The author hopes that some day someone with an inclination for combinatorics and algebra will take up and complete the explicit construction of g_{kl} and its properties.

CHAPTER 6

Three Topological Tools

Chapter 6 is roughly divided into three units of unequal length. We begin with a whirlwind tour through p-primary homotopy theory, to establish notation and to review basic facts. The second unit reviews Adams-Hilton models and examines a certain extension problem. This culminates in an existence/uniqueness theorem for mod p^t homotopy classes. Under suitable hypotheses, the theorem enables the construction of a canonical mod p^t homotopy class using only information about homology and about mod p^{t-1} homotopy. Since this theorem and its corollaries are the only tools in the entire monograph which construct new mod p^t homotopy classes, its importance cannot be overstated.

The third and longest unit explores a certain homotopy inverse limit for three spaces, along with its algebraic analog for Hopf algebras. Given three spaces, consider the diagram formed by taking all three bouquets of two of them, along with the projections onto the three individual spaces. The homotopy inverse limit of this diagram is called the "thin product" of the three spaces. We study thin products because they lead us to the correct place to prove each of three essential relationships, all analogs of the Jacobi identity, among certain homotopy classes. The Hopf algebra analog, which we call a "cyclic cotensor," describes the loop space homology of the thin product. We include a corollary showing that the dgL N^{klm} defined in 5.15 has a linear differential and computing an indecomposable submodule.

Review of p-Primary Homotopy Theory

We assume that the reader is somewhat acquainted with p-primary homotopy theory, and we include this review for completeness and convenience only. Proofs for nearly everything in this unit can be found in [N1].

Let p be a prime, $r \geq 1$, $m \geq 2$. The mod p^r *Moore space* of dimension m is the cofiber of the map of Brouwer degree p^r on S^{m-1}, denoted $P^m(p^r) = S^{m-1} \cup_{p^r} e^m$. Given a (pointed) space X (recall that for us all spaces are pointed), the mth mod p^r *homotopy set* of X is the set of homotopy classes,

$$\pi_m(X; \mathbb{Z}_{p^r}) = [P^m(p^r); X].$$

When $m \geq 3$ or X is an H-space it is a group; if $p \neq 2$ this group is abelian and it is annihilated by p^r.

There are several correlations and operations among mod p^r homotopy groups. Certain maps among Moore spaces and spheres will crop up very often, so we standardize our names for them once and for all. The inclusion map of S^{m-1} into $P^m(p^r)$ is denoted ω^r, and the pinch map from $P^m(p^r)$ onto S^m is called ρ^r. Together with the map of Brouwer degree p^r, they comprise the coexact sequence

$$\text{(6A)} \qquad \cdots \xrightarrow{p^r} S^{m-1} \xrightarrow{\omega^r} P^m(p^r) \xrightarrow{\rho^r} S^m \xrightarrow{p^r} S^m \xrightarrow{\omega^r} \cdots$$

Sequence (6A) yields the famous universal coefficient short exact sequence:

$$\text{(6B)} \qquad 0 \to \pi_m(X) \otimes \mathbb{Z}_{p^r} \to \pi_m(X; \mathbb{Z}_{p^r}) \to \operatorname{Tor}(\pi_{m-1}(X), \mathbb{Z}_{p^r}) \to 0.$$

For $p \geq 3$ (6B) always splits, but not functorially.

A second coexact sequence involves Moore spaces only. For $t \geq s \geq r \geq 1$, let ω_s^t denote the map from $P^m(p^s)$ to $P^m(p^t)$ that has Brouwer degree p^{t-s} on the included $(m-1)$ sphere and is a relative homeomorphism on the m-cells. Let $\rho_t^r : P^m(p^t) \to P^m(p^r)$ be the map from $P^m(p^t)$ to the cofiber of the ω_{t-r}^t. It is a homeomorphism on the included S^{m-1}'s. These two maps participate in the coexact sequence

$$\text{(6C)} \qquad \cdots \to P^m(p^s) \xrightarrow{\omega_s^t} P^m(p^t) \xrightarrow{\rho_t^{t-s}} P^m(p^{t-s}) \xrightarrow{\omega^s \rho^{t-s}} P^{m+1}(p^s) \xrightarrow{\omega_s^t} \cdots \quad .$$

All of these maps induce natural transformations via composition on the right, e.g., the natural transformation

$$(\omega_s^t)^\# : \pi_*(-; \mathbb{Z}_{p^t}) \to \pi_*(-; \mathbb{Z}_{p^s}).$$

The coexactness of (6C) means that $\operatorname{im}(\rho_t^{t-s})^\# = \ker(\omega_s^t)^\#$, and so on, and similar relations are derivable from (6A).

The isomorphism relating the mod p^r homotopy of X to that of ΩX is denoted

$$\text{(6D)} \qquad (\,)^a : \pi_{m+1}(X; \mathbb{Z}_{p^r}) \xrightarrow{\approx} \pi_m(\Omega X; \mathbb{Z}_{p^r}),$$

where $(\,)^a$ is our notation for "adjoint." We denote the inverse of (6D) by $(\,)^{\bar{a}}$.

We will frequently prefer to work with the mod p^r homotopy of ΩX rather than of X because the former carries the additional structure of Samelson products when $p \geq 3$. The *(absolute) Samelson product*

$$[\ \ ,\ \ ] : \pi_m(\Omega X; \mathbb{Z}_{p^r}) \otimes \pi_q(\Omega X; \mathbb{Z}_{p^r}) \to \pi_{m+q}(\Omega X; \mathbb{Z}_{p^r})$$

is an antisymmetric bilinear pairing. When $p \geq 5$ it satisfies the (graded) Jacobi identity, making $\pi_*(\Omega X; \mathbb{Z}_{p^r})$ into a graded Lie algebra over $\mathbb{Z}_{p^r}$. When $p = 3$ the Jacobi identity is satisfied modulo a certain perturbation term. The fact that $\pi_*(\Omega X; \mathbb{Z}_{p^r})$ is a true Lie algebra only for $p \geq 5$ is one of the principal reasons why most of the geometric results in this monograph are stated for $p \geq 5$ only.

As defined in the previous paragraph, the Lie algebra $\pi_*(\Omega X; \mathbb{Z}_{p^r})$ only makes sense for $* \geq 2$. For loop spaces we can make the special definition

$$\pi_1(\Omega X; \mathbb{Z}_{p^r}) = \pi_1(\Omega X) \otimes \mathbb{Z}_{p^r},$$

and then $\pi_{\geq 1}(\Omega X; \mathbf{Z}_{p^r})$ becomes a natural graded lie $\mathbf{Z}_{p^r}$-algebra, for $p \geq 5$. Whenever we write $\pi_*(\Omega X; \mathbf{Z}_{p^r})$, the component for $* = 1$ will be assumed to be part of this object.

Closely related to Samelson products are Whitehead products. To avoid extra signs we define the ***Whitehead product***

$$[\ ,\ ] : \pi_m(X; \mathbf{Z}_{p^r}) \otimes \pi_q(X; \mathbf{Z}_{p^r}) \to \pi_{m+q-1}(X; \mathbf{Z}_{p^r})$$

by $[\hat{x}, \hat{y}] = ([\hat{x}^a, \hat{y}^a])^a$. Also closely related to Samelson products are ***relative Samelson products***. Given a fibration

$$X \xrightarrow{f} Y \to Z,$$

there is a functorial action of $\pi_*(\Omega Y; \mathbf{Z}_{p^r})$ on $\pi_*(\Omega X; \mathbf{Z}_{p^r})$,

$$\text{(6E)} \qquad [\ ,\ ] : \pi_*(\Omega Y; \mathbf{Z}_{p^r}) \otimes \pi_*(\Omega X; \mathbf{Z}_{p^r}) \to \pi_*(\Omega X; \mathbf{Z}_{p^r}).$$

The relative and absolute Samelson products are connected via the formulas

$$(\Omega f)_\#[\hat{y}, \hat{x}] = [\hat{y}, (\Omega f)_\#(\hat{x})] \quad , \quad [(\Omega f)_\#(\hat{x}), \hat{x}'] = [\hat{x}, \hat{x}'],$$

where in each formula the bracket on the left is a relative Samelson product and the bracket on the right is an absolute Samelson product. Relative Samelson products will be important to us, since they are the starting point for the study of homotopy operations under a space (Chapter 8). When $p \geq 5$, (6E) makes $\pi_*(\Omega X; \mathbf{Z}_{p^r})$ into an "extended Lie ideal" of $\pi_*(\Omega Y; \mathbf{Z}_{p^r})$. In particular, $\pi_*(\Omega X; \mathbf{Z}_{p^r})$ is a left module over the graded algebra $\mathcal{U}\pi_*(\Omega Y; \mathbf{Z}_{p^r})$.

Another essential structure on $\pi_*(-; \mathbf{Z}_{p^r})$ is the p^r-*Bockstein*. We denote it by ${}_\pi\delta^r : \pi_m(-; \mathbf{Z}_{p^r}) \to \pi_{m-1}(-; \mathbf{Z}_{p^r})$, and its definition is

$${}_\pi\delta^r = (\omega^r \rho^r)^\# = (\rho^r)^\#(\omega^r)^\#.$$

Since $\rho^r\omega^r = 0$ (see (6A)), $({}_\pi\delta^r)^2 = 0$, and ${}_\pi\delta^r$ makes the mod p^r homotopy into a functor from spaces to chain complexes. Furthermore, for $p \geq 3$ it is a derivation with respect to Samelson and relative Samelson products. For $p \geq 5$ ${}_\pi\delta^r$ makes $\pi_*(\Omega-; \mathbf{Z}_{p^r})$ into a functor from spaces to dgL's. This is the dgL we mean when we refer to the "homotopy dgL" of a space. For $p \geq 5$, the relative Samelson product (6E) makes $\pi_*(\Omega X; \mathbf{Z}_{p^r})$ into a differential graded left module over the dga $\mathcal{U}\pi_*(\Omega Y; \mathbf{Z}_{p^r})$.

The ***Hurewicz homomorphism*** h is a natural transformation between mod p^r homotopy and mod p^r homology. It is defined by $h(\hat{x}) = x_*(z) \in H_m(X; \mathbf{Z}_{p^r})$, where $x : P^m(p^r) \to X$ represents the homotopy class $\hat{x} \in \pi_m(X; \mathbf{Z}_{p^r})$, and z denotes the canonical generator of $H_m(P^m(p^r); \mathbf{Z}_{p^r})$. The Hurewicz homomorphism is a chain map; i.e., $h_\pi\delta^r = \delta^r h$, where $\delta^r : H_m(X; \mathbf{Z}_{p^r}) \to H_{m-1}(X; \mathbf{Z}_{p^r})$ is the Bockstein associated to the short exact sequence of coefficient groups,

$$0 \to \mathbf{Z}_{p^r} \to \mathbf{Z}_{p^{2r}} \to \mathbf{Z}_{p^r} \to 0.$$

The connection with the integral Hurewicz homomorphism, also denoted h, is best summarized by stating that h induces a natural transformation between the respective universal coefficient sequences.

When $p \geq 3$, $h : \pi_*(\Omega X; \mathbf{Z}_{p^r}) \to H_*(\Omega X; \mathbf{Z}_{p^r})$ satisfies the beautiful formula

$$h[\hat{x}, \hat{y}] = [h(\hat{x}), h(\hat{y})],$$

where the bracket on the left is the Samelson product and the bracket on the right is the (graded) commutator in the Pontrjagin ring. The image of h consists entirely of primitives in $H_*(\Omega X; \mathbb{Z}_{p^r})$. The mod p Hurewicz theorem states that, if X is $(m-1)$-connected, $m \geq 2$, then

$$h : \pi_q(X; \mathbb{Z}_{p^r}) \to H_q(X; \mathbb{Z}_{p^r})$$

is an isomorphism for $q \leq \min(m+2p-4, 2m-2)$, and an epimorphism at $\min(m+2p-3, 2m-1)$.

We summarize some further formulas in Lemma 6.1.

Lemma 6.1. *Let $t \geq s \geq r \geq 1$.*

(a) $(\rho^s)^\#$, $(\omega^s)^\#$, $(\rho^s_t)^\#$,$(\omega^s_t)^\#$ *all commute with* $(\,)^a$ *and* $(\,)^{\bar{a}}$.
(b) $\rho^r_s\rho^s_t = \rho^r_t$ and $\omega^t_s\omega^s_r = \omega^t_r$.
(c) $\rho^r\rho^r_t = p^{t-r}\rho^t$; $\rho^t\omega^t_s = \rho^s$; $\rho^r_t\omega^t = \omega^r$; $\omega^t_s\omega^s = p^{t-s}\omega^t$.
(d) $\omega^t_s\rho^s_t = p^{t-s} : P^m(p^t) \to P^m(p^t)$.
(e) $\rho^s_t\omega^t_s = p^{t-s} : P^m(p^s) \to P^m(p^s)$.
(f) *(A consequence of (c))* $(\omega^t_s)^\# {}_\pi\delta^t(\rho^s_t)^\# = (\rho^s_t\omega^t\rho^t\omega^t_s)^\# = {}_\pi\delta^s$.

For (g)-(i), p is assumed to be odd. The bracket can denote a Whitehead product, a Samelson product, or a relative Samelson product.

(g) *For $\hat{x}, \hat{y} \in \pi_*(\,;\mathbb{Z}_{p^t})$,*

$$(\omega^t_s)^\#[\hat{x}, \hat{y}] = [(\omega^t_s)^\#(\hat{x}), (\omega^t_s)^\#(\hat{y})],$$

where the first bracket is for $\pi_*(\,;\mathbb{Z}_{p^t})$ and the second is for $\pi_*(\,;\mathbb{Z}_{p^s})$.

(h) *For $\hat{x} \in \pi_*(\,;\mathbb{Z}_{p^s})$ and $\hat{y} \in \pi_*(\,;\mathbb{Z}_{p^t})$,*

$$(\rho^s_t)^\#[\hat{x}, (\omega^t_s)^\#(\hat{y})] = [(\rho^s_t)^\#(\hat{x}), \hat{y}],$$

where the first bracket is for $\pi_*(\,;\mathbb{Z}_{p^s})$ and the second is for $\pi_*(\,;\mathbb{Z}_{p^t})$.

(i) *For $\hat{x} \in \pi_*(\,;\mathbb{Z}_{p^t})$ and $\hat{y} \in \pi_*(\,;\mathbb{Z}_{p^s})$,*

$$(\rho^s_t)^\#[(\omega^t_s)^\#(\hat{x}), \hat{y}] = [\hat{x}, (\rho^s_t)^\#(\hat{y})],$$

where the first bracket is for $\pi_*(\,;\mathbb{Z}_{p^s})$ and the second is for $\pi_*(\,;\mathbb{Z}_{p^t})$.

Notation/Convention. We are going to be sloppy about distinguishing between a map $\hat{x}$ and its homotopy class $[\hat{x}]$, i.e., we will generally employ the same notation for both. In particular, when $\hat{x} \in \pi_*(X; \mathbb{Z}_{p^s})$, we will frequently write $\hat{x}\rho^s_t$ or $\hat{x}\omega^s_r$ instead of $(\rho^s_t)^\#(\hat{x})$ or $(\omega^s_r)^\#(\hat{x})$. Another example will be a statement like "$(1)^a \in \pi_{m-1}(\Omega P^m(p^r); \mathbb{Z}_{p^r})$" to indicate the adjoint of the identity map. In the few instances where it matters, e.g., that a square commute on the nose rather than merely up to homotopy, we will be more careful.

The Extension Problem for Adams-Hilton Models

An *Adams-Hilton model* for a space X is a free dga that is quasi-isomorphic as a dga with the 0th Eilenberg subcomplex of the singular chain complex $C_*(\Omega X)$. For our next task, we review some of the basic properties of Adams-Hilton models. We then look at the question of when a mod p^{t-1} homotopy class can be extended to a mod p^t homotopy class (i.e., extending through ω_{t-1}^t), and we discern what this question means for Adams-Hilton models. We prove a theorem giving sufficient conditions for the solvability of the extension problem, and we consider some useful corollaries of the theorem.

We assume the reader has at least a passing acquaintence with the category of differential graded algebras, in particular, with the notion of dga homotopy. References for dga theory are [Mu], [A3]; for Adams-Hilton models, [AH] and [A3].

Let X be a simply-connected space. The space ΩX of Moore loops on X is a path-connected topological monoid, and the 0th Eilenberg subcomplex of the singular chains on ΩX, denoted $C_*(\Omega X)$, is a dga. Any coefficient ring may be used; for us the default coefficient ring will be $\mathbb{Z}_{(p)}$, so that what we call $C_*(\Omega X)$ would normally be denoted $C_*^0(\Omega X; \mathbb{Z}_{(p)})$. By elementary dga theory there must exist a free dga over $\mathbb{Z}_{(p)}$, (TV, d), which is a *free model* for $C_*(\Omega X)$, i.e., (TV, d) comes with a quism $\theta_X : (TV, d) \xrightarrow{\simeq} C_*(\Omega X)$. An *Adams-Hilton model* (henceforth *AH model*) for X is any free dga (TV, d) which comes with a quism $\theta_X : (TV, d) \to C_*(\Omega X)$. We typically denote an AH model for X by $\mathcal{A}(X)$. Similarly, given a map $f : X \to Y$ and given AH models $\mathcal{A}(X)$ for X and $\mathcal{A}(Y)$ for Y, there is always a dga homomorphism, typically denoted $\mathcal{A}(f)$, which makes this diagram commute up to dga homotopy:

$$\begin{array}{ccc} \mathcal{A}(X) & \xrightarrow{\mathcal{A}(f)} & \mathcal{A}(Y) \\ {\scriptstyle \theta_X}\downarrow & & \downarrow{\scriptstyle \theta_Y} \\ C_*(\Omega X) & \xrightarrow{C_*(\Omega f)} & C_*(\Omega Y) \end{array}$$

We call $\mathcal{A}(f)$ an *Adams-Hilton (AH) model* for the map f. (Note carefully: even though the quism is with the chains on ΩX, we call $\mathcal{A}(X)$ a model for X, not a model for ΩX.)

Suppose $X' = * \cup \bigcup_{i \in I} e^{m_i}$, $m_i \geq 2$, is a CW approximation to X, i.e., it is a CW complex for which there is a weak homotopy equivalence from X' to X. (Here $\{m_i\}$ are the dimensions of the positive-dimensional cells of X'.) Adams and Hilton proved in their original paper [AH] that there is an AH model for X that has the form (TV, d), where V is the free graded $\mathbb{Z}_{(p)}$-module with $\mathbb{Z}_{(p)}$-basis $\{a_i\}_{i \in I}$, and $|a_i| = m_i - 1$. In other words, the generators of $\mathcal{A}(X)$ are in one-to-one correspondence with the cells of X', and the dimension of each generator is one less than the dimension of the corresponding cell. The linear component d^1 of the differential d makes (V, d^1) into a chain complex isomorphic with the desuspension of the reduced cellular chain complex for X'. This by itself is enough to determine $\mathcal{A}(X)$ for some extremely simple spaces. Since S^{q+1} has a CW structure with just one positive-dimensional cell, $\mathcal{A}(S^{q+1})$ is free on a single q-dimensional generator, with zero differential. Similarly, $P^{q+1}(p^r)$ is free on two generators that are connected via $d(v) = p^r u$.

One property of AH models that is simultaneously a great strength and a great weakness is their flexibility. For any space X there will be many free dga's that satisfy the definition to serve as $\mathcal{A}(X)$. However, any two models for X must have the same dga homotopy type. Similarly, even when $\mathcal{A}(X)$ and $\mathcal{A}(Y)$ are fixed, a map $f : X \to Y$ will typically have more than one AH model. If $\phi : \mathcal{A}(X) \to \mathcal{A}(Y)$ is a model for f, then any dga homomorphism that is homotopic to ϕ may also serve as $\mathcal{A}(f)$. (Conversely, any two models for the same map f must be homotopic as dga homomorphisms.) If $\mathcal{A}(X)$ and $\mathcal{A}(Y)$ exhibit even a moderate degree of complexity it can be difficult to discern whether two dga homomorphisms from $\mathcal{A}(X)$ to $\mathcal{A}(Y)$ are homotopic. Lemma 6.2 will shortly give a criterion for two dga homomorphisms to be homotopic when the source is $\mathcal{A}(S^{q+1})$ or $\mathcal{A}(P^{q+1}(p^s))$.

The flexibility implicit in the notation $\mathcal{A}(f)$ also leads to a potentially tricky situation for compositions, and we alert the reader to this anomoly now. If $X \xrightarrow{f} Y \xrightarrow{g} Z$, then regardless of how $\mathcal{A}(f)$ and $\mathcal{A}(g)$ are selected (from among their allowable values, of course) it is permissible to set $\mathcal{A}(gf)$ equal to the composition $\mathcal{A}(g)\mathcal{A}(f)$. However, if $\mathcal{A}(gf)$ is known in advance for some other reason, then the most that can be said in general is that $\mathcal{A}(gf)$ is homotopic to $\mathcal{A}(g)\mathcal{A}(f)$. Furthermore, an arbitrary AH model $\mathcal{A}(gf)$ for gf need not be factorable at all as an $\mathcal{A}(g)$ composed with an $\mathcal{A}(f)$.

In view of the flexibility of AH models, we find it useful to fix once and for all our models for Moore spaces and spheres, and for certain maps among Moore spaces and spheres. Our AH model for the Moore space $P^{q+1}(p^m)$ will always be the free dga over $\mathbb{Z}_{(p)}$ that has two generators v_q and u_{q-1}, subscript signifying dimension, and differential given by $d(u_{q-1}) = 0$, $d(v_q) = p^m u_{q-1}$. Occasionally we shall not want to suppress the Moore space's order from the notation, and on these occasions we write $v_q^{(m)}$ for v_q and $u_{q-1}^{(m)}$ for u_{q-1}. For S^{q+1} the AH model will always be the polynomial ring over $\mathbb{Z}_{(p)}$ on a single generator v_q of dimension q, with zero differential.

Let $t \geq s \geq 1$. For the q-sphere inclusion map $\omega^s : S^q \to P^{q+1}(p^s)$, we always take our AH model to be the homomorphism defined by $\mathcal{A}(\omega^s)(v_{q-1}) = u_q^{(s)}$. For ω_s^t our model is always the homomorphism described by $\mathcal{A}(\omega_s^t)(u_{q-1}^{(s)}) = p^{t-s}u_{q-1}^{(t)}$ and $\mathcal{A}(\omega_s^t)(v_q^{(s)}) = v_q^{(t)}$. For the pinch map ρ^t, we adopt the model $\mathcal{A}(\rho^t)(v_{q-1}^{(t)}) = 0$, $\mathcal{A}(\rho^t)(v_q^{(t)}) = v_q$. And for ρ_t^s, use $\mathcal{A}(\rho_t^s)(u_{q-1}^{(t)}) = u_{q-1}^{(s)}$, $\mathcal{A}(v_q^{(t)}) = v^{(s)}$. All of these models can be deduced easily from the corresponding facts about the maps between cellular chain complexes. Among these models for these particular maps, the reader can check that $\mathcal{A}(gf)$ does always equal $\mathcal{A}(g)\mathcal{A}(f)$ whenever f and g are composable.

With Lemma 6.2 we explore what it means for the AH models of two integral or mod p^s homotopy classes to be homotopic. Recall that two dga homomorphisms are homotopic if and only if they are the restrictions to the two "ends" of a dga homomorphism out of the dga cylinder on their common source. When the source is $\mathcal{A}(S^{q+1}) = (\mathbb{Z}_{(p)}\langle v_q\rangle, 0)$, this cylinder is $(\mathbb{Z}_{(p)}\langle v_q, v_q', v_{q+1}''\rangle, d(v_{q+1}'') = v_q - v_q')$. Thus two dga homomorphisms ϕ and ϕ' from $\mathcal{A}(S^{q+1})$ to $\mathcal{A}(X)$ are dga homotopic if and only if the difference element $\phi(v_q) - \phi'(v_q)$ is a boundary in $\mathcal{A}(X)$. A similar analysis can be performed for dga homomorphisms out of $\mathcal{A}(P^{q+1}(p^s))$. The results are summarized in Lemma 6.2, whose proof is left as an exercise.

Lemma 6.2.

(a) *Let $\hat{a}_1, \hat{a}_2 : S^{q+1} \to X$. Fix $\mathcal{A}(X)$, and choose models $\mathcal{A}(\hat{a}_1)$, $\mathcal{A}(\hat{a}_2) : \mathcal{A}(S^{q+1}) \to \mathcal{A}(X)$. (It is possible to choose $\mathcal{A}(\hat{a}_1)$ different from $\mathcal{A}(\hat{a}_2)$ even if $\hat{a}_1 = \hat{a}_2$.) A necessary and sufficient condition for $\mathcal{A}(\hat{a}_1)$ to be homotopic to $\mathcal{A}(\hat{a}_2)$ is*

$$\mathcal{A}(\hat{a}_1)(v_q) - \mathcal{A}(\hat{a}_2)(v_q) \in d(\mathcal{A}(X)),$$

and this condition is necessary for $\hat{a}_1 \simeq \hat{a}_2$.

(b) *Let* $\hat{a}_1, \hat{a}_2 : P^{q+1}(p^s) \to X$. *Fix* $\mathcal{A}(X)$ *and choose models* $\mathcal{A}(\hat{a}_1), \mathcal{A}(\hat{a}_2) : \mathcal{A}(P^{q+1}(p^s)) \to \mathcal{A}(X)$. *Put* $y_1 = \mathcal{A}(\hat{a}_1)(v_q^{(s)})$ *and* $y_2 = \mathcal{A}(\hat{a}_2)(v_q^{(s)})$. *The following are equivalent, and are necessary conditions for* $\hat{a}_1 \simeq \hat{a}_2$:

(i) $\mathcal{A}(\hat{a}_1) \simeq \mathcal{A}(\hat{a}_2)$

(ii) *There exist* $w_q \in \mathcal{A}(X)$ *and* $w_{q+1} \in \mathcal{A}(X)$, *subscripts signifying dimension, such that* $d(y_1) - d(y_2) = p^s d(w_q)$ *and*

(6F)
$$y_1 - y_2 = p^s w_q + d(w_{q+1}).$$

(iii) $y_1 - y_2 \in p^m \mathcal{A}(x)_q + d(\mathcal{A}(X)_{q+1})$.

(iv) *(if one assumes further that* $\mathcal{A}(\hat{a}_1)(u_{q-1}^{(s)}) = \mathcal{A}(\hat{a}_2)(u_{q-1}^{(s)})$*)*

$$y_1 - y_2 \in (p^s)\ker(d) + \text{im } d.$$

Conversely, if $\mathcal{A}(\hat{a}_1)$ *is any model for* $\hat{a}_1$, *and we put* $y_1 = \mathcal{A}(\hat{a}_1)(v_q^{(s)})$, *and if* y_2 *is any other element of* $\mathcal{A}(X)$ *for which (6F) holds, then the homomorphism* $\phi : \mathcal{A}(P^{q+1}(p^s)) \to \mathcal{A}(X)$ *defined by* $\phi(v_q^{(s)}) = y_2$, $\phi(u_{q-1}^{(s)}) = \mathcal{A}(\hat{a}_1)(u_{q-1}^{(s)}) - w_q$, *is also an AH model for* $\hat{a}_1$.

We are ready to begin discussing the extension problem for AH models. Suppose $\hat{a} : P^{q+1}(p^s) \to X$ satisfies $\mathcal{A}(\hat{a})(v_q^{(s)}) = x \in \mathcal{A}(X)$, and suppose we wish to extend $\hat{a}$ over $P^{q+1}(p^{s+1})$, i.e., suppose we wish to find an $\hat{a}_0 : P^{q+1}(p^{s+1}) \to X$ for which $\hat{a}_0\omega_s^{s+1} = \hat{a}$. Geometrically, by (6C), we are asking whether $\hat{a} \in \text{im}(\omega_s^{s+1})^\# = \ker(\omega^s\rho^1)^\#$, i.e., whether $\hat{a}\omega^s\rho^1$ is null-homotopic. However, let us make the problem harder by requiring in addition that $\mathcal{A}(\hat{a}_0)(v^{(s+1)}) = x$. Now there will be an algebraic obstruction as well as a geometric obstruction (the geometric obstruction is $\hat{a}\omega^s\rho^1$) to the existence of such an extension.

At first glance it might appear that $\mathcal{A}(\hat{a}_0)(v_q^{s+1})$ equals x whenever $\hat{a}_0$ exists. After all, we have

$$x = \mathcal{A}(\hat{a})(v_q^{(s)}) = \mathcal{A}(\hat{a}_0\omega_s^{s+1})(v_q^{(s)}) = \mathcal{A}(\hat{a}_0)\mathcal{A}(\omega_s^{s+1})(v_q^{(s)}) = \mathcal{A}(\hat{a}_0)(v_q^{(s+1)}).$$

The fallacy in this "proof" is in the innocent-looking step $\mathcal{A}(\hat{a}_0\omega_s^{s+1}) = \mathcal{A}(\hat{a}_0)\mathcal{A}(\omega_s^{s+1})$. While it would be possible to choose $\mathcal{A}(\hat{a}_0\omega_s^{s+1})$ to be $\mathcal{A}(\hat{a}_0)\mathcal{A}(\omega_s^{s+1})$ if $\mathcal{A}(\hat{a}_0)$ were chosen first, that is not what happens here. Here we have chosen $\mathcal{A}(\hat{a}) = \mathcal{A}(\hat{a}_0\omega_s^{s+1})$ first, and we are comparing possible models $\mathcal{A}(\hat{a}_0)$ for $\hat{a}_0$ with the given $\mathcal{A}(\hat{a})$. Of course $\hat{a}_0$ must have some model (call it $\mathcal{A}(\hat{a}_0)$), but there is no guarantee that $\mathcal{A}(\hat{a}_0)$ composed with the canonical $\mathcal{A}(\omega_s^{s+1})$ should coincide with the prescribed homomorphism $\mathcal{A}(\hat{a})$. Reverting to the level of geometry may or may not help: there might be a different extension $\hat{a}_1$ of $\hat{a}$ for which $\mathcal{A}(\hat{a}_1)$ can be chosen so that $\mathcal{A}(\hat{a}_1)\mathcal{A}(\omega_s^{s+1})$ equals the prescribed $\mathcal{A}(\hat{a})$, but such an $\hat{a}_1$ need not exist in general.

Let us look at this problem in the context of Lemma 6.2. We can pick some model $\mathcal{A}(\hat{a}_0)$ for $\hat{a}_0$, and then $\mathcal{A}(\hat{a}_0)\mathcal{A}(\omega_s^{s+1})$ will be homotopic to $\mathcal{A}(\hat{a})$. Thus the difference element

$$w = \mathcal{A}(\hat{a}_0)(v_q^{(s+1)}) - \mathcal{A}(\hat{a})(v_q^{(s)}) = \mathcal{A}(\hat{a}_0)(v_q^{(s+1)}) - x$$

will lie in $p^s\mathcal{A}(X) + d(\mathcal{A}(X))$, i.e., it will be a multiple of p^s plus a boundary. Still trying to get $\mathcal{A}(\hat{a}_0)(v_q^{(s+1)})$ to equal x, we may think to replace the model $\mathcal{A}(\hat{a}_0)$ by another dga homomorphism that is homotopic to it. This replacement will cause $\mathcal{A}(\hat{a}_0)(v_q^{(s+1)})$ to be altered by the addition of some perturbation term y, where $y \in p^{s+1}\mathcal{A}(X) + d(\mathcal{A}(X))$. If w belongs to $p^{s+1}\mathcal{A}(X)+d(\mathcal{A}(X))$ we may choose y to be $-w$ and then the new $\mathcal{A}(\hat{a}_0)$ will equal x. Thus the algebraic obstruction to $\mathcal{A}(\hat{a}_0)$ being choosable so that $\mathcal{A}(\hat{a}_0)(v_q^{(s+1)})$ equals x can be thought of as the coset $w+(p^{s+1}\mathcal{A}(X)+d(\mathcal{A}(X))$. Since w lies in $p^s\mathcal{A}(X)+d(\mathcal{A}(X))$, the algebraic obstruction group is $(p^s\mathcal{A}(X) + \mathrm{im}(d))/(p^{s+1}\mathcal{A}(X) + \mathrm{im}(d))$.

We may summarize the discussion as follows. Lemma 6.2 says roughly that the indeterminancy in $\mathcal{A}(\hat{a})(v_q^{(s)})$ for a map $\hat{a}$ on a mod p^s Moore space is $p^s\mathcal{A}(X) + \mathrm{im}(d)$. Once we choose an extension $\hat{a}_0$ there will be less indeterminancy because we are now mapping out of a mod p^{s+1} Moore space instead. If we still want $\mathcal{A}(\hat{a})(v_q^{(s+1)})$ to equal the prescribed element x, we will need to make some special argument as to why the algebraic obstruction vanishes.

Theorem 6.4 will provide us with precisely such a special argument. Under suitable hypotheses it will even assure us that the extension $\hat{a}_0$ is unique. We first introduce some notation to dispense with clumsy expressions like $\mathcal{A}(\hat{a}_0)(v_q^{(s+1)})$, and we look at a lemma for transferring information about AH models from one map to another. Then comes a remark about the Hurewicz homomorphism, followed by the theorem.

Notation. Given $f : X \to Y$ and a homotopy class $\hat{a} \in \pi_{q+1}(X)$, define $\underline{a}_f(\hat{a})$ to be the element $\mathcal{A}(f\hat{a})(v_q)$ of $\mathcal{A}(Y)$. Notice that $\underline{a}_f(\hat{a})$ is a cycle in $\mathcal{A}(Y)$ and that the symbol $\underline{a}_f(\hat{a})$ has built into it indeterminancy equal to $d(\mathcal{A}(Y)_{q+1})$. Similarly, for $\hat{a} \in \pi_{q+1}(X;\mathbb{Z}_{p^s})$, define $\underline{a}_f(\hat{a})$ to be $\mathcal{A}(f\hat{a})(v_q^{(s)})$. Then p^s divides $d(\underline{a}_f(\hat{a}))$ and $\underline{a}_f(\hat{a})$ has indeterminancy built into it equal to $p^s\mathcal{A}(Y)_q + d(\mathcal{A}(Y)_{q+1})$. Notice that $\underline{a}_f(\hat{a}\omega^s) = p^{-s}d(\underline{a}_f(\hat{a}))$.

More Notation. When $\hat{x} \in \pi_q(\Omega X)$ or $\hat{x} \in \pi_q(\Omega X;\mathbb{Z}_{p^s})$, we shall also write $\underline{a}_f(\hat{x})$ for $\mathcal{A}(f \circ (\hat{x})^{\overline{a}})(v_q)$ or for $\mathcal{A}(f \circ (\hat{x})^{\overline{a}})(v_q^{(s)})$. I.e., the notation can be applied either to a homotopy class or to its adjoint. It will always be clear what it intended because f is a map from X to Y, so if $\hat{a} \in \pi_*(X)$ we do $\mathcal{A}(f\hat{a})$ directly but if $\hat{a} \in \pi_*(\Omega X)$ we take $(\hat{a})^{\overline{a}}$ first.

Lemma 6.3. *Let $q \geq 2$, $s \geq 1$. Let*

$$\begin{array}{ccc} X & \xrightarrow{\ f\ } & Y \\ {\scriptstyle g_X}\downarrow & & \downarrow{\scriptstyle g_Y} \\ X' & \xrightarrow{\ f'\ } & Y' \end{array}$$

be a homotopy-commutative square, and fix a model $\mathcal{A}(g_Y) : \mathcal{A}(Y) \to \mathcal{A}(Y')$. Suppose $\hat{a} \in \pi_{q+1}(X;\mathbb{Z}_{p^s})$ has $\underline{a}_f(\hat{a}) = y \in \mathcal{A}(Y)$. Then $\underline{a}_{f'}(g_X\hat{a}) = \mathcal{A}(g_Y)(y)$.

Proof. Straightforward.

Remark. The definition of the mod p^s Hurewicz homomorphism required that a choice be made of a generator for $H_{q+1}(P^{q+1}(p^s);\mathbb{Z}_{p^s})$. We also made a choice when we picked out the generator $v_q^{(s)}$ for $\mathcal{A}(P^{q+1}(p^s))$. We may assume that these choices are made compatibly, in the following sense. When $H_*(\Omega P^{q+1}(p^s);\mathbb{Z}_{p^s})$ is identified (via the quism) with the homology of $\mathcal{A}(P^{q+1}(p^s)) \otimes \mathbb{Z}_{p^s}$, $h((1)^a)$ is the homology class represented by

$v_q^{(s)}$. Since $P^{q+1}(p^s)$ serves as a universal example, it follows that when $H_*(\Omega X; \mathbf{Z}_{p^s})$ is identified with the homology of $\mathcal{A}(X) \otimes \mathbf{Z}_{p^s}$, the homology class $h(\hat{x})$ is represented by $\mathcal{A}(\hat{x}^{\overline{a}})(v_q^{(s)})$, for any $\hat{x} \in \pi_q(\Omega X; \mathbf{Z}_{p^s})$. As a result the equation

$$(\Omega f)_* h(\hat{x}) = \overline{\underline{a}_f(\hat{x})}$$

is true when $f : X \to Y$, the right-hand side denoting the homology class in $H_q(\Omega Y; \mathbf{Z}_{p^s})$ represented by $\underline{a}_f(\hat{x})$. Notice that the indeterminancy in $\underline{a}_f(\hat{x})$ is exactly the group that we divide out by when we pass from $\mathcal{A}(Y)$ to $H_*(\Omega Y; \mathbf{Z}_{p^s})$, so this equation can be taken to be an alternate definition of $\underline{a}_f(\hat{x})$. A similar equation holds when $\hat{x} \in \pi_q(\Omega X)$ instead.

Theorem 6.4 will be the critical tool that permits us to build mod p^{s+1} homotopy classes out of mod p^s homotopy classes while maintaining control over their Adams-Hilton images. In spite of the seeming plethora of hypotheses, we will find a variety of situations in which all of the hypotheses are fulfilled. Its proof is reminiscent of a diagram chase.

Theorem 6.4. *Let $q \geq 3$, $t \geq 1$. Let*

(6G)
$$X \xrightarrow{f} Y \xrightarrow{g} Z$$

be a fibration up to homotopy, or more generally suppose merely that gf is null-homotopic. Suppose the sequence

(6H)
$$0 \longrightarrow p^t\pi_j(\Omega X) \xrightarrow{(\Omega f)_* h} p^t H_j(\Omega Y) \xrightarrow{(\Omega g)_*} p^t H_j(\Omega Z)$$

is exact when $q-2 \leq j \leq q$. Suppose further that

(6I)
$$p^{t+1}H_q(\Omega Y) = 0 \quad \text{and} \quad p^{t+1}H_q(\Omega Z) = 0.$$

Let $\hat{y} \in \pi_{q+1}(X; \mathbf{Z}_{p^t})$ and $\hat{z} \in \pi_q(X; \mathbf{Z}_{p^t})$ be homotopy classes that satisfy ${}_\pi\delta^t(\hat{z}) = 0$ and ${}_\pi\delta^t(\hat{y}) = p\hat{z}$. Fix a model $\mathcal{A}(g) : \mathcal{A}(Y) \to \mathcal{A}(Z)$, and suppose further that we have elements $y = \underline{a}_f(\hat{y})$ and $z = \underline{a}_f(\hat{z})$ such that $d(y) = p^{t+1}z$ (hence z is a cycle) and p^{t+1} divides $\mathcal{A}(g)(y)$. Then there exists a unique homotopy class $\hat{y}_2 \in \pi_{q+1}(X; \mathbf{Z}_{p^{t+1}})$ such that $\underline{a}_f(\hat{y}_2) = y$ and $\hat{y}_2\omega_t^{t+1} = \hat{y}$ and $\hat{y}_2\omega^{t+1}\rho^t = \hat{z}$. (The last equation can also be written $({}_\pi\delta^{t+1}(\hat{y}_2))\omega_t^{t+1} = \hat{z}$.)

Proof. We begin by establishing that $\hat{z}\omega^t = 0$. Let $\hat{w} = \hat{z}\omega^t \in \pi_{q-1}(X)$. Because $0 = {}_\pi\delta^t(\hat{z}) = \hat{w}\rho^t$, we have $\hat{w} \in \ker(\rho^t)^\# = p^t\pi_{q-1}(X)$. By the exactness of (6H) at $p^t\pi_{q-2}(\Omega X)$ it suffices to see that $(\Omega f)_* h(\hat{w}^a) = 0$. But $\underline{a}_f(\hat{w}) = p^{-t}d(\underline{a}_f(\hat{z})) = p^{-t}d(z) = 0$ because z is a cycle.

Because $\hat{z}\omega^t = 0$ we have $\hat{z} \in \ker(\omega^t)^\# = \operatorname{im}(\rho^t)^\#$, and we may therefore write $\hat{z} = \hat{z}_0\rho^t$, where $\hat{z}_0 \in \pi_q(X)$. Put $z_0 = \underline{a}_f(\hat{z}_0)$, and then $z - z_0$ is a cycle lying in $p^t\mathcal{A}(Y) + d(\mathcal{A}(Y))$ by Lemma 6.2. Let $z_1 = z - z_0$; the homology class of the cycle z_1, denoted $\overline{z}_1$, belongs to $p^tH_*(\Omega Y)$. We also know that $\overline{z}_1$ lies in $\ker(\Omega g)_*$ because $\mathcal{A}(g)(z_1) = \mathcal{A}(g)(z) - \mathcal{A}(g)(z_0)$, which is a boundary in $\mathcal{A}(Z)$. (To justify this assertion, $\mathcal{A}(g)(z) = d(p^{-t-1}\mathcal{A}(g)(y))$, and $\mathcal{A}(g)(z_0)$ is a boundary by Lemma 6.2(a) because gfz_0 is null-homotopic.) Applying the exactness of (6H) at $p^tH_{q-1}(\Omega Y)$, there is a homotopy class $\hat{z}_1 \in \pi_q(X)$ for which $(\Omega f)_* h(p^t\hat{z}_1^a) = \overline{z}_1$. We can choose $\underline{a}_f(p^t\hat{z}_1)$ to be z_1. Let $\hat{z}_2 = \hat{z}_0 + p^t\hat{z}_1 \in \pi_q(X)$. Then $\underline{a}_f(\hat{z}_2) = z_0 + z_1 = z$, and simultaneously $\hat{z}_2p^t = \hat{z}_0\rho^t + p^t(\hat{z}_1\rho^t) = \hat{z}_0\rho^t + 0 = \hat{z}$.

Consider $\hat{x} = \hat{y}\omega^t - p\hat{z}_2 \in \pi_q(X)$. We have $\hat{x}\rho^t = {}_\pi\delta^t(\hat{y}) - p\hat{z} = 0$, so $\hat{x} \in \ker(\rho^t)^\# = p^t\pi_q(X)$. But $(\Omega f)_* h(\hat{x}^a)$ can be computed as $\underline{a}_f(\hat{x}) = p^{-t}d\underline{a}_f(\hat{y}) - p\underline{a}_f(\hat{z}_2) = p^{-t}d(y) - pz = 0$. Using the exactness of (6H) at $p^t\pi_{q-1}(\Omega X)$, we see that $\hat{x} = 0$, i.e., $\hat{y}\omega^t = p\hat{z}_2$.

The homotopy push-out diagram

$$\text{(6J)} \qquad \begin{array}{ccc} S^q & \xrightarrow{\omega^t} & P^{q+1}(p^t) \\ {\scriptstyle p}\big\downarrow & & \big\downarrow{\scriptstyle \omega_t^{t+1}} \\ S^q & \xrightarrow{\omega^{t+1}} & P^{q+1}(p^{t+1}) \end{array}$$

shows that there exists $\hat{y}_0 \in \pi_{q+1}(X; \mathbb{Z}_{p^{t+1}})$ such that $\hat{y}_0\omega^{t+1} = \hat{z}_2$ and $\hat{y}_0\omega_t^{t+1} = \hat{y}$. We may choose a model for $f\hat{y}_0$ that sends the generator $u_{q-1}^{(t+1)}$ of $\mathcal{A}(P^{q+1}(p^{t+1}))$ to $\underline{a}_f(\hat{z}_2) = z$, i.e., it is possible to choose $\underline{a}_f(\hat{y}_0)$ so that $d(\underline{a}_f(\hat{y}_0)) = p^{t+1}z$. Make such a choice and set $y_0 = \underline{a}_f(\hat{y}_0) \in \mathcal{A}(Y)_q$. Then

$$y - y_0 = \underline{a}_f(\hat{y} - \hat{y}_0\omega_t^{t+1}) = \underline{a}_f(0),$$

i.e., $y - y_0 = y_1$ for some cycle $y_1 \in p^t\mathcal{A}(Y) + d(\mathcal{A}(Y))$. (We know y_1 is a cycle because our choice of $\mathcal{A}(fy_0)$ makes $d(y - y_0) = p^{t+1}z - d\underline{a}_f(\hat{y}_0) = p^{t+1}z - d\mathcal{A}(fy_0)(v_{q-1}^{(t+1)}) = p^{t+1}z - p^{t+1}\mathcal{A}(fy_0)(u_{q-1}^{(t+1)}) = 0$.) The homology class $\overline{y}_1$ represented by y_1 belongs to $p^tH_q(\Omega Y)$.

Consider the cycle $\mathcal{A}(g)(y_1) = \mathcal{A}(g)(y) - \mathcal{A}(g)(y_0)$. We have $\mathcal{A}(g)(y) \in p^{t+1}\mathcal{A}(Z)$ by hypothesis. The fact that gfy_0 is null-homotopic means by Lemma 6.2(b) that $\mathcal{A}(g)(y_0) \in p^{t+1}\mathcal{A}(Z) + d(\mathcal{A}(Z))$. Consequently $\mathcal{A}(g)(y_1)$ represents a homology class in $p^{t+1}H_q(\Omega Z)$, which is zero by (6I). Using the exactness of (6H) at $p^tH_q(\Omega Y)$, we may write $\overline{y}_1 = (\Omega f)_* h(\hat{y}_1^a)$ for some $\hat{y}_1 \in p^t\pi_{q+1}(X)$, and we may choose $\underline{a}_f(\hat{y}_1)$ to equal y_1. Put $\hat{y}_2 = \hat{y}_0 + \hat{y}_1\rho^{t+1}$. Then $\hat{y}_2\omega_t^{t+1} = \hat{y}_0\omega_t^{t+1} = \hat{y}$ and $\underline{a}_f(\hat{y}_2) = y_0 + y_1 = y$ and $\hat{y}_2\omega^{t+1} = \hat{y}_0\omega^{t+1} = \hat{z}_2$. From this last equation we deduce the desired relation,

$$\hat{z} = \hat{z}_2\rho^t = \hat{y}_2\omega^{t+1}\rho^t.$$

Now let us verify the uniqueness of $\hat{y}_2$. Suppose $\hat{y}_3$ also satisfies $\underline{a}_f(\hat{y}_3) = y$ and $\hat{y}_3\omega_t^{t+1} = \hat{y}$ and $\hat{y}_3\omega^{t+1}\rho^t = \hat{z}$. Put $\hat{z}_3 = \hat{y}_3\omega^{t+1}$. We show first that $\hat{z}_3 = \hat{z}_2$. To begin with, $\underline{a}_f(\hat{z}_3) = p^{-t-1}d(\underline{a}_f(\hat{y}_3)) = z$. Since $\hat{z}_3\rho^t = \hat{z} = \hat{z}_2\rho^t$, we have $(\hat{z}_3 - \hat{z}_2) \in \ker(\rho^t)^\# = p^t\pi_q(X)$. But $\underline{a}_f(\hat{z}_3 - \hat{z}_2) = 0$, so the exactness of (6H) at $p^t\pi_{q-1}(\Omega X)$ tells us that $\hat{z}_3 = \hat{z}_2$.

Given that $\hat{z}_3 = \hat{z}_2$, let us see why $\hat{y}_3$ must equal $\hat{y}_2$. Consider the difference $\hat{y}_3 - \hat{y}_2$. We have $(\hat{y}_3 - \hat{y}_2)\omega^{t+1} = \hat{z}_3 - \hat{z}_2 = 0$, so $\hat{y}_3 - \hat{y}_2 \in \ker(\omega^{t+1})^\# = \operatorname{im}(\rho^{t+1})^\#$, and we may write $\hat{y}_3 = \hat{y}_2 + \hat{y}_4\rho^{t+1}$ for some $\hat{y}_4 \in \pi_{q+1}(X)$. Then

$$0 = \hat{y} - \hat{y} - (\hat{y}_3 - \hat{y}_2)\omega_t^{t+1} = \hat{y}_4\rho^{t+1}\omega_t^{t+1} = \hat{y}_4\rho^t,$$

so $\hat{y}_4 \in \ker(\rho^t)^\# = p^t\pi_{q+1}(X)$. Finally, the fact that

$$0 = \underline{a}_f(\hat{y}_3 - \hat{y}_2) = \underline{a}_f(\hat{y}_4\rho^{t+1}) \in \underline{a}_f(\hat{y}_4) + p^{t+1}\mathcal{A}(Y) + d(\mathcal{A}(Y))$$

shows us that $(\Omega f)_* h(\hat{y}_4^a) \in p^{t+1}H_q(\Omega Y) = 0$, using (6I). The exactness of (6H) at $p^t\pi_q(\Omega X)$ gives $\hat{y}_4 = 0$, i.e., $\hat{y}_3 = \hat{y}_2$.

We follow up Theorem 6.4 with three results that attempt to simplify the hypotheses of 6.4, or to extend their range of applicability.

Lemma 6.5. *Equations (6I) and the exactness of (6H) will hold, as required for Theorem 6.4, if these four conditions are satisfied:*

(i) *(6G) is a fibration that splits when looped once, i.e., Ωg has a section.*

(ii) *The Hurewicz homomorphism*

$$h : p^t\pi_j(\Omega X) \to p^t H_j(\Omega X) \tag{6K}$$

is bijective for $q-2 \le j \le q$.

(iii) $p^{t+1}H_q(\Omega Y) = 0$.

(iv) *For some $m \ge 1$, $p^t\overline{H}_{\le q-m}(\Omega X) = 0$ and $p^t\overline{H}_{<m}(\Omega Z) = 0$.*

Proof. By (i) $H_*(\Omega Z)$ is a retract of $H_*(\Omega Y)$, so (6I) follows from (iii). Since both (6K) and $(\Omega f)_*$ are monomorphic, (6H) is exact at $p^t\pi_j(\Omega X)$. The Kunneth formula along with (iv) proves the exactness of (6H) at $p^t H_j(\Omega Y)$.

One common complaint against AH models is that, because they are by definition free (i.e., tensor algebras), they are sometimes too big and unwieldy for calculations. We sometimes prefer to replace $\mathcal{A}(X)$ by an *AH quotient*. By this we mean a dga (Q, d) which comes with a surjective quism $\zeta_X : \mathcal{A}(X) \to (Q, d)$. A good example is that $\mathcal{A}(X) \otimes \mathcal{A}(Y)$ is an AH quotient for the product space $X \times Y$ whenever $\mathcal{A}(X)$ and $\mathcal{A}(Y)$ are AH models for X and Y (see [AH, Section 4] or [A3, 8.1(m,n)]).

For theoretical purposes AH quotients are far less convenient than AH models, because the extensive body of knowledge about free dga's does not apply to them. For instance, given a map $f : X \to Y$ and AH quotients (Q_X, d_X), and (Q_Y, d_Y), there need not exist a dga homomorphism Q_f making this square commute up to dga homotopy:

$$\begin{array}{ccc} \mathcal{A}(X) & \xrightarrow{\mathcal{A}(f)} & \mathcal{A}(Y) \\ \zeta_X \downarrow & & \downarrow \zeta_Y \\ (Q_X, d_X) & \overset{Q_f}{\dashrightarrow} & (Q_Y, d_Y). \end{array}$$

Of course, Q_f might exist for a particular application. The next corollary adapts Theorem 6.4 to the setting of AH quotients. It merely restates the theorem, substituting $\zeta_Y \underline{a}_f$ and Q_g for $\underline{a}_f$ and $\mathcal{A}(g)$. Its proof also merely restates the proof of 6.4, so we omit it.

Corollary 6.6. *Let q, t, X, Y, Z, f, g, $\hat{y}$, $\hat{z}$ be as in Theorem 6.4 Suppose there are AH quotients for Y and Z and for the map g making this square commute on the nose:*

$$\begin{array}{ccc} \mathcal{A}(Y) & \xrightarrow{\mathcal{A}(g)} & \mathcal{A}(Z) \\ \zeta_Y \downarrow & & \downarrow \zeta_Z \\ (Q_Y, d_Y) & \xrightarrow{Q_g} & (Q_Z, d_Z) \end{array} \tag{6L}$$

Suppose we have elements $y = \zeta_Y \underline{a}_f(\hat{y})$ and $z = \zeta_Y \underline{a}_f(\hat{z})$ such that $d_Y(y) = p^{t+1}z$ (hence z is a cycle) and p^{t+1} divides $Q_g(y)$. Then there exists a unique homotopy class $\hat{y}_2 \in \pi_{q+1}(X; \mathbb{Z}_{p^{t+1}})$ such that $\zeta_Y \underline{a}_f(\hat{y}_2) = y$ and $\hat{y}_2\omega_t^{t+1} = \hat{y}$ and $\hat{y}_2\omega^{t+1}\rho^t = \hat{z}$.

A very convenient device for showing that two mod p^{t+1} homotopy classes coincide is the following.

Corollary 6.7. *Let $X \xrightarrow{f} Y \xrightarrow{g} Z$ be a fibration that satisfies conditions (i)-(iv) of Lemma 6.5 for a certain value of q. Let $\hat{y}_0 \in \pi_{q+1}(Y; \mathbb{Z}_{p^{t+1}})$ satisfy these four equations:*

$$g_{\#}(\hat{y}_0) = 0; \quad h(\hat{y}_0^a) = 0; \quad (\omega_t^{t+1})^{\#}(\hat{y}_0) = 0; \quad (\omega_t^{t+1})^{\#}({}_{\pi}\delta^{t+1}(\hat{y}_0)) = 0.$$

Then $\hat{y}_0 = 0$.

Proof. Condition 6.5(i) tells us that $f_{\#}$ is monomorphic. We have $\hat{y}_0 \in \ker(g_{\#}) = \mathrm{im}(f_{\#})$, so there is a unique $\hat{y}_1 \in \pi_{q+1}(X; \mathbb{Z}_{p^{t+1}})$ for which $f_{\#}(\hat{y}_1) = \hat{y}_0$. In Theorem 6.4, put $\hat{y} = 0$, $\hat{z} = 0$, $y = 0$, $z = 0$. Then $\hat{y}_2 = 0$ obviously satisfies the three equations in the conclusion of the theorem. However, $\hat{y}_2 = \hat{y}_1$ also satisfies all three equations since

$$\underline{a}_f(\hat{y}_1) = (\Omega f)_* h(\hat{y}_1^a) = h((f\hat{y}_1)^a) = h(\hat{y}_0) = 0,$$

and $\hat{y}_1\omega_t^{t+1} = f_{\#}^{-1}(\hat{y}_0\omega_t^{t+1}) = 0$ and $\hat{y}_1\omega^{t+1}\rho^t = f_{\#}^{-1}(\hat{y}_0\omega^{t+1}\rho^t) = 0$. By the uniqueness aspect of Theorem 6.4 we must have $\hat{y}_1 = 0$, hence $\hat{y}_0 = 0$.

Thin Products and Cyclic Cotensors

This third unit of Chapter 6 we devote to the study of a certain homotopy limit, called the "thin product" of three spaces, and to its algebraic analog, called a "cyclic cotensor." To begin, we apply some classical constructions of Eilenberg and Moore to the diagram defining the thin product, and we see how the cyclic cotensor construction arises naturally. We embark on a long digression to study the cyclic cotensor. Using combinatorial methods, we find explicit presentations, both for the cyclic cotensor and for a certain Hopf algebra kernel associated with it. As a corollary we spin off the useful fact that N^{klm} has a linear differential.

Returning to topology, we exploit our knowledge about the cyclic cotensor to determine an AH quotient for the thin product. This procedure raises the issue of AH models being dgH's, which we address with the concept of a dgH-AH model. We prove a proposition about dgH-AH models that is unrelated to thin products. To finish, we apply the presentation of the Hopf algebra kernel mentioned above to give a wedge decomposition for a certain homotopy-theoretic fiber related to the thin product.

Let $C_*(\ )$ denote the 0th Eilenberg subcomplex of the coalgebra of singular chains on a space, with coefficients in $\mathbb{Z}_{(p)}$. Let

$$\text{(6M)} \qquad \begin{array}{ccc} E & \longrightarrow & X \\ \big\downarrow & & \big\downarrow f \\ Y & \underset{g}{\longrightarrow} & Z \end{array}$$

be a pull-back diagram in which either f of g is a fibration. Eilenberg and Moore [EM] expressed $C_*(E)$ as chain equivalent to the differential graded cotensor product

$$C_*(X) \underset{C_*(Z)}{\square} C_*(Y).$$

Suppose in (6M) that X, Y, Z are path-connected topological groups and that f and g are homomorphisms of topological groups, so that $C_*(X)$, $C_*(Y)$, $C_*(Z)$ are dgH's. Then E is a topological group, and there is a natural quism

$$\text{(6N)} \qquad C_*(X) \underset{C_*(Z)}{\square} C_*(Y) \xrightarrow{\simeq} C_*(E)$$

as dga's.

Definition/Notation 6.8. . Let X^1, X^2, X^3 denote 1-connected spaces. Consider the diagram

(6O)
$$\begin{array}{ccccc} X^1 \vee X^2 & \longrightarrow & X^2 & \longleftarrow & X^2 \vee X^3 \\ \downarrow & & & & \downarrow \\ X^1 & \longleftarrow & X^1 \vee X^3 & \longrightarrow & X^3, \end{array}$$

in which every arrow is a collapsing map onto a wedge summand. The homotopy inverse limit of (6O) is called the ***thin product*** of the three spaces, and we denote it by

$$R = \mathcal{R}(X^1, X^2, X^3).$$

Since there are compatible projections from $X^1 \vee X^2 \vee X^3$ to each of the six spaces in (6O), the thin product comes with a natural map, denoted

$$\tilde{\lambda} = \tilde{\lambda}(X^1, X^2, X^3) : X^1 \vee X^2 \vee X^3 \to \mathcal{R}(X^1, X^2, X^3).$$

Remark. The term "thin product" arises from the fact that the thin product is the Hilton-Eckmann dual to the fat wedge. The term was first introduced in [Ho], where the relationship between the thin product and various notions of cocategory were studied.

One of our goals is to describe Adams-Hilton models for R and for $\tilde{\lambda}$. Another is to understand the homotopy-theoretic fiber of $\tilde{\lambda}$, since this fiber will serve as the "home" where three Jacobi identities are proved.

Let us try to get a better understanding of $C_*(\Omega R)$. Notice first that R is also the homotopy limit of the expanded diagram

(6P)
$$\begin{array}{ccccc} X^1 \vee X^2 & \longrightarrow & X^1 & \longleftarrow & X^1 \vee X^3 \\ \downarrow & & \downarrow & & \downarrow \\ X^2 & \longrightarrow & * & \longleftarrow & X^3 \\ \uparrow & & \uparrow & & \uparrow \\ X^2 \vee X^3 & \xrightarrow{=} & X^2 \vee X^3 & \xleftarrow{=} & X^2 \vee X^3. \end{array}$$

We may apply the functor Ω to the diagram (6P) to obtain a diagram whose limit is ΩR. We may replace ΩX^i in this looped diagram by topological groups G^i, and we may replace $\Omega(X^i \vee X^j)$ by the free product of groups $G^i * G^j$. Then ΩR has the homotopy type of the limit of

(6Q)
$$\begin{array}{ccccc} G^1 * G^2 & \longrightarrow & G^1 & \longleftarrow & G^1 * G^3 \\ \downarrow & & \downarrow & & \downarrow \\ G^2 & \longrightarrow & * & \longleftarrow & G^3 \\ \uparrow & & \uparrow & & \uparrow \\ G^2 * G^3 & \xrightarrow{=} & G^2 * G^3 & \xleftarrow{=} & G^2 * G^3. \end{array}$$

Taking (6Q) one row at a time, and applying (6N) repeatedly, we obtain a quism

$$(C_*(G^2 * G^1) \underset{C_*(G^1)}{\square} C_*(G^1 * G^3)) \underset{C_*(G^2 \times G^3)}{\square} C_*(G^2 * G^3) \xrightarrow{\simeq} C_*(\Omega R).$$

There are quisms that permit us to replace $C_*(G^i * G^j)$ by $C_*(G^i) \amalg C_*(G^j)$ and $C_*(G^2 \times G^3)$ by $C_*(G^2) \otimes C_*(G^3)$ as dgH's. Writing A^i for $C_*(G^i)$, we obtain the quism of dga's,

(6R) $$((A^2 \amalg A^1) \underset{A^1}{\square} (A^1 \amalg A^3)) \underset{A^2 \otimes A^3}{\square} (A^2 \amalg A^3) \xrightarrow{\simeq} C_*(\Omega R).$$

The left-hand side of (6R) as written obscures the intrinsic symmetry among A^1, A^2, and A^3. We offer for it instead the notation

(6S) $$(A^2 \amalg A^1) \underset{A^1}{\square} (A^1 \amalg A^3) \underset{A^3}{\square} (A^3 \amalg A^2) \underset{A^2}{\square},$$

where $\underset{A^2}{\square}$ is invoking both the right A^2-comodule structure of $A^3 \amalg A^2$ and the left A^2-comodule structure of $A^2 \amalg A^1$. In other words, (6S) equals the sub-dga of

$$P(A^1, A^2, A^3) = (A^2 \amalg A^1) \otimes (A^1 \amalg A^3) \otimes (A^3 \amalg A^2)$$

that is the intersection of the equalizers of the three pairs of maps:

$$P(A^1, A^2, A^3) \rightrightarrows (A^2 \amalg A^1) \otimes A^1 \otimes (A^1 \amalg A^3) \otimes (A^3 \amalg A^2),$$
$$P(A^1, A^2, A^3) \rightrightarrows (A^2 \amalg A^1) \otimes (A^1 \amalg A^3) \otimes A^3 \otimes (A^3 \amalg A^2),$$

$$\begin{array}{ccc} P(A^1, A^2, A^3) & \longrightarrow & P(A^1, A^2, A^3) \otimes A^2 \\ \downarrow & & \downarrow = \\ A^2 \otimes P(A^1, A^2, A^3) & \xrightarrow{\tau} & P(A^1, A^2, A^3) \otimes A^2. \end{array}$$

To simplify notation even further, we denote (6S) by $\mathrm{CC}(A^1, A^2, A^3)$ and call it the *cyclic cotensor product* of the three Hopf algebras (or dgH's) A^1, A^2, and A^3. There is an inclusion of Hopf algebras or dgH's

(6T) $$\mathrm{CC}(A^1, A^2, A^3) \to P(A^1, A^2, A^3).$$

Lemma 6.9. *Let A^1, A^2, and A^3 be dgH's over $\mathbb{Z}_{(p)}$. Let $\varepsilon : A^i \to \mathbb{Z}_{(p)}$ denote the augmentations, and let*

$$\varepsilon^1 = \varepsilon \amalg 1 \amalg 1 : A^1 \amalg A^2 \amalg A^3 \to \mathbb{Z}_{(p)} \amalg \amalg A^2 \amalg A^3 = A^2 \amalg A^3,$$

and similarly for ε^2 and ε^3. With $P(A^1, A^2, A^3)$ and $\mathrm{CC}(A^1, A^2, A^3)$ as above, there is a natural commuting diagram

(6U) $$\begin{array}{ccc} A^1 \amalg A^2 \amalg A^3 & \xrightarrow{(\Delta \otimes 1)\Delta \amalg (\Delta \otimes 1)\tau\Delta \amalg (\Delta \otimes 1)\Delta} & (A^1 \amalg A^2 \amalg A^3)^{\otimes 3} \\ {\scriptstyle \lambda = \lambda(A^1, A^2, A^3)} \downarrow & & \downarrow {\scriptstyle \varepsilon^3 \otimes \varepsilon^2 \otimes \varepsilon^1} \\ \mathrm{CC}(A^1, A^2, A^3) & \longrightarrow & P(A^1, A^2, A^3) \end{array}$$

in which the bottom arrow is the dgH inclusion (6T). Furthermore, when $A^i = C_*(G^i)$ *and* $G^i \approx \Omega X^i$ *as above,* $i = 1, 2, 3$, *then the square*

$$\text{(6V)}\qquad \begin{array}{ccc} A^1 \amalg A^2 \amalg A^3 & \xrightarrow{\lambda=\lambda(A^1,A^2,A^3)} & \mathrm{CC}(A^1, A^2, A^3) \\ \simeq\big\downarrow & & \big\downarrow\simeq \\ C_*(G^1 * G^2 * G^3) & \xrightarrow{C_*(\Omega\bar{\lambda})} & C_*(\Omega R) \end{array}$$

commutes, up to dga homotopy. The vertical arrows in (6V) are the quisms that we have mentioned earlier.

Proof. Part of what we are doing in this lemma is defining $\lambda = \lambda(A^1, A^2, A^3)$. Since the bottom arrow of (6U) is the monomorphism (6T), λ is a unique and natural dgH homomorphism if it exists. To see that λ exists, use the definition given above for CC as an intersection of three equalizers. It suffices to check that the images under $(\varepsilon^3 \otimes \varepsilon^2 \otimes \varepsilon^1)((\Delta \otimes 1)\Delta \amalg (\Delta \otimes 1)\tau\Delta \amalg (\Delta \otimes 1)\Delta)$ of a set of generators for $A^1 \amalg A^2 \amalg A^3$ all lie in each equalizer. The set $A^1_+ \cup A^2_+ \cup A^3_+$ serves well as a set of generators, and the check is straightforward.

For (6V) it also suffices to check that (6V) commutes up to homotopy when restricted to each A^i. But then it becomes nearly trivial, since we are only considering one space X^i at a time.

We now digress from the geometric ideas we have been considering to pursue an algebraic and combinatorial understanding of the cyclic cotensor. We will obtain explicit presentations for CC and for HAK (λ). We do this through a series of ten lemmas, followed by two corollaries relevant to dgL's.

Conventions. For the next ten lemmas, an algebra or Lie algebra or Hopf algebra is assumed to be connected and graded, and free as a module over the ground ring, which may be any principal ideal domain. By "module" we now mean a non-negatively graded module over this ground ring. Dispensing with our usual convention, a Hopf algebra need not be cocommutative for these lemmas. All of the lemmas are stated only for modules and algebras and Lie algebras and Hopf algebras, but the results and the proofs remain valid in the presence of a differential, i.e., for chain complexes, dga's, dgL's, and dgH's.

Definition. Given an algebra (resp. dga) M and module homomorphisms (resp. chain maps)

$$f^{(i)} : M^{(i)} \to M, \quad 1 \le i \le m,$$

their ***sequential product***

$$f = f^{(1)} \otimes \cdots \otimes f^{(m)} : M^{(1)} \otimes \cdots \otimes M^{(m)} \to M$$

is defined by

$$f(x_1 \otimes \cdots \otimes x_m) = f^{(1)}(x_1)f^{(2)}(x_2)\cdots f^{(m)}(x_m),$$

i.e., multiply in order in M.

Definition. Given algebras (resp. dga's) D and C, call a homomorphism (resp. chain map)

$$f : D_+ \otimes C_+ \to D \amalg C$$

principal if and only if $(\varepsilon \amalg 1)f = 0 = (1 \amalg \varepsilon)f$ and $\overline{\pi}_{21} f\tau : C_+ \otimes D_+ \to C_+ \otimes D_+$ is an isomorphism, where $\overline{\pi}_{21} : D \amalg C \to C_+ \otimes D_+$ is the projection defined in Lemma 5.2.

Lemma 6.10. *Let C and D be algebras. Let $f : D_+ \otimes C_+ \to D \amalg C$ be principal, and extend f to the multiplicative homomorphism $f : T(D_+ \otimes C_+) \to D \amalg C$. Then the squential product*

$$1 \otimes f \otimes 1 : D \otimes T(D_+ \otimes C_+) \otimes C \to D \amalg C$$

is an isomorphism.

Proof. Use the high term argument in the proof of Lemma 5.1 to show that $1 \otimes f \otimes 1$ is surjective, and then use a Hilbert series argument to show that it is bijective.

Now we specialize to the setting where $D = A \amalg B$.

Notation. Consider the direct sum decomposition,

$$A \amalg B = (\text{ground ring}) \oplus A_+ \oplus B_+ \oplus (A_+ \otimes B_+) \oplus (B_+ \otimes A_+) \oplus (A_+ \otimes B_+ \otimes A_+) \oplus \cdots .$$

Let $A_+ \amalg B_+$ denote the direct sum of all summands other than the first three in this list, i.e.,

$$A_+ \amalg B_+ = (A_+ \otimes B_+) \oplus (B_+ \otimes A_+) \oplus (A_+ \otimes B_+ \otimes A_+) \oplus (B_+ \otimes A_+ \otimes B_+) \oplus \cdots .$$

Equivalently, $A_+ \amalg B_+$ is the right or the left ideal in $A \amalg B$ generated by $(A_+ B_+) \oplus (B_+ A_+)$. Notice that the augmentation ideal of $A \amalg B$ has the decomposition

$$(A \amalg B)_+ = A_+ \oplus (A_+ \amalg B_+) \oplus B_+.$$

More notation. In a Hopf algebra, $\mathrm{Ad}_H(x)(-)$ denotes $[x, -]_H$. Since $\mathrm{Ad}_H(x)(z) = \mathrm{ad}(x)(z)$ for primitive z, we have $\mathrm{Ad}_H(xy) = \mathrm{ad}(xy) = \mathrm{ad}(x)\,\mathrm{ad}(y) = \mathrm{Ad}_H(x)\,\mathrm{Ad}_H(y)$ when acting on a primitive. In general however $\mathrm{Ad}_H(xy)$ does not equal $\mathrm{Ad}_H(x)\,\mathrm{Ad}_H(y)$.

We do not try to define ad_H in an arbitrary Hopf algebra, but for $w \in A \amalg B$ we define $\mathrm{ad}_H(w)$ as follows. If $w = a_1 b_1 \cdots a_m b_m$ where $a_j \in A_+$ and $b_j \in B_+$, let

$$\mathrm{ad}_H(w) = \mathrm{Ad}_H(a_1)\,\mathrm{Ad}_H(b_1) \cdots \mathrm{Ad}_H(a_m)\,\mathrm{Ad}_H(b_m),$$

and make a similar definition if the alternating string of elements from A_+ and B_+ starts with a $b \in B_+$ or ends in an $a \in A_+$. Putting $\mathrm{ad}_H(1)(z) = z$ and extending linearly, we have defined $\mathrm{ad}_H(\ )$ on all of $A \amalg B$. This notion of ad_H occurs in Lemma 6.11.

Lastly, if our Hopf algebra is identified with a tensor algebra, say TV, we can define $\mathrm{ad}_H(w)$ for $w = v_1 \cdots v_m \in T^m V$ by

$$\mathrm{ad}_H(w) = \mathrm{Ad}_H(v_1)\,\mathrm{Ad}_H(v_2) \cdots \mathrm{Ad}_H(v_m)$$

and extending linearly. This notion of ad_H occurs in Lemma 6.12.

Lemma 6.11. *Let A, B, and C be Hopf algebras. Suppose $f_{AC} : A_+ \otimes C_+ \to A \amalg C$ and $f_{BC} : B_+ \otimes C_+ \to B \amalg C$ are principal. Define a module homomorphism*

$$f_0 : (A_+ \amalg B_+) \otimes C_+ \to (A \amalg B) \amalg C$$

by

$$f_0(wa \otimes c) = \mathrm{ad}_H(w) f_{AC}(a \otimes c), \quad \textit{if } a \in A_+,\ w \in (A \amalg B)B_+;$$
$$f_0(wb \otimes c) = \mathrm{ad}_H(w) f_{BC}(b \otimes c), \quad \textit{if } b \in B_+,\ w \in (A \amalg B)A_+.$$

Define

(6W) $$f : (A \amalg B)_+ \otimes C_+ \to (A \amalg B) \otimes C$$

by identifying $(A \amalg B)_+ \otimes C_+$ *with*

$$(A_+ \oplus (A_+ \amalg B_+) \oplus B_+) \otimes C_+ = (A_+ \otimes C_+) \oplus (A_+ \amalg B_+) \otimes C_+ \oplus (B_+ \otimes C_+)$$

and setting $f = f_{AC} \oplus f_0 \oplus f_{BC}$. *Then* f *is principal.*

Proof. It boils down to checking that

$$\overline{\pi}_{21} f_0 \tau : C_+ \otimes (A_+ \amalg B_+) \to C_+ \otimes (A_+ \amalg B_+)$$

is an isomorphism. Writing $\theta = \overline{\pi}_{21} f_{AC}$, an isomorphism by hypothesis, we find

$$\overline{\pi}_{21} f_0(wa \otimes c) = (\overline{\pi}_{21} f_{AC}(a \otimes c))\chi(w)$$

for $w \in (A \amalg B)B_+$, so $\overline{\pi}_{21} f_0 \tau$ is an isomorphism of $C_+ \otimes (A \amalg B)B_+A_+$ with $C_+ \otimes A_+B_+(A \amalg B)$. Similarly, $\overline{\pi}_{21} f_0 \tau$ is an isomoprhism of $C_+ \otimes (A \amalg B)A_+B_+$ with $C_+ \otimes B_+A_+(A \amalg B)$.

Lemma 6.12. *Let* M_1, M_2, *and* M_3 *denote positively graded modules that are free as modules over the ground ring. Let*

$$g_{13} : M_1 \otimes M_3 \to T(M_1 \oplus M_3) = TM_1 \amalg TM_3$$

be principal. Let $U' = M_2 \oplus (M_1 \otimes M_3)$, *and define*

$$g' = 1 \oplus g_{13} : U' \to T(M_1 \oplus M_2 \oplus M_3).$$

Let

$$U = TM_3 \otimes U' \otimes TM_1,$$

and define $g : U \to T(M_1 \oplus M_2 \oplus M_3)$ *by*

(6X) $$g(z \otimes u' \otimes x) = \mathrm{ad}_H(x)\,\mathrm{ad}_H(z)(g'(u')).$$

Extend g *multiplicatively to*

$$g : TU \to T(M_1 \oplus M_2 \oplus M_3).$$

Then the sequential product

$$1 \otimes g \otimes 1 : TM_1 \otimes TU \otimes TM_3 \to T(M_1 \oplus M_2 \oplus M_3)$$

is an isomorphism.

Proof. This is simply a more sophisticated version of Lemma 5.1. The algebra $T(M_1 \oplus M_2 \oplus M_3)$ has three gradings, one coming from each M_i. If

$$x = a_{11}a_{12}a_{13}a_{21}a_{22}a_{23} \cdots a_{m1}a_{m2}a_{m3} \quad ,$$

where $a_{ij} \in TM_j$, define the **rank** of x to be

$$\sum_{i<j}(|a_{i3}|\,|a_{j1}| + |a_{i3}|\,|a_{j2}| + |a_{i2}|\,|a_{j1}|).$$

The ***high term*** of a non-zero trihomogenous $x \in T(M_1 \oplus M_2 \oplus M_3)$, denoted $HT(x)$, is the non-zero component of maximal rank. From (6X) we see that $HT(g(z \otimes u' \otimes x)) = zy\chi(x)$ if $u' = y \in M_2$, while if $u' \in M_1 \otimes M_3$, $HT(g(z \otimes u' \otimes x)) = z(\overline{\pi}_{21}g_{13}(u'))\chi(x)$.

We leave the remainder of the proof as an exercise; it follows the same pattern as the proof of Lemma 5.1. Notice that if $M_2 = 0$ we recover a special case of Lemma 6.10.

We are leading up to a description of $\mathrm{CC}(A, B, C)$, so we let A, B, and C be Hopf algebras. Let

$$\begin{aligned} K_{AC} &= \mathrm{HAK}(A \amalg C \to A \otimes C), \\ K_{BC} &= \mathrm{HAK}(B \amalg C \to B \otimes C), \\ K_{(A\amalg B)C} &= \mathrm{HAK}((A \amalg B) \amalg C \to (A \amalg B) \otimes C). \end{aligned}$$

Suppose

$$f_{AC} : A_+ \otimes C_+ \to K_{AC} \subseteq A \amalg C$$

and

$$f_{BC} : B_+ \otimes C_+ \to K_{BC} \subseteq B \amalg C$$

are principal. In Lemma 6.12 put $M_1 = A_+ \otimes C_+$ and $M_3 = B_+ \otimes C_+$. Combining Lemma 6.10 with the Hilbert series argument of 5.1, we see that the multiplicative extension of f_{AC},

$$f_{AC} : TM_1 \to K_{AC},$$

is an isomorphism of algebras, so there is a unique Hopf algebra structure on TM_1 for which f_{AC} is an isomorphism of Hopf algebras. Similarly, there is a unique Hopf algebra structure on TM_3 that makes the multiplicative extension

$$f_{BC} : TM_3 \to K_{BC}$$

of f_{BC} into an isomorphism of Hopf algebras.

Let $M_2 = (A_+ \amalg B_+) \otimes C_+$ in Lemma 6.12, so that $M_1 \oplus M_2 \oplus M_3 = (A \amalg B)_+ \otimes C_+$, and define f_0 and f as in Lemma 6.11. Observe that $\mathrm{im}(f) \subseteq K_{(A\amalg B)C}$, since f_{AB} and f_0 and f_{BC} all have their images in $K_{(A\amalg B)C}$. Then f has a multiplicative extension

$$f : T(M_1 \oplus M_2 \oplus M_3) \to K_{(A\amalg B)C}$$

which is an isomorphism of algebras, and which determines a Hopf algebra structure on $T(M_1 \oplus M_2 \oplus M_3)$. With respect to this Hopf algebra structure, the Hopf algebras TM_1

and TM_3 as defined above are sub-Hopf algebras. Putting $g_{13} = [\ ,\ ]_H$, we have verified the hypotheses of Lemma 6.12 for this $T(M_1 \oplus M_2 \oplus M_3)$.

Combining Lemmas 6.10–6.12, we find that the following composite sequential product is an isomorphism:

$$(A \amalg B) \otimes TM_1 \otimes TU \otimes TM_3 \otimes C \xrightarrow{1\otimes 1\otimes g\otimes 1\otimes 1} (A \amalg B) \otimes T(M_1 \oplus M_2 \oplus M_3) \otimes C$$
$$\xrightarrow{=} (A \amalg B) \otimes T((A \amalg B)_+ \otimes C_+) \otimes C \xrightarrow{1\otimes f\otimes 1} (A \amalg B) \amalg C.$$

Here f is the multiplicative extension of (6W) over the tensor algebra, as mentioned above, and g is the multiplicative extension of (6X). Since $f|_{TM_1} = f_{AC}$ and $f|_{TM_3} = f_{BC}$, we have shown

Lemma 6.13. *The sequential product* $1 \otimes f_{AC} \otimes fg \otimes f_{BC} \otimes 1$:

(6Y) $$(A \amalg B) \otimes T(A_+ \otimes C_+) \otimes TU \otimes T(B_+ \otimes C_+) \otimes C \to (A \amalg B) \amalg C$$

is an isomorphism.

Notation/Definition. Let A, B, and C be Hopf algebras (resp. dgH's). Let $J = J(A, B, C)$ be the two-sided ideal of $A \amalg B \amalg C$ generated by

(6Z) $$[[C, B]_H, A]_H + [[C, A]_H, B]_H + [[C, A]_H, [C, B]_H]_H.$$

Lemma 6.14. *Let A, B, and C be Hopf algebras. Let f_{AC} and f_{BC} be given by $[\ ,\ ]_H$. Let $J = J(A, B, C)$, and let*

$$q = 1 \otimes f_{AC} \otimes f_{BC} \otimes 1 : K = (A \amalg B) \otimes T(A_+ \otimes C_+) \otimes T(B_+ \otimes C_+) \otimes C \to A \amalg B \amalg C$$

be the sequential product. Then the composition

$$K \xrightarrow{q} A \amalg B \amalg C \longrightarrow A \amalg B \amalg C/J$$

is surjective.

Proof. We want to show that $q(K) + J$ equals all of $A \amalg B \amalg C$.

By Lemma 6.13, as modules $A \amalg B \amalg C$ equals $q(K)$ plus the image of

$$(A \amalg B) \otimes T(A_+ \otimes C_+) \otimes T^{\geq 1}U \otimes T(B_+ \otimes C_+) \otimes C.$$

Since J is a two-sided ideal, it suffices to see that the image of U under fg is contained in J. From the way U is defined, it suffices to see that $fg'(U') \subseteq J$. This is trivial, using (5D).

Notation. If M is free as a (graded) module and of finite type, its Hilbert series $\sum_{m=0}^{\infty} \dim(M_m)t^m$ is also denoted $M(t)$.

Lemma 6.15. *Let A, B, and C be Hopf algebras. Then $J = J(A,B,C)$ is the kernel of the homomorphism*

$$\lambda = \lambda(A,B,C) : A \amalg B \amalg C \to \mathrm{CC}(A,B,C)$$

defined by (6U), and λ is surjective. I.e., λ induces a natural isomorphism of $A \amalg B \amalg C/J$ with $\mathrm{CC}(A,B,C)$.

Proof. First, check that $J \subseteq \ker(\lambda)$, by using (5D) to see that everything in (6Z) is sent to zero by $(\varepsilon^3 \otimes \varepsilon^2 \otimes \varepsilon^1)((\Delta \otimes 1)\Delta \amalg (\Delta \otimes 1)\tau\Delta \amalg (\Delta \otimes 1)\Delta)$ in (6U). The right-hand arrow is defined in the diagram

$$\text{(6AA)} \qquad \begin{array}{ccc} K & \xrightarrow{\overline{q}} & A \amalg B \amalg C/J \\ {\scriptstyle q'}\downarrow & & \downarrow{\scriptstyle \overline{\lambda}} \\ P(A,B,C) & \longleftarrow & \mathrm{CC}(A,B,C). \end{array}$$

In (6AA), the bottom arrow is (6T), $\overline{q}$ is the surjective composition of Lemma 6.14, and q' is chosen so as to make (6AA) commute. The homomorphism q' can be identified as a sequential product, but K has to be decomposed into six factors (not just four) first. Write K as

$$K = A \otimes T(A_+ \otimes B_+) \otimes B \otimes T(A_+ \otimes C_+) \otimes T(B_+ \otimes C_+) \otimes C,$$

and let $f_{AB} : T(A_+ \otimes B_+) \to A \amalg B$ be the multiplicative extension of $[\ ,\ ]_H$. Let Δ_A be the composition

$$\Delta_A : A \xrightarrow{\Delta} A \otimes A \to (A \amalg B) \otimes (A \amalg C) = (A \amalg B) \otimes (A \amalg C) \otimes 1 \subseteq P(A,B,C),$$

and likewise for Δ_B and Δ_C. Then $q' = \Delta_A \otimes f_{AB} \otimes \Delta_B \otimes f_{AC} \otimes f_{BC} \otimes \Delta_C$. The important thing about this formula for q' is that it becomes easy to see that q' has a left inverse, i.e., q' is one-to-one.

By a direct limit argument we can reduce Lemma 6.15 to the case where A, B, and C have finite type. We shall utilize a Hilbert series argument to show that $\overline{\lambda}$ is an isomorphism. Observe first that, whenever M_1 is right cofree or M_2 is left cofree as comodules over a coalgebra N, then the Hilbert series of $M_1 \square_N M_2$ equals the quotient of the Hilbert series of $M_1 \otimes M_2$ divided by the Hilbert series for N, i.e., $(M_1 \square_N M_2)(t) = M_1(t)M_2(t)/N(t)$. From the fact that

$$\mathrm{CC}(A,B,C) = ((B \amalg A) \underset{A}{\square} (A \amalg C)) \underset{B \otimes C}{\square} (B \amalg C),$$

we find that

$$\text{(6BB)} \qquad \mathrm{CC}(A,B,C)(t) = \frac{(B \amalg A)(t) \cdot (A \amalg C)(t) \cdot (B \amalg C)(t)}{A(t)\,B(t)\,C(t)}.$$

The right-hand side of (6BB) equals $P(A,B,C)(t)/A(t)\,B(t)\,C(t)$. Because

$$K \otimes A \otimes B \otimes C \approx (A \amalg B) \otimes [A \otimes T(A_+ \otimes C_+) \otimes C] \otimes [B \otimes T(B_+ \otimes C_+) \otimes C]$$

as modules, we have $K(t)\,A(t)\,B(t)\,C(t) = P(A,B,C)(t)$, i.e., $K(t)$ also equals the right-hand side of (6BB).

We are almost done. Because q' is monomorphic, $\overline{q}$ in (6AA) is monomorphic. Since $\overline{q}$ is also surjective (Lemma 6.14), it is bijective. This means that

$$(6CC) \qquad (A \amalg B \amalg C/J)(t) = K(t).$$

It also means that $\overline{\lambda}$ is monomorphic (again, because q' is monomorphic). But the source and target of $\overline{\lambda}$ have the same Hilbert series. Since this argument is valid over any field, $\overline{\lambda}$ is an isomorphism.

Lemma 6.16. *Let A, B, and C be Hopf algebras, and let*

$$\lambda = \lambda(A,B,C) : A \amalg B \amalg C \to \mathrm{CC}(A,B,C).$$

The homomorphism $fg : TU \to A \amalg B \amalg C$ of (6Y) is an isomorphism of TU with $\mathrm{HAK}(\lambda)$.

Proof. As is easily checked, $fg(U)$ is contained in the Hopf ideal generated by (6Z), and (6Z) is clearly contained in $\mathrm{HAK}(\lambda)$, so $fg(TU) \subseteq \mathrm{HAK}(\lambda)$. Since (6Y) is an isomorphism, $fg|_{TU}$ is an embedding.

We know by [MM] that an upper bound on the Hilbert series of $\mathrm{HAK}(\lambda)$ is the quotient of Hilbert series,

$$(6DD) \qquad \mathrm{HAK}(\lambda)(t) \le \frac{(A \amalg B \amalg C)(t)}{\mathrm{CC}(A,B,C)(t)}.$$

Since $TU \otimes K \approx A \amalg B \amalg C$ as modules, however, we have $TU(t) = (A \amalg B \amalg C)(t)/K(t)$, which equals the right-hand side of (6DD) by (6CC) and Lemma 6.15. Since TU is embedded in $\mathrm{HAK}(\lambda)$ but $\mathrm{HAK}(\lambda)(t) \le TU(t)$, we have $\mathrm{HAK}(\lambda) = fg(TU)$.

Lemma 6.17. *(cf. 5.16) Let $L^{(1)}$, $L^{(2)}$, and $L^{(3)}$ be Lie algebras, and let $A^{(i)} = \mathcal{U}L^{(i)}$. Let $L^{(123)}$ denote the Lie ideal of $L^{(1)} \amalg L^{(2)} \amalg L^{(3)}$ consisting of all elements which involve something from each factor, i.e., $L^{(123)} = \{x \in L^{(1)} \amalg L^{(2)} \amalg L^{(3)} |$ the projections of x onto the submodules $L^{(1)} \amalg L^{(2)}$ and $L^{(1)} \amalg L^{(3)}$ and $L^{(2)} \amalg L^{(3)}$ are all zero$\}$. Then the two-sided ideal in $A^{(1)} \amalg A^{(2)} \amalg A^{(3)}$ generated by $L^{(123)}$ is $J = J(A^{(1)}, A^{(2)}, A^{(3)})$, and*

$$\mathcal{U}(L^{(1)} \amalg L^{(2)} \amalg L^{(3)}/L^{(123)}) = \mathrm{CC}(A^{(1)}, A^{(2)}, A^{(3)}).$$

Proof. Let $L^{(12)}$ denote the Lie algebra kernel of $L^{(1)} \amalg L^{(2)} \to L^{(1)} \oplus L^{(2)}$, and likewise for $L^{(13)}$ and $L^{(23)}$. As modules,

$$(6EE) \qquad L^{(1)} \amalg L^{(2)} \amalg L^{(3)} = L^{(1)} \oplus L^{(2)} \oplus L^{(3)} \oplus L^{(12)} \oplus L^{(13)} \oplus L^{(23)} \oplus L^{(123)}.$$

We will show that $L^{(123)} \subseteq J$, hence the two-sided ideal $\langle L^{(123)} \rangle$ that $L^{(123)}$ generates in $A^{(1)} \amalg A^{(2)} \amalg A^{(3)}$ is contained in J. Then $\mathrm{CC}(A^{(1)}, A^{(2)}, A^{(3)}) = A^{(1)} \amalg A^{(2)} \amalg A^{(3)}/J$ is a quotient of

$$A^{(1)} \amalg A^{(2)} \amalg A^{(3)}/\langle L^{(123)} \rangle = \mathcal{U}(L^{(1)} \amalg L^{(2)} \amalg L^{(3)}/L^{(123)}).$$

Apply the Poincare-Birkhoff-Witt theorem to (6EE) to see that $\mathcal{U}(L^{(1)} \amalg L^{(2)} \amalg L^{(3)}/L^{(123)})$ and $\mathrm{CC}(A^{(1)}, A^{(2)}, A^{(3)})$ have the same Hilbert series. Since one is a quotient of the other, they are equal.

To get $L^{(123)} \subseteq J$, note that $L^{(123)}$ is spanned as a module by iterated brackets of the form

$$\text{(6FF)} \qquad [x_m, [x_{m-1}, \cdots [x_2, x_1] \cdots]], \quad m \geq 3,$$

where each x_i belongs to $L^{(1)} \cup L^{(2)} \cup L^{(3)}$ and at least one of $x_1, \ldots, x_m$ belongs to each of $L^{(1)}$, $L^{(2)}$, and $L^{(3)}$. Using the Jacobi identities we see easily that it suffices to consider brackets (6FF) for which $x_1 \in L^{(3)}$ and $x_2 \in L^{(1)} \cup L^{(2)}$. Associated to (6FF) is a function $e : \{1, \cdots, m\} \to \{1, 2, 3\}$ defined by $x_i \in L^{(e(i))}$, and we may restrict our attention to those having $e(1) = 3$, $e(2) \in \{1, 2\}$, and $e(\{1, \cdots, m\}) = \{1, 2, 3\}$.

Assume inductively that brackets (6FF) of length $m - 1$ or shorter are known to belong to J (this is trivial for brackets of length 3). Denote (6FF) by z; we want $z \in J$. Supposing $e(2) = 2$ (the case $e(2) = 1$ is symmetrical) consider four cases. If $e(\{1, \cdots, m-1\}) = \{1, 2, 3\}$ then $[x_{m-1}, \cdots [x_2, x_1] \cdots] \in J$ by our inductive assumption. The remaining cases have $e(\{1, \cdots, m-1\}) = \{2, 3\}$ and $e(m) = 1$.

If $e(3) = \cdots = e(m-1) = 2$ then

$$z = [x_m, \mathrm{ad}(x_{m-1} \cdots x_2)(x_1)] = [x_m, [x_{m-1} \cdots x_2, x_1]_H]_H \subseteq J.$$

If $e(3) = \cdots = e(m-1) = 3$ then $z = \pm[x_m, \mathrm{ad}(x_{m-1} \cdots x_3 x_1)(x_2)] \in J$. For the last case, the maximal index j for which $e(\{x_j, \cdots, x_{m-1}\}) = \{2, 3\}$ satisfies $j \geq 3$. Let $u = \mathrm{ad}(x_{j-1} \cdots x_2)(x_1)$ and write

$$x_m x_{m-1} \cdots x_j = \pm\,\mathrm{ad}(x_j \cdots x_{m-1})(x_m) + w,$$

where w is a sum of $\pm x_{\sigma(m)} \cdots x_{\sigma(j)}$ for various permutations σ on the set $\{j, \cdots, m\}$, all of which satisfy $\sigma(m) \neq m$ (hence $e(\sigma(s)) = 1$ for some $s = \sigma^{-1}(m) < m$). Let $v = \pm\,\mathrm{ad}(x_j \cdots x_{m-1})(x_m) \in L^{(123)}$, which belongs to J by our inductive assumption. Then

$$z = \mathrm{ad}(v + w)(u) = [v, u] + \Sigma(\pm)\,\mathrm{ad}(x_{\sigma(m)})\,\mathrm{ad}(x_{\sigma(m-1)} \cdots x_{\sigma(j)})(u).$$

Since $v \in J$, and since each $\mathrm{ad}(x_{\sigma(m-1)} \cdots x_{\sigma(j)})(u)$ belongs to J by our inductive assumption, we get $z \in J$.

In the above proof we never needed the term $[[C, A]_H, [C, B]_H]_H$ of (6Z). If J' is the ideal generated by the submodule

$$[[A^{(3)}, A^{(2)}]_H, A^{(1)}]_H + [[A^{(3)}, A^{(1)}]_H, A^{(2)}]_H$$

of (6Z), then $\langle L^{(123)} \rangle \subseteq J' \subseteq J$, so $\langle L^{(123)} \rangle = J'$. Since the modules $[[A^{(3)}, A^{(2)}]_H, A^{(1)}]_H$ and $\mathrm{ad}(A_+^{(1)})[A^{(3)}, A^{(2)}]_H$ generate the same two-sided ideal, we have shown

Lemma 6.18. *For Lie algebras $L^{(1)}$, $L^{(2)}$, and $L^{(3)}$ with $L^{(123)}$ as above, both $\langle L^{(123)} \rangle$ and J equal the two-sided ideal in $A^{(1)} \amalg A^{(2)} \amalg A^{(3)}$ generated by*

$$\mathrm{ad}(A_+^{(1)})[A^{(3)}, A^{(2)}]_H + \mathrm{ad}(A_+^{(2)})[A^{(3)}, A^{(1)}]_H.$$

Lemma 6.19. *Let $L^{(1)}$, $L^{(2)}$, and $L^{(3)}$ be Lie algebras, and put $A^{(i)} = \mathcal{U}L^{(i)}$. Suppose $f_{13} : A_+^{(1)} \otimes A_+^{(3)} \to L^{(13)}$ and $f_{23} : A_+^{(2)} \otimes A_+^{(3)} \to L^{(23)}$ are principal (i.e., they become principal when viewed as homomorphisms into $\mathcal{U}L^{(13)} \subseteq A^{(1)} \amalg A^{(3)}$ and $\mathcal{U}L^{(23)} \subseteq A^{(2)} \amalg A^{(3)}$). Then Lemma 6.16 remains true if f_{13} and f_{23} are used as f_{AC} and f_{BC} in defining the isomorphism (6Y). Furthermore, $fg(U) \subseteq L^{(123)}$, so fg induces an isomorhism of $\mathcal{L}(U)$ with $L^{(123)}$.*

Proof. Working carefully through the definitions shows directly that $fg(U) \subseteq L^{(123)}$. The only fact we needed about fg for Lemma 6.16 was that $fg(U) \subseteq \mathrm{HAK}(\lambda)$, but it is clear that $L^{(123)}$ consists of primitives belonging to $\ker(\lambda)$, i.e., $L^{(123)} \subseteq \mathrm{HAK}(\lambda)$.

Proposition 6.20. *Let $L^{(1)}$, $L^{(2)}$, and $L^{(3)}$ be dgL's that are free as modules, and suppose that any two out of three of $L^{(12)}$, $L^{(13)}$, and $L^{(23)}$ have a linear differential. Then $L^{(123)}$ has a linear differential.*

Proof. We saw in Chapter 5 that $L^{(12)}$, $L^{(13)}$, and $L^{(23)}$ are automatically free as Lie algebras, since their enveloping algebras are free. Without loss of generality we assume that $L^{(13)}$ and $L^{(23)}$ have linear differentials. Let

$$f_{13} : (V^{(13)}, d) \to L^{(13)}$$

be a chain map that is an isomorphism onto indecomposables. Since $L^{(13)}/([L^{(13)}, L^{(13)}] + (L_{\mathrm{odd}}^{(13)})^2) \approx \mathcal{U}L_+^{(13)}/(\mathcal{U}L_+^{(13)})^2 = \mathcal{U}L_+^{(1)} \otimes \mathcal{U}L_+^{(3)}$ as chain complexes, $(V^{(13)}, d)$ is isomorphic with $\mathcal{U}L_+^{(1)} \otimes \mathcal{U}L_+^{(3)}$. We may simply replace $(V^{(13)}, d)$ by $\mathcal{U}L_+^{(1)} \otimes \mathcal{U}L_+^{(3)}$, and then f_{13} is principal. Likewise there is a principal f_{23} with image in $L^{(23)}$, as needed for Lemma 6.19. Since the isomorphism of $\mathcal{L}(U)$ with $L^{(123)}$ is a dgL isomorphism (since it extends the differential module inclusion $fg : U \to L^{(123)}$), we have shown that $L^{(123)}$ has a linear differential.

Combining this with Theorem 5.12, we obtain the following corollary.

Corollary 6.21. *Let $k, l, m \geq 0$. The dgL N^{klm} of 5.15 has a linear differential, and $fg(U)$ describes an indecomposable submodule.*

Having expressed $\mathrm{CC}(A, B, C)$ in a recognizable form, our detour is finished, and we return to the matter of $C_*(\Omega R)$ when $R = \mathcal{R}(X^1, X^2, X^3)$. Let $\mathcal{A}(X^i)$ be an AH model for X^i. Because there are quisms $\theta_{X^i} : \mathcal{A}(X^i) \to A^i = C_*(G^i) \approx C_*(\Omega X^i)$, it would seem that the "obvious" thing to do is to write down the diagram of quisms,

$$\begin{array}{ccccc}
\mathrm{CC}(\mathcal{A}(X^1), \mathcal{A}(X^2), \mathcal{A}(X^3)) & \xrightarrow{\simeq} & \mathrm{CC}(A^1, A^2, A^3) & & \\
{\scriptstyle =}\uparrow & & \downarrow{\scriptstyle \simeq} & & \\
\mathcal{A}(X^i) \amalg \mathcal{A}(X^2) \amalg \mathcal{A}(X^3)/J & & C_*(\Omega R) & \underset{\theta_R}{\xleftarrow{\simeq}} & \mathcal{A}(R),
\end{array}$$

and then use the lifting lemma [A3, Prop. 2.9] to obtain the quism

(6GG) $$\mathcal{A}(R) \xrightarrow{\simeq} \mathcal{A}(X^i) \amalg \mathcal{A}(X^2) \amalg \mathcal{A}(X^3)/J.$$

Since $\mathcal{A}(X^1 \vee X^2 \vee X^3) = \mathcal{A}(X^1) \amalg \mathcal{A}(X^2) \amalg \mathcal{A}(X^3)$, we obtain using (6V) the homotopy-commutative square

$$\text{(6HH)}\qquad \begin{array}{ccc} \mathcal{A}(X^1 \vee X^2 \vee X^3) & \xrightarrow{\mathcal{A}(\bar{\lambda})} & \mathcal{A}(R) \\ =\big\downarrow & & \big\downarrow\simeq \\ \mathcal{A}(X^i) \amalg \mathcal{A}(X^2) \amalg \mathcal{A}(X^3) & \xrightarrow{\lambda=\lambda(\mathcal{A}(X^i))} & \mathcal{A}(X^i) \amalg \mathcal{A}(X^2) \amalg \mathcal{A}(X^3)/J, \end{array}$$

in which the bottom arrow is the quotient homomorphism. By altering $\mathcal{A}(R)$ if necessary, we can get each $\mathcal{A}(X^i)$ to be a retract of it (because each X^i is a retract of R), and then (6HH) can be assumed to commute on the nose.

Unfortunately, the scheme we have outlined will only work if $\mathcal{A}(X^i)$ are dgH's, not just dga's, and if the quisms θ_{X^i} preserve the dgH information. This is not always the case. [A3, Section 8] discusses the fact that an AH model is always a "Hopf algebra up to homotopy" but it is not necessary a dgH. The idea is that the diagonal map $\Delta_X : X \to X \times X$ induces a dga homomorphism

$$\text{(6II)}\qquad \psi_X : \mathcal{A}(X) \xrightarrow{\mathcal{A}(\Delta_X)} \mathcal{A}(X \times X) \xrightarrow{\simeq} \mathcal{A}(X) \otimes \mathcal{A}(X)$$

for which the square

$$\text{(6JJ)}\qquad \begin{array}{ccc} \mathcal{A}(X) & \xrightarrow{\theta_X} & C_*(\Omega X) \\ \psi_X\big\downarrow & & \big\downarrow\Delta_* \\ \mathcal{A}(X) \otimes \mathcal{A}(X) & \xrightarrow{\theta_X \otimes \theta_X} & C_*(\Omega X) \otimes C_*(\Omega X), \end{array}$$

with Δ_* denoting the Alexander-Whitney diagonal approximation, commutes up to homotopy. We call ψ_X the ***topological coproduct*** for $\mathcal{A}(X)$. For many spaces X it is possible to make a judicious choice of ψ_X so that $(\mathcal{A}(X), \psi_X)$ is a cocommutative dgH. In such cases we call $(\mathcal{A}(X), \psi_X)$ a ***dgH-AH model*** for X. We have sketched the proof of the following.

Proposition 6.22. *Suppose each of X^1, X^2, and X^3 has a dgH-AH model $\mathcal{A}(X^i)$. Then it is possible to choose $\mathcal{A}(R)$ for $R = R(X^1, X^2, X^3)$ such that the quism (6GG) exists and the diagram (6HH) commutes. Furthermore, if*

$$(\mathcal{A}(X^i), \psi_{X^i}) \to (B^i, e^i, \psi^i)$$

are dgH quisms, $i = 1, 2, 3$, then there is a commuting square

$$\text{(6KK)}\qquad \begin{array}{ccc} \mathcal{A}(X^1 \vee X^2 \vee X^3) & \xrightarrow{\mathcal{A}(\bar{\lambda})} & \mathcal{A}(R) \\ \simeq\big\downarrow & & \big\downarrow\simeq \\ B^1 \amalg B^2 \amalg B^3 & \longrightarrow & (B^1 \amalg B^2 \amalg B^3)/J(B^1, B^2, B^3) \end{array}$$

in which the vertical arrows are quisms and the bottom arrow is the quotient homomorphism.

DgH-AH models are a fascinating topic, whose study could easily take us far afield from our intended path. For instance, an open question is whether or not every 1-connected space has a dgH-AH model. The property of having a dgH-AH model can sometimes be transferred from a total space to a fiber, as the next result shows. Lemma 6.23 really should have been part of [A4], but the author omitted it due to inadequate foresight. Again, we merely sketch the proof rather than spend too long on what is for us a tangential matter.

Lemma 6.23.

Let $F \xrightarrow{\mu} E \xrightarrow{\rho} S^{m+1}$ be a fibration over a sphere, $m \geq 1$, F 1-connected. Suppose that $\mathcal{A}(E)$ is a dgH-AH model for E, and that $\mathcal{A}(\rho) : \mathcal{A}(E) \to \mathcal{A}(S^{m+1}) = (\mathbb{Z}_{(p)}\langle b\rangle, 0)$ is a surjective dgH homomorphism, where $|b| = m$.

(a) *The Hopf algebra kernel $\mathrm{HAK}(\mathcal{A}(\rho))$ serves as an AH model for F, and the inclusion of $\mathrm{HAK}(\mathcal{A}(\rho))$ into $\mathcal{A}(E)$ is an AH model for the map μ.*

(b) *Suppose further that $\phi : (N, d) \to (\mathbb{L}\langle b\rangle, 0)$ is a surjective dgL homomorphism, where $|b| = m$, and suppose that there is a commuting diagram*

$$\text{(6LL)}\qquad \begin{array}{ccc} \mathcal{A}(E) & \xrightarrow{\mathcal{A}(\rho)} & \mathcal{A}(S^{m+1}) \\ \simeq\downarrow & & \downarrow = \\ \mathcal{U}(N,d) & \xrightarrow{\mathcal{U}\phi} & \mathcal{U}(\mathbb{L}\langle b\rangle, 0) \end{array}$$

in which the left vertical arrow is a quism. Then there is a dgH quism $\mathrm{HAK}(\mathcal{A}(\rho)) \to \mathcal{U}(\ker(\phi))$. If $\ker(\phi)$ is a free dgL then $\mathcal{U}(\ker(\phi))$ serves as a dgH-AH model for F.

Proof. (a) This follows via the techniques of [A4]. The model $\mathcal{A}(F)$ constructed in [A4] can be recognized as having the same dga homotopy type as $\mathrm{HAK}(\mathcal{A}(\rho))$. The one fact that is needed that is not readily apparent in [A4] is that the derivation Q of [A4] (which mirrors the action of S^m on F) commutes with coproducts, i.e., $\psi_F Q(x) = (Q \otimes 1 + 1 \otimes Q)\psi_F(x)$ for $x \in \mathcal{A}(F)$.

We provide here the barest outline of why Q commutes with ψ_F. We must extend the ideas of [A4] to the setting of a fibration over $S^{m+1} \times S^{m+1}$, where there will be two operations Q' and Q'' on the Adams-Hilton model of the fiber. A commuting diagram of fibrations

$$\begin{array}{ccccc} F & \longrightarrow & E & \xrightarrow{\rho} & S^{m+1} \\ f\downarrow & & \downarrow & & \downarrow\Delta \\ F' & \longrightarrow & E' & \xrightarrow{\rho'} & S^{m+1} \times S^{m+1} \end{array}$$

will have an induced map f on fibers whose AH model satisfies $\mathcal{A}(f)Q = (Q' + Q'')\mathcal{A}(f)$. When ρ' is itself a product, $\rho' = \rho_1 \times \rho_2 : E_1 \times E_2 \to S^{m+1} \times S^{m+1}$, then Q' and Q'' commute on the tensor product $\mathcal{A}(F_1) \otimes \mathcal{A}(F_2)$ of the two fibers' AH models, and $Q' = Q \otimes 1$ while $Q'' = 1 \otimes Q$. Replacing $\mathcal{A}(f)$ by ψ_F (which is, by (6II), $\mathcal{A}(\Delta_F)$ followed by a quism) gives the needed formula.

(b) Straightforward, using (a).

We have one remaining task for Chapter 6: we want to understand the homotopy-theoretic fiber of the map $\tilde{\lambda} = \tilde{\lambda}(X^1, X^2, X^3)$. As we shall see, the combinatorial algebra that led to the identification of $\mathrm{HAK}(\lambda)$ with TU will also lead to a description of this fiber as a certain bouquet. We denote the fiber by $G = \mathcal{G}(X^1, X^2, X^3)$ so that our fibration is

$$\text{(6MM)}\qquad G = \mathcal{G}(X^1, X^2, X^3) \longrightarrow X^1 \vee X^2 \vee X^3 \xrightarrow{\tilde{\lambda}(X^1,X^2,X^3)} R = \mathcal{R}(X^1, X^2, X^3)$$

As motivation, we will need to know something about the homotopy of G, because (6MM) is going to serve as the fibration (6G) in Theorem 6.4, on three separate occasions.

We begin with a quick review of the analogous fibration for two spaces, which is

$$G^{12} \to X^1 \vee X^2 \to X^1 \times X^2, \tag{6NN}$$

where we denote by G^{12} the homotopy-theoretic fiber of the inclusion map from the wedge into the product. The fibration (6NN) splits when looped once, i.e.,

$$\Omega(X^1 \vee X^2) = \Omega X^1 \times \Omega G^{12} \times \Omega X^2. \tag{6OO}$$

Actually, (6OO) is the topological analog of Lemma 6.10, when the homomorphism f of Lemma 6.10 is $[\ ,\ ]_H$. To explain this statement, we need to review external Whitehead products.

Given two maps,

$$f_1 : \Sigma X^1 \to Y, \quad f_2 : \Sigma X^2 \to Y,$$

the *commutator* of their adjoints, denoted

$$[f_1^a, f_2^a] : X^1 \times X^2 \to \Omega Y,$$

is defined by $[f_1^a, f_2^a](x_1, x_2) = f_1^a(x_1)f_2^a(x_2)f_1^a(x_1)^{-1}f_2^a(x_2)^{-1}$, where we are using the loop multiplication in ΩY to define this product. The commutator has a canonical homotopy factorization through $X^1 \wedge X^2$, and the induced extension is called the *external Samelson product* and is denoted $\langle f_1, f_2 \rangle^a$:

$$\begin{array}{ccc} X_1 \times X_2 & \xrightarrow{[f_1^a, f_2^a]} & \Omega Y \\ \downarrow & & \downarrow = \\ X^1 \wedge X^2 & \xrightarrow{\langle f_1, f_2 \rangle^a} & \Omega Y. \end{array}$$

The coadjoint of the external Samelson product is the *external Whitehead product*,

$$\langle f_1, f_2 \rangle : \Sigma X^1 \wedge X^2 \to Y.$$

Let $e_i : \Sigma\Omega X^i \xrightarrow{(\Omega 1)^a} X^i \longrightarrow X^1 \vee X^2$ for $i = 1, 2$. Then the map $e_{12} = \langle e_1, e_2 \rangle : \Sigma\Omega X^1 \wedge \Omega X^2 \to X^1 \vee X^2$ turns out to be the fiber inclusion map $G^{12} \to X^1 \vee X^2$ in (6NN). The reason for this is as follows. Let $H_*(\ )$ denote homology with coefficients in a field, and let A^i be the Hopf algebra $H_*(\Omega X^i)$. We have $H_*(\Omega(X^1 \vee X^2)) = A^1 \amalg A^2$ as Hopf algebras, and $H_*(\Omega X^1 \wedge \Omega X^2) = A_+^1 \otimes A_+^2$. By the Bott-Samelson theorem, $H_*(\Omega(\Sigma\Omega X^1 \wedge \Omega X^2)) = T(A_+^1 \otimes A_+^2)$. By [N1, p. 49], this diagram commutes:

$$\begin{array}{ccc} H_*(\Omega X^1 \wedge \Omega X^2) & \xrightarrow{(e_{12}^a)_*} & H_*(\Omega(X^1 \vee X^2)) \\ = \downarrow & & \downarrow = \\ A_+^1 \otimes A_+^2 & \xrightarrow{[\ ,\ ]_H} & A^1 \amalg A^2. \end{array} \tag{6PP}$$

Consequently, $(\Omega e_{12})_* : H_*(\Omega(\Sigma\Omega X^1 \wedge \Omega X^2)) \to H_*(\Omega(X^1 \vee X^2))$ is precisely the map $f : T(A_+^1 \otimes A_+^2) \to A^1 \amalg A^2$ of Lemma 6.10. In particular, $(\Omega e_{12})_*$ is one-to-one.

Suppose X^1 and X^2 are 1-connected and of finite type. Denote the Poincaré series of a space Y by $Y(t)$, and retain the notation $M(t)$ for the Hilbert series of a graded module. If we did not already know that $G^{12} = \Sigma\Omega X^1 \wedge \Omega X^2$, one way to prove it would be to make four observations. First, $\Omega G^{12}(t)\Omega(X^1 \times X^2)(t) = \Omega(X^1 \vee X^2)(t)$ because the fibration (6NN) splits when looped once, so

$$\begin{aligned}\Omega G^{12}(t) &= \Omega(X^1 \vee X^2)(t)/\Omega X^1(t)\Omega X^2(t) = (A^1 \amalg A^2)(t)/A^1(t)A^2(t)\\ &= [A^1(t) + A^2(t) - A^1(t)A^2(t)]^{-1} = [1 - A^1_+(t)A^2_+(t)]^{-1}.\end{aligned}$$

Second,

$$\begin{aligned}(\Omega(\Sigma\Omega X^1 \wedge \Omega X^2))(t) &= [1 - (\Omega X^1 \wedge \Omega X^2)(t)]^{-1}\\ &= [1 - A^1_+(t)A^2_+(t)]^{-1} = \Omega G^{12}(t).\end{aligned}$$

Third, the homomorphism $(\Omega e_{12})_*$ is one-to-one, as we have already noted. Fourth, e_{12} factors through G^{12} because e_{12} becomes trivial when continued into $X^1 \times X^2$ (because the commutator becomes trivial.)

Here is why the four observations suffice. If $\tilde{e}_{12} : \Sigma\Omega X^1 \wedge \Omega X^2 \to G^{12}$ denotes the lift of e_{12}, then $(\Omega\tilde{e}_{12})_*$ must also be one-to-one. Since it is a monomorphism between graded modules having the same Hilbert series, $(\Omega\tilde{e}_{12})_*$ is an homology isomorphism between 0-connected H-spaces, with coefficients in any field, hence $\Omega\tilde{e}_{12}$ is a homotopy equivalence. It follows that $\tilde{e}_{12}$ is a homotopy equivalence, i.e., $G^{12} \approx \Sigma\Omega X^1 \wedge \Omega X^2$. We shall imitate the pattern of this proof, for (6MM) instead of (6NN).

Lemma 6.24. *Let X^1, X^2, and X^3 be 1-connected. Let $R = \mathcal{R}(X^1, X^2, X^3)$. Let G^{12} be as in (6NN), and define G^{13} and G^{23} likewise.*

(a) *ΩR has the homotopy type of*

(6QQ) $$\Omega X^1 \times \Omega G^{12} \times \Omega X^2 \times \Omega G^{13} \times \Omega G^{23} \times \Omega X^3,$$

hence also the homotopy type of

(6RR) $$\Omega(X^1 \vee X^2) \times \Omega G^{13} \times \Omega G^{23} \times \Omega X^3.$$

(b) *The fibration (6MM) splits when looped once,*

$$\Omega G \longrightarrow \Omega(X^1 \vee X^2 \vee X^3) \underset{\longleftarrow}{\longrightarrow} \Omega R,$$

hence $\Omega(X^1 \vee X^2 \vee X^3) \approx \Omega G \times \Omega R$ as spaces (not as loop spaces).

(c) *Taking homology with coefficients in any field, let $A^i = H_*(\Omega X^i)$, and let K of Lemma 6.14 and U of Lemma 6.12 be these modules as defined for the triple of Hopf algebras A^1, A^2, and A^3. If X^1, X^2, and X^3 have finite type, then*

$$\begin{aligned}&\Omega R(t) = \mathrm{CC}(A^1, A^2, A^3)(t) = K(t);\\ &\Omega G(t) = TU(t).\end{aligned}$$

Proof. (a) Applying Ω to (6O), ΩR is the homotopy limit of

$$\begin{array}{ccccc} \Omega(X^1 \vee X^2) & \longrightarrow & \Omega X^2 & \longleftarrow & \Omega(X^2 \vee X^3) \\ \downarrow & & & & \downarrow \\ \Omega X^1 & \longleftarrow & \Omega(X^1 \vee X^3) & \longrightarrow & \Omega X^3, \end{array}$$

which is the same as the diagram

$$\text{(6SS)}\qquad \begin{array}{ccccc} \Omega X^1 \times \Omega G^{12} \times \Omega X^2 & \longrightarrow & \Omega X^2 & \longleftarrow & \Omega X^2 \times \Omega G^{23} \times \Omega X^3 \\ \downarrow & & & & \downarrow \\ \Omega X^1 & \longleftarrow & \Omega X^1 \times \Omega G^{13} \times \Omega X^3 & \longrightarrow & \Omega X^3. \end{array}$$

The homotopy limit of (6SS) is clearly (6QQ).

(b) By the Hilton-Milnor theorem, $\Omega(X^1 \vee X^2 \vee X^3)$ is a (weak infinite) product of a list of factors of the form $\Omega\Sigma(\Omega X^1)^{\wedge a_1} \wedge (\Omega X^2)^{\wedge a_2} \wedge (\Omega X^3)^{\wedge a_3}$ for a certain list of integer triples (a_1, a_2, a_3). The first six entries in the list give the six factors (6QQ), and $\Omega\tilde{\lambda}$ is simply the projection onto these six factors (this is evident from (6SS)). So $\Omega\tilde{\lambda}$ has a section.

(c) Since we have already seen that $\Omega G^{12}(t) = T(A^1_+ \otimes A^2_+)(t)$ and likewise for ΩG^{13} and ΩG^{23}, we need only to compare (6RR) and the definition of K in 6.14 to see that their series coincide. Since Lemma 6.16 shows (6DD) to be an equality, we have

$$\begin{aligned} \Omega G(t) &= \Omega(X^1 \vee X^2 \vee X^3)(t)/\Omega R(t) \\ &= (A^1 \amalg A^2 \amalg A^3)(t)/K(t) \quad = \quad TU(t). \end{aligned}$$

Using the explicit description of U in Lemma 6.12 as our guide, we will now give a list of iterated brackets which will index a wedge decomposition for G. Since we are now working with three spaces, we denote by e_i the composition

$$e_i : \Sigma\Omega X^i \xrightarrow{(\Omega 1)^{\overline{\ }}} X^i \to X^1 \vee X^2 \vee X^3$$

for $i = 1, 2, 3$. Let $e_{12} = \langle e_1, e_2 \rangle$, and likewise for e_{13} and e_{23}. Let $\mathrm{Ad}(f)(-)$ be another notation for $\langle f, - \rangle$, and following our usual convention let $\mathrm{ad}(w_1 \cdots w_m)$ denote the composition $\mathrm{Ad}(w_1)\,\mathrm{Ad}(w_2) \cdots \mathrm{Ad}(w_m)$.

Let $\widetilde{U}' = \{\langle e_{13}, e_{23} \rangle\} \cup \{\mathrm{ad}(w)(e_3) | w$ is an alternating string of e_1's and e_2's of length at least two $\}$. Let

$$\widetilde{U} = \{\mathrm{Ad}^q(e_{13})\,\mathrm{Ad}^s(e_{23})(u') | q \geq 0,\ s \geq 0,\ u' \in \widetilde{U}'\}.$$

For each expression $u \in \widetilde{U}$, let $a_j(u)$ denote the total number of e_j's that appear in it, $1 \leq j \leq 3$, when it is fully expanded as an iterated bracket of e_i's. Let $(\Omega X)^{\wedge a(u)}$ be a shorthand for $(\Omega X)^{\wedge a_1(u)} \wedge (\Omega X)^{\wedge a_2(u)} \wedge (\Omega X)^{\wedge a_3(u)}$, and let

$$\gamma_u : \Sigma(\Omega X)^{\wedge a(u)} \to X^1 \vee X^2 \vee X^3$$

be the actual iterated external Whitehead product of e_i's described by u. Let

$$\text{(6TT)}\qquad V = \bigvee_{u \in U} (\Omega X)^{\wedge a(u)}.$$

Proposition 6.25. *Let X^1, X^2, and X^3 be 1-connected spaces.*

(a) *The cardinality of $\{u \in \widetilde{U} | a_i(u) = m_i\}$ for a prescribed triple (m_1, m_2, m_3) is described by the generating function*

$$f(t_1,t_2,t_3) = \sum_{u\in U} t_1^{a_1(u)} t_2^{a_2(u)} t_3^{a_3(u)} = \frac{t_1t_2t_3(2+t_1+t_2+t_3-t_1t_2t_3)}{(1-t_1t_2)(1-t_1t_3)(1-t_2t_3)}. \tag{6UU}$$

(b) *If X^1, X^2, and X^3 have finite type, then $V(t) = U(t)$ and $(\Omega\Sigma V)(t) = TU(t) = \Omega G(t)$, where V is given by (6TT).*

(c) *If $\gamma = \bigvee_{u\in U} \gamma_u : \Sigma V \to X^1 \vee X^2 \vee X^3$, then this diagram commutes, where $H_*(\ )$ denotes homology with field coefficients as above:*

$$\begin{array}{ccc} H_*(V) & \xrightarrow{(\gamma^a)_*} & H_*(\Omega(X^1 \vee X^2 \vee X^3)) \\ {\scriptstyle =}\big\downarrow & & \big\downarrow{\scriptstyle =} \\ U & \xrightarrow{fg} & A^1 \amalg A^2 \amalg A^3. \end{array}$$

Here fg is the homomorphism appearing in (6Y). Consequently

$$\begin{array}{ccc} H_*(\Omega\Sigma V) & \xrightarrow{(\Omega\gamma)_*} & H_*(\Omega(X^1 \vee X^2 \vee X^3)) \\ {\scriptstyle =}\big\downarrow & & \big\downarrow{\scriptstyle =} \\ TU & \xrightarrow{fg} & A^1 \amalg A^2 \amalg A^3 \end{array}$$

commutes; in particular, $(\Omega\gamma)_*$ is one-to-one.

(d) *For each $u \in \widetilde{U}$, $\tilde{\lambda}\gamma_u$ is null-homotopic, hence γ_u factors through G. Denote a lifting by*

$$\tilde{\gamma}_u : \Sigma(\Omega X)^{\wedge a(u)} \to G.$$

(e) *$\tilde{\gamma} = \bigvee_{u\in U} \tilde{\gamma}_u : \Sigma V \to G$ is a homotopy equivalence.*

Proof. Part (a) is an elementary counting argument. For (b), substitute $A^i_+(t)$ for t_i in (6UU) to get $V(t)$. The calculation of $U(t)$ from its definition is the same, practically step for step. For (c), again we compare the definitions of γ_u and fg, keeping in mind that $(\langle\ ,\ \rangle^a)_* = [\ ,\ ]_H$. For (d), use the fact that any iterated external Whitehead product that involves all three of e_1, e_2, e_3 will vanish when projected onto any one of $X^1 \vee X^2$, $X^1 \vee X^3$, $X^2 \vee X^3$. Since every $u \in \widetilde{U}$ involves all three e_i's in this way, $\tilde{\lambda}\gamma_u$ is null-homotopic. We have now completed the "four observations," so (e) follows as before, using a direct limit argument if necessary to reduce to the case of finite type.

CHAPTER 7

Coherent Sequences of Algebras

In Chapter 7 we explore the category CSA of "coherent sequences of algebras" and we prove several results of a formal nature about them. A csa is a device for handling simultaneously the homology or homotopy groups with coefficients in $\mathbb{Z}_p$, in $\mathbb{Z}_{p^2}$, in $\mathbb{Z}_{p^3}$, and so on. The language and properties of the csa's will be essential for stating and proving many of the properties of the spaces D_k and E_k.

The standard tool for working simultaneously with torsion of all orders is the Bockstein spectral sequence. However, it turns out that the BSS has too much indeterminacy built into it for the applications we have in mind. We are led to seek a structure that has all the information in the BSS but which also keeps some the information about how to mulitply and divide by p. We especially need it to accommodate easily to work with mod p^t homotopy classes, for $t > 1$. The device that meets these requirements is the coherent sequence of algebras, or, as appropriate, the coherent sequence of Lie algebras.

The csa and csL structures will have all of the following properties: both homology and homotopy carry the structure functorially, and it is preserved by the Hurewicz homomorphism; we can recognize what it is explicitly for the important chain complexes $\mathcal{U}M^k$ and $\overline{L}^k$; and it is "natural" in the sense of being the "obvious" thing to define. The drawback is that it requires the introduction of a number of technical axioms that are reminiscent in their interaction to the axioms of semi-simplicial set theory. The reader is encouraged to peak ahead to Lemmas 7.2 and 7.3, which provide the motivating examples.

In the first of three units comprising Chapter 7, we introduce coherent sequences of chain complexes, of algebras, of Lie algebras, and of modules. Using the important concept of a Bockstein subalgebra, we demonstrate how checking only a few equations can suffice to guarantee the existence of a homomorphism which preserves all the rich structure on these objects, and which extends a given homomorphism. The second unit of Chapter 7 examines a setting where a coherent Lie algebra structure and a coherent module structure interact. This culminates in Theorem 7.15, which again presents sufficient conditions for the construction of structure-preserving homomorphisms. The underlying insight for Theorem 7.15 is that such a homomorphism has very few degrees of freedom, and the freedom that it does have mirrors closely the minimal presentation for a certain (typically very small) module over a certain associative $\mathbb{Z}_p$-algebra. We call the structure on the csL corresponding to the module presentation a "virtual structure". The third unit of Chapter

7 analyzes three specific examples from the virtual structure viewpoint, namely $\overline{L}^k, N^{kl}$, and N^{klm}.

Definitions and Basic Properties

Definition 7.1. Fix a prime p. A *coherent sequence of chain complexes* over $\mathbb{Z}_{(p)}$, or *csc*, is a finite or infinite list

$$(A_{[1]}, d_{[1]}),\ (A_{[2]}, d_{[2]}),\ \dots\ ,\ (A_{[j]}, d_{[j]})\ \dots \tag{7AA}$$

of chain complexes over $\mathbb{Z}_{(p)}$, together with $\mathbb{Z}_{(p)}$-homomorphisms

$$\theta^t_s :\ A_{[t]} \to A_{[s]} \text{ and } \zeta^s_t :\ A_{[s]} \to A_{[t]} \text{ for } s \leq t,$$

that satisfy the following axioms for $q \leq s \leq t$. ($p:\ A \to A$ denotes multiplication by p):

(a) θ^s_s and ζ^s_s are the identity on $A_{[s]}$.

(b) $\theta^s_q \theta^t_s = \theta^t_q$ and $\zeta^s_t \zeta^q_s = \zeta^q_t$.

(c) $\zeta^s_t \theta^t_s = p^{t-s} :\ A_{[t]} \to A_{[t]}$, and $\theta^t_s \zeta^s_t = p^{t-s} :\ A_{[s]} \to A_{[s]}$.

(d) $d_{[s]} = \theta^t_s d_{[t]} \zeta^s_t$.

If in addition each $(A_{[s]} d_{[s]})$ is a dga, we call the csc together with the multiplications a *coherent sequence of algebras* or *csa* if

(e) $\theta^t_s(xy) = \theta^t_s(x)\theta^t_s(y)$ for $x, y \in A_{[t]}$;

(f1) $\zeta^s_t(x\theta^t_s(y)) = \zeta^s_t(x)y$ for $x \in A_{[s]},\ y \in A_{[t]}$;

(f2) $\zeta^s_t(\theta^t_s(x)y) = x\zeta^s_t(y)$ for $x \in A_{[t]},\ y \in A_{[s]}$.

If each $(A_{[s]}, d_{[s]})$ is instead a dgL, we call the csc together with the brackets a *coherent sequence of Lie algebras* or *csL* if

(e′) $\theta^t_s([x,y]) = [\theta^t_s(x), \theta^t_s(y)]$ for $x, y \in A_{[t]}$;

(f′) $\zeta^s_t([x, \theta^t_s(y)]) = [\zeta^s_t(x), y]$ for $x \in A_{[s]},\ y \in A_{[t]}$.

A csc (or csa or csL) is called *proper* if it satisfies

(g) each ζ^s_t is monomorphic; and

(h) Whenever $\zeta^{t-1}_t(x) = d_{[t]}\zeta^{t-1}_t(y)$, then there exists $z \in A_{[t]}$ such that $y = \theta^t_{t-1}(z)$ and $x = \theta^t_{t-1} d_{[t]}(z)$.

A *morphism* of csc's (resp. csa's, csL's) is a sequence of chain maps (resp. dga, dgL homomorphisms) commuting with all the θ^t_s's and ζ^s_t's. The respective categories are denoted CSC, CSA, CSL; the full subcategories consisting of proper objects are called PCSC, PCSA, PCSL. Occasionally we shall find it convenient to have a csc begin at $(A_{[m]}, d_{[m]})$ for some $m > 1$ instead of at $(A_{[1]}, d_{[1]})$ as in (7AA), or at $m = 0$.

The next two lemmas provide motivating examples for csc's, csa's, and csL's. The proof of the first is left as an exercise, while the second largely recapitulates Lemma 6.1.

Lemma 7.2. *Let (A,d) be an chain complex (resp. dga, dgL) over $\mathbb{Z}_{(p)}$.*

(a) *$\{H_*(A,d;\mathbb{Z}_{pt}),\delta^t\}_{1\le t<\infty}$ is a csc (resp. csa, csL), where θ^t_s and ζ^s_t are the natural transformations induced by the evident coefficient group homomorphisms $\mathbb{Z}_{p^t}\to\mathbb{Z}_{p^s}$ and $\mathbb{Z}_{p^s}\to\mathbb{Z}_{p^t}$. Here δ^t is the Bockstein associated to the short exact sequence*

$$0\to\mathbb{Z}_{p^t}\to\mathbb{Z}_{p^{2t}}\to\mathbb{Z}_{p^t}\to 0.$$

Furthermore, $\{H_*(\ ;\mathbb{Z}_{p^t})\}$ is a functor from chain complexes (resp. dga's, dgL's) to CSC (resp. CSA, CSL).

(b) *If A is $\mathbb{Z}_{(p)}$-free, let $\mathcal{F}(A)$ denote the seqence $\{A_{[t]},\delta^t\}$, where $A_{[t]}=\{x\in A | p^t \text{divides } d(x)\}$ and $\delta^t(x)$ is defined as $p^{-t}d(x)$. Note that δ^t is well-defined because A is torsion-free. For $s\le t$, assign to $\mathcal{F}(A)$ the homomorphisms θ^t_s and ζ^s_t defined by $\theta^t_s(x)=x$ (makes sense since p^s divides $d(x)$ if p^t does) and $\zeta^t_s(x)=p^{t-s}x$. Then $\mathcal{F}$ is a functor from $\mathbb{Z}_{(p)}$-free chain complexes (resp. $\mathbb{Z}_{(p)}$-free dga's, $\mathbb{Z}_{(p)}$-free dgL's) to PCSC (resp. PCSA, PCSL).*

(c) *If A is $\mathbb{Z}_{(p)}$-free, the quotient homomorphisms*

(7AB) $$q_{[t]}:\ A_{[t]}\to A_{[t]}/(d(A)+p^tA)=H_*(A,d;\mathbb{Z}_{p^t})$$

comprise a morphism of csc's (resp. csa's, csL's), and this quotient homomorphism is a natural transformation between the functors $\mathcal{F}$ and $\{H_*(\ ;\mathbb{Z}_{p^t}),\delta^t\}$ on $\mathbb{Z}_{(p)}$-free chain complexes (resp. $\mathbb{Z}_{(p)}$-free dga's, $\mathbb{Z}_{(p)}$-free dgL's).

(d) *Referring to (7AB), the kernel of $q_{[t]}$ can be expressed as $\zeta^1_t(A'_{[1]})+d_{[t]}\zeta^1_t(A'_{[1]})$ for a certain submodule $A'_{[1]}$ of $A_{[1]}$. $A'_{[1]}$ always contains $pA_{[1]}$, and $A'_{[1]}=pA_{[1]}$ if and only if $d(A)\subseteq pA$. When $A_{[1]}$ is viewed as $\{x\in A\mid p \text{ divides } d(x)\}$ as in part (b), then $A'_{[1]}=pA$.*

Notation. Recall from the beginning of Chapter 6 the maps ω^t_s and ρ^s_t and the operation ${}_\pi\delta^t=(\rho^t)^\#(\omega^t)^\#$. Define

$$\theta^t_s=(\omega^t_s)^\#:\ \pi_*(X;\mathbb{Z}_{p^t})\to\pi_*(X;\mathbb{Z}_{p^s});$$
$$\zeta^s_t=(\rho^s_t)^\#:\ \pi_*(X;\mathbb{Z}_{p^s})\to\pi_*(X,\mathbb{Z}_{p^t});$$
$$d_{[t]}=\ {}_\pi\delta^t:\ \pi_*(X;\mathbb{Z}_{p^t})\to\pi_{*-1}(X;\mathbb{Z}_{p^t}).$$

Lemma 7.3.

(a) *$\{H_*(\ ;\mathbb{Z}_{p^t})\}$ is a functor from spaces to CSC.*

(b) *$\{\pi_*(\ ;\mathbb{Z}_{p^t})\}$ is a functor from 1-connected spaces to CSC, if p is odd, where θ^t_s and ζ^s_t and $d_{[t]}$ are described above.*

(c) *$\{\pi_*(\ ;\mathbb{Z}_{2^t})\}$ is a functor from 2-connected spaces to CSC.*

(d) *For either* (b) *or* (c)*, the Hurewicz homomorphism is a natural homomorphism of csc's.*

(e) *$\{H_*(\Omega-;\mathbb{Z}_{p^t})\}$ is a functor from 1-connected spaces to CSA.*

(f) *$\{\pi_*(\Omega-;\mathbb{Z}_{p^t})\}$ is a functor from 1-connected spaces to CSL, if $p\ge 5$.*

(g) *View $\{H_*(\Omega-;\mathbb{Z}_{p^t})\}$ as a csL, with the Pontrjagin commutator as bracket. For $p\ge 5$, the Hurewicz homomorphism*

$$h:\ \pi_*(\Omega-;\mathbb{Z}_{p^t})\to H_*(\Omega-;\mathbb{Z}_{p^t})$$

is a natural csL homomorphism, for 1-connected spaces.

In general it can be hard to construct morphisms between csa's or csL's. One reason for this is that the sources $A_{[t]}$ tend to be very far from being free algebras or free Lie algebras. We will circumvent this difficulty by exploiting the coherence implicit in a csc. Proposition 7.5 and its variations will be our principal tools for constructing morphisms out of proper csa's or csL's. Its purpose is to reduce the amount of work needed in order to verify the axioms for a morphism in CSA or CSL. It is clearly intended to serve as the inductive step in a recursive construction.

Definition. Let $\{A_{[t]}, d_{[t]}\}_{t_0 \leq t \leq m}$ be a proper csc (resp. csa, csL). Let $\overline{A}_{[t_0]} = pA_{[t_0]}$, and for $t > t_0$, write $\overline{A}_{[t]}$ for $\zeta_t^{t-1}(A_{[t-1]}) + d_{[t]}\zeta_t^{t-1}(A_{[t-1]})$. We call a submodule (resp. subalgebra, Lie subalgebra) $N_{[t]}$ of $A_{[t]}$ a *Bockstein submodule* (resp. *Bockstein subalgebra*) of $A_{[t]}$ if

$$A_{[t]} = N_{[t]} + \overline{A}_{[t]} \tag{7AC}$$

as $\mathbb{Z}_{(p)}$-modules, and

$$N_{[t]} \cap \overline{A}_{[t]} = pN_{[t]}. \tag{7AD}$$

If $N_{[t]}$ is also closed under $d_{[t]}$, we call $N_{[t]}$ a *Bockstein subcomplex* (resp. *Bockstein sub-dga*, *Bockstein sub-dgL*) of $A_{[t]}$.

Lemma 7.4. *Let (A, d) be a $\mathbb{Z}_{(p)}$-free chain complex. Put $\{A_{[t]}, d_{[t]}\} = \mathcal{F}(A)$ and let $\overline{A}_{[t]}$ be as above. Suppose $t > 1$ or $t = 1$ with $d(A) \subseteq pA$.*

(a) *$pA_{[t]} \subseteq \overline{A}_{[t]}$, and $A_{[t]}/\overline{A}_{[t]}$ may be identified with the tth term $E^t(A)$ of the BSS for (A, d).*

(b) *A submodule $N_{[t]} \subseteq A_{[t]}$ is a Bockstein submodule if and only if the composite homomorphism*

$$N_{[t]} \otimes \mathbb{Z}_p \to A_{[t]} \otimes \mathbb{Z}_p = A_{[t]}/(pA_{[t]}) \to A_{[t]}/\overline{A}_{[t]} = E^t(A) \tag{7AE}$$

is an isomorphism.

Proof. (a) $pA_{[t]} = \zeta_t^{t-1}\theta_{t-1}^t(A_{[t]}) \subseteq \overline{A}_{[t]}$. When $\{A_{[t]}, d_{[t]}\} = \mathcal{F}(A)$, then $E^t(A) = A_{[t]}/(pA_{[t-1]} + d_{[t-1]}(A_{[t-1]})) = A_{[t]}/\overline{A}_{[t]}$.

(b) The kernel of $N_{[t]} \to N_{[t]} \otimes \mathbb{Z}_p$ is $pN_{[t]}$, while the kernel of $N_{[t]} \to A_{[t]} \to A_{[t]}/\overline{A}_{[t]}$ is $N_{[t]} \cap \overline{A}_{[t]}$, so (7AE) is one-to-one if and only if (7AD) holds. Likewise, (7AE) is onto if and only if (7AC) holds.

Proposition 7.5. *Let $\{A_{[s]}, d_{[s]}\}_{s_0 \leq s \leq m}$ be a proper csc (resp. csa, csL). Let $\{C_{[s]}, e_{[s]}\}$ be any csc (resp. csa, csL). Let $s_0 < t \leq m$, and suppose that*

$$\{f_{[s]}\}_{s_0 \leq s \leq t-1} : \{A_{[s]}, d_{[s]}\}_{s_0 \leq s \leq t-1} \to \{C_{[s]}, e_{[s]}\}_{s_0 \leq s \leq t-1}$$

is a morphism in CSC (resp. CSA, CSL). Let $N_{[t]}$ be a Bockstein subcomplex (resp. Bockstein sub-dga, sub-dgL) of $A_{[t]}$ and suppose there is a homomorphism (resp. algebra, Lie algebra homomorphism)

$$g : N_{[t]} \to C_{[t]} \tag{7AF}$$

which satisfies

(7AG) $$\theta^t_{t-1}g(x) = f_{[t-1]}\theta^t_{t-1}(x)$$

and

(7AH) $$gd_{[t]}(x) = e_{[t]}g(x),$$

as x runs through a $\mathbb{Z}_{(p)}$-basis (resp. through a generating set as an algebra, Lie algebra) for $N_{[t]}$. Then there is a unique extension of g to a homomorphism

$$f_{[t]} : A_{[t]} \to C_{[t]}$$

for which $\{f_{[s]}\}_{s_0 \le s \le t}$ is a morphism in CSC (resp. CSA, CSL).

Proof. Since (7AG) holds when x is a generator and since g and θ^t_{t-1} and $f_{[t-1]}$ all preserve products or brackets as appropriate, (7AG) holds for any $x \in N_{[t]}$. Similarly, (7AH) holds for any $x \in N_{[t]}$.

We establish next that the formula

(7AI) $$e_{[t]}\zeta^{t-1}_t f_{[t-1]}\theta^t_{t-1}(z) = \zeta^{t-1}_t f_{[t-1]}\theta^t_{t-1}d_{[t]}(z)$$

is valid, for any $z \in A_{[t]}$. Because of (7AC) it suffices to check (7AI) in each of the three separate cases $z \in N_{[t]}$, $z = \zeta^{t-1}_t(x)$, and $z = d_{[t]}\zeta^{t-1}_t(y)$. When $z \in N_{[t]}$, use (7AG) and 7.1(c) and (7AH) to see that

$$\begin{aligned} e_{[t]}\zeta^{t-1}_t f_{[t-1]}\theta^t_{t-1}(z) &= (p)e_{[t]}g(z) = (p)gd_{[t]}(z) \\ &= \zeta^{t-1}_t\theta^t_{t-1}gd_{[t]}(z) = \zeta^{t-1}_t f_{[t-1]}\theta^t_{t-1}d_{[t]}(z). \end{aligned}$$

When $z = \zeta^{t-1}_t(x)$, we use 7.1(c) and 7.1(d) and the fact that $f_{[t-1]}$ is a chain map:

$$\begin{aligned} e_{[t]}\zeta^{t-1}_t f_{[t-1]}\theta^t_{t-1}\zeta^{t-1}_t(x) &= (p)e_{[t]}\zeta^{t-1}_t f_{[t-1]}(x) = \zeta^{t-1}_t\theta^t_{t-1}e_{[t]}\zeta^{t-1}_t f_{[t-1]}(x) \\ &= \zeta^{t-1}_t e_{[t-1]}f_{[t-1]}(x) = \zeta^{t-1}_t f_{[t-1]}d_{[t-1]}(x) = \zeta^{t-1}_t f_{[t-1]}\theta^t_{t-1}d_{[t]}\zeta^{t-1}_t(x). \end{aligned}$$

When $z = d_{[t]}\zeta^{t-1}_t(y)$, we find similarly that both sides of (7AI) are zero.

Next, we define the homomorphism

(7AJ) $$g_1 : \overline{A}_{[t]} \to C_{[t]}$$

by setting $g_1(\zeta^{t-1}_t(x)) = \zeta^{t-1}_t f_{[t-1]}(x)$ and $g_1(d_{[t]}\zeta^{t-1}_t(y)) = e_{[t]}\zeta^{t-1}_t f_{[t-1]}(y)$. We must check that g_1 is well-defined. There are three possibilities to consider. If $\zeta^{t-1}_t(x) = \zeta^{t-1}_t(x')$ then by 7.1(g) $x = x'$, so $g_1(\zeta^{t-1}_t(x)) = g_1(\zeta^{t-1}_t(x'))$, i.e., the definition of g_1 was unambigous. If $\zeta^{t-1}_t(x) = d_{[t]}\zeta^{t-1}_t(y)$, then by 7.1(h) there exits z such that $y = \theta^t_{t-1}(z)$ while $x = \theta^t_{t-1}d_{[t]}(z)$. We find using (7AI) that

$$\begin{aligned} \zeta^{t-1}_t f_{[t-1]}(x) &= \zeta^{t-1}_t f_{[t-1]}\theta^t_{t-1}d_{[t]}(z) = e_{[t]}\zeta^{t-1}_t f_{[t-1]}\theta^t_{t-1}(z) \\ &= e_{[t]}\zeta^{t-1}_t f_{[t-1]}(y), \end{aligned}$$

so $g_1(\zeta_t^{t-1}(x)) = g_1(d_{[t]}\zeta_t^{t-1}(y))$ as needed. Lastly, if $d_{[t]}\zeta_t^{t-1}(y) = d_{[t]}\zeta_t^{t-1}(y')$, then $d_{[t]}\zeta_t^{t-1}(y-y') = 0 = \zeta_t^{t-1}(0)$, so by the previous calculation we have $e_{[t]}\zeta_t^{t-1}f_{[t-1]}(y-y') = \zeta_t^{t-1}f_{[t-1]}(0) = 0$, i.e., $e_{[t]}\zeta_t^{t-1}f_{[t-1]}(y) = e_{[t]}\zeta_t^{t-1}f_{[t-1]}(y')$.

Having defined g_1, we observe that g_1 and g coincide on the intersection of their domains, which is, by (7AD), precisely $pN_{[t]}$. To see this, let $x \in N_{[t]}$. We find that

$$g_1(px) = g_1(\zeta_t^{t-1}(\theta_{t-1}^t(x))) = \zeta_t^{t-1}f_{[t-1]}(\theta_{t-1}^t(x)) = \zeta_t^{t-1}\theta_{t-1}^t g(x) = g(px).$$

Consequently, g_1 and g together determine a well-defined homomorphism

(7AK) $$f_{[t]} : A_{[t]} = N_{[t]} + \overline{A}_{[t]} \to C_{[t]}.$$

This $f_{[t]}$ satisfies

(7AL) $$\theta_{t-1}^t f_{[t]}(z) = f_{[t-1]}\theta_{t-1}^t(z),$$

as needed for $f_{[t]}$ to be a term in a CSC (CSA, CSL) morphism. To see that (7AL) holds, we check it separately for $z \in N_{[t]}$ (evident by (7AH)), for $z \in \mathrm{im}(\zeta_t^{t-1})$ (trivial by 7.1(c)), and for $z \in \mathrm{im}(d_{[t]}\zeta_t^{t-1})$ (by 7.1(d)). Likewise we find that $f_{[t]}d_{[t]} = e_{[t]}f_{[t]}$ and (by definition) that $\zeta_t^{t-1}f_{[t-1]} = f_{[t]}\zeta_t^{t-1}$. Thus $\{f_{[s]}\}_{s_0 \le s \le t}$ is a morphism of csc's.

The only remaining axiom to check is that $f_{[t]}$ preserves multiplication (resp. bracket) in the case of csa's (csL's). To confirm the equation

$$f_{[t]}(xy) = f_{[t]}(x)f_{[t]}(y)$$

there are nine cases, depending upon whether each of x and y belongs to $N_{[t]}$ or to $\mathrm{im}(\zeta_t^{t-1})$ or to $\mathrm{im}(d_{[t]}\zeta_t^{t-1})$. We omit the verification, noting only that the already-established properties of $f_{[t]}$ are essential in the proof, and that some of the nine cases require establishing other cases first and then applying $e_{[t]}$ to both sides.

Corollary 7.6. *Weaken the hypotheses of Proposition 7.5 for csL's as follows. First, instead of assuming that $N_{[t]}$ is a Bockstein sub-dgL, assume merely that it is a Bockstein subalgebra. Second, replace (7AH) by the hypothesis that*

(7AM) $$e_{[t]}g(px) = \zeta_t^{t-1}f_{[t-1]}\theta_{t-1}^t d_{[t]}(x)$$

as x runs through a generator set for $N_{[t]}$. (We continue to suppose that (7AF) preserves brackets and that (7AG) holds on some generator set for $N_{[t]}$.) Then a unique Lie algebra homomorphism $f_{[t]} : A_{[t]} \to C_{[t]}$ exists satisfying the axioms

(7AN) $$f_{[t-1]}\theta_{t-1}^t = \theta_{t-1}^t f_{[t]},\ f_{[t]}\zeta_t^{t-1} = \zeta_t^{t-1}f_{[t-1]},\ f_{[t]}d_{[t]}\zeta_t^{t-1} = e_{[t]}\zeta_t^{t-1}f_{[t-1]},$$

but this $f_{[t]}$ is not necessarily a chain map. If this $f_{[t]}$ satisfies

$$f_{[t]}d_{[t]}(x) = e_{[t]}f_{[t]}(x)$$

as x runs through a generator set for $N_{[t]}$, then $f_{[t]}$ is a chain map and $\{f_{[s]}\}_{s_0 \le s \le t}$ is a csL homomorphism.

Note. It is tempting to rewrite the right-hand side of (7AM), with the help of (7AG), as $\zeta_t^{t-1}\theta_{t-1}^t g d_{[t]}(x)$, which would equal $gd_{[t]}(px)$. This is valid if $N_{[t]}$ happens to be a Bockstein sub-dgL, but in general $d_{[t]}(x)$ will not belong to $N_{[t]}$, so an expression containing $gd_{[t]}(x)$ may have no meaning.

Proof. As before, because (7AG) holds for generators, it holds for any $x \in N_{[t]}$. Similarly, (7AM) holds for any $x \in N_{[t]}$ (a routine but delicate exercise). Having (7AM) on $N_{[t]}$ is enough to make (7AI) true on all of $A_{[t]}$, and (7AI) is enough for the existence of (7AK). Formula (7AL) is true as before, and the other two axioms of (7AN) reiterate how $f_{[t]}$ is constructed. That $f_{[t]}$ preserves brackets is similarly straightforward. The last claim is trivial.

Remark. If (7AG) and (7AM) only hold in dimensions $\le m+1$, then $f_{[t]}$ is well-defined up to and including in dimension m. Also, $f_{[t]}$ will be defined on $N_{[t]} + \mathrm{im}(\zeta_t^{t-1})$ in dimension $m+1$, and we will have

$$e_{[t]} f_{[t]} \zeta_t^{t-1}(x) = f_{[t]} d_{[t]} \zeta_t^{t-1}(x)$$

for $x \in A_{[t-1]}$ when $|x| \le m+1$.

Discussion 7.7. Although csc's, csa's, csL's all work equally well in Proposition 7.5, csL's in particular present us with a subtle difficulty. The problem arises in trying to find a good Bockstein sub-dgL $N_{[t]}$. For this discussion, let us restrict our attention to csL's of the form $\mathcal{F}(L,d)$ where L is a free dgL, $L = \mathcal{L}(V)$ for some $\mathbb{Z}_{(p)}$-free graded module V. If elements x and y in L satisfy $d(y) = p^{t-1}x$, $p \nmid x$, then as we noted in Chapter 5 there are elements $\tau_j(y)$ and $\sigma_j(y)$ satisfying $d(\tau_j(y)) = p^t\sigma_j(y)$. Thus $\tau_j(y)$ and $\sigma_j(y)$ belong to $\mathcal{F}(L,d)_{[t]}$ even though y does not belong to $\mathcal{F}(L,d)_{[t]}$. If y and x survive to $E^{t-1}(L)$, then typically $\tau_j(y)$ and $\sigma_j(y)$ survive to $E^t(L)$, and they are typically Lie algebra generators in $E^t(L)$, so they would (in most cases) need to belong to any Bockstein sub-dgL $N_{[t]}$.

Being forced to include the elements $\tau_j(y)$ and $\sigma_j(y)$ in our $N_{[t]}$ presents two kinds of difficulties. One difficulty is that the $N_{[t]}$'s we get this way are very far from being free algebras. In all cases we shall consider, $N_{[t]}$ will be a free Lie algebra "except for" $\tau_j(y)$'s and $\sigma_j(y)$'s. An $N_{[t]}$ that deviates significantly from being free will cause us trouble at the stage where we are trying to define the Lie algebra homomorphism (7AF) of Proposition 7.5, since once g is defined on the generators of $N_{[t]}$ there will still be lots of relators (i.e. the defining relators in a presentation for $N_{[t]}$) whose g-images must be shown to vanish. At the very least, constructing such a g could be time-consuming and inconvenient. A second difficulty is more subtle, and it arises when the target csL is $\{\pi_*(\Omega X; \mathbb{Z}_{p^t})\}$ for a space X. We shall discuss this issue at some length in Chapter 11.

The way we shall get around these two difficulties is with a compromise. We will not work with the full Lie algebra L, but with a submodule of L which is not necessarily closed under the brackets and differential. For instance, for $L = \mathcal{L}(V)$, if we considered the submodule $\mathcal{L}(V)^{\ell<p}$ consisting only of iterated brackets whose length ℓ is $p-1$ or less, no $\tau_j(y)$'s nor $\sigma_j(y)$'s would occur. Because ζ_t^s and θ_s^t preserve bracket length on

$\mathcal{F}(L,d)$, the notations $\mathcal{F}(L,d)_{[t]}^{\ell<p}$ make sense. Of course, $\mathcal{L}(V)^{\ell<p}$ and $\mathcal{F}(L,d)_{[t]}^{\ell<p}$ are not dgL's, but they do satisfy the dgL axioms whenever those axioms make sense, i.e., whenever both sides of an axiom equation are defined then they are equal. Similarly, a list of homomorphisms $\{f_{[t]}\}$ with sources $\mathcal{F}(L,d)_{[t]}^{\ell<p}$ can be a homomorphism of csL's, in the sense that all axioms are satisfied for which both sides exist. We shall indeed refer to such homomorphisms as csL homomorphisms, leaving it up to the context to clarify what we truly intend. In particular — and this is the main point of this discussion — Proposition 7.5 and Corollary 7.6 are still valid for these submodules of (proper) csL's.

Before departing from the topic of how to work effectively with csL's in spite of their difficulties, let us examine one csL which is of particular interest to us, namely $\mathcal{F}(\overline{L}^k, d)$. Recall that $\overline{L}^k$ and L^k and M^k have a second grading, called $*$-degree. On generators of $\overline{L}^k$ it satisfies $|y_i|_* = 1$ and $|z_i'|_* = 2$ for each $i \geq 2$, and d has $*$-degree $+1$. Since each $|y_i|$ is odd and each $|z_i'|$ is even, we see that even-dimensional elements $\overline{L}^k$ always have $*$-degree ≥ 2. As a result, any $\tau_j(y)$ has $*$-degree $p^j|y|_*+1 \geq 2p+1$ and $|\sigma_j(y)|_* = p^j|y|_*+2 \geq 2p+2$. If we restrict ourselves to $*$-degrees at or below $2p$ the aforementioned problems do not arise. Our notation for the corresponding submodules is $(\overline{L}^k)^{*\leq 2p}$, $\mathcal{F}(\overline{L}^k)^{*\leq 2p}$. In $*$-degrees $\leq 2p$, the free Lie algebra $\overline{L}_s^k = \mathbb{L}\langle y_i, z_i' | i \in I_s\rangle$ is a Bockstein sub-dgL of $\mathcal{F}(\overline{L}^k)$. This fact follows from Proposition 2.7, since the BSS functor commutes with $\mathcal{U}$, in $*$-degree $\leq 2p$.

We turn our attention next to coherent sequences of modules over a csa. It is necessary to have a separate definition and proposition and corollaries for this setting, because the termwise tensor product of two csc's is not in general a csc, and because the termwise endomorphism ring of a csc need not in general be a csa.

Definition 7.8. A csc $\{M_{[t]}, e_{[t]}\}$ is a *coherent sequence of modules* or *csm* over the csa $\{A_{[t]}, d_{[t]}\}$ if there is for each t a homomorphism

$$\psi_{[t]} : A_{[t]} \otimes M_{[t]} \to M_{[t]}$$

satisfying the following axioms, for $s \leq t$:

(a) $\psi_{[t]}$ is a module structure homomorphism, i.e.,

$$\psi_{[t]}(xy \otimes z) = \psi_{[t]}(x \otimes \psi_{[t]}(y \otimes z))$$

(b) $\psi_{[t]}$ is a chain map, i.e., $\psi_{[t]}\varepsilon_{[t]} = e_{[t]}\psi_{[t]}$, where

$$\varepsilon_{[t]}(x \otimes z) = d_{[t]}(x) \otimes z + (-1)^{|x|} x \otimes e_{[t]}(z).$$

(c) $\theta_s^t \psi_{[t]}(x \otimes z) = \psi_{[s]}(\theta_s^t(x) \otimes \theta_s^t(z))$

(d) $\zeta_t^s \psi_{[s]}(\theta_s^t(x) \otimes z) = \psi_{[t]}(x \otimes \zeta_t^s(z))$

(e) $\zeta_t^s \psi_{[s]}(x \otimes \theta_s^t(z)) = \psi_{[t]}(\zeta_t^s(x) \otimes z)$

A homomorphisms of csc's

$$\{f_{[t]}\} : \{M_{[t]}, e_{[t]}\} \to \{M'_{[t]}, e'_{[t]}\}$$

in which both source and target are csm's over a csa $\{A_{[t]}, d_{[t]}\}$ is called a ***homomorphism of csm's*** over $\{A_{[t]}, d_{[t]}\}$ if for each t

$$f_{[t]}\psi_{[t]} = \psi'_{[t]}(1 \otimes f_{[t]}) : \; A_{[t]} \otimes M_{[t]} \to M'_{[t]}.$$

For a fixed csa $\{A_{[t]}, d_{[t]}\}$, the csm's and csm homomorphisms over it obviously form a category.

Proposition 7.9. *Let $\{A_{[j]}, d_{[j]}\}$ be a proper csa, and let $N_{[t]}$ be a Bockstein sub-dga of $A_{[t]}$. Let $\{M_{[j]}, e_{[j]}\}$ be a csc, and suppose for $1 \le s < t$ that there are module structure homomorphisms*

$$\psi_{[s]} : \; A_{[s]} \otimes M_{[s]} \to M_{[s]}$$

that make $\{M_{[j]}, e_{[j]}\}_{1\le j<t}$ into a csm over $\{A_{[j]}, d_{[j]}\}_{1\le j<t}$. Suppose further that there is a module structure homomorphism

$$g : \; N_{[t]} \otimes M_{[t]} \to M_{[t]}$$

which satisfies these three formulas, as w runs through a set of algebra generators for $N_{[t]}$ and $z \in M_{[t]}$, $y \in M_{[t-1]}$:

(7AO) $$g\varepsilon_{[t]}(w \otimes z) = e_{[t]}g(w \otimes z);$$

(7AP) $$\theta^t_{t-1}g(w \otimes z) = \psi_{[t-1]}(\theta^t_{t-1}(w) \otimes \theta^t_{t-1}(z));$$

(7AQ) $$g(w \otimes \zeta^{t-1}_t(y)) = \zeta^{t-1}_t\psi_{[t-1]}(\theta^t_{t-1}(w) \otimes y).$$

Then g extends uniquely to a module structure homomorphism

$$\psi_{[t]} : \; A_{[t]} \otimes M_{[t]} \to M_{[t]}$$

which makes $\{M_{[s]}, e_{[s]}\}_{1\le s\le t}$ into a csm over $\{A_{[s]}, d_{[s]}\}_{1\le s\le t}$.

Proof. The proof mirrors very closely the proof of Proposition 7.5, so only a scant outline will be provided.

First, observe that (7AO)-(7AQ) hold for any $w \in N_{[t]}$. Second, establish the formula which is an analog of (7AI),

(7AR) $$e_{[t]}\zeta^{t-1}_t\psi_{[t-1]}(\theta^t_{t-1} \otimes \theta^t_{t-1}) = \zeta^{t-1}_t\psi_{[t-1]}(\theta^t_{t-1} \otimes \theta^t_{t-1})\varepsilon_{[t]},$$

as homomorphisms from $A_{[t]} \otimes M_{[t]}$ to $M_{[t]}$. Third, use properness and (7AR) to establish that

$$g_1 : \; (\zeta^{t-1}_t(A_{[t-1]}) + d_{[t]}\zeta^{t-1}_t(A_{[t-1]})) \otimes M_{[t]} \to M_{[t]}$$

is well-defined, where

$$g_1(\zeta^{t-1}_t(w) \otimes z) = \zeta^{t-1}_t\psi_{[t-1]}(w \otimes \theta^t_{t-1}(z)),$$

and

$$g_1(d_{[t]}\zeta_t^{t-1}(w)\otimes z) = e_{[t]}\zeta_t^{t-1}\psi_{[t-1]}(w\otimes\theta_{t-1}^t(z))$$
$$-(-1)^{|w|}\zeta_t^{t-1}\psi_{[t-1]}(w\otimes\theta_{t-1}^t e_{[t]}(z)).$$

Fourth, g and g_1 together determine a well-defined $\psi_{[t]}$. Fifth, prove casewise that $\psi_{[t]}$ is a chain map, and note that axiom 7.8(e) holds by definition. Sixth, verify that 7.8(c) and 7.8(d) hold (three cases for x). Finally, check 7.8(a) (nine cases for x and y).

Remark. Sometimes the csc $M_{[t]}$ is functorial on some category $\mathcal{X}$, i.e., $M_{[t]} = M_{[t]}(X)$ is a functor, for each t. For instance, this is the case if $M_{[t]} = \pi_*(X;\mathbb{Z}_{p^t})$ and $\mathcal{X} = \text{TOP}$. If $e_{[t]}$ and ζ_t^{t-1} and θ_{t-1}^t are natural transformations, and if $g(w\otimes -)$ is a natural transformation for each generator w of $N_{[t]}$, then the csm structure constructed in Proposition 7.9 is also functorial on $\mathcal{X}$.

Lemma 7.10. *Let $\{M_{[t]}, e_{[t]}\}$ be csc having the property that $p^t M_{[t]} = 0$ for each t. For example, suppose $\{M_{[t]}, e_{[t]}\}$ is $\{H_*(M, e;\mathbb{Z}_{p^t})\}$ for some chain complex (M, e) or that p is odd and $\{M_{[t]}, e_{[t]}\}$ is $\{\pi_*(X;\mathbb{Z}_{p^t})\}$ for some 1-connected space X.*

(a) *Let (C, d) be a $\mathbb{Z}_{(p)}$-free chain complex, and let $\{f_{[t]}\} : \mathcal{F}(C, d) \to \{M_{[t]}, e_{[t]}\}$ be a csc homomorphism. If $d(C) \subseteq pC$, or more generally if the module $C'_{[1]}$ of Lemma 7.2(d) belongs to $\ker(f_{[1]})$, then $\{f_{[t]}\}$ may be factored in CSC,*

$$\mathcal{F}(C,d) \xrightarrow{q} \{H_*(C,d;\mathbb{Z}_{p^t})\} \xrightarrow{\{\overline{f}_{[t]}\}} \{M_{[t]}, e_{[t]}\},$$

where $\underline{q}$ is the surjection of (7B).

(b) *Let (A, d) be a $\mathbb{Z}_{(p)}$-free dga. If $d(A) \subseteq pA$, or more generally if the submodule $A'_{[1]}$ of Lemma 7.2(d) annihilates $M_{[1]}$, then the module structure homomorphisms factor through homomorphisms*

$$\overline{\psi}_{[t]} : H_*(A, d;\mathbb{Z}_{p^t})\otimes M_{[t]} \to M_{[t]},$$

i.e., there is an induced structure on $\{M_{[t]}, e_{[t]}\}$ as a csm over $\{H_*(A, d;\mathbb{Z}_{p^t})\}$.

Proof. Straightforward, by Lemma 7.2(d) and some easy calculations.

Remark. The condition on $\{M_{[t]}, e_{[t]}\}$ in Lemma 7.10, namely $p^t M_{[t]} = 0$ for all t, is equivalent to the simpler-looking condition $pM_{[1]} = 0$. If $pM_{[1]} = 0$ then $p^t M_{[t]} = \zeta_t^1\theta_1^t(pM_{[t]}) \subseteq \zeta_t^1(pM_{[1]}) = 0$.

Virtual Structures

The second unit of Chapter 7 examines the setting where a csL $\{M_{[t]}, e_{[t]}\}$ also happens to be a csm over a csa $\{A_{[t]}, d_{[t]}\}$. Frequently the two structures interact in an interesting way. The interaction is of such a nature that it gives a formula for an action on a bracket in terms of brackets of actions. Although Theorem 7.15 and the work leading up to it are rather technical, the payoff is an extremely efficient machine for constructing homomorphisms that are simultaneously homomorphisms of csL's and of csm's.

Definition 7.11. A *diagonalized csa*, or *Dcsa*, is a csa $\{A_{[t]}, d_{[t]}\}$ together with $\mathbb{Z}_{(p)}$-homomorphisms $\chi_{imt} : A_{[t]} \to A_{[m]} \otimes A_{[m]}$ for $i = 0$ or 1 and $1 \le m \le t-i$. The degree of each χ_{omt} is zero and the degree of each χ_{1mt} is $+1$. We impose no conditions on χ_{imt} except that when $w \in (A_{[t]})_+$, i.e., $|w| > 0$, $\overline{\chi}_{0tt}(w) = \chi_{0tt}(w) - (w \otimes 1 + 1 \otimes w)$ should lie in $(A_{[t]})_+ \otimes (A_{[t]})_+$. Another notation for $\chi_{imt}(w)$ will be $\Sigma'_{i,m,t} w' \otimes w''$. With this notation we are suppressing both the index over which the summation runs and the dependence of w' and w'' upon i and m. Let $\{A_{[t]}, d_{[t]}, \chi_{imt}\}$ be a Dcsa, and suppose $\{M_{[t]}, e_{[t]}\}$ is a csL which is also a csm over $\{A_{[t]}, d_{[t]}\}$. For $w \in A_{[t]}$, $x \in M_{[t]}, y \in M_{[t]}$, write

$$\text{(7BA)} \qquad S_{it}(w,x,y) = \sum_{m=0}^{t-i} \Sigma'_{i,m,t} (-1)^{|w''||x|} \zeta_t^m([\psi_{[m]}(w' \otimes \theta_m^t(x)), \psi_{[m]}(w'' \otimes \theta_m^t(y))]).$$

We call $\{M_{[t]}, e_{[t]}\}$ a *csL over* $\{A_{[t]}, d_{[t]}, \chi_{imt}\}$ if for all $w \in A_{[t]}$, $x \in M_{[t]}, y \in M_{[t]}$,

$$\begin{aligned} \psi_{[t]}(w \otimes [x,y]) =& S_{0t}(w,x,y) + d_{[t]} S_{1t}(w,x,y) \\ \text{(7BB)} \qquad & - (-1)^{|w|} S_{1t}(w, d_{[t]}(x), y) - (-1)^{|w|+|x|} S_{1t}(w,x,d_{[t]}(y)). \end{aligned}$$

Remark. If a Dcsa is going to permit non-trivial csL's over it, the homomorphisms χ_{imt} will have to satisfy a great many formulas. For instance, $\chi_{0mt}(w_2 w_1)$ is completely determined by $\chi_{0mt}(w_2)$ and $\chi_{0mt}(w_1)$, the Jacobi identities in $M_{[t]}$ impose further constraints, and so on. By saying "diagonalized csa" we are actually dancing around the edges of a concept for "coherent sequence of Hopf algebras", but a good definition for a csH should really have all of these constraints on the χ_{imt}'s built into it. For our purposes, only the form of (7BB) will matter at all; the properties of w' and w'' or the relations among them will be irrelevant. In particular, for each fixed $w \in A_{[t]}$, (7BB) is an identity, reminiscent of a polynomial identity on an assocative ring, that holds simultaneously for all elements x and y of $M_{[t]}$.

Let us describe an important source of Dcsa's and csL's over them.

Lemma 7.12.

(a) *Let (A,d) and (A',d') be $\mathbb{Z}_{(p)}$-free chain complexes. Then*

$$\mathcal{F}(A \otimes A')_{[t]} = B_{[t]} + \sum_{m=1}^{t-1} (p^{t-m} B_{[m]} + \delta^t(p^{t-m} B_{[m]})),$$

where $B_{[m]} = \mathcal{F}(A)_{[m]} \otimes \mathcal{F}(A')_{[m]}$. Here δ^t is $p^{-t}d$ as in Lemma 7.2(b).

(b) *Let (A,d,Δ) be a $\mathbb{Z}_{(p)}$-free dgH. We associate to $\mathcal{F}(A)$ the csa homomorphism $\mathcal{F}(\Delta) : \mathcal{F}(A) \to \mathcal{F}(A \otimes A)$. There are homomorphisms $\chi_{imt} : \mathcal{F}(A)_{[t]} \to \mathcal{F}(A)_{[m]} \otimes \mathcal{F}(A)_{[m]}$ that diagonalize $\mathcal{F}(A)$ satisfying*

$$\mathcal{F}(\Delta)_{[t]}(w) = w \otimes 1 + 1 \otimes w + \bar{\chi}_{0tt}(w) + \sum_{m=1}^{t-1} (p^{t-m}\chi_{0mt}(w) + \delta^t(p^{t-m}\chi_{1mt}(w))).$$

(c) *Let (M,d) be a $\mathbb{Z}_{(p)}$-free dgL and let $L \subseteq M$ be a differential Lie ideal (i.e., $d(L) \subseteq L$ and $[L,M] \subseteq L$). Then $\mathcal{F}(L,d)$ is a csL over $(\mathcal{F}(\mathcal{U}M,d), \chi_{imt})$, where the χ_{imt} are given in part (b). The action of $w \in \mathcal{F}(\mathcal{U}M)_{[t]}$ on $\mathcal{F}(L)_{[t]}$ is $\psi_{[t]}(w \otimes x) = \mathrm{ad}(w)(x)$.*

Proof. (a) Both sides are biadditive functors of A and A'. Check the formula when each of A and A' is merely $(0 \to \mathbb{Z}_{(p)} \to 0)$ or $(\mathbb{Z}_{(p)} \xrightarrow{p^s} \mathbb{Z}_{(p)})$. Because A and A' are $\mathbb{Z}_{(p)}$-free, they are direct sums of the two types of building blocks, by Lemma 5.4.

(b) Follows from (a).

(c) The action of $w \in \mathcal{U}M$ on $[x,y] \in L$ is given by going either way around the commuting diagram

$$\begin{array}{ccccc} \mathcal{U}M \otimes L \otimes L & \xrightarrow{\Delta\otimes 1\otimes 1} & \mathcal{U}M \otimes \mathcal{U}M \otimes L \otimes L & \xrightarrow{1\otimes\tau\otimes 1} & \mathcal{U}M \otimes L \otimes \mathcal{U}M \otimes L \\ \Big\downarrow {\scriptstyle 1\otimes[\ ,\ ]} & & & & \Big\downarrow {\scriptstyle \psi\otimes\psi} \\ \mathcal{U}M \otimes L & \xrightarrow{\psi=\mathrm{ad}} & L & \xleftarrow{[\ ,\ ]} & L \otimes L \end{array}$$

Comparing the two ways gives the formula (7BB).

Example 7.13. Putting $M = M^k$ in Lemma 7.12(c), the diagonal satisfies $\Delta(b^{p^s}) = \sum_{j=0}^{p^s} (p^s|j) b^j \otimes b^{p^s - j}$ for $0 \le s \le k$. From this we can derive what (7BB) says for $w = b^{p^s}$ when $t \le r+s$ and $x \in \mathcal{F}(L)_{[t]}$, $y \in \mathcal{F}(L)_{[t]}$. Letting a raised dot denote the action of $\mathcal{F}(\mathcal{U}M^k)$ on $\mathcal{F}(L)_{[t]}$, we have

$$\text{(7BC)} \qquad b^{p^s} \cdot [x,y] = [b^{p^s} \cdot x, y] + [x, b^{p^s} \cdot y] + \sum_{j=1}^{p^s-1} ((p^s|j)p^{m-t})\zeta_t^m [b^j \cdot x, b^{p^s-j} \cdot y].$$

In (7BC) the m in the summation stands for the integer such that $p^m = \gcd(p^r j, p^t)$. The corresponding formulas for $c_s \cdot [x,y]$ are a little more complicated. To obtain them, apply $d_{[t]}$ to both sides of (7BC) and then solve for the term $c_s \cdot [x,y]$. In our applications, the differential ideal L of M^k will always be $\overline{L}^k$.

In general, to give a diagonal structure on a csa $\{A_{[t]}, d_{[t]}\}$, it suffices to tell for each t what $\chi_{imt}(w)$ are as w runs through a generating set for a Bockstein subalgebra of $A_{[t]}$. For the example of $\mathcal{F}(\mathcal{U}M^k)$, a Bockstein sub-dga of $\mathcal{F}(\mathcal{U}M^k)_{[t]}$ is R_s^k, where $s = \max(0, t-r)$. (This follows from Corollary 4.9). A set of generators for R_s^k is $\{b^{p^s}, c_s, c_{s+1}, \dots, c_k\}$. In view of (4F), knowing χ_{imt} on the set $\{b^{p^s}, c_s, a_{s+1}, \dots, a_k\}$ would also determine $\chi_{imt}(w)$ uniquely for all $w \in \mathcal{F}(\mathcal{U}M^k)_{[t]}$. It is more convenient to tell or check the χ_{imt} on a a_q than on c_q, because each a_q for $s < q \le k$ is *primitive*, i.e., $\chi_{imt}(a_q) = 0$ for $m < t$ and $\bar{\chi}_{0tt}(a_q) = 0$. Equivalently, $\psi_{[t]}(a_q \otimes -)$ is a derivation on any csL over it, i.e.,

$$\text{(7BD)} \qquad a_q \cdot [x,y] = [a_q \cdot x, y] + (-1)^{|x|}[x, a_q \cdot y], \quad s < q \le k.$$

Suppose we have a csL homomorphism between two csL's and we want to know whether it is a csm homomorphism over a certain csa. If the csa is a Dcsa and both csL's are csL's over it, then it suffices to check on generators that f commutes with the actions. We make this precise with our next proposition.

Proposition 7.14. *Let $\{f_{[s]}\}_{1\le s\le t} : \{M_{[s]}, e_{[s]}\}_{1\le s\le t} \to \{M'_{[s]}, e'_{[s]}\}_{1\le s\le t}$ be a homomorphism of csL's and suppose that each of these csL's is a csL over some Dcsa $\{A_{[s]}, d_{[s]}, \chi_{im_s}\}_{1\le s\le t}$. Suppose further that $\{f_{[s]}\}_{1\le s\le t-1}$ is a homomorphism of csm's over $\{A_{[s]}, d_{[s]}\}$. Suppose $A_{[t]}$ has a Bockstein subalgebra $P_{[t]}$ and $M_{[t]}$ has a Bockstein subalgebra $N_{[t]}$, and that*

$$\text{(7BE)} \qquad \psi'_{[t]}(w\otimes f_{[t]}(x)) = f_{[t]}\psi_{[t]}(w\otimes x)$$

as w and x run through generator sets for $P_{[t]}$ and $M_{[t]}$ respectively. Then $f_{[t]}$ is a module homomorphism, so $\{f_{[s]}\}_{1\le s\le t}$ is a homomorphism of csm's.

Proof. As usual we consider nine cases for w and x. For this proof it is convenient to embed the proof in an induction on the dimension of w. We assume inductively that (7BE) holds whenever $|w| < l$, and we prove that (7BE) is true for w of dimension l. The first of the nine cases deals with $w \in P_{[t]}$ and $x \in N_{[t]}$. If w is decomposable in $P_{[t]}$ it is easy, so suppose w is in our generator set. Using (7BA) we find

$$f_{[t]}S_{it}(w,x,y) = S_{it}(w, f_{[t]}(x), f_{[t]}(y))$$

because $\{f_{[m]}\}_{m<t}$ is a csm homomorphism and a csL homomorphism. Then (7BB), together with $f_{[t]}$ being a chain map, tells us that (7BE) holds for $[x,y]$ if it holds separately for x and for y. This explains the inductive step that permits us to increment the bracket length of x in (7BE) while holding w fixed. This completes the first case. The remaining eight cases are routine.

We move now into a discussion and proof of Theorem 7.15, which will occupy the remainder of this unit of Chapter 7. Let $\{M_{[s]}, e_{[s]}\}_{1\le s\le t}$ and $\{M'_{[s]}, e'_{[s]}\}_{1\le s\le t}$ be two csL's over a Dcsa $\{A_{[s]}, d_{[s]}, \chi_{im_s}\}_{1\le s\le t}$. Suppose that

$$\{f_{[s]}\}_{1\le s\le t-1} : \{M_{[s]}, e_{[s]}\} \to \{M'_{[s]}, e'_{[s]}\}$$

is a csL homomorphism that is also a csm homomorphism over $\{A_{[s]}, d_{[s]}\}_{1\le s\le t-1}$. We are going to give sufficient conditions for the existence and uniqueness of a homomorphism $f_{[t]}$ that makes $\{f_{[s]}\}_{1\le s\le t}$ into both a csL and a csm homomorphism (we shall sometimes write this as "$f_{[t]}$ is a csL/csm homomorphism").

We assume that both $A_{[t]}$ and $M_{[t]}$ are locally finite, i.e., finitely generated as $\mathbb{Z}_{(p)}$-modules in each dimension. We also assume that $A_{[t]}$ is $\mathbb{Z}_{(p)}$-free and has a Bockstein sub-dga $(P_{[t]}, d_{[t]})$, while $M_{[t]}$ has a Bockstein sub-dgL $(N_{[t]}, e_{[t]})$ which is free as a Lie algebra, i.e., $N_{[t]} = \mathcal{L}(V)$ for some $\mathbb{Z}_{(p)}$-free positively graded $\mathbb{Z}_{(p)}$-module V. We shall frequently think of $V = \mathcal{L}^1(V)$ as a (non-differential) submodule of $N_{[t]}$.

We can write a presentation for $P_{[t]}$ in the form $P_{[t]} = TU/ < r_\beta >$, where U is a free graded $\mathbb{Z}_{(p)}$-module, TU is the tensor algebra on U, and $< r_\beta >$ denotes the two-sided ideal of TU generated by a set $\{r_\beta\}$ of relators. Of course, $\psi_{[t]}(r_\beta\otimes x) = 0$ for any $x \in M_{[t]}$, for any r_β in this set. An idea that is central to Theorem 7.15 is that there may be further elements of $P_{[t]}$ that are "virtual relators" or "relators modulo higher terms." Instead of $\psi_{[t]}(w\otimes -)$ being identically zero, as it is when w is a true relator, there is a reasonable formula for $\psi_{[t]}(w\otimes -)$.

To make this precise, we digress to present a definition. Call a *formal linear expression* or *fle* for $M_{[t]}$ (of degree l) any (homogeneous of degree l) linear combination of terms of the form

$$\text{(7BF)} \qquad \mathrm{ad}(y_t)\psi_{[t]}(z_t \otimes -) \quad \text{with} \quad |y_t| > 0$$

and of the form

$$\text{(7BG)} \qquad ((e_{[t]}))\zeta_t^m \,\mathrm{ad}(y_m)\psi_{[m]}(z_m \otimes \theta_m^t((e_{[t]}))(-)), \quad 1 \le m \le t-1,$$

where $y_j \in \mathcal{U}M_{[j]}$ and $z_j \in A_{[j]}$. In (7BG), the notation $((e_{[t]}))$ means that, at each spot in each such term, $e_{[t]}$ may or may not be part of the formula. A formal linear expression is a linear homomorphism defined on $x \in M_{[t]}$ and we can denote it as $F(x)$. Given a fle $F(x)$ and a csL/csm homomorphism $\{g_{[t]}\} : \{M_{[s]}, e_{[s]}\} \to \{M'_{[s]}, e'_{[s]}\}$, let $(g_{[*]}F)(x)$ denote the corresponding fle for $M'_{[t]}$, where we obtain $g_{[*]}F$ by replacing each y_j by $\mathcal{U}g_{[j]}(y_j)$ and each $e_{[t]}$ by $e'_{[t]}$. Define a *virtual relator* for $M_{[t]}$ to be any $w \in A_{[t]}$ for which $\psi_{[t]}(w \otimes -)$ is a formal linear expression, i.e., for which there exists a fle $F(x)$ such that

$$\psi_{[t]}(w \otimes x) = F(x) \quad \text{for all} \quad x \in M_{[t]}.$$

It should be obvious that $F(\psi_{[t]}(w \otimes -))$ is a fle whenever $w \in A_{[t]}$ and F is a fle, since the $\theta_j^t(w)$'s become absorbed into the z_j's. Somewhat less obvious is the fact that $\psi_{[t]}(w \otimes F(-))$ is a fle; we must use (7BB) to derive this. These two facts show that the virtual relators form a two-sided ideal in $A_{[t]}$. Also, $e_{[t]}F(-) - (-1)^l F(e_{[t]}(-))$ is a fle of degree $l-1$ if F is a fle of degree l. As a result, the ideal of virtual relators is a differential ideal.

Another property of the ideal of virtual relators (henceforth *VR*'s) is that it contains all of $\mathrm{im}(\zeta_t^{t-1}(w)) + \mathrm{im}(d_{[t]}\zeta_t^{t-1})$, if $t > 1$. Actually, this is trivial. The csm axioms tell us that

$$\psi_{[t]}(\zeta_t^{t-1}(w) \otimes x) = \zeta_t^{t-1}\psi_{[t-1]}(w \otimes \theta_{t-1}^t(x)),$$

i.e., $\psi_{[t]}(\zeta_t^{t-1}(w) \otimes -)$ coincides with the fle $\zeta_t^{t-1}\psi_{[t-1]}(w \otimes \theta_{[t-1]}^t(-))$. So $\zeta_t^{t-1}(w)$ is a VR. Since the ideal of VR's is closed under $d_{[t]}$, $d_{[t]}\zeta_t^{t-1}(w)$ is a VR too. By special convention all of $pA_{[1]}$ is considered to consist of VR's as well, since $t = 1$ is not covered by the above argument.

When w is a VR, notice that

$$\psi_{[t]}(w \otimes M_{[t]}) \subseteq \mathcal{L}^{\ge 2}(V) + \overline{M}_{[t]},$$

where the overbar notation means $\overline{M}_{[t]} = \zeta_t^{t-1}(M_{[t-1]}) + e_{[t]}\zeta_t^{t-1}(M_{[t-1]})$ like it does in (7AC) and (7AD). To prove this, consider separately the three cases for $x \in M_{[t]}$; we omit this routine check, but we note that the condition $|y_t| > 0$ in (7BF) is crucial for this step. Let N denote $V \otimes \mathbb{Z}_p$ and notice by (7AC) and (7AD) and the modular law that $M_{[t]}/(\mathcal{L}^{\ge 2}(V) + \overline{M}_{[t]}) = N_{[t]}/(\mathcal{L}^{\ge 2}(V) + pN_{[t]}) = V/pV = N$. What this means is that $\psi_{[t]}(w \otimes M_{[t]})$ is zero when projected down to N, for any VR w. Since $\psi_{[t]}(pP_{[t]} \otimes M_{[t]}) \subseteq pM_{[t]} = \zeta_t^{t-1}\theta_{t-1}^t(M_{[t]}) \subseteq \overline{M}_{[t]}$ and since

$$\psi_{[t]}(P_{[t]} \otimes (\mathcal{L}^{\ge 2}(V) + \overline{M}_{[t]})) \subseteq \mathcal{L}^{\ge 2}(V) + \overline{M}_{[t]}$$

((7BB) is used to show this), $\psi_{[t]}$ induces an action of the quotient algebra $P = P_{[t]}/(pP_{[t]} + R_{[t]})$ on N, where $R_{[t]}$ is the two-sided ideal consisting of all VR's in $P_{[t]}$.

Theorem 7.15 will relate the existence of a csL/csm homomorphism $f_{[t]}$ as above to certain constructions associated with a presentation for N as a graded P-module.

First, we consider a presentation for P as a $\mathbf{Z}_p$-algebra. Recalling the surjection $TU \to P_{[t]}$ with kernel $< r_\beta >$, we know there is a presentation for P of the form $T\overline{U}/\overline{R}$, where $\overline{U} = U \otimes \mathbf{Z}_p$ and $\overline{R}$ is a two-sided ideal. The mod p reductions $\overline{r}_\beta$ of the elements r_β obviously belong to $\overline{R}$. Let $\{\overline{r}_\alpha\}$ denote a minimal set of elements that, together with $\{\overline{r}_\beta\}$, generate $\overline{R}$ as a two-sided ideal in $T\overline{U}$. In other words, $\{\overline{r}_\alpha\}$ is minimal having the property that $\overline{R} =< \overline{r}_\alpha > + < \overline{r}_\beta >$. Then each $\overline{r}_\alpha$ is the mod p reduction of some element $r_\alpha \in TU$ whose image in $P_{[t]}$ is a virtual relator. We denote the image of r_α in $P_{[t]}$ also as r_α, and we denote the corresponding fle as F_{r_α}. Thus

$$\psi_{[t]}(r_\alpha \otimes x) = F_{r_\alpha}(x) \tag{7BH}$$

for every $x \in M_{[t]}$. Clearly, F_{r_α} has degree $|r_\alpha|$.

Consider next the start of a minimal free left resolution for N as a P-module,

$$0 \leftarrow N \overset{\overline{\delta}^{(0)}}{\longleftarrow} \bigoplus_{i\in I} P\overline{x}_i \overset{\overline{\delta}^{(1)}}{\longleftarrow} \bigoplus_{j\in J} P\overline{y}_j, \tag{7BI}$$

where $P\overline{x}_i$ denotes a free cyclic left P-module generated by the (graded) element $\overline{x}_i$, and likewise for $P\overline{y}_j$. As i runs through the indexing set I for the minimal P-module generators for N, we choose (any) lifts $x_i \in M_{[t]}$ for $\overline{\delta}^{(0)}(\overline{x}_i) \in N = V \otimes \mathbf{Z}_p$. We refer to a set $\{x_i\} \subseteq M_{[t]}$, whose projection to N generates N as a P-module, as a set of ***virtual generators*** (sometimes *VG*'s) for $M_{[t]}$. Similarly, as j runs through the indexing set J for the minimal syzygies of N, we write

$$\overline{\delta}^{(1)}(\overline{y}_j) = \sum_{i\in I} \overline{b}_{ij}\overline{x}_i$$

with $\overline{b}_{ij} \in P$, and we choose lifts $b_{ij} \in A_{[t]}$ for the coefficients $\overline{b}_{ij}$. For each index $j \in J$, the element

$$\sum_{i\in I} \psi_{[t]}(b_{ij} \otimes x_i) \in M_{[t]}$$

projects to zero in $N = M_{[t]}/(\mathcal{L}^{\geq 2}(V) + \overline{M}_{[t]})$. Consequently there is for each $j \in J$ an equation that holds in $M_{[t]}$,

$$\sum_{i\in I} \psi_{[t]}(b_{ij} \otimes x_i) = z_j^0 + \zeta_t^{t-1}(z_j^1) + e_{[t]}\zeta_t^{t-1}(z_j^2), \tag{7BJ}$$

where $z_j^0 \in \mathcal{L}^{\geq 2}(V)$ and $z_j^1 \in M_{[t-1]}$ have the same dimension $|\overline{y}_j|$ as $\overline{y}_j$, and $z_j^2 \in M_{[t-1]}$ with $|z_j^2| = |\overline{y}_j| + 1$. We call an equation of the form (7BJ) a ***virtual syzygy*** (sometimes *VS*) for $M_{[t]}$.

Taken together, the virtual relators, virtual generators, and virtual syzygies comprise a ***virtual structure*** for $M_{[t]}$ as an $A_{[t]}$-module. A ***virtual structure*** for a csL over a Dcsa provides a virtual structure at the tth stage, for each t.

Let $\{\tilde{x}_i\} \in M'_{[t]}$ be artibrary elements in the target csL whose dimensions match up with those of $\{\overline{x}_i\}$, i.e., $|\tilde{x}_i| = |\overline{x}_i|$ for $i \in I$. We think of the x_i's and b_{ij}'s as fixed; even though some choice was exercised in defining them, this choice "comes out in the wash." The choice of $\{\tilde{x}_i\}$ however will be highly deliberate, and we are finally ready to explain what Theorem 7.15 is all about. Theorem 7.15 will answer the following: What conditions must $\{\tilde{x}_i\}$ satisfy, if there is to be a unique csL/csm homomorphism $f_{[t]}$ that satisfies

$$f_{[t]}(x_i) = \tilde{x}_i?$$

It turns out that $f_{[t]}$ is automatically unique if it exists. The best way to describe the sufficient conditions for $f_{[t]}$ to exist is to suppose that $f_{[t]}$ has been defined and is unique in dimensions below m for some m, and to present the conditions that must be fulfilled in order for $f_{[t]}$ to have a unique csL/csm extension over the m-dimensional component $(M_{[t]})_m$ of $M_{[t]}$. Another way to say this is that we are giving the obstructions to extending $f_{[t]}$ uniquely over $(M_{[t]})_m$.

Summary of Notations and Conventions. Before stating Theorem 7.15, let us summarize the notations and conventions that it deals with. We begin with a csL/csm homomorphism

$$\{f_{[s]}\}_{1\leq s\leq t-1} : \{M_{[s]}, e_{[s]}\} \to \{M'_{[s]}, e'_{[s]}\},$$

where both source and target are csL's over a Dcsa $\{A_{[s]}, d_{[s]}, \chi_{im_s}\}_{1\leq s\leq t}$. The source csL is assumed to be proper and the Dcsa is assumed to be both $\mathbb{Z}_{(p)}$-free and proper. We have a Bockstein sub-dga $P_{[t]}$ of $A_{[t]}$, and we have a quotient algebra P of $P_{[t]} \otimes \mathbb{Z}_p$ which is presented as $P = T\overline{U}/ < \overline{r}_\alpha, \overline{r}_\beta >$, with $\overline{U}$ locally finite. The $\overline{r}_\beta$'s are mod p reductions of elements that vanish in $P_{[t]}$, hence each r_β annihilates any module over $A_{[t]}$, while the $\overline{r}_\alpha$'s are mod p reductions of virtual relators r_α for $M_{[t]}$. We have a Bockstein sub-dgL $N_{[t]} = \mathcal{L}(V)$ of $M_{[t]}$ which is assumed to be free as a Lie $\mathbb{Z}_{(p)}$-algebra on the $\mathbb{Z}_{(p)}$-free graded module V. We observe that $M_{[t]}/(\mathcal{L}^{\geq 2}(V) + \overline{M}_{[t]})$ equals $V \otimes \mathbb{Z}_p$, a module we denote as N, and that N is a left P-module. We choose minimal generators $\{\overline{x}_i\}_{i\in I}$ for N as a left P-module, and we choose minimal primary syzygies $\{\overline{y}_j\}_{j\in J}$ as well. We lift the $\overline{x}_i$'s to virtual generators $x_i \in M_{[t]}$ and we lift the primary syzygies to the virtual syzygies (7BJ). Lastly, we have elements $\{\tilde{x}_i\}_{i\in I}$ in the target csL $\{x_i\}$, and we are wondering under what conditions an $f_{[t]}$ that works below some dimension m can be extended to dimension m, such that $f_{[t]}(x_i) = \tilde{x}_i$.

Theorem 7.15. *With notations and conventions as above, suppose a csL/csm homomorphism $f_{[t]} : (M_{[t]})_{<m} \to (M'_{[t]})$ having $f_{[t]}(x_i) = \tilde{x}_i$ has been defined below dimension m. By this we mean that $f_{[t]}$ commutes with everything it should, provided that the element that $f_{[t]}$ is being applied to has dimension less than m. Then $f_{[t]}$ extends to a unique csL/csm homomorphism below dimension $m+1$, still satisfying $f_{[t]}(x_i) = \tilde{x}_i$, if the following four conditions are fulfilled.*

(i) $\theta^t_{t-1}(\tilde{x}_i) = f_{[t-1]}\theta^t_{t-1}(x_i)$ *for any index $i \in I$ having $|x_i| \leq m+1$, i.e., the elements $\tilde{x}_i$ should be compatible with $f_{[t-1]}$.*

(ii) $e'_{[t]}(\tilde{x}_i) = f_{[t]}e_{[t]}(x_i)$ *for any index $i \in I$ having $|x_i| \leq m$ and*

$$(p)e'_{[t]}(\tilde{x}_i) = \zeta_t^{t-1} f_{[t-1]}\theta^t_{t-1}e_{[t]}(x_i) \tag{7BK}$$

for indices i having $|x_i| = m+1$, i.e., the elements $\tilde{x}_i$ should have the correct boundaries.

(iii) *For each index $j \in J$ for which $|\overline{y}_j| \le m$, the equation corresponding to (7BJ)*

(7BL) $$\sum_{i\in I} \psi'_{[t]}(b_{ij} \otimes \tilde{x}_i) = f_{[t]}(z_j^0) + \zeta_t^{t-1} f_{[t-1]}(z_j^1) + e'_{[t]}\zeta_t^{t-1} f_{[t-1]}(z_j^2)$$

is true in $M'_{[t]}$. Here z_j^0 has dimension m, and $f_{[t]}(z_j^0)$ refers to the obvious extenstion of $f_{[t]}|_{V_{<m}}$ over $\mathcal{L}^{\ge 2}(V_{<m})$, which contains z_j^0.

(iv) *For each index α having $|r_\alpha| < m$, r_α is a virtual relator for $M'_{[t]}$. Specifically,*

(7BM) $$\psi'_{[t]}(r_\alpha \otimes x) = (f_{[*]}F_{r_\alpha})(x),$$

an identity that holds for all $x \in M'_{[t]}$. (F_{r_α} is defined in (7BH).)

Proof. The plan for the proof is as follows. We begin by choosing "good" $\mathbb{Z}_p$-bases for P and for N. Combinatorial ring theory techniques are utilized for this purpose. We lift the basis for N to $M_{[t]}$, obtaining a $\mathbb{Z}_{(p)}$-module W. The Lie algebra $\mathcal{L}(W)$ generated by W is shown to be a Bockstein subalgebra of $M_{[t]}$, and the hypotheses of Corollary 7.6 are verified for it. Corollary 7.6 constructs a csL homomorphism $f_{[t]}$ in dimensions $\le m$ and a "partial" $f_{[t]}$ in dimension $m+1$. This much of the proof uses conditions (i) and (ii). Using (iii) and (iv), we show that this $f_{[t]}$ is a csm homomorphism below dimension $m+1$.

As just mentioned, we apply some standard ring-theoretic methods to choose $\mathbb{Z}_p$-bases for P and for N. These methods are described in [A2] and [Be]. We are doing this because we will need to work explicitly with basis elements, and we need a good way to keep track of them. A good place to start is our presentation $P_{[t]} = TU/R_{[t]}$ for $P_{[t]}$ as a $\mathbb{Z}_{(p)}$-module, where $R_{[t]}$ consists of all virtual relators in TU, and its mod p reduction, the presentation $P = T\overline{U}/\overline{R}$ for P. For convenience we think of $M_{[t]}$ and $M'_{[t]}$ as modules over TU rather than over $P_{[t]}$, so the notation $\psi_{[t]}(u \otimes x)$ will make sense whenever $x \in M_{[t]}$ and $u \in TU$. Let U^1 be a homogeneous $\mathbb{Z}_{(p)}$-basis for U and let $\tilde{U}$ denote the free associative monoid on U^1. We call the elements of $\tilde{U}$ "monomials." Clearly, $\tilde{U}$ sits inside of TU as a $\mathbb{Z}_{(p)}$-basis.

The set $\tilde{U}$ also bijects with a $\mathbb{Z}_p$-basis for $T\overline{U}$, and we denote by $\overline{u}$ the basis element obtained by projecting $u \in \tilde{U}$ to $T\overline{U}$. We impose a total ordering on U^1 and we give to $\tilde{U}$ the corresponding lexicographic ordering. Because U is locally finite, there are only finitely many monomials to be compared lexicographically, in each dimension. Let $q : TU \to P_{[t]}$ and $\overline{q} : T\overline{U} \to P$ be the quotient homomorphisms.

Let $\tilde{P} = \{x \in \tilde{U} | \overline{q}(\overline{x})$ is not in the span of the $\overline{q}$-images of the monomials lexicographically lower than $x\}$. By [A1, Lemma 1.1] $\tilde{P}$ is a $\mathbb{Z}_p$-basis for P. Similarly, $\tilde{U} \times I$ is a $\mathbb{Z}_{(p)}$-basis for the free left TU-module $\bigoplus_{i\in I}(TU)x_i$, which surjects onto the $\mathbb{Z}_p$-module N via $\delta^{(0)}$ (see (7BI)). We choose a linear ordering for I and we let $\tilde{U} \times I$ be lexicographically ordered. Put $\delta_0(u,i) = \overline{\delta}_0(\overline{q}(\overline{u})\overline{x}_i)$ and define

$$\tilde{N} = \{\text{pairs } (u,i) \in \tilde{U} \times I | \delta_0(u,i) \text{is not in the span of stuff lower than } (u,i)\}.$$

Then $\tilde{N} \subseteq \tilde{P} \times I$, and we let $\overline{a}_{u,i}$ denote $\delta_0(u,i)$. The set $\{\overline{a}_{u,i} | (u,i) \in \tilde{N}\}$ is a $\mathbb{Z}_p$-basis for N. Let

(7BN) $$a_{u,i} = \psi_{[t]}(u \otimes x_i) \in M_{[t]};$$

the image of $a_{u,i}$ upon projecting to N is of course $\overline{a}_{u,i}$. There is no reason to expect $a_{u,i}$ to belong to V in general. Let $W = \mathrm{Span}\,\{a_{u,i} | (u,i) \in \tilde{N}\}$, a $\mathbb{Z}_{(p)}$-free submodule of $M_{[t]}$.

The relationship between $\mathcal{L}(W)$ and $\mathcal{L}(V) = N_{[t]}$ is rather interesting. Because $\mathcal{L}(V)$ is a Bockstein sub-dgL, the surjection $\mathcal{L}(V) \to M_{[t]}/\overline{M}_{[t]}$ has kernel $\mathcal{L}(V) \cap \overline{M}_{[t]} = p\mathcal{L}(V)$, i.e.,

$$\mathcal{L}(N) = \mathcal{L}(V) \otimes \mathbb{Z}_p \to M_{[t]}/\overline{M}_{[t]}$$

is an isomorphism. The inclusion of W into $M_{[t]}$ induces $\mathcal{L}(W) \to M_{[t]}$, and the resulting Lie algebra homomorphism

$$\mathcal{L}(W) \to M_{[t]}/\overline{M}_{[t]}$$

must be onto, because N is in its image. Since the image is $\mathcal{L}(N)$, however, the kernel $\mathcal{L}(W) \cap \overline{M}_{[t]}$ must be precisely $p\mathcal{L}(W)$. We have proved (7AC) and (7AD) for $\mathcal{L}(W)$, i.e., $\mathcal{L}(W)$ is a Bockstein subalgebra (but not in general a sub-dgL) of $M_{[t]}$.

We are finally ready to begin the construction of $f_{[t]}$ in dimension m. To do so we shall apply Corollary 7.6, using $\mathcal{L}(W)$ as our Bockstein subalgebra. We must define $g : \mathcal{L}(W) \to M'_{[t]}$ and check (7AG) and (7AM) for the generating set $\{a_{u,i} | (u,i) \in \tilde{N}\}$. Define g by $g = f_{[t]}$ below dimension m and

(7BO) $$g(a_{u,i}) = \psi'_{[t]}(u \otimes \tilde{x}_i) \text{ for } (u,i) \in \tilde{N},\ |a_{u,i}| \geq m.$$

This clearly satisfies $g(x_i) = \tilde{x}_i$, since the minimality of (7BI) guarantees that $(1,i) \in \tilde{N}$ for each $i \in I$. Using (i), we have

$$\begin{aligned}\theta g(a_{u,i}) &= \psi'_{[t-1]}(\theta(u) \otimes \theta(\tilde{x}_i)) = \psi'_{[t-1]}(\theta(u) \otimes f_{[t-1]}\theta(x_i)) \\ &= f_{[t-1]}\psi_{[t-1]}(\theta(u) \otimes \theta(x_i)) = f_{[t-1]}\theta(a_{u,i}),\end{aligned}$$

where we are writing θ for θ^t_{t-1}. This establishes (7AG) in all dimensions where condition (i) holds, in particular, in dimensions $\leq m+1$.

Next we check that $e'_{[t]}g(a_{u,i}) = f_{[t]}e_{[t]}(a_{u,i})$ if $|a_{u,i}| = m$. To simplify the calculation we omit all subscripts of $[t]$, and we write $\pm$ for $(-1)^{|u|}$. Using (ii), we find for $|a_{u,i}| = m$ that

$$\begin{aligned}e'g(a_{u,i}) &= \psi'(d(u) \otimes \tilde{x}_i) \pm \psi'(u \otimes e'(\tilde{x}_i)) = \psi'(d(u) \otimes f(x_i) \pm u \otimes fe(x_i)) \\ &= f\psi(d(u) \otimes x_i \pm u \otimes e(x_i)) = fe\psi(u \otimes x_i) = fe(a_{u,i}),\end{aligned}$$

where we have exploited the inductive assumption that $f\psi$ does equal $\psi'(1 \otimes f)$ when applied to something of dimension $m-1$. Since we do not yet know this for dimension m, the result when $|a_{u,i}| = m+1$ is much weaker. Fortunately, (7AM) is true in dimension $m+1$, as we now verify. It is true by hypothesis (ii) for $a_{1,i}$, so assume $|u| \geq 1$. Writing ζ for ζ^{t-1}_t and skipping several steps,

$$\begin{aligned}\zeta f_{[t-1]}\theta e(a_{u,i}) &= \zeta\psi_{[t-1]}(\theta d(u) \otimes f_{[t-1]}\theta(x_i) \pm \theta(u) \otimes f_{[t-1]}\theta e(x_i)) \\ &= \zeta\psi_{[t-1]}(\theta \otimes \theta)(1 \otimes g)(d(u) \otimes x_i \pm u \otimes e(x_i)) \\ &= \zeta\theta\psi_{[t]}(1 \otimes g)(d(u) \otimes x_i \pm u \otimes e(x_i)) = (p)e'g(a_{u,i}).\end{aligned}$$

By Corollary 7.6 and its subsequent Remark, this suffices to construct a well-defined bracket-preserving $f_{[t]}$ satisfying the axioms (7AN). Since $f_{[t]}$ does commute with differentials in dimension m, it is a dgL homomorphism below dimension $m+1$, and we also have some results about it in dimension $m+1$. Specifically, $f_{[t]}$ is defined on $\mathcal{L}^{\geq 2}(W)+\mathrm{im}(\zeta_t^{t-1})$ in dimension $m+1$. We shall need this in order to apply $f_{[t]}$ to the m-dimensional expressions of the form $e_{[t]}\zeta_t^{t-1}(\ )$ that are coming up.

It remains only to show that

$$f_{[t]}\psi_{[t]}(w\otimes x)=\psi'_{[t]}(w\otimes f_{[t]}(x)),\ \text{if } |w|+|x|=m. \tag{7BP}$$

By the argument of Proposition 7.14, we already know that (7BP) is true if $x\in\mathcal{L}^{\geq 2}(W)+\overline{M}_{[t]}$. In particular, it is true on any $w\otimes x$ that is a multiple of p, since $pM_{[t]}=\zeta_t^{t-1}\theta_{t-1}^t(M_{[t]})\subseteq \mathrm{im}(\zeta_t^{t-1})\subseteq \overline{M}_{[t]}$. Only the case $x\in W$ remains to be done. Proving (7BP) when $x=a_{u,l}$ reduces easily to showing

$$f_{[t]}\psi_{[t]}(v\otimes x_l)=\psi'_{[t]}(v\otimes \tilde{x}_l)\ \text{for } (v,l)\in\tilde{U}\times I,\ |v|+|x_l|=m. \tag{7BQ}$$

So we shall focus on proving (7BQ).

It is time to examine what conditions (iii) and (iv) are telling us. Suppose $s\in R_{[t]}$, i.e., s is a virtual relator. Then s is a linear combination of summands of the form $ur_\alpha v$ or $ur_\beta v$. The fle $F_s(x)$ corresponding to s equals the same linear combination of the corresponding fle's for these summands, which are compositions $\psi_{[t]}(u\otimes -)\circ F_{r_\alpha}(-)\circ\psi_{[t]}(v\otimes -)$. (Of course, the F_{r_β}'s are zero.) Since these compositions involve elements of $A_{[t]}$ only, they will commute with $f_{[*]}$, and we can conclude from (7BM) that s is a virtual relator for $M'_{[t]}$. Specifically,

$$\psi'_{[t]}(s\otimes x)=(f_{[*]}F_s)(x) \tag{7BR}$$

for any $x\in M'_{[t]}$. In particular, for any $i\in I$ we have $\psi_{[t]}(s\otimes x_i)=F_s(x_i)\in\mathcal{L}^{\geq 2}(W)+\overline{M}_{[t]}$, whence

$$f_{[t]}\psi_{[t]}(s\otimes x_i)=f_{[t]}(F_s(x_i))=(f_{[*]}F_s)(\tilde{x}_i)=\psi'_{[t]}(s\otimes\tilde{x}_i). \tag{7BS}$$

In a similar fashion, (7BL) tells us that

$$\sum_{i\in I}\psi'_{[t]}(b_{ij}\otimes\tilde{x}_i)=f_{[t]}(z_j),$$

where we can afford to write z_j for the right-hand side of (7BJ) now that the notation $f_{[t]}(\ )$ has been defined for m-dimensional elements of $\mathcal{L}^{\geq 2}(W)+\overline{M}_{[t]}$. We are also taking the liberty of allowing b_{ij} to denote lifts of $\overline{b}_{ij}$ to TU and not just to $P_{[t]}$. (Although in the definition of a virtual syzygy b_{ij} is allowed to be any lift of $\overline{b}_{ij}$ in $A_{[t]}$, without loss of generality we can now suppose $b_{ij}\in P_{[t]}$, because $P_{[t]}+\overline{A}_{[t]}=A_{[t]}$ by (7AC) and when $b'\in\overline{A}_{[t]}$, $\psi_{[t]}(b'\otimes x_i)$ can be absorbed into the terms $\zeta_t^{t-1}(z_j^1)$ and $e_{[t]}\zeta_t^{t-1}(z_j^2)$ of (7BJ).) For any $w\in TU$ we have

$$\begin{aligned} f_{[t]}\psi_{[t]}(w\sum_{i\in I}b_{ij}\otimes x_i)&=f_{[t]}\psi_{[t]}(w\otimes z_j)=\psi'_{[t]}(w\otimes f_{[t]}(z_j))\\ &=\psi'_{[t]}(w\sum_{i\in I}b_{ij}\otimes\tilde{x}_i)\ \text{if } |w|+|\overline{y}_j|=m.\end{aligned} \tag{7BT}$$

Now we will pull it all together. The module N has a presentation (7BI) as

$$N = (T\overline{U} \otimes \mathbb{Z}_p(I))/(\overline{R} \otimes \mathbb{Z}_p(I) + \mathrm{im}(\overline{\delta}^1)),$$

where $\mathbb{Z}_p(I)$ means the $\mathbb{Z}_p$-span of the graded set $\{x_i\}_{i\in I}$. Let $(v,l) \in \tilde{U} \times I$ have dimension m, and write

$$\overline{a}_{v,l} = \Sigma \overline{c}_{u,h}\overline{a}_{u,h},$$

where the sum ranges over all $(u,h) \in \tilde{N}$ and $\overline{c}_{u,h} \in \mathbb{Z}_p$. Lifting to $TU \otimes \mathbb{Z}_{(p)}(I)$, we have

$$\text{(7BU)} \qquad v \otimes x_l = (p)z(v,l) + \sum c_{u,h}(u \otimes x_h) + \sum_{i\in I} s_i \otimes x_i + \sum_{j\in J} w_j \sum_{i \in I} b_{ij} \otimes x_i,$$

where $s_i \in R_{[t]}$ and $w_j \in TU$ and the $c_{u,h} \in \mathbb{Z}_{(p)}$ project to $\overline{c}_{u,h}$ and we have no control whatsoever over $z(v,l) \in TU \otimes \mathbb{Z}_{(p)}(I)$. Apply $f_{[t]}\psi_{[t]}$ to both sides of (7BU). The terms $c_{u,h}(u \otimes x_l)$ are carried to $c_{u,h}\psi'_{[t]}(u \otimes \tilde{x}_h)$ by (7BO), and we have already seen that $f_{[t]}\psi_{[t]}$ and $\psi'_{[t]}(1 \otimes f_{[t]})$ coincide on the other three terms, by (7BP), (7BS), (7BT). The conclusion is that (7BQ) is true. This completes the proof.

Three Examples with Discussion

We give a complete description of a virtual structure for $\mathcal{F}(L)$, when L is $\overline{L}^k$, N^{kl}, or N^{klm}. Assume for the remainder of Chapter 7 that $p \geq 3$.

Example 7.16. An important example of a csL over a Dcsa that does have virtual relators is $\mathcal{F}(\overline{L}^k)$ as a csL over $\mathcal{F}(\mathcal{U}M^k)$. Let $1 \leq t \leq r+k$ and put $s = \max(0, t-r)$, so that our Bockstein sub-dga for the Dcsa is

$$R^k_s = \mathbb{Z}_{(p)}\langle b^{p^s}, c_s, c_{s+1}, \ldots, c_k\rangle / \langle pc_{j+1} - \epsilon(b^{p^j}, c_j)\rangle.$$

Our Bockstein sub-dgL for $\mathcal{F}(\overline{L}^k)$, below $*$-degree $2p+1$, is

$$\overline{L}^k_s = \mathbb{Z}_{(p)}\langle y_i, z'_i | i \in I_s\rangle.$$

The element $w = b^{p^s}c_s - c_s b^{p^s}$ is a virtual relator because, by Lemma 4.4(c),

$$\text{(7CA)} \qquad \psi_{[t]}(w \otimes -) = [y_{2p^s}, -] + \sum_{i=2}^{2p^s-1} (\mu_i p^{m-s})\zeta_t^{r+m}[y_i, \psi_{[r+m]}(b^{2p^s-i} \otimes \theta^t_{r+m}(-))],$$

where in the summation m denotes the largest integer not exceeding s for which $i \in I_m$. (The summation in (7CA) is vacuously zero if $s = 0$). The element $(c_s)^2$ is a virtual relator too. If $s = 0$ (i.e. if $t \leq r$) this is trivial since $c_0^2 = a_0^2$ and $\psi_{[t]}(c_0^2 \otimes -) = [z_2, -]$. If $s > 0$, $d_{[r+s]}(b^{p^s}c_s - c_s b^{p^s}) = -2(c_s)^2$ is a VR because the set of VR's is closed under $d_{[t]}$. Since p is odd, $(c_s)^2$ is a VR.

Similarly, $w_q = c_{q+1} - c_q b^{p^q(p-1)}$ is a VR for q in the range $s \leq q \leq k-1$, by Lemma 4.4(d). When $s = 0$ Lemma 4.4(d) tell us that

$$\text{(7CB)} \qquad \psi_{[t]}(w_q \otimes -) = [y_{p^{q+1}}, -] + \sum_{i=2}^{p^{q+1}-1} \nu_i[y_i, \psi_{[t]}(b^{p^{q+1}-i} \otimes -)]$$

When $s > 0$ some of the terms in the summation in (7CB) would not make sense, but Lemma 4.4(d) assures us that the coefficients $(\nu_i p^{m-s})$ are integers in

$$\text{(7CC)} \quad \psi_{[t]}(w_q \otimes -) = [y_{p^{q+1}}, -] + \sum_{i=2}^{p^{q+1}-1} (\nu_i p^{m-s}) \zeta_i^{r+m} [y_i, \psi_{[r+m]}(b^{p^{q+1}-i} \otimes \theta^t_{r+m}(-))],$$

where m again denotes the largest integer not exceeding s for which $i \in I_m$. Our quotient ring is

$$P = (R^k_s / \langle [b^{p^s}, c_s], (c_s)^2, c_{q+1} - c_q b^{p^q(p-1)} \rangle) \otimes \mathbb{Z}_p = \mathbb{Z}_p[b^{p^s}] \otimes \Lambda(c_s),$$

an extremely simple commutative ring.

The module N associated to $\mathcal{F}(\overline{L}^k)_{[t]}$ has $\{y_i, z'_i | i \in I_s\}$ as a $\mathbb{Z}_p$-basis, where we allow y_j and z'_j to denote the projections of these elements to N as well as the elements themselves. The action of P on N is given by

$$b^{p^s} \cdot y_i = \begin{cases} 0 & \text{if } i + p^s = p^m \text{ for some } m \le k \\ y_{i+p^s} & \text{otherwise} \end{cases}$$

$$b^{p^s} \cdot z'_i = \begin{cases} 0 & \text{if } i = p^m \text{ for some } m \le k \\ z'_{i+p^s} & \text{otherwise} \end{cases}$$

$$c_s \cdot y_i = z'_{i+p^s}, \; c_s \cdot z'_i = 0.$$

An easy calculation shows that the minimal generating set for N as a P-module is

$$\text{(7CD)} \qquad \{y_{2p^s}, \; z'_{2p^s}, \; y_{p^s+1}, \; \ldots \;, \; y_{p^k}\},$$

so (7CD) also describes a set of virtual generators for $\mathcal{F}(\overline{L}^k)_{[t]}$. The minimal syzygies for $\mathcal{F}(\overline{L}^k)_{[t]}$ (we give the $\overline{\delta}^{(1)}$-images) are

$$\text{(7CE)} \qquad \begin{array}{l} (c_s)z'_{2p^s}, \; (c_s)y_{2p^s} - (b^{p^s})z'_{2p^s}, \; (b^{p^s})^{p-2} y_{2p^s}, \\ (b^{p^s})^{p^2-p} y_{p^s+1}, \; \ldots \;, \; (b^{p^s})^{p^{k-s}-p^{k-s-1}} y_{p^k-1}. \end{array}$$

This shows that very few equations need to be checked in order to construct a csL/csm homomorphism out of $\mathcal{F}(\overline{L}^k)$, below $*$-degree $2p+1$. We shall construct precisely such a homomorphism, with target $\{\pi_*(\Omega E_k; \mathbb{Z}_{p^t})\}$, in Chapters 10 and 13.

Discussion/Example 7.17. We discuss at some length the example of (N^{kl}, d) from Chapter 5. By (5R) and Lemma 7.12(c), $\mathcal{F}(N^{kl})$ is a csL over $\mathcal{F}(\mathcal{U}M^k \amalg \mathcal{U}M^k)$. We will describe a virtual structure for $\mathcal{F}(N^{kl})$ over $\mathcal{F}(\mathcal{U}M^k \amalg \mathcal{U}M^l)$. In the process, we shall illustrate some tricks that can be used to obtain or describe a virtual structure.

We continue to employ the abbreviations VR, VG, VS for virtual relator, virtual generator, and virtual syzygy.

Let us begin with a quite general observation. Suppose

$$\text{(7DA)} \qquad 0 \to N \to M \xrightarrow{\phi} M/N \to 0$$

is a short exact sequence of $\mathbb{Z}_{(p)}$-free dgL's, which induces a structure for $\mathcal{F}(N)$ as a csL over $\mathcal{F}(\mathcal{U}M)$ by Lemma 7.12(c). Then anything belonging to the image of the composition

$$\text{(7DB)} \qquad \bar{e}_{[t]} : \mathcal{F}(N)_{[t]} \to \mathcal{F}(M)_{[t]} \to \mathcal{F}(\mathcal{U}M)_{[t]}$$

is automatically a VR, for the following reason. Let $u \in \mathcal{F}(N)_{[t]}$. Then $\psi_{[t]}(\bar{e}_{[t]}(u) \otimes w) = \mathrm{ad}(\bar{e}_{[t]}(u))(w) = [u, w]$ for $w \in \mathcal{F}(N)_{[t]}$. This says that $\psi_{[t]}(\bar{e}_{[t]}(u) \otimes -)$ coincides with the fle $[u, -]$ on $\mathcal{F}(N)_{[t]}$, i.e., $\bar{e}_{[t]}(u)$ is a VR.

Now let us take up the specific example of $\mathcal{F}(N^{kl})_{[t]}$. Let s always denote $\max(0, t-r)$. We restrict our attention to t in the range $1 \le t \le r + \min(k, l)$. Observe that $P_{[t]} = R^k_s \amalg R^l_s$ is a Bockstein sub-dga of $\mathcal{F}(\mathcal{U}M^k \amalg \mathcal{U}M^l)_{[t]}$. Let $\langle VR's \rangle$ be our informal notation for the two-sided ideal of $\mathcal{F}(\mathcal{U}M^k \amalg \mathcal{U}M^l)_{[t]}$ consisting of all VR's for $\mathcal{F}(N^{kl})_{[t]}$.

Let $g = g_{kl} : \mathcal{L}(\mathcal{U}M^k_+ \otimes \mathcal{U}M^l_+) \to N^{kl}$ be the isomorphism constructed in Theorem 5.12. Let V be the submodule $V = (R^k_s)_+ \otimes (R^l_s)_+$ of $\mathcal{U}M^k_+ \otimes \mathcal{U}M^l_+$. Let $N_{[t]} = g(\mathcal{L}(V))$; it is a free sub-dgL of $\mathcal{F}(N^{kl})_{[t]}$. The left action of $P_{[t]}$ on V is the "obvious" one, i.e., $e^{(1)}(x)$ for $x \in R^k_s$ acts as $x \otimes 1$ on $(R^k_s)_+ \otimes (R^l_s)_+$, and $e^{(2)}(y)$ for $y \in R^l_s$ acts as $1 \otimes y$.

We have not said that $N_{[t]}$ is a Bockstein sub-dgL of $\mathcal{F}(N^{kl})_{[t]}$, which is one of the assumptions going into Theorem 7.15. When $s = 0$, $N_{[t]}$ is N^{kl}, and N^{kl} is a Bockstein sub-dgL of $\mathcal{F}(N^{kl})_{[t]} = N^{kl}$. Unfortunately, when $s > 0$, $N_{[t]}$ fails to meet the definition of a Bockstein sub-dgL. Formula (7AD) holds for this $N_{[t]}$, but (7AC) misses due to $\tilde{\tau}_j$'s and $\tilde{\sigma}_j$'s. When it comes to N^{kl}, unlike $\bar{L}^k$, we cannot simply duck the issue by declaring it doesn't matter because we are only interested in low $*$-degrees anyhow. The dgL N^{kl} has important even-dimensional elements in $*$-degree zero, so it has $\tilde{\tau}_j$'s and $\tilde{\sigma}_j$'s in the very relevant $*$-degrees 1 and 2. For now we are going to act as if $N_{[t]}$ is a Bockstein sub-dgL of $\mathcal{F}(N^{kl})_{[t]}$. We will spend considerable energy clearing up this issue in Chapter 11 before we begin to use the virtual structure in Chapter 12. For now we ask the reader to proceed on faith that this apparent gap in our reasoning is bridgeable.

The first thing we shall compute is a minimal set of virtual generators for $N_{[t]}$. We are considering the action of $P_{[t]} \otimes \mathbb{Z}_p$ on $V \otimes \mathbb{Z}_p$, as described above. The minimal generators for $(R^k_s)_+ \otimes (R^l_s)_+ \otimes \mathbb{Z}_p$ as a left $P_{[t]} \otimes \mathbb{Z}_p$-module are given by $S^{(1)} \otimes S^{(2)}$, where $S^{(1)} = \{b_s, c_s, c_{s+1}, \ldots, c_k\}$ is a minimal set of generators for $(R^k_s)_+$ as a left (R^k_s)-module, and $S^{(2)} = \{b_s, c_s, \ldots, c_l\}$ is a minimal set of generators for $(R^l_s)_+$ as a left (R^l_s)-module. We proceed to describe a set of VG's in $\mathcal{F}(N^{kl})_{[t]}$.

For $s = \max(0, t-r)$, $1 \le t \le r + \min(k, l)$, the list of VG's for $N_{[t]}$ always contains the four elements, which we dub the *basic* VG's:

$$\text{(7DC)} \qquad g(c_s \otimes c_s),\ g(b_s \otimes c_s),\ g(c_s \otimes b_s),\ g(b_s \otimes b_s),$$

where $g = g_{kl}$ as above. In addition to the basic VG's, the list contains the following elements, where $s < m \le k$ and $s < q \le l$ (if one or both of these ranges is empty the corresponding VG's are simply absent from the list):

$$\begin{aligned} &\mathrm{ad}^{(2)}(b_s)(a^{(1)}_m),\ \mathrm{ad}^{(2)}(c_s)(a^{(1)}_m),\ \mathrm{ad}^{(1)}(b_s)(a^{(2)}_q), \\ \text{(7DD)} \qquad &\mathrm{ad}^{(1)}(c_s)(a^{(2)}_q),\ [a^{(1)}_m, a^{(2)}_q]. \end{aligned}$$

We look at the VR's next. Return momentarily to the general case where (7DA) induces our virtual structure, and recall the homomorphism $\bar{e}_{[t]}$ of (7DB). Suppose (7DA) has the property that $E^j(\mathcal{U}\phi)$ is a surjection of BSS terms, for $j < t$. We saw in Lemma 2.3 that $E^j(\mathcal{U}N)$ is the HAK of $E^j(\mathcal{U}\phi)$, for $j \le t$. Suppose further that there is some dgL $N_{[t]} \subseteq \mathcal{F}(N)_{[t]}$ having the property that the composition along the middle row of

$$\text{(7DE)}\qquad \begin{array}{ccccc} N_{[t]} \otimes \mathbf{Z}_p & \longrightarrow & \mathcal{F}(N)_{[t]} \otimes \mathbf{Z}_p & & \\ \downarrow & & \downarrow & & \\ \mathcal{U}(N_{[t]}) \otimes \mathbf{Z}_p & \longrightarrow & \mathcal{F}(\mathcal{U}N)_{[t]} \otimes \mathbf{Z}_p & \longrightarrow & E^t(\mathcal{U}N) \\ & & \downarrow & & \downarrow \\ & & \mathcal{F}(\mathcal{U}M)_{[t]} \otimes \mathbf{Z}_p & \longrightarrow & E^t(\mathcal{U}M) \end{array}$$

is an isomorphism. Then the diagram (7DE) shows that all of $\mathrm{HAK}(E^t(\mathcal{U}\phi))$ is generated as an algebra by $\bar{e}_{[t]}$-images, hence all of $\mathrm{HAK}(E^t(\mathcal{U}\phi))_+$ consists of VR's.

This general situation does apply to $N = N^{kl}$ for the particular $N_{[t]}$ we have been considering. All of

$$\mathrm{HAK}(E^t(\mathcal{U}M^k \amalg \mathcal{U}M^l \to \mathcal{U}M^k \otimes \mathcal{U}M^l))_+$$

consists of VR's. Within this ideal are all the commutators $[x^{(1)}, y^{(2)}]$ as $x^{(1)}$ runs through the generators $S^{(1)}$ of R^k_s and $y^{(2)}$ runs through the generators $S^{(2)}$ of R^l_s. Consequently $\mathcal{F}(\mathcal{U}M^k \amalg \mathcal{U}M^l)_{[t]}/\langle VR's\rangle$ equals either $R^k_s \otimes R^l_s \otimes \mathbf{Z}_p$ or a quotient of $R^k_s \otimes R^l_s \otimes \mathbf{Z}_p$. The latter possibility is ruled out because $V \otimes \mathbf{Z}_p = (R^k_s)_+ \otimes (R^l_s)_+ \otimes \mathbf{Z}_p$ has to be a module over this ring. Deduce that we have found all the VR's, and that

$$\text{(7DF)}\qquad \mathcal{F}(\mathcal{U}M^k \amalg \mathcal{U}M^l)_{[t]}/\langle VR's\rangle = R^k_s \otimes R^l_s \otimes \mathbf{Z}_p,$$

a ring we henceforth denote as P. In particular,

$$\psi_{[t]}([b^{(1)}_s, b^{(2)}_s] \otimes -) = [g(b_s \otimes b_s), -] + (\text{terms of the form (7BG)}),$$

and a similar formula holds for each of the other three basic VG's.

The virtual syzygies are found by resolving $V \otimes \mathbf{Z}_p$ as a module over $P = R^k_s \otimes R^l_s \otimes \mathbf{Z}_p$. Since $V = (R^k_s)_+ \otimes (R^l_s)_+$, one way to find a resolution is to tensor together, a resolution for $(R^k_s)_+$ as a left (R^k_s)-module, and a resolution for $(R^l_s)_+$ as a left (R^l_s)-module. We omit the calculation and give next an explicit description of all the minimal VS's.

There is one minimal VS for each occurence of $a^{(1)}_m$ and one for each occurrence of $a^{(2)}_q$ among the VG's (7DD). To obtain the VS corresponding to an occurence of $a^{(i)}_m$ in a VG, where $i = 1$ or 2, multiply the VG by p and simplify, using one of these two equations:

$$\text{(7DG)}\qquad pa^{(i)}_m = \mathrm{ad}^{(i)}(b^{p^{m-s-1}(p-1)})(a^{(i)}_{m-1}) \text{ if } m \ge s+2 \text{ or } s = 0, i = 1,2 \quad ;$$

$$\text{(7DH)}\qquad pa^{(i)}_m = \mathrm{ad}^{(i)}(b^{p-2}_s)(y^{(i)}_{2p^s}) \text{ if } m-1 = s > 0,\ i = 1,2 \quad .$$

Here $y^{(i)}_{2p^s}$ denotes the $e^{(i)}$-image of $y_{2p^s} \in \bar{L}^k$ in $M^k \amalg M^l$. By "simplify" we mean using the csL axioms, the VS's for $\bar{L}^k$, and the VR's, to get the expression into the form of a true VS.

For example, if $m \geq s+2$, the first non-basic VG gives us

$$\begin{aligned} 0 &\equiv (p)\,\mathrm{ad}^{(2)}(b_s)(a_m^{(1)}) = \mathrm{ad}^{(2)}(b_s)\,\mathrm{ad}^{(1)}(b_s^{p^{m-s-1}(p-1)})(a_{m-1}^{(1)}) \\ &\equiv \mathrm{ad}^{(1)}(b^{p^{m-s-1}(p-1)})\,\mathrm{ad}^{(2)}(b_s)(a_{m-1}^{(1)}), \end{aligned}$$

where the congruences are modulo

$$\text{(7DI)} \qquad \mathrm{ad}^{(1)}(R_s^k)g(b_s \otimes b_s) + \mathrm{im}(\zeta_t^{t-1}) + \mathrm{im}(d_{[t]}\zeta_t^{t-1}).$$

The congruence is true as a result of the already-established VR $[b_s^{(1)}, b_s^{(2)}]$, which allows us to commute $\mathrm{ad}^{(1)}(b_s)$ and $\mathrm{ad}^{(2)}(b_s)$, modulo (7DI).

When $m = s+1$ the simplifying process needs one more equation in order to get to a true VS. It is the identity

$$\text{(7DJ)} \qquad \mathrm{ad}^{(2)}(b_s)(y_{2p^s}^{(1)}) \equiv \mathrm{ad}^{(1)}(c_s)g(b_s \otimes b_s) - \mathrm{ad}^{(1)}(b_s)g(c_s \otimes b_s),$$

$$\text{modulo } \mathrm{im}(\zeta_t^{t-1}) + \mathrm{im}(d_{[t]}\zeta_t^{t-1}),$$

i.e., for a certain $x_s^{12} \in \mathcal{F}(N^{kl})_{[t-1]}$ and $w_s^{12} \in \mathcal{F}(N^{kl})_{[t-1]}$ we have

$$\text{(7DK)} \qquad \begin{aligned} &-\mathrm{ad}^{(1)}(c_s)g(b_s \otimes b_s) + \mathrm{ad}^{(1)}(b_s)g(c_s \otimes b_s) + \mathrm{ad}^{(2)}(b_s)(y_{2p^s}^{(1)}) \\ &= \zeta_t^{t-1}(x_s^{12}) + d_{[t]}\zeta_t^{t-1}(w_s^{12}) \end{aligned}$$

Similarly, for a certain $y_s^{12}, z_s^{12} \in \mathcal{F}(N^{kl})_{[t-1]}$ we have

$$\text{(7DL)} \qquad \begin{aligned} &\mathrm{ad}^{(1)}(c_s)g(b_s \otimes c_s) - \mathrm{ad}^{(1)}(b_s)g(c_s \otimes c_s) + \mathrm{ad}^{(2)}(c_s)(y_{2p^s}^{(1)}) \\ &= \zeta_t^{t-1}(z_s^{12}) + d_{[t]}\zeta_t^{t-1}(y_s^{12}) \end{aligned}$$

When $t \leq r$ (i.e. $s = 0$) we may take $w_s^{12} = x_s^{12} = y_s^{12} = z_s^{12} = 0$, and then (7DK) and (7DL) merely restate a familiar Jacobi identity. We postpone the proofs of (7DK) and (7DL) for $t > r$ until the end of Discussion 7.19.

Lastly, the VS that arises from an occurence of $a_q^{(2)}$ in the list (7DD) is simplified with the help of the counterparts of (7DK) and (7DL) obtained by switching (1) and (2). There exist $w_s^{21}, x_s^{21}, y_s^{21}, z_s^{21} \in \mathcal{F}(N^{kl})_{[t-1]}$ for which

$$\text{(7DM)} \qquad \begin{aligned} &\mathrm{ad}^{(2)}(c_s)g(b_s \otimes b_s) - \mathrm{ad}^{(2)}(b_s)g(b_s \otimes c_s) + \mathrm{ad}^{(1)}(b_s)(y_{2p^s}^{(2)}) \\ &= \zeta_t^{t-1}(x_s^{21}) + d_{[t]}\zeta_t^{t-1}(w_s^{21}) \end{aligned}$$

and

$$\text{(7DN)} \qquad \begin{aligned} &\mathrm{ad}^{(2)}(c_s)g(c_s \otimes b_s) + \mathrm{ad}^{(2)}(b_s)g(c_s \otimes c_s) - \mathrm{ad}^{(1)}(c_s)(y_{2p^s}^{(2)}) \\ &= \zeta_t^{t-1}(z_s^{21}) + d_{[t]}\zeta_t^{t-1}(y_s^{21}). \end{aligned}$$

Summary 7.18. We summarize the main points of Example 7.17 as follows. The minimal VG's for $\mathcal{F}(N^{kl})_{[t]}$ as a module over $\mathcal{F}(\mathcal{U}M^k \amalg \mathcal{U}M^l)_{[t]}$ are given by (7DC) and the "crossed conjugates" (7DD). All of the minimal VR's are the images of these VG's; this is very convenient, since when u is such an image the VR has the particularly simple form, $\psi_{[t]}(u\otimes -) = [u,-]$. Lastly, there are no VS's when $k = l$ and $t = r + k$, and in general all the VS's are consequences of the equations (7DG), (7DH), and (7DK)-(7DN).

Example 7.19. We describe next, a virtual structure for $\mathcal{F}(N^{klm})$ as a csL over the Dcsa $\mathcal{F}(\mathcal{U}M^k \amalg \mathcal{U}M^l \amalg \mathcal{U}M^m)$. We give the answer only; proofs are omitted save occasional hints. For t in the range $1 \le t \le r + \min(k,l,m)$, let $s = \max(0, t-r)$. A Bockstein sub-dga of $\mathcal{F}(\mathcal{U}M^k \amalg \mathcal{U}M^l \amalg \mathcal{U}M^m)_{[t]}$ is $P_{[t]} = R^k_s \amalg R^l_s \amalg R^m_s$.

First we describe the module V and the corresponding dgL $N_{[t]} = \mathcal{L}(V)$ which will be "almost" a Bockstein sub-dgL of $\mathcal{F}(N^{klm})_{[t]}$. In Proposition 6.21, the submodule $fg(U) \subseteq N^{klm}$ is identified as submodule of indecomposables. We quickly retrace the construction of U in order to extract a suitable V. Recall that U is built out of the principal homomorphisms $g_{km} : \mathcal{U}M^k_+ \otimes \mathcal{U}M^m_+ \to N^{km}$ and $g_{lm} : \mathcal{U}M^l_+ \otimes \mathcal{U}M^m_+ \to N^{lm}$. Define V to be the subalgebra of N^{klm} obtained by everywhere replacing A,B,C in the construction of U by R^k_s, R^l_s, R^m_s, respectively. Put $N_{[t]} = \mathcal{L}(V)$. This $N_{[t]}$ is "almost" a Bockstein sub-dgL of $\mathcal{F}(N^{klm})_{[t]}$ in the sense that (7AD) holds while (7AC) fails only on account of $\tilde{\tau}_j$'s and $\tilde{\sigma}_j$'s. As in Example 7.17, we proceed as if $N_{[t]}$ is a Bockstein sub-dgL, promising to make this precise in Chapter 11.

We do not need to give a minimal set of VG's. Any spanning set of VG's has some minimal subset, and we will not need to know the list explicitly. By Lemma 6.18, $V \otimes \mathbb{Z}_p$ is generated as a $P_{[t]} \otimes \mathbb{Z}_p$-module by the two sets

$$\mathrm{ad}^{(1)}((R^k_s)_+)g^{(23)}_{lm}((R^l_s)_+ \otimes (R^m_s)_+), \quad \mathrm{ad}^{(2)}((R^l_s)_+)g^{(13)}_{km}((R^k_s)_+ \otimes (R^m_s)_+).$$

Here $g^{(23)}_{lm}$ is the composite

$$\mathcal{U}M^l_+ \otimes \mathcal{U}M^m_+ \xrightarrow{g_{lm}} N^{lm} \to M^l \amalg M^m \to M^k \amalg M^l \amalg M^m,$$

and similarly for $g^{(12)}_{kl}$ and $g^{(13)}_{km}$.

We can choose a minimal set of VG's from among the elements of the form

$$\text{(7EA)} \qquad \mathrm{ad}^{(1)}(w)g^{(23)}_{lm}(u); \; \mathrm{ad}^{(2)}(w')g^{(13)}_{km}(u').$$

where $w \in \{b_s, c_s, a_{s+1}, \dots, a_k\}$, $w' \in \{b_s, c_s, a_{s+1}, \dots, a_l\}$, $u \in (R^l_s)_+ \otimes (R^m_s)_+$, and $u' \in (R^k_s)_+ \otimes (R^m_s)_+$.

The VR's are just the $\bar{e}_{[t]}$-images of VG's (notation $\bar{e}_{[t]}$ as in (7DB)), and the same reasoning that we used in 7.17 can be used to show this for $\mathcal{F}(N^{klm})_{[t]}$. Because the cyclic cotensor construction of Chapter 6 commutes with the homology functor for dgH's over $\mathbb{Z}_p$, the natural homomorphism $\lambda : \; A \amalg B \amalg C \to \mathrm{CC}(A,B,C)$ induces a surjection on every term of the BSS; specifically,

$$E^j(\lambda(A,B,C)) = \lambda(E^j(A), E^j(B), E^j(C)) :$$
$$E^j(A) \amalg E^j(B) \amalg E^j(C) \to \mathrm{CC}(E^j(A), E^j(B), E^j(C)).$$

We chose $N_{[t]}$ for N^{klm} so that $\mathcal{U}N_{[t]} \otimes \mathbb{Z}_p \to E^t(\mathcal{U}N^{klm})$ would be bijective. As we observed in connection with (7DE), this is enough to guarantee that the images of the minimal VG's, together with the "trivial" VR's $\operatorname{im}(\zeta_t^{t-1}) + \operatorname{im}(d_{[t]}\zeta_t^{t-1})$, generate the ideal of all VR's for $\mathcal{F}(N^{klm})_{[t]}$.

In other words, when checking condition 7.15(iv) for $\mathcal{F}(N^{klm})_{[t]}$, it suffices to do this for the $\bar{e}_{[t]}$-images in $\mathcal{F}(\mathcal{U}M^k \amalg \mathcal{U}M^l \amalg \mathcal{U}M^m)_{[t]}$ of elements of the form (7EA).

Now we come to the VS's. Like in Example 7.17, it is more useful to describe the source equations from which the VS's can be derived than to try to list them individually. The minimal VS's for $\mathcal{F}(N^{klm})$ are all deducible from a certain set of equations, and we recognize three types of equations in this set. The first are (7DG), (7DH) and (7DK)-(7DN). The second are the VR's for $\mathcal{F}(\bar{L}^k)$ (described in Example 7.16) and for $\mathcal{F}(N^{kl})_{[t]}$, $\mathcal{F}(N^{km})_{[t]}$, $\mathcal{F}(N^{lm})_{[t]}$ (described in Example 7.17).

The third group of equations consists of eight equations that are similar to (7DK) and (7DL). We call them "Jacobi identities" for N^{klm}. We postpone our remarks about how to prove the Jacobi identities until Outline 7.21. The first of the eight equations is

$$\text{(7EB)} \qquad \operatorname{ad}^{(1)}(b_s)g_{lm}^{(23)}(b_s\otimes b_s) - \operatorname{ad}^{(2)}(b_s)g_{km}^{(13)}(b_s\otimes b_s) + \operatorname{ad}^{(3)}(b_s)g_{kl}^{(12)}(b_s\otimes b_s) \equiv 0,$$

modulo $\operatorname{im}(\zeta_t^{t-1}) + \operatorname{im}(d_{[t]}\zeta_t^{t-1})$. Because we happen to be in $*$-degree zero there is no $d_{[t]}\zeta_t^{t-1}$ term, so (7EB) actually says that

$$\text{(7EC)} \quad \operatorname{ad}^{(1)}(b_s)g_{lm}^{(23)}(b_s\otimes b_s) - \operatorname{ad}^{(2)}(b_s)g_{km}^{(13)}(b_s\otimes b_s) + \operatorname{ad}^{(3)}(b_s)g_{kl}^{(12)}(b_s\otimes b_s) = \zeta_t^{t-1}(w_s^0)$$

for a certain $w_s^0 \in \mathcal{F}(N^{klm})_{[t-1]}$.

The other seven Jacobi identities are the congruences modulo $\operatorname{im}(\zeta_t^{t-1}) + \operatorname{im}(d_{[t]}\zeta_t^{t-1})$ that are obtained from (7EB) by applying one of the seven differentiation operations:

$$d^{(1)}, d^{(2)}, d^{(3)}, d^{(2)}d^{(1)}, d^{(3)}d^{(1)}, d^{(3)}d^{(2)}, d^{(3)}d^{(2)}d^{(1)}.$$

(Note: this is a description of how to write down the equations with the proper signs, not of how to prove them!) For example, applying $-d^{(1)}$ to (7EB) gives

$$\text{(7ED)} \qquad \operatorname{ad}^{(1)}(c_s)g_{lm}^{(23)}(b_s\otimes b_s) - \operatorname{ad}^{(2)}(b_s)g_{km}^{(13)}(c_s\otimes b_s) + \operatorname{ad}^{(3)}(b_s)g_{kl}^{(12)}(c_s\otimes b_s) \equiv 0,$$

to which we may apply $d^{(3)}$ to produce

$$\text{(7EE)} \qquad \operatorname{ad}^{(1)}(c_s)g_{lm}^{(23)}(b_s\otimes c_s) - \operatorname{ad}^{(2)}(b_s)g_{km}^{(13)}(c_s\otimes c_s) - \operatorname{ad}^{(3)}(c_s)g_{kl}^{(12)}(c_s\otimes b_s) \equiv 0,$$

and so on.

This completes a description of the virtual structure for $\mathcal{F}(N^{klm})_{[t]}$ as a module over $\mathcal{F}(\mathcal{U}M^k \amalg \mathcal{U}M^l \amalg \mathcal{U}M^m)_{[t]}$. However, we wish to make one more point regarding the Jacobi identities, and that is their connection with (7DK)-(7DN). Congruence (7ED) says that there exist elements $w_s^1, x_s^1 \in \mathcal{F}(N^{klm})_{[t-1]}$ such that

$$\text{(7EF)} \qquad \text{(left-hand side of (7ED))} = \zeta_t^{t-1}(x_s^1) + d_{[t]}\zeta_t^{t-1}(w_s^1).$$

Similarly, there exist $y_s^{13}, z_s^{13} \in \mathcal{F}(N^{klm})_{[t-1]}$ such that

$$\text{(7EG)} \qquad \text{(left-hand side of (7EE))} = \zeta_t^{t-1}(z_s^{13}) + d_{[t]}\zeta_t^{t-1}(y_s^{13}).$$

To make the connection with (7DK) and (7DL), recall from 5.13 that the folding homomorphism $\nabla : M^k \amalg M^k \to M^k$ induces a homomorphism, also denoted ∇, from N^{kk} to $\bar{L}^k$. In a similar manner,

$$\nabla^{(12)} = \nabla \amalg 1 : \ (M^k \amalg M^k) \amalg M^k \to M^k \amalg M^k$$

induces a homomorphism, also denoted $\nabla^{(12)}$, between dgL kernels in

(7EH)

$$\begin{array}{ccccccccc} 0 & \longrightarrow & N^{kkk} & \longrightarrow & M^k \amalg M^k \amalg M^k & \longrightarrow & M^k \amalg M^k \amalg M^k / N^{kkk} & \longrightarrow & 0 \\ & & \Big\downarrow{\scriptstyle \nabla^{(12)}} & & \Big\downarrow{\scriptstyle \nabla^{(12)}} & & \Big\downarrow & & \\ 0 & \longrightarrow & N^{kk} & \longrightarrow & M^k \amalg M^k & \longrightarrow & M^k \oplus M^k & \longrightarrow & 0. \end{array}$$

Define $\nabla^{(13)}$ and $\nabla^{(23)}$ likewise.

Now suppose $k = l = m$. What happens if we apply $\nabla^{(12)}$ to both sides of (7EF)? The coordinates labeled (1) and (2) are both collapsed to (1), and (3) becomes relabeled as (2). By Lemma 5.14(c), the result is

$$\begin{aligned} &\mathrm{ad}^{(1)}(c_s)g(b_s \otimes b_s) - \mathrm{ad}^{(1)}(b_s)(c_s \otimes b_s) - \mathrm{ad}^{(2)}(b_s)(y^{(1)}_{2p^s}) \\ &= \zeta_t^{t-1}(\nabla^{(12)}(x_s^1)) + d_{[t]}\zeta_t^{t-1}(\nabla^{(12)}(w_s^1)) \end{aligned} \tag{7EI}$$

If we define

$$w_s^{12} = \nabla^{(12)}(w_s^1), \ x_s^{12} = \nabla^{(12)}(x_s^1), \tag{7EJ}$$

we have just derived (7DK) from (7EF). Likewise, (7DL) is a consequence of (7EG). Lastly, (7DM) and (7DN) are likewise deducible from the appropriate Jacobi identities through use of a suitable folding homomorphism.

Summary 7.20. For $\mathcal{F}(N^{klm})_{[t]}$ as a module over $\mathcal{F}(\mathcal{U}M^k \amalg \mathcal{U}M^l \amalg \mathcal{U}M^m)_{[t]}$, with $1 \leq t \leq r + \min(k, l, m)$ and $s = \max(0, t - r)$, a virtual structure is as follows. There is a minimal set of VG's whose members all have the form (7EA). A minimal set of VR's is obtained by taking the images of the minimal VG's in $\mathcal{F}(\mathcal{U}M^k \amalg \mathcal{U}M^l \amalg \mathcal{U}M^m)_{[t]}$. The minimal VS's are all deducible from the following list of equations: (7DG), (7DH), and (7DK)-(7DN); the VR's for $\mathcal{F}(\bar{L}^i)$ and for $\mathcal{F}(N^{ij})$ as i and j run through $\{k, l, m\}$; the eight Jacobi identities, including (7EC), (7EF), and (7EG).

Remark. In the notation of Chapter 6, for non–trivial Hopf algebras A^1, A^2, A^3, there are always some "generic" syzygies for the indecomposables of $TU = \mathrm{HAK}(\lambda)$ as a module over $CC(A^1, A^2, A^3)$. When A^1, A^2, A^3 are free algebras these are the only syzygies. One can give a three–variable Hilbert series (like (6UU)) for these generic syzygies in terms of the Hilbert series t_i of A^i. The series is $(t_1)^2(t_2)^2(t_3)^2 + \ldots$. For the case of $A^1 = A^2 = A^3 = E^{r+s}(\mathcal{U}M^k)$ we have for $0 \leq s \leq k$

$$t_1 \ = \ t_2 \ = \ t_3 \ = \ \text{Hilbert series of } E^{r+s}(\mathcal{U}M^k) \ = \ t^{2np^s-1} \ + \ t^{2np^s} \ + \ \ldots,$$

so the lowest–dimensional generic VS for $\mathcal{F}(N^{kkk})_{[r+s]}$ occurs in dimension $6(2np^s - 1) = 12np^s - 6$. Because $E^{r+s}(\mathcal{U}M^k) = R_s^k$ coincides with the free algebra $\mathbb{Z}_p\langle b_s, c_s\rangle$ at least up to dimension $2np^{s+1} - 1$, the first non-generic syzygy occurs in dimension at least $(2np^{s+1} - 1) + (2np^s - 1) + (2np^s - 1) \geq 2np^s(p+2) - 3$. When $s = k$ then $E^{r+k}(\mathcal{U}M^k)$ is free, so the only syzygies are the generic ones.

Outline 7.21. We wish to outline how the Jacobi identities for N^{klm} are proved. We start with some general discussion of N^{klm} and its BSS. Then we focus on just one equation, since the technique is the same for all eight. Specifically, we prove (7ED).

To begin with, if $s = 0$ (i.e. $t \leq r$), then $g^{(13)}_{km}(c_0 \otimes b_0)$ is the commutator $[a_0^{(1)}, b^{(3)}]$, and an analogous statement holds for all the other expressions $g^{(ij)}(\;)$ appearing in any of the eight equations. When $s = 0$ the eight Jacobi identities reduce to instances of the ordinary Jacobi identity for Lie algebras. The equations hold in N^{klm} because N^{klm} embeds in $M^k \amalg M^l \amalg M^m$, and the Jacobi identities are certainly true in the Lie algebra $M^k \amalg M^l \amalg M^m$. When $s = 0$, therefore, the eight equations are true as "pure" equations, not only as congruences modulo $\mathrm{im}(\zeta_t^{t-1}) + \mathrm{im}(d_{[t]}\zeta_t^{t-1})$. In particular, when $s = 0$ we have $x_0^1 = 0$ and $w_0^1 = 0$ in (7EF), $w_0^0 = 0$ in (7EC), and so on. For the remainder of this discussion we assume $s \geq 1$.

The entire BSS for N^{klm} can be computed from 6.21 and 5.6(d). Corollary 6.21 permits us to write $N^{klm} = \mathcal{L}(V, d)$ and

$$(V, d) = V^{(r)} \oplus V^{(r+1)} \oplus \ldots \oplus V^{(q)} \tag{7EK}$$

as in (5M), where $q = r + \min(k, l, m)$. One key consequence of 6.21 is that the $V^{(j)}$'s may be computed explicitly. The free $\mathbb{Z}_{(p)}$-module V also serves as a module of indecomposables for $\mathcal{U}N^{klm}$, which is (by 6.16 and 6.19) precisely

$$\mathrm{HAK}(\lambda : \mathcal{U}M^k \amalg \mathcal{U}M^l \amalg \mathcal{U}M^m \to \mathrm{CC}(\mathcal{U}M^k, \mathcal{U}M^l, \mathcal{U}M^m)).$$

Its Hilbert series and BSS can be computed, as in Chapter 6.

Here is what the computation tells us. Just as (N^{kl}, d) is trigraded, with a two-component $\#$-grade and a $*$-degree, so (N^{klm}, d) has four linearly independent gradings, which we can take to be $\#$-trigrade, i.e. $(|x|^{(1)}_{\#}, |x|^{(2)}_{\#}, |x|^{(3)}_{\#})$, and (total) $*$-degree. For $s \leq q - r = \min(k, l, m)$, $V^{(r+s)}$ is zero in $\#$-trigrade (i_1, i_2, i_3), if any i_j is less than p^s. $\#$-Trigrade (p^s, p^s, p^s) is where all eight Jacobi identities live. In $\#$-trigrade (p^s, p^s, p^s), V^{r+s} has precisely sixteen generators (as a free $\mathbb{Z}_{(p)}$-module). To describe all sixteen, recall the notation Γ^s_s for the two-element set $\{b_s, c_s\}$. The sixteen generators are the sixteen elements in the two sets

$$\mathrm{ad}^{(1)}(\Gamma^s_s) g^{(23)}_{lm}(\Gamma^s_s \otimes \Gamma^s_s); \;\; \mathrm{ad}^{(2)}(\Gamma^s_s) g^{(13)}_{km}(\Gamma^s_s \otimes \Gamma^s_s). \tag{7EL}$$

In $*$-degrees, $0, 1, 2, 3$ the number of generators is respectively $2, 6, 6, 2$. We give special names to the two that have $*$-degree zero, namely

$$u_s = \mathrm{ad}^{(1)}(b_s) g^{(23)}_{lm}(b_s \otimes b_s); \;\; v_s = \mathrm{ad}^{(2)}(b_s) g^{(13)}_{km}(b_s \otimes b_s). \tag{7EM}$$

Now consider what Lemma 5.6(d) tells us. Let a superscript $(\;)^{\%}$ denote the component in $\#$-trigrade (p^s, p^s, p^s) and in $*$-degree 1. Recalling (5P), we compute

$$Q = \tilde{W}^{r+s-1}(V^{(\geq r+s-1)})^{\%} = \mathrm{Span}(\tilde{\tau}_1(u_{s-1}), \tilde{\tau}_1(v_{s-1})), \tag{7EN}$$

where u_{s-1} and v_{s-1} are defined in (7EM), and (7EN) is our definition of Q. We just described $(V^{(\geq r+s)})^{\%}$, and it equals $(\mathcal{L}(V^{(\geq r+s)}))^{\%}$ because $V^{(\geq r+s)}$ is zero below #-trigrade (p^s, p^s, p^s). Let R denote the $\mathbb{Z}_{(p)}$-span of the six elements in the list (7EL) which have $*$-degree 1. Then Lemma 5.6(d) tells us that

(7EO) $$E^{r+s}(N^{klm})^{\%} = (Q \oplus R) \otimes \mathbb{Z}_p$$

a $\mathbb{Z}_p$-vector space of rank eight.

Let y denote the left-hand side of (7ED). The element y belongs to $(N^{klm})^{\%}$ and can also be viewed as belonging to $E^{r+s}(N^{klm})^{\%}$. By (7EO) we may uniquely write

(7EP) $$y = y' + y'' \text{ in } E^{r+s}(N^{klm})^{\%},$$

where $y' \in Q \otimes \mathbb{Z}_p$ and $y'' \in R \otimes \mathbb{Z}_p$. Our plan is to show that both y' and y'' are zero.

To determine y'', use the inclusion

$$\bar{e}: \; N^{klm} \to \mathcal{U}M^k \amalg \mathcal{U}M^l \amalg \mathcal{U}M^m,$$

and consider $E^{r+s}(\bar{e})$. We know from Corollary 4.9 that

$$E^{r+s}(\mathcal{U}M^k \amalg \mathcal{U}M^l \amalg \mathcal{U}M^m) = (R^k_s \amalg R^l_s \amalg R^m_s) \otimes \mathbb{Z}_p.$$

We see easily that

$$E^{r+s}(\bar{e})(\mathrm{ad}^{(1)}(c_s)g^{(23)}_{lm}(b_s \otimes b_s)) = [c^{(1)}_s, [b^{(2)}_s, b^{(3)}_s]]$$

in $E^{r+s}(\mathcal{U}M^k \amalg \mathcal{U}M^l \amalg \mathcal{U}M^m)$, and likewise $E^{r+s}(\bar{e})$ converts each of our other five basis elements for R into a three-fold bracket. Also, $Q \otimes \mathbb{Z}_p \subseteq \ker(E^{r+s}(\bar{e}))$, but the six basis elements of R remain linearly independent in $E^{r+s}(\mathcal{U}M^k \amalg \mathcal{U}M^l \amalg \mathcal{U}M^m)$, hence $Q \otimes \mathbb{Z}_p = \ker(E^{r+s}(\bar{e}))^{\%}$.

The ordinary Jacobi identity tells us that $E^{r+s}(\bar{e})(y) = 0$. It follows that $y'' = 0$, hence $y = y'$.

We need a different trick in order to determine y'. The trick is the method of "detection" inroduced during the proof of Theorem 5.12. Let $\hat{N}^{lm}$ be the Lie ideal of $\mathbb{L}^{\mathrm{ab}}[b] \amalg \mathcal{U}M^l \amalg \mathcal{U}M^m$ consisting of all elements x for which $|x|^{(i)}_{\#} > 0$ for $i = 1, 2, 3$. The surjection $\gamma : M^k \to \mathbb{L}^{\mathrm{ab}}[b]$ induces the dgL homomorphism $\gamma_0 : \; N^{klm} \to \hat{N}^{lm}$. Recalling the definition of y as the left-hand side of (7ED), direct computation shows that $E^{r+s}(\gamma_0)(y) = 0$, hence $E^{r+s}(\gamma_0)(y') = 0$ since $y = y'$.

However, $E^{r+s}(\gamma_0)$ is one-to-one on $Q \otimes \mathbb{Z}_p$. To see this, first observe that $\hat{N}^{lm}$ has a linear differential (by 6.20) and write $\hat{N}^{lm} = \mathcal{L}(\hat{V})$. Let $U' = \mathrm{Span}(u_{s-1}, v_{s-1})$; there is a chain complex retract $\hat{U}$ of $\hat{V}$ such that the composite

$$L(U') \subseteq N^{klm} \xrightarrow{\gamma_0} \hat{N}^{lm} = \mathcal{L}(\hat{V}) \to \mathcal{L}(\hat{U})$$

is an isomorphism of dgL's. So $E^{r+s}(\gamma_0)$ is one-to-one on the image in $E^{r+s}(N^{klm})$ of $E^{r+s}(\mathcal{L}(U'))$, which contains $Q \otimes \mathbb{Z}_p$. It follows that $y' = 0$, hence $y = 0$.

This proves that (7ED) is true if viewed as an equation in $E^{r+s}(N^{klm})$ rather than a congruence. But in general

$$E^t(N^{klm}) = \mathcal{F}(N^{klm})_{[t]}/(\mathrm{im}(\zeta^{t-1}_t) + \mathrm{im}(d_{[t]}\zeta^{t-1}_t)),$$

so (7ED) is true in $\mathcal{F}(N^{klm})_{[t]}$ as a congruence modulo $\mathrm{im}(\zeta^{t-1}_t) + \mathrm{im}(d_{[t]}\zeta^{t-1}_t)$. This completes Outline 7.21.

CHAPTER 8

Stable Homotopy Operations Under a Space

Let A be a topological space, and let $t \geq 1$. In Chapter 8 we define and examine a certain dga $\mathcal{O}_A$ associated to A and t, which we call the ring of stable homotopy operations under A. We may think of $\mathcal{O}_A = \mathcal{O}_A[t]$ as a simultaneous generalization, of the usual ring of stable mod p^t homotopy operations on one hand, and of iterated Whitehead or Samelson products on the other. There is a strong connection between $\mathcal{O}_A$ and $H_*(\Omega A; \mathbb{Z}_{p^t})$, and we present some results and some open questions about this connection. The dga of stable homotopy operations will later provide the correct framework within which to understand and prove many of the important properties of our spaces D_k.

We begin with a longish but elementary detour to define two operations, denoted $\mathfrak{f}$ and $\mathfrak{g}$, that convert homotopy classes in a fiber to maps between fibers.

Notation. Let $\{Y_i, Z_i, A_i\}_{i=1,2,3}$ denote spaces. Let CA_i denote the reduced cone on A_i, i.e., $CA_i = (A_i \times I)/(A_i \times 0 \cup * \times I)$. Points of CA_i are denoted via coordinate pairs in $A_i \times I$, e.g., (a,t). Let $1_i : A_i \to CA_i$ be $1_i(a) = (a,1)$. Suppose that

$$\begin{array}{ccccc} Y_1 & \xrightarrow{f_1} & Y_2 & \xrightarrow{f_2} & Y_3 \\ \downarrow{\scriptstyle g_1} & & \downarrow{\scriptstyle g_2} & & \downarrow{\scriptstyle g_3} \\ Z_1 & \xrightarrow{h_1} & Z_2 & \xrightarrow{h_2} & Z_3 \end{array} \tag{8A}$$

is a strictly commutative diagram. Let X_i denote the homotopy–theoretic fiber of $(g_i \vee 1_i) : Y_i \vee A_i \to Z_i \vee CA_i$. Then

$$X_i = \{\text{pairs } (\phi, e) \in (Z_i \vee CA_i)^I \times (Y_i \vee A_i) | \phi(0) = z_{i0} \text{ and } \phi(1) = (g_i \vee 1_i)(e)\},$$

where z_{i0} denotes the base point of Z_i. Let $j_i : X_i \to Y_i \vee A_i$ be the fiber inclusion, i.e., $j_i(\phi, e) = e$, and let $w_i : A_i \to X_i$ denote the canonical lifting of A_i through j_i, i.e., $w_i(a) = (\phi_a, a)$, where $\phi_a(t) = (a,t) \in CA_i \subseteq Z_i \vee CA_i$.

Lemma 8.1. *Refer to diagram (8A) and its associated notation. Given any map $\hat{a}_1 : A_1 \to X_2$, there is a natural map $\mathfrak{f}(\hat{a}_1) : X_1 \to X_2$ for which*

(8B) $$j_2\mathfrak{f}(\hat{a}_1) = (f_1 \vee j_2\hat{a}_1)j_1 \quad ,$$

and

(8C) $$\mathfrak{f}(\hat{a}_1)w_1 = \hat{a}_1.$$

(To define $\mathfrak{f}(\hat{a}_1)$ requires only the left square of (8A), and the naturality of $\mathfrak{f}$ is with respect to such commuting squares). If $\hat{a}_2 : A_2 \to X_3$ is another map, then

(8D) $$\mathfrak{f}(\mathfrak{f}(\hat{a}_2)\hat{a}_1) = \mathfrak{f}(\hat{a}_2)\mathfrak{f}(\hat{a}_1).$$

If $\hat{b}_1 : A_1 \to A_2$ is any map, let $\mathfrak{g}(\hat{b}_1)$ denote $\mathfrak{f}(w_2\hat{b}_1)$. Then $\mathfrak{g}$ is functorial (with respect to commuting squares) and satisfies

(8E) $$\mathfrak{f}(\hat{a}_2)\mathfrak{g}(\hat{b}_1) = \mathfrak{f}(\hat{a}_2\hat{b}_1).$$

If in addition we have $\hat{b}_2 : A_2 \to A_3$, then

(8F) $$\mathfrak{g}(\hat{b}_2)\mathfrak{f}(\hat{a}_1) = \mathfrak{f}(\mathfrak{g}(\hat{b}_2)\hat{a}_1) \quad ,$$

and

(8G) $$\mathfrak{g}(\hat{b}_2)\mathfrak{g}(\hat{b}_1) = \mathfrak{g}(\hat{b}_2\hat{b}_1).$$

Lastly, viewed as a map from $X_2^{A_1}$ to $X_2^{X_1}$ (resp. $A_2^{A_1}$ to $X_2^{X_1}$), $\mathfrak{f}$ (resp. $\mathfrak{g}$) is continuous, hence $\mathfrak{f}$ (resp. $\mathfrak{g}$) preserves homotopy.

Proof. There is associated to $\hat{a}_1$ a functorial commuting square

(8H) $$\begin{array}{ccc} Y_1 \vee A_1 & \xrightarrow{f_1 \vee j_2\hat{a}_1} & Y_2 \vee A_2 \\ {\scriptstyle g_1 \vee 1_1}\big\downarrow & & \big\downarrow{\scriptstyle g_2 \vee 1_2} \\ Z_1 \vee CA_1 & \xrightarrow{h_1 \vee \alpha(\hat{a}_1)} & Z_2 \vee CA_2, \end{array}$$

where $\alpha(\hat{a}_1)(a,t) = \phi_2(t)$ when we write $\hat{a}_1(a) = (\phi_2, e_2) \in X_2$. We leave it as an exercise to check that α is well–defined on CA_1 and that (8H) commutes. Define $\mathfrak{f}(\hat{a}_1)$ to be the canonical induced map between homotopy–theoretic fibers, for (8H). A formula for it is

$$\mathfrak{f}(\hat{a}_1)(\phi_1, e_1) = ((h_1 \vee \alpha(\hat{a}_1)) \circ \phi_1,\ (f_1 \vee j_2\hat{a}_1)(e_1)),$$

which is clearly continuous as a map on the function space $X_2^{A_1}$. Property (8B) follows at once, and (8C) is easily checked. Formulas (8E), (8F), and (8G) follow from (8C) and (8D). Formula (8D) requires verifying that

$$(f_2 \vee j_3\hat{a}_2) \circ (f_1 \vee j_2\hat{a}_1) = (f_2f_1) \vee (j_3\mathfrak{f}(\hat{a}_2)\hat{a}_1)$$

and that

$$(h_2 \vee \alpha(\hat{a}_2)) \circ (h_1 \vee \alpha(\hat{a}_1)) = (h_2 h_1) \vee (\alpha(\mathfrak{f}(\hat{a}_2)\hat{a}_1)),$$

and then the two sides of (8D) both equal the induced map on fibers for the same commuting square. We omit the verification.

Before proceeding we examine one extreme specialization of the setting (8A). Let A be a space, and suppose in (8A) that $Y_i = Z_i = A$ for $i = 1, 2, 3$ and that all the arrows in (8A) are identity maps. Then X_i has the homotopy type of the half–smash $(\Omega A) \ltimes A_i$. We shall view the half–smash construction as a cofiber, i.e.,

$$X \ltimes Y = (X \times Y) \underset{X}{\cup} C(X \times y_0).$$

Of course, the quotient map $X \ltimes Y \to X_+ \wedge Y$ is a homotopy equivalence, where $(\ )_+$ denotes "adjoin a disjoint point."

Note that ΩZ always acts on the left on X_i via

$$\phi_1 * (\phi_2, e) = (\phi_1 * \phi_2, e),$$

so in this case we have ΩA acting on the left on X_i. The map $\mathfrak{f}(\hat{a}_i)$ will be a map of left ΩA–modules. The homotopy equivalence

$$(\Omega A) \ltimes A_i = (\Omega A \times A_i) \underset{\Omega A}{\cup} C(\Omega A \times a_{i0}) \to X_i$$

can be given by the explicit formulas

$$\begin{array}{rll} (\phi, a) \longmapsto \phi * w_1(a) & \text{for} & (\phi, a) \in \Omega A \times A_i, \\ (\phi \times a_{i0}, s) \longmapsto (\phi^s, \phi(s)) & \text{for} & (\phi \times a_{i0}, s) \in C(\Omega A \times a_{i0}), \end{array}$$

where $\phi^s : [0,1] \to A^I$ is given by $\phi^s(t) = \phi(ts)$. This homotopy equivalence is natural in the sense that

$$\text{(8I)} \qquad \begin{array}{ccccc} \Omega A_+ \wedge A_1 & \xleftarrow{\simeq} & (\Omega A) \ltimes A_1 & \xrightarrow{\simeq} & X_1 \\ \downarrow{\scriptstyle 1_{\Omega A_+} \wedge b_1} & & \downarrow{\scriptstyle 1_{\Omega A} \ltimes b_1} & & \downarrow{\scriptstyle \mathfrak{g}(\hat{b}_1)} \\ \Omega A_+ \wedge A_2 & \xleftarrow{\simeq} & (\Omega A) \ltimes A_2 & \xrightarrow{\simeq} & X_2 \end{array}$$

commutes on the nose, for any $\hat{b}_1 : A_1 \to A_2$. Up to homotopy, then, we shall identify naturally $\mathfrak{g}(\hat{b}_1)$ with $1_{\Omega A_+} \wedge \hat{b}_1$, when all the f_i's, g_i's, and h_i's of (8A) are identity maps.

Continuing with the case where (8A) consists of identity maps, let X_i^Σ denote the homotopy–theoretic fiber associated to ΣA_i rather than A_i. This space X_i^Σ has the homotopy type of $(\Omega A_+) \wedge \Sigma A_i$, which has the homotopy type of $\Sigma(\Omega A_+) \wedge A_i$ and hence of ΣX_i. Denoting the homotopy equivalence by $\gamma_{X_i} : \Sigma X_i \to X_i^\Sigma$, we observe that γ_{X_i} is natural, in the following sense. Given any map $\hat{a}_1 : A_1 \to X_2$, the diagram

$$\text{(8J)} \qquad \begin{array}{ccc} \Sigma X_1 & \xrightarrow{\Sigma \mathfrak{f}(\hat{a}_1)} & \Sigma X_2 \\ {\scriptstyle \gamma_{X_1}}\downarrow & & \downarrow{\scriptstyle \gamma_{X_2}} \\ X_1^\Sigma & \xrightarrow{\mathfrak{f}(\gamma_{X_2} \circ \Sigma \hat{a}_1)} & X_2^\Sigma \end{array}$$

commutes. We omit the proof as tedious. In general we shall suppress $\gamma_{(\ )}$ from our notation and simply identify X_i with $(\Omega A_+) \wedge A_i$ and X_i^{Σ} with ΣX_i.

With this background material established, let us move on to the main business of this chapter, namely, the construction and characterization of the ring $\mathcal{O}_A$ of stable homotopy operations under A.

Notation. Given a space A, let $A'(q)$ denote the homotopy–theoretic fiber of the inclusion map $A \vee S^q \to A \vee CS^q$, and let $A'(q,p^t)$ be the homotopy–theoretic fiber of $A \vee P^q(p^t) \to A \vee CP^q(p^t)$. We shall sometimes write A' for $A'(q)$ or for $A'(q,p^t)$ when the meaning is clear from the context. Notice that $A'(q)$ (resp. $A'(q,p^t)$) has the homotopy type of $S^q \vee \Sigma^q\Omega A$ (resp. $P^q(p^t) \vee (P^q(p^t) \wedge \Omega A)$) if $q \geq 2$. We denote the homotopy class of the lift $w_1 : S^q \to A'(q)$ or $w_1 : P^q(p^t) \to A'(q,p^t)$ by $\hat{v}_q$ or simply $\hat{v}$.

In view of our earlier remarks, we shall identify $A'(q)$ with $(\Omega A_+) \wedge S^q$ and $A'(q,p^t)$ with $(\Omega A_+) \wedge P^q(p^t)$. In particular, there are maps
(cf. (8I))

$$\mathfrak{g}(\rho^t) = 1_{\Omega A_+} \wedge \rho^t : A'(q,p^t) \to A'(q),$$
$$\mathfrak{g}(\omega^t) = 1_{\Omega A_+} \wedge \omega^t : A'(q-1) \to A'(q,p^t),$$

where ρ^t and ω^t are the pinch and inclusion maps described at the beginning of Chapter 6.

Notation. Define $\tilde{\pi}_q(X;\mathbb{Z}_{p^t})$ to be the zero group if $q < 2$, and let $\tilde{\pi}_q(X;\mathbb{Z}_{p^t}) = \pi_{q-1}(\Omega X;\mathbb{Z}_{p^t})$ for $q \geq 2$. Thus $\tilde{\pi}_q$ is identical with π_q if $q \geq 3$, but $\tilde{\pi}_2(X;\mathbb{Z}_{p^t}) = \pi_2(X) \otimes \mathbb{Z}_{p^t}$. An equivalent definition is that $\tilde{\pi}_q$ is π_q applied to the universal covering space. For any X, $\tilde{\pi}_*(X;\mathbb{Z}_{p^t})$ is a graded $\mathbb{Z}_{p^t}$–module. We may reinterpret the natural transformation ${}_\pi\delta^t$ as the transformation

$${}_\pi\delta^t : \tilde{\pi}_q(\ ;\mathbb{Z}_{p^t}) \to \tilde{\pi}_{q-1}(\ ;\mathbb{Z}_{p^t})$$

given by

(8K) $${}_\pi\delta^t = \begin{cases} (\rho^t\omega^t)^\#, & \text{if } q \geq 4 \\ (\omega^t)^\# \text{ followed by reduction } \bmod p^t, & \text{if } q = 3 \\ 0, & \text{if } q \leq 2. \end{cases}$$

Notation. Let A be a topological space, and let $\mathcal{M}_A$ denote the category of morphisms in the category of spaces under A. Objects in $\mathcal{M}_A$ are thus diagrams

(8L) $$A \xrightarrow{f} Y \xrightarrow{g} Z.$$

We typically denote (8L) as (f,g) and call it a "composable pair of maps with initial source A" or ***map pair*** for short. Morphisms of map pairs are strictly commuting diagrams

(8M) $$\begin{array}{ccccc} A & \xrightarrow{f_1} & Y_1 & \xrightarrow{g_1} & Z_1 \\ {=}\downarrow & & \downarrow & & \downarrow \\ A & \xrightarrow{f_2} & Y_2 & \xrightarrow{g_2} & Z_2. \end{array}$$

On the cateory $\mathcal{M}_A$ of map pairs consider the forgetful functor (it forgets f and keeps g) composed with the homotopy–theoretic fiber functor. Our notation for this composite functor applied to (f,g) is K_g. Then the $\mathbf{Z}_{p^t}$–modules $\tilde{\pi}_q(K_g;\mathbf{Z}_{p^t})$ are viewed as functors on $\mathcal{M}_A$. A *homotopy operation under A of type* $(m,p^s;q,p^t)$ is any natural transformation between the two functors

$$(f,g) \longmapsto \tilde{\pi}_m(K_g;\mathbf{Z}_{p^s}) \text{ and } (f,g) \longmapsto \tilde{\pi}_q(K_g;\mathbf{Z}_{p^t})$$

from $\mathcal{M}_A$ to $\mathbf{Z}_{(p)}$–modules. The set of homotopy operations under A of type $(m,p^s;q,p^t)$ is denoted $S(m,p^s;q,p^t)$. It is a $\mathbf{Z}_{(p)}$–module via $(\chi+c\omega)(\hat{x}) = \chi(\hat{x}) + c\omega(\hat{x})$, $c \in \mathbf{Z}_{(p)}$.

There is a universal example for homotopy operations under A. It is the space $A'(m,p^s)$ when $m \geq 3$, and it is $A'(2)$ when $m = 2$. When $m < 2$ or $q < 2$ we obviously have $S(m,p^s;q,p^t) = 0$. To see why these spaces serve as the respective universal examples, define $F: S(m,p^s;q,p^t) \to \tilde{\pi}_q(A'(m,p^s);\mathbf{Z}_{p^t})$ if $m \geq 3$ (resp. $\to \tilde{\pi}_q(A'(2);\mathbf{Z}_{p^t})$ if $m = 2$) by $F(\chi) = \chi(\hat{v}_m)$.

More notation. Write $X \in \mathcal{W}$ if and only if X has the p–local homotopy type of a 1–connected finite–type bouquet of Moore spaces and spheres.

Lemma 8.2. *F is an injection of $\mathbf{Z}_{(p)}$–modules. It is a bijection if $m \geq 3$, or if $m = 2$ and $s \geq t$ and $\Sigma^2\Omega A \in \mathcal{W}$.*

Proof. Let $A \xrightarrow{f} Y \xrightarrow{g} Z$ be a map pair. We apply the ideas of Lemma 8.1 by setting $Y_2 = Y$, $Z_2 = Z$; $A_2 = *$, whence $X_2 = K_g$. Set $Y_1 = Z_1 = A$ and $A_1 = P^m(p^s)$ if $m \geq 3$ (resp. S^2 if $m = 2$), so that $X_1 = A'(m,p^s)$ (resp. $A'(2)$). The left square of (8A) is now

$$\begin{array}{ccc} A & \xrightarrow{f} & Y \\ {\scriptstyle =}\downarrow & & \downarrow{\scriptstyle g} \\ A & \longrightarrow & Z. \end{array}$$

Given an element $\hat{x} \in \tilde{\pi}_m(K_g;\mathbf{Z}_{p^s})$, take $\hat{a}_1 = P^m(p^s) \to X_2 = K_g$ (resp. $\hat{a}_1 : S^2 \to X_2$) to be any representative. Then $\mathfrak{f}(\hat{a}_1) : A' \to K_g$ satisfies $\mathfrak{f}(\hat{a}_1)w_1 = \hat{a}_1$ by (8C), whence $\hat{x} = \mathfrak{f}(\hat{a}_1)_\#(\hat{v})$. The naturality of any $\chi \in S(m,p^s;q,p^t)$ guarantees that

$$\chi(\hat{x}) = \mathfrak{f}(\hat{a}_1)_\#\chi(\hat{v}) = \mathfrak{f}(\hat{a}_1)_\#(F(\chi)),$$

i.e., χ is determined by $F(\chi)$, i.e., F is one–to–one.

To see that F is onto when $m \geq 3$, let $\hat{z} \in \pi_q(A'(m,p^s);\mathbf{Z}_{p^t})$, and let $\hat{x} \in \pi_m(K_g;\mathbf{Z}_{p^s})$ be arbitrary. Put $\chi(\hat{x}) = \mathfrak{f}(\hat{a}_1)_\#(\hat{z})$, where $\hat{a}_1$ is a map representing $\hat{x}$. If χ belongs to $S(m,p^s;q,p^t)$ then $F(\chi) = \mathfrak{f}(w_1)_\#(\hat{z}) = \hat{z}$ since $\mathfrak{f}(w_1)$ is the identity map on X_1. So we are done if χ is natural. We proceed to prove this.

To maximize the consistency of notation between (8A) and this setting, we denote our morphism of map pairs as

$$\begin{array}{ccccc} A & \xrightarrow{f_0} & Y_2 & \xrightarrow{g_2} & Z_2 \\ {\scriptstyle =}\downarrow & & {\scriptstyle f_2}\downarrow & & {\scriptstyle h_2}\downarrow \\ A & \xrightarrow{f_3} & Y_3 & \xrightarrow{g_3} & Z_3. \end{array}$$

Given any element $\hat{x} \in \tilde{\pi}_m(K_{g_2}; \mathbb{Z}_{p^s})$, we want to show that $g_{\#}\chi(\hat{x}) = \chi(g_{\#}(\hat{x}))$, where $g : K_{g_2} \to K_{g_3}$ now denotes the induced map between homotopy–theoretic fibers. Put $A_1 = P^m(p^s)$ and $A_2 = A_3 = *$ in Lemma 8.1. Let $\hat{a}_1$ represent $\hat{x}$ and let $\hat{b}_2$ be the trivial map. We find that $g = \mathfrak{g}(\hat{b}_2)$ and that (8F) says $\mathfrak{f}(g\hat{a}_1) = g\mathfrak{f}(\hat{a}_1)$, whence

$$g_{\#}\chi(\hat{x}) = g_{\#}\mathfrak{f}(\hat{a}_1)_{\#}(\hat{z}) = \mathfrak{f}(g\hat{a}_1)_{\#}(\hat{z}) = \chi(g_{\#}(\hat{x})),$$

as desired.

When $m = 2$ we use the same proof, except that A_1 is S^2 instead of $P^m(p^s)$. We have something additional to prove, however, because the map $\hat{a}_1$ representing an element $\hat{x}$ of $\tilde{\pi}_2(K_{g_2}; \mathbb{Z}_{p^s})$ has indeterminacy equal to $p^s\tilde{\pi}_2(K_{g_2})$. We assert that the indeterminacy is irrelevant to the above argument. To be precise, we assert that if $\hat{a}_1$ and $\hat{a}_2$ both represent $\hat{x}$, i.e. if $\hat{a}_2 = \hat{a}_1 + p^s\hat{a}_0$ for some $\hat{a}_0$: $S^2 \to K_{g_2}$, then $\mathfrak{f}(\hat{a}_1)$ and $\mathfrak{f}(\hat{a}_2)$ induce the same homomorphism on $\tilde{\pi}_*(\ ; \mathbb{Z}_{p^t})$. Granting this assertion, the above proof works at $m = 2$.

The assertion requires a rather technical proof for which we provide only an outline. It clearly suffices to do the case $s = t$, since taking $s > t$ can only reduce the amount of indeterminacy. First we show that $\Sigma^2 X \xrightarrow{p} \Sigma^2 X$ induces the zero homomorphism on $\pi_*(\ ; \mathbb{Z}_p)$, whenever $\Sigma^2 X \in \mathcal{W}$ (the map "p" means: add p copies of the identity map together, using the co–H structure on the source space $\Sigma^2 X$). Given this fact, we prove by induction on t that $\Sigma^2 X \xrightarrow{p^t} \Sigma^2 X$ induces the zero homomorphism on $\pi_*(\ ; \mathbb{Z}_{p^t})$, and then we prove the assertion (take X to be ΩA_+).

The first step is the hardest. Let $X^{\vee p}$ denote the bouquet of p copies of X, and note that $(\Sigma^2 X)^{\vee p} \approx \Sigma^2(X^{\vee p})$. There is a factorization of $\Sigma^2 X \xrightarrow{p^t} \Sigma^2 X$ as $\Sigma^2 X \xrightarrow{\varphi} (\Sigma^2 X)^{\vee p} \xrightarrow{\nabla} \Sigma^2 X$, where φ is the iterated comultiplication and ∇ is the folding map. Because $\Sigma^2 X$ is a double suspension, the comultiplication is coassociative and cocommutative, hence $\operatorname{im}(\varphi)_{\#}$ is invariant under the action of the symmetric group Σ_p on $\pi_*((\Sigma^2 X)^{\vee p}; \mathbb{Z}_p)$. Careful use of the Hilton–Milnor decomposition for $(\Sigma^2 X)^{\vee p}$ (recall that $\Sigma^2 X \in \mathcal{W}$ even though $\Sigma^2 X$ might not be a wedge of spheres and Moore spaces as a co–H space) enables us to show that any element of $\pi_*((\Sigma^2 X)^{\vee p}; \mathbb{Z}_p)^{(\Sigma_p)}$ is sent to zero by $\nabla_{\#}$. A key ingredient in the proof is the fact that, in the free product $\coprod_{i=1}^p B$ of p copies of a free $\mathbb{Z}_p$–algebra B, the (Σ_p)–invariant elements all lie in the kernel of the folding homomorphism $\coprod_{i=1}^p B \to B$ ([BC] is relevant to determining the invariant subalgebra).

Corollary 8.3. (a) $S(m, p^s; q, p^t) = 0$ *if* $q < m-1$. (b) *If A is 1-connected then* $S(q, p^t; q-1, p^t) = \mathbb{Z}_{p^t}$ *for* $q \geq 3$, *with a generator being the transformation* ${}_\pi\delta^t$ *of (8K).*

Proof. $(\Omega A_+) \wedge S^2$ is 1-connected, and $(\Omega A_+) \wedge P^m(p^s)$ is $(m-2)$–connected. If A is 1-connected then $\tilde{\pi}_{q-1}((\Omega A_+) \wedge P^q(p^t); \mathbb{Z}_{p^t}) = \tilde{\pi}_{q-1}(P^q(p^t); \mathbb{Z}_{p^t}) = \mathbb{Z}_{p^t}$.

Definition. Fix the space A. A homotopy operation χ of type $(m, p^s; q, p^t)$ under A is called ***additive*** if $\chi(\hat{x}_1 + c\hat{x}_2) = \chi(\hat{x}_1) + c\chi(\hat{x}_2)$ for $c \in \mathbb{Z}_{(p)}$ and for $\hat{x}_1, \hat{x}_2 \in \tilde{\pi}_m(K_g; \mathbb{Z}_{p^s})$.

Lemma 8.4. *χ is additive if and only if $F(\chi)$ is a co–H map. In particular, χ is additive if $F(\chi)$ is a suspension.*

Proof. To begin with, the coefficient $c \in \mathbb{Z}_{(p)}$ is irrelevant since over a cyclic coefficient group c must lie in the image of $\mathbb{Z}_+$, so it suffices to take $c = 1$.

The cases $m = 2$ and $m \geq 3$ can be handled simultaneously if we allow $\tilde{A}'$ to denote either $A'(2)$ or $A'(m,p^s)$ as needed, and if we likewise allow $\tilde{P}^m(p^s)$ to denote either S^2 or $P^m(p^s)$. Let $\lambda^m : \tilde{P}^m(p^s) \to \tilde{P}^m(p^s) \vee \tilde{P}^m(p^s)$ be the standard co–H structure map (arising from $\tilde{P}^m(p^s)$ being a suspension.) The co–H structure map for $\tilde{A}'$ is expressible as

$$\begin{aligned}\tilde{A}' = (\Omega A_+) \wedge \tilde{P}^m(p^s) &\xrightarrow{1\wedge\lambda^m} (\Omega A_+) \wedge (\tilde{P}^m(p^s) \vee \tilde{P}^m(p^s)) \\ &= ((\Omega A_+) \wedge \tilde{P}^m(p^s)) \vee ((\Omega A_+) \wedge \tilde{P}^m(p^s)) = \tilde{A}' \vee \tilde{A}'.\end{aligned}$$

Let $\tilde{A}''$ be the homotopy–theoretic fiber of $A \vee (\tilde{P}^m(p^s) \vee \tilde{P}^m(p^s)) \to A$. Then $\tilde{A}'' = (\Omega A_+) \wedge (\tilde{P}^m(p^s) \vee \tilde{P}^m(p^s))$, and the co–H structure map for $\tilde{A}'$ can be identified as $\mathfrak{g}(\lambda^m) : \tilde{A}' \to \tilde{A}''$.

The universal example for the additivity formula is the map pair $A \xrightarrow{f} A \vee (\tilde{P}^m(p^s) \vee \tilde{P}^m(p^s)) \xrightarrow{g} A$, for which $K_g = \tilde{A}''$. This fiber $\tilde{A}''$ has two canonical m–dimensional mod p^s homotopy classes (both are $(w_1)_\#$–images), denoted $\hat{v}_{(1)}$ and $\hat{v}_{(2)}$. χ will be additive if and only if $\chi(\hat{v}_{(1)} + \hat{v}_{(2)}) = \chi(\hat{v}_{(1)}) + \chi(\hat{v}_{(2)})$ in $\tilde{\pi}_q(\tilde{A}''; \mathbb{Z}_{p^t})$. It is easily checked that $\chi(\hat{v}_{(1)} + \hat{v}_{(2)}) = \chi((\lambda^m)_\#(\hat{v})) = \lambda^m_\# F(\chi)$ while (using (8G)) $\chi(\hat{v}_{(1)}) + \chi(\hat{v}_{(2)}) = (F(\chi) \vee F(\chi))_\#(\lambda^q)$. Thus the additivity of χ boils down to the formula $\lambda^m \circ F(\chi) \simeq (F(\chi) \vee F(\chi))\lambda^q$, which is precisely the co–H map criterion.

Definition. Let A be a space. A natural transformation which has degree m as a homomorphism of graded $\mathbb{Z}_{(p)}$–modules, from the functor $\tilde{\pi}_*(K_g; \mathbb{Z}_{p^t})$ on $\mathcal{M}_A$ to itself, may be identified with a list $\{\chi_q\}$ of homotopy operations under A, where χ_q has type $(q, p^t; q+m, p^t)$. Such a natural transformation or its associated list is called a ***stable homotopy operation of dimension*** m ***(under*** A***)*** if each χ_q is additive and if the diagrams

$$\text{(8N)}\qquad \begin{array}{ccc} P^{q+m+1}(p^t) & \xrightarrow{=} & \Sigma P^{q+m}(p^t) \\ {\scriptstyle F(\chi_{q+1})}\downarrow & & \downarrow{\scriptstyle \Sigma F(\chi_q)} \\ A'(q+1, p^t) & \xrightarrow{=} & \Sigma A'(q, p^t) \end{array}$$

all commute, for $q \geq 2$. (When $q = 2$ the lower right corner of (8N) must be replaced by $\Sigma A'(2)$ and the bottom arrow by $\mathfrak{g}(\rho^t)$; when $m + q = 2$ the upper right corner of (8N) must become ΣS^2, with the upper arrow being ρ^t.)

Remark. Because F is one–to–one, if $\{\chi_q\}$ is a stable operation then χ_q for $q > 3$ is determined by χ_3, and there is a compatibility condition between χ_2 and χ_3. Since χ_q must be zero for $q < 2$ (since $S(q, p^t; m+q, p^t) = 0$ there), we generally view the list $\{\chi_q\}$ as starting with χ_2, i.e., as $\{\chi_q\}_{q \geq 2}$.

Lemma 8.5. *Fix* $t \geq 1$. *Let* $\mathcal{O}_A = \mathcal{O}_A[t]$ *denote the* $\mathbb{Z}$*–graded* $\mathbb{Z}_{(p)}$*–module whose* m^{th} *component is precisely the* $\mathbb{Z}_{(p)}$*–module of* m*–dimensional stable homotopy operations under* A.

(a) $\mathcal{O}_A$ *is bounded below, with* $(\mathcal{O}_A)_m = 0$ *for* $m \leq -2$.
(b) $\mathcal{O}_A$ *is a graded ring under composition.*
(c) *There is a differential* d *of degree* -1 *on* $\mathcal{O}_A$ *which is a derivation, making* $\mathcal{O}_A$ *into a (non–connected) dga over* $\mathbb{Z}_{(p)}$.

Proof. Part (a) merely repeats Corollary 8.3(a). To define composition, let $\chi = \{\chi_q\}$ and $\omega = \{\omega_q\}$ be stable operations of dimensions m_1 and m_2, respectively. Define $\omega\chi = \{(\omega\chi)_q\}$ by $(\omega\chi)_q = \omega_{q+m_1}\chi_q$. Then

$$\begin{aligned} F((\omega\chi)_q) &= F(\omega_{q+m_1}\chi_q) = (\omega_{q+m_1}\chi_q)(\hat{v}_q) = \mathfrak{f}(\chi_q(\hat{v}_q))_{\#}\omega_{q+m_1}(\hat{v}_{q+m_1}) \\ &= \mathfrak{f}(F(\chi_q))_{\#}F(\omega_{q+m_1}), \end{aligned}$$

so

$$\begin{aligned} \Sigma F((\omega\chi)_q) &= \Sigma\mathfrak{f}(F(\chi_q))_{\#}(\Sigma f(\omega_{q+m_1})) = (\text{by (8J)})\ \mathfrak{f}(\Sigma F(\chi_q))_{\#}(\Sigma F(\omega_{q+m_1})) \\ &= \mathfrak{f}(F(\chi_{q+1}))_{\#}(F(\omega_{q+1+m_1})) = F((\omega\chi)_{q+1}). \end{aligned}$$

This verifies stability, except when $q = 2$ or $m_1 + m_2 + q = 2$. We leave these cases as exercises. Thus $\omega\chi \in \mathcal{O}_A$. Two–sided distributivity of composition over addition follows from the additivity of each ω_q and χ_q. The identity natural transformation is clearly a unit for $\mathcal{O}_A$.

To insert the differential into $\mathcal{O}_A$, let $\chi = \{\chi_q\}$ be stable of dimension m, and define $d\chi$ by

$$(d\chi)_q = {}_{\pi}\delta^t \circ \chi_q - (-1)^m \chi_{q-1} \circ_{\pi} \delta^t,$$

where ${}_{\pi}\delta^t$ is given by (8K). The formula $d(d\chi) = 0$ is easily checked. It is obvious that $(d\chi)_q$ is additive whenever both χ_q and χ_{q-1} are additive, since ${}_{\pi}\delta^t$ is additive. The derivation formula

$$(d(\omega\chi))_q = ((d\omega)\chi)_q + (-1)^{|\omega|}(\omega(d\chi))_q$$

is a straightforward calculation. The calculation which shows $\{(d\chi)_q\}$ to be stable begins with the formula

$$\begin{aligned} F(d\chi)_q &= \mathfrak{f}(F(\chi_q))_{\#}F({}_{\pi}\delta^t) - (-1)^m\mathfrak{f}(F({}_{\pi}\delta^t))_{\#}F(\chi_{q-1}) \\ &= \mathfrak{f}(F(\chi_q))_{\#}(w_1 \circ_{\pi} \delta^t(1)) - (-1)^m\mathfrak{f}(w_1 \circ_{\pi} \delta^t(1))_{\#}F(\chi_{q-1}) \\ &= (\text{by (8C)})\ F(\chi_q)_{\#}\ {}_{\pi}\delta^t(1) - (-1)^m\mathfrak{g}({}_{\pi}\delta^t(1))_{\#}F(\chi_{q-1}). \end{aligned}$$

When $q \geq 4$,

$$\begin{aligned} \Sigma F(d\chi)_q &= \Sigma F(\chi_q)_{\#}(\Sigma_{\pi}\delta^t(1)) - (-1)^m\Sigma\mathfrak{g}({}_{\pi}\delta^t(1))_{\#}(\Sigma F(\chi_{q-1})) \\ &= F(\chi_{q+1})_{\#}({}_{\pi}\delta^t(1)) - (-1)^m\mathfrak{g}({}_{\pi}\delta^t(1))_{\#}F(\chi_q) = F(d\chi)_{q+1}. \end{aligned}$$

We leave the remaining cases, $q = 3$ and $q = 2$, as exercises.

We consider next the detection of stable homotopy operations under A via a Hurewicz homomorphism. Note that

$$\overline{H}_{m+2}(A'(2); \mathbb{Z}_{p^t}) = \overline{H}_{m+2}(\Sigma^2(\Omega A_+); \mathbb{Z}_{p^t}) = H_m(\Omega A; \mathbb{Z}_{p^t}) \quad ,$$

and

$$\overline{H}_*(A'(q,p^t); \mathbb{Z}_{p^t}) = \overline{H}_*((\Omega A_+) \wedge P^q(p^t); \mathbb{Z}_{p^t}) = H_*(\Omega A; \mathbb{Z}_{p^t}) \otimes \mathbb{Z}_{p^t}(u,v),$$

where $\mathbf{Z}_{p^t}(u,v)$ denotes the free graded $\mathbf{Z}_{p^t}$–module on generators u of dimension $q-1$ and v of dimension q. Consider the compositions

$$S(q,p^t;q+m,p^t) \xrightarrow{F} \tilde{\pi}_{q+m}(A'(q,p^t);\mathbf{Z}_{p^t}) \xrightarrow{h} \overline{H}_{q+m}(A'(q,p^t);\mathbf{Z}_{p^t})$$
$$\text{(8O)} \qquad \xrightarrow{\approx} H_m(\Omega A;\mathbf{Z}_{p^t}) \oplus H_{m+1}(\Omega A;\mathbf{Z}_{p^t}), \text{ for } q \geq 3;$$

$$S(2,p^t;2+m,p^t) \xrightarrow{F} \tilde{\pi}_{m+2}(A'(2);\mathbf{Z}_{p^t}) \xrightarrow{h} \overline{H}_{m+2}(A'(2);\mathbf{Z}_{p^t})$$
$$\text{(8P)} \qquad \xrightarrow{\approx} H_m(\Omega A;\mathbf{Z}_{p^t}), \text{ for } q = 2.$$

The stability condition for elements of $\mathcal{O}_A$ leads immediately to the following result, whose proof we omit as trivial.

Lemma 8.6. *Let $\chi = \{\chi_q\} \in \mathcal{O}_A[t]$ have dimension m. Then the image of χ_q under the composition (8O) is independent of q, for $q \geq 3$. As a result there is a well–defined homomorphism $\mathcal{H}_0$ of degree zero and a second homomorphism $\mathcal{H}_1$ of degree +1,*

$$\mathcal{H} = (\mathcal{H}_0, \mathcal{H}_1) : \mathcal{O}_A[t] \to H_*(\Omega A;\mathbf{Z}_{p^t}) \oplus H_{*+1}(\Omega A;\mathbf{Z}_{p^t}),$$

for which $\mathcal{H}(\chi)$ equals the image under (8O) of any χ_q, $q \geq 3$. The component $\mathcal{H}_0(\chi)$ also equals the image under (8P) of χ_2.

Lemma 8.7. *Let δ^t denote the Bockstein differential on $H_*(\Omega A;\mathbf{Z}_{p^t})$, and recall that it makes the Pontrjagin ring $H_*(\Omega A;\mathbf{Z}_{p^t})$ into a dga. Then $\mathcal{H} = (\mathcal{H}_0, \mathcal{H}_1)$ satisfies the following properties.*

(a) $\mathcal{H}(1) = (1,0)$
(b) $\mathcal{H}(\chi \circ {}_\pi\delta^t) = (0, \mathcal{H}_0(\chi))$
(c) $\mathcal{H}d = \delta^t\mathcal{H}$, i.e., $\mathcal{H}_0(d\chi) = \delta^t\mathcal{H}_0(\chi)$ and $\mathcal{H}_1(d\chi) = \delta^t\mathcal{H}_1(X)$
(d) $\mathcal{H}_0(\omega\chi) = \mathcal{H}_0(\omega)\mathcal{H}_0(\chi) + \mathcal{H}_1(\omega)\delta^t\mathcal{H}_0(\chi)$
(e) $\mathcal{H}_1(\omega\chi) = \mathcal{H}_0(\omega)\mathcal{H}_1(\chi) + (-1)^{|\chi|}\mathcal{H}_1(\omega)\mathcal{H}_0(\chi) + \mathcal{H}_1(\omega)\delta^t\mathcal{H}_1(\chi)$.

Proof. Return to the notation $av + bu$ for $(a,b) \in H_*(\Omega A;\mathbf{Z}_{p^t}) \oplus H_{*+1}(\Omega A;\mathbf{Z}_{p^t})$. All of these can be deduced readily from the observations that

$$\begin{aligned}
\mathfrak{f}(F(\chi_q))_*(v) &= \mathcal{H}_0(\chi)v + \mathcal{H}_1(\chi)u;\\
\mathfrak{f}(F(\chi_q))_*(u) &= \mathfrak{f}(F(\chi_q))_*\delta^t(v) = \delta^t\mathfrak{f}(F(\chi_q))_*(v)\\
&= (\delta^t\mathcal{H}_0(\chi))v + ((-1)^{|\chi|}\mathcal{H}_0(\chi) + \delta^t\mathcal{H}_1(\chi))u; \text{ and}\\
\mathfrak{f}(F(\chi_q))_*(av+bu) &= a\mathfrak{f}(F(\chi_q))_*(v) + b\mathfrak{f}(F(\chi_q))_*(u).
\end{aligned}$$

The last formula here comes from the fact that $\mathfrak{f}(F(\chi_q))$ is a map of left ΩA–modules.

Corollary 8.8. *Fix $t \geq 1$. $\mathrm{Ker}(\mathcal{H}_1)$ is a sub–dga of $\mathcal{O}_A$, and $\mathcal{H}_0 : \ker(\mathcal{H}_1) \to H_*(\Omega A;\mathbf{Z}_{p^t})$ is a dga homomorphism.*

Let us examine next how $\mathcal{O}_A$ and $\ker(\mathcal{H}_1)$ can serve as generalizations of the relative Samelson product. For $m \geq 1$, let $\hat{z} \in \pi_m(\Omega A;\mathbf{Z}_{p^t})$. Then $\hat{z}$ determines an m–dimensional stable homotopy operation under A, denoted $\Phi(\hat{z})$, as follows. Given a map pair (f,g), let $\Phi(\hat{z})$ be the relative Whitehead product $[f_\#(\hat{z}^{\bar{a}}), -]$, which determines a homotopy operation of type $(q,p^t;q+m,p^t)$ under A, for any $q \geq 2$. The well–known properties of the Whitehead product assure us that the operation on $\tilde{\pi}_*(K_g;\mathbf{Z}_{p^t})$ defined this way is stable. The Jacobi identity and the derivation formula for Samelson products lead us to the following delightful lemma.

Lemma 8.9. *Fix $t \geq 1$ and the space A. $\Phi : \pi_*(\Omega A; \mathbb{Z}_{p^t}) \to \mathcal{O}_A$ is a chain map, and Φ preserves brackets in the sense that*

$$\Phi([\hat{x}, \hat{y}]) = \Phi(\hat{x})\Phi(\hat{y}) - (-1)^{|\hat{x}||\hat{y}|}\Phi(\hat{y})\Phi(\hat{x}).$$

As a result, Φ extends to a dga homomorphism

$$\Phi : \mathcal{U}\pi_*(\Omega A; \mathbb{Z}_{p^t}) \to \mathcal{O}_A \tag{8Q}$$

Furthermore, $\mathrm{im}(\Phi) \subseteq \ker(\mathcal{H}_1)$, *and $\mathcal{H}_0\Phi$ coincides with the usual* $\bmod p^t$ *Hurewicz homomorphism for ΩA.*

So far we have been treating A as fixed, but everything we have done is actually functorial in A. To see this, let $\alpha : A \to B$ be any map. Let (f, g) be a map pair in $\mathcal{M}_B$ and let χ_q be an operation under A of type $(q, p^s; q+m; p^t)$. Because $(f\alpha, g) \in \mathcal{M}_A$, χ_q takes $\tilde{\pi}_q(K_g; \mathbb{Z}_{p^s})$ naturally to $\tilde{\pi}_{q+m}(K_g; \mathbb{Z}_{p^t})$. Thus χ_q determines an operation of the same type under B. We leave it as an exercise to check that stability of a list $\{\chi_q\}$ is preserved this way, whence α determines $\mathcal{O}_{(\alpha)} : \mathcal{O}_A \to \mathcal{O}_B$. The verification that $\mathcal{O}_{(\ )}$ is now a functor is straightforward. We summarize and extend this information in the following lemma.

Lemma 8.10. *Fix $t \geq 1$. $\mathcal{O}_{(\ )}[t]$ is a functor from spaces to dga's over $\mathbb{Z}_{p^t}$, and $\mathcal{H}_0, \mathcal{H}_1, \Phi$ are natural transformations.*

Remark. When $A = *$ is a point, then the universal object $A'(q, p^t)$ is simply $P^q(p^t)$, and $\mathcal{O}_{(*)}$ reduces to the ring of "absolute" stable $\bmod p^t$ homotopy operations. The ring $\mathcal{O}_{(*)}$ differs from what is usually understood by the term "ring of stable homotopy operations" in the following ways. In the latter ring, lists $\{\chi_q\}$ are acceptable if they are eventually stable (i.e. if (8N) commutes for $q >> 0$), and two lists are identified if they coincide for almost all q, whereas we have required stability starting at χ_2 and different lists are never collapsed. It is possible to view the starting dimension of 2 as an adjustable parameter for $\mathcal{O}_{(\ )}$, or to define a "classically stable" version of $\mathcal{O}_{(\ )}$ for which (8N) is only required to commute for $q >> 0$ and lists that coincide for $q >> 0$ are set equal. We leave these variations for the interested reader to explore. For now we note merely that the functoriality of $\mathcal{O}_{(\ )}$ makes $\mathcal{O}_{(*)}$ a retract of $\mathcal{O}_A$, for any A. For instance, the operation ${}_\pi\delta^t$ mentioned in Corollary 8.3(b) occurs in every $\mathcal{O}_A$ because it exists in $(\mathcal{O}_{(*)})_{-1}$.

We proceed next to give a necessary and sufficient condition for $\mathcal{H}_0$ to be surjective. This result is followed by a number of remarks and open problems. We first detour to prove an elementary topological fact.

Lemma 8.11. *Let X be simply-connected and of finite type. The following are equivalent.*

(i) *X has the p-local homotopy type of a wedge of spheres and Moore spaces;*
(ii) *The Hurewicz homomorphism $h : \tilde{\pi}_*(X; \mathbb{Z}_{p^t}) \to \overline{H}_*(X; \mathbb{Z}_{p^t})$ surjects, for all $t \geq 1$.*

Proof. That (i) implies (ii) is easy, since h is onto for all t for an individual sphere or Moore space, and $\overline{H}_*(\ ; \mathbb{Z}_{p^t})$ converts wedges into direct sums.

Assuming (ii), we now prove (i). Since each $H_m(X)$ is a finitely generated abelian group, write $H_m(X; \mathbb{Z}_{(p)}) = F_m \oplus T_m$, where $F_m = \mathbb{Z}_{(p)}^{r_m}$ is $\mathbb{Z}_{(p)}$-free and the torsion component is

$T_m = \bigoplus_{j=1}^{s_m}(\mathbb{Z}/p^{a_{mj}}\mathbb{Z})$. Let $W = \bigvee_{m=2}^{\infty}\bigvee_{j=1}^{s_m} P^{m+1}(p^{a_{mj}})$, and use condition (ii) to choose a map $f:\ W \to X$ such that $H_*(f;\mathbb{Z}_{(p)})$ equals the inclusion of T_* into $H_*(X;\mathbb{Z}_{(p)})$. We will show inductively that for each $m \geq 1$ there is a map $f_m:\ W_m \to X$, where W_m is a 1–connected finite type wedge of spheres and Moore spaces, for which $H_*(f_m;\mathbb{Z}_{(p)})$ equals the inclusion of $F_{\leq m} \oplus T_*$ into X. This statement holds true at $m = 1$, with $W_1 = W$ and $f_1 = f$. Let us see how f_{m+1} is obtained from f_m.

Since $H_{\leq m}(X, W_m;\mathbb{Z}_{(p)}) = 0$ but $H_{m+1}(X, W_m;\mathbb{Z}_{(p)}) = \mathbb{Z}_{(p)}^{r_{m+1}}$, (X, W_m) must be p–locally an m–connected relative CW complex having precisely r_{m+1} $(m+1)$–cells. Thus there is for some attaching map α a cofibration sequence and a factorization of f_m,

$$\begin{array}{ccccccc}
\bigvee_{j=1}^{r_{m+1}} S^m & \xrightarrow{\alpha} & W_m & \longrightarrow & W_{m+1}(=\text{cofiber}) & \longrightarrow & \bigvee_{j=1}^{r_{m+1}} S^{m+1} \\
 & & & f_m \searrow & \downarrow f_{m+1} & & \\
 & & & & X & &
\end{array}$$

Here f_{m+1} makes (X, W_{m+1}) $(m+1)$–connected, and $H_*(f_{m+1};\mathbb{Z}_{(p)})$ is indeed the inclusion of $F_{\leq m+1} \oplus T_*$ into $H_*(X;\mathbb{Z}_{(p)})$. So the inductive step will be complete if we can show that $\alpha = 0$ (because then W_{m+1} will be a wedge of spheres and Moore spaces). Actually, when $m = 1$ α is automatically null because W_1 is 1–connected, so we can assume that $m \geq 2$.

To get α to be null-homotopic when $m \geq 2$, let $t \geq 1$ and use hypothesis (ii) to choose r_{m+1} maps $P^{m+1}(p^t) \to X$ whose Hurewicz images are the mod p^t reductions of a basis for F_{m+1}, sitting in $H_{m+1}(X;\mathbb{Z}_{p^t})$. Because f_{m+1} is $(m+1)$–connected these maps will lift through f_{m+1} to W_{m+1}, and the homomorphism induced on $H_{m+1}(\ ;\mathbb{Z}_{p^t})$ by the wedge sum of all these maps will be an isomorphism when continued to $H_{m+1}(\bigvee_{j=1}^{r_{m+1}} S^{m+1};\mathbb{Z}_{p^t})$. We have constructed the diagram

$$\text{(8R1)}\qquad \begin{array}{ccccccc}
\bigvee S^m & \xrightarrow{p^t} & \bigvee S^m & \longrightarrow & \bigvee P^{m+1}(p^t) & \longrightarrow & \bigvee S^{m+1} \\
\vdots\downarrow & & \vdots\downarrow & & \downarrow & & \downarrow \simeq \text{(p-locally)} \\
\bigvee S^m & \xrightarrow{\alpha} & W_m & \longrightarrow & W_{m+1} & \longrightarrow & \bigvee S^{m+1}, \\
 & & & f_m \searrow & \downarrow f_{m+1} & & \\
 & & & & X & &
\end{array}$$

in which all wedges run from $j = 1$ to $j = r_{m+1}$.

The dotted arrows of (8R1) can be filled in, because in dimensions $\leq m$ both rows are fibrations as well as cofibrations. When they are filled in, the left–most vertical arrow of (8R1) is a p–local equivalence because the right–most arrow is. Conclude that α factors through p^t. But (8R1) exists for every t, so α must be divisible by p^t, for every t. Because W_m is 1–connected and of finite type, the only element of $[\bigvee S^m; W_m] = \bigoplus_{j=1}^{r_{m+1}} \pi_m(W_m)$ that is divisible by p^t for all t is $\alpha = 0$. This finishes the inductive step. Letting $f_\infty = \lim_{\rightarrow}\{f_m\}:\ W_\infty \to X$, we have that $H_*(f_\infty;\mathbb{Z}_{(p)})$ is an isomorphism, hence f_∞ is a p–local homotopy equivalence, while W_∞ is a wedge of spheres and Moore spheres.

Corollary 8.12. *Let A be 1–connected. The following are equivalent.*

(i) $\Sigma^2\Omega A$ *is p–locally a wedge of spheres and Moore spaces.*
(ii) $h: \tilde{\pi}_*(\Sigma^2\Omega A;\mathbb{Z}_{p^t}) \to \overline{H}_*(\Sigma^2\Omega A;\mathbb{Z}_{p^t})$ *surjects, for all* $t \geq 1$.
(iii) $\mathcal{H}_0 : \mathcal{O}_A[t] \to H_*(\Omega A;\mathbb{Z}_{p^t})$ *surjects, for all* $t \geq 1$.
(iv) $\mathcal{H}_0 : \ker(\mathcal{H}_1) \to H_*(\Omega A;\mathbb{Z}_{p^t})$ *surjects, for all* $t \geq 1$.

Proof. That (i) $\Leftrightarrow$ (ii) is an application of Lemma 8.11. That (iii) $\Rightarrow$ (ii) is really just Lemma 8.6 (specifically, the image of χ_2 under (8P)), along with the fact that $A'(2) = S^2 \vee \Sigma^2\Omega A$. That (iv) implies (iii) is trivial, and for (iii) $\Rightarrow$ (iv) let $x \in H_m(\Omega A;\mathbb{Z}_{p^t})$ and pick $\chi \in (\mathcal{O}_A)_m$ having $\mathcal{H}_0(\chi) = x$. Then pick $\omega \in (\mathcal{O}_A)_{m+1}$ for which $\mathcal{H}_0(\omega) = \mathcal{H}_1(\chi)$. By Lemma 8.7(b), $\chi - \omega_\pi\delta^t$ lies in $\ker(\mathcal{H}_1)$ while $\mathcal{H}_0(\chi - \omega_\pi\delta^t) = x$.

To see that (i) implies (iii), let

$$(\bigvee_\beta w_\beta) \vee (\bigvee_\alpha y_\alpha) : (\bigvee_\beta S^{m(\beta)}) \vee (\bigvee_\alpha P^{m(\alpha)}(p^{t(\alpha)})) \to \Sigma^2\Omega A \vee S^2 = A'(2)$$

be a p–local homotopy equivalence. The Hurewicz images of the set

$$\text{(8R2)} \qquad \hat{W} = \{w_\beta\rho^t\} \cup \{y_\alpha\omega^{t(\alpha)}\rho^t\} \cup \{y_\alpha\rho_t^{t(\alpha)} \mid t(\alpha) \leq t\} \cup \{y_\alpha\omega_t^{t(\alpha)} \mid t(\alpha) > t\}$$

span $\overline{H}_*(A'(2);\mathbb{Z}_{p^t})$. Letting $\hat{w}$ be a typical element of $\hat{W}$, it suffices to show that the typical spanning element $h(\hat{w})$ belongs to $\mathrm{im}(\mathcal{H}_0)$. Let $\chi_2 = F^{-1}(\hat{w})$, which exists by Lemma 8.2. We claim that χ_3 exists such that (8N) commutes (after setting $q = 2$ in (8N)). Granting this claim, put $\chi_j = F^{-1}\Sigma^{j-3}F(\chi_3)$ for $j \geq 4$. Then $\chi = \{\chi_j\}_{j\geq 2}$ belongs to $\mathcal{O}_A[t]$ and has $\mathcal{H}_0(\chi) = hF(\chi_2) = h(\hat{w})$.

To prove the claim, let $P(\hat{w})$ denote the sphere or Moore space on which $\hat{w}$ lives (i.e. $P(\hat{w})$ is $S^{m(\beta)}$ or $P^{m(\alpha)}(p^{t(\alpha)})$) and let $\iota(\hat{w})$: $P(\hat{w}) \to A'(2)$ be its inclusion map into $A'(2)$. The relevant diagram is

$$\text{(8R3)} \qquad \begin{array}{ccccccccc} P(\hat{w}) & \xrightarrow{p^t} & P(\hat{w}) & \longrightarrow & G(\hat{w}) & \longrightarrow & \Sigma P(\hat{w}) & & \\ {\scriptstyle \iota(\hat{w})}\downarrow & & \vdots{\scriptstyle \iota'(\hat{w})} & & \vdots{\scriptstyle \iota''(\hat{w})} & & \downarrow{\scriptstyle \Sigma\iota(\hat{w})} & & \\ A'(2) & \xrightarrow{p^t} & A'(2) & \longrightarrow & A'(3,p^t) & \longrightarrow & \Sigma A'(2) & \xrightarrow{=} & A'(3)\,. \end{array}$$

There exists a map $\iota'(\hat{w})$ which makes the left square of (8R3) commute up to homotopy. To construct $\iota'(\hat{w})$, we use the fact, which we saw in connection with Lemma 8.2, that $A'(2) \xrightarrow{p^t} A'(2)$ induces zero on $\pi_*(A'(2);\mathbb{Z}_{p^t})$. If $P(\hat{w})$ is an S^2 or a Moore space of order $\leq t$, we can take $\iota'(\hat{w}) = \iota(\hat{w})$. If $P(\hat{w})$ is an S^q with $q \geq 3$, then the composite $(p^t) \circ \iota(\hat{w}) \circ \omega^t$ is null, and we let $\iota'(\hat{w})$ be the induced map between cofibers. The case where $P(\hat{w})$ is a Moore space of order $> t$ is handled in a similar manner.

Having constructed $\iota'(\hat{w})$, let $G(\hat{w})$ be the cofiber of $P(\hat{w}) \xrightarrow{p^t} P(\hat{w})$; there exists a map $\iota''(\hat{w})$: $G(\hat{w}) \to A'(3,p^t)$ making the right–hand square of (8R3) commute up to homotopy. To construct χ_3 we again consider cases, depending upon which component of the union (8R2) contains $\hat{w}$. If $\hat{w}$ has the form $w_\beta\rho^t$ we simply put $\chi_3 = F^{-1}(\iota''(\hat{w}))$. If $P(\hat{w})$ is the

α^{th} Moore space $P^{m(\alpha)}(p^{t(\alpha)})$ and $t(\alpha) \le t$, then the arrow from $G(\hat{w})$ to $\Sigma P(\hat{w})$ in (8R3) has a right inverse, call it $\sigma(\hat{w})$. Take $F(\chi_3)$ to be the composite

$$\Sigma P^{|\hat{w}|}(p^t) \xrightarrow{\Sigma\hat{w}} \Sigma P(\hat{w}) \xrightarrow{\sigma(\hat{w})} G(\hat{w}) \xrightarrow{\iota''(\hat{w})} A'(3, p^t) .$$

Lastly, if $P(\hat{w}) = P^{m(\alpha)}(p^{t(\alpha)})$ and $t(\alpha) > t$, then $G(\hat{w}) \approx P^{m(\alpha)}(p^t) \vee P^{m(\alpha)+1}(p^t)$. If $\hat{w} = y_\alpha \omega^{t(\alpha)} \rho^t$ we take $F(\chi_3)$ to be the composition

$$P^{m(\alpha)}(p^t) \to G(\hat{w}) \xrightarrow{\iota''(\hat{w})} A'(3, p^t) ,$$

and if $\hat{w} = y_\alpha \omega_t^{t(\alpha)}$ use instead the composition

$$P^{m(\alpha)+1}(p^t) \to G(\hat{w}) \xrightarrow{\iota''(\hat{w})} A'(3, p^t) .$$

Remark 8.13. Recall that $X \in \mathcal{W}$ means that X has the p–local homotopy type of a finite type 1–connected wedge of Moore spaces and spheres. Let us contrast the criterion $\Sigma^2\Omega A \in \mathcal{W}$ with the stronger criterion $\Sigma\Omega A \in \mathcal{W}$. The latter condition was studied in [A5], where two equivalent conditions were found. The statement "$\Sigma\Omega A \in \mathcal{W}$" is equivalent to ΩA factoring into certain kinds of pieces, and it is also equivalent to (expressed in our terms) $\mathcal{H}_0 \circ \Phi$ being surjective (Φ as in (8Q)). Thus the change from $\Sigma\Omega A$ splitting to $\Sigma^2\Omega A$ splitting is precisely the change from $\mathcal{H}_0\Phi$ being onto to $\mathcal{H}_0$ being onto.

Spaces for which $\Sigma\Omega A \notin \mathcal{W}$ but $\Sigma^2\Omega A \in \mathcal{W}$ seem particularly interesting from two perspectives. First, the maximal wedge decomposition of $\Sigma\Omega A$ must contain some intriguing co–H–spaces that are not in $\mathcal{W}$ but whose suspensions are. The attaching maps in such co–H–spaces must all be unstable (zero after one suspension) co–H–maps. (We are allowing the attaching maps here to have as source either a sphere or a Moore space.) Second, the ring $\mathcal{O}_A$ for these spaces will contain some stable homotopy operations that are genuinely "new" in the sense that they cannot be written in terms of iterated Samelson products and compositions. Map pairs out of A permit one or more operations on $\tilde{\pi}_*(K_g; \mathbb{Z}_{p^t})$ which "come from nowhere" in the sense that these operations have no interpretation in terms of $\pi_*(A; \mathbb{Z}_{p^t})$. The author thinks of these as secondary homotopy operations, by analogy with secondary cohomology operations. Two examples of finite complexes for which $\Sigma^2\Omega A \in \mathcal{W}$ while $\Sigma\Omega A \notin \mathcal{W}$ are the $(p-1)^{\text{st}}$ stage of the James construction on an even sphere, denoted $J_{p-1}(S^{2n})$ or $\widehat{S}^{2n}$, and the spaces D_k that originally motivated this monograph.

The space $\widehat{S}^{2n}$ is illustrative. $H_*(\Omega\widehat{S}^{2n}; \mathbb{Z}_{(p)})$ is the commutative ring $\Lambda(a_{2n-1}) \otimes \mathbb{Z}_{(p)}[c_{2np-2}]$. When $n \ge 2$ the p–local analog of Hopf invariant one says that $c_{2np-2} \notin \text{im}(h)$, whence $\Sigma\Omega\widehat{S}^{2n} \notin \mathcal{W}$. However, the composition

$$\Sigma^2 S^{2n-1} \to \Sigma^2\Omega\widehat{S}^{2n} \to \Sigma^2\Omega^2 S^{2n+1} \to S^{2n+1}$$

shows S^{2n+1} to be a retract, and hence a wedge summand, of the co–H space $\Sigma^2\Omega\widehat{S}^{2n}$. The complementary wedge summand is $(2np-1)$–connected, so the twice–suspended element $\Sigma^2 c_{2np-2}$ (in $H_{2np}(\Sigma^2\Omega\widehat{S}^{2n})$) does belong to $\text{im}(h)$. When $n > 1$ we have immediately that all of $H_{\le 2np+1}(\Sigma^2\Omega\widehat{S}^{2n}; \mathbb{Z}_{p^t})$ is contained in $\text{im}(h)$ for any t, whence

$H_{\leq 2np-2}(\Omega\widehat{S}^{2n};\mathbb{Z}_{p^t})$ lies in $\mathcal{H}_0(\ker(\mathcal{H}_1))$ by the argument of 8.12(ii) $\Rightarrow$ 8.12(iv). Since $\mathcal{H}_0$ is multiplicative on $\ker(\mathcal{H}_1)$ and both multiplicative generators of $H_*(\Omega\widehat{S}^{2n};\mathbb{Z}_{p^t})$ have been shown to lie in $(\mathcal{H}_0)$, deduce that $\mathcal{H}_0$ is onto. By Corollary 8.12, $\Sigma^2\Omega\widehat{S}^{2n} \in \mathcal{W}$. This fact was already well-known, but with this concise proof we hope to illustrate the power and versatility of our techniques. There is a stable homotopy operation under $\widehat{S}^{2n}$ whose $\mathcal{H}_0$-image is c_{2np-2}, call it $\gamma \in (\mathcal{O}_{\widehat{S}^{2n}})_{2np-2}$, and γ may be useful in the study of the EHP sequence. While we are discussing $\widehat{S}^{2n}$, let us further remark (cf. the previous paragraph) that $\Sigma\Omega\widehat{S}^{2n}$ does have as a retract the interesting two-cell complex $\Sigma(S^{2n-1}\cup_{w_n} e^{2np-2})$, whose attaching map is the first non-zero element in the kernel of the double suspension.

When $\Sigma^2\Omega A \in \mathcal{W}$, it is obvious that $H^*(\Omega A;\mathbb{Z}_p)$ cannot have non-zero $\mathcal{P}^i$ Steenrod operations. A conjecture of McGibbon and Wilkerson [MW] asserts that $\mathcal{P}^i$ is trivial on $H^*(\Omega A;\mathbb{Z}_p)$ for almost all p when A is finite and 1-connected, so Corollary 8.12 may be relevant to proving the McGibbon-Wilkerson conjecture. Recalling our earlier remarks that there is nothing sacred about starting with $A'(2)$, we note that for each $l \geq 2$ an analog of Corollary 8.12 will hold, equating the condition $\Sigma^l\Omega A \in \mathcal{W}$ with the surjectivity of $\mathcal{H}_0$ for the analog of $\mathcal{O}_A$ obtained by requiring the commutativity of (8N) only for $q \geq l$. Or we could consider the "classical stability" analog of $\mathcal{O}_A$ and $\mathcal{H}_0$, and relate it to splitting $\Sigma^\infty\Omega A$. Any of these could also yield an approach to proving the McGibbon-Wilkerson conjecture.

We proceed next to list several open questions regarding the ring of stable homotopy operations under A. Then we consider $\pi_*(\Omega K_g;\mathbb{Z}_{p^t})$ as a module over $\mathcal{O}_A$, and we finish up this chapter by examining the case of $A = P^{l+1}(p^r)$.

Questions 8.14. First, what is the kernel of Φ? We shall see at the end of Chapter 11 that $\ker(\Phi)$ can be non-zero. Next, suppose a space A does satisfy $\Sigma^2\Omega A \in \mathcal{W}$ and hence $\mathcal{H}_0$ is surjective. Must $\mathcal{H}_0|_{\ker(\mathcal{H}_1)}$ have a section as dga's? For spaces A that allow $\mathcal{H}_0|\ker(\mathcal{H}_1)$ to have a section, we obtain a new interpretation of the Pontrjagin ring $H_*(\Omega A;\mathbb{Z}_{p^t})$, since we can view it as a ring of homotopy operations on $\tilde{\pi}_*(K_g;\mathbb{Z}_{p^t})$. For $A = D_k$ such a section does exist. For $A = \widehat{S}^{2n}$ it is equivalent to the question of whether $\gamma \in (\mathcal{O}_{\widehat{S}^{2n}})_{2np-2}$ can be chosen in such a way that it commutes with $\Phi(\hat{a}_{2n-1})$, where $h(\hat{a}_{2n-1}) = a_{2n-1}$ (notation is being continued from our previous discussion about $\widehat{S}^{2n}$). When such a section does exist for all $t \geq 1$, we can observe that $\{H_*(\Omega A;\mathbb{Z}_{p^t})\}$ is a csa that acts termwise on the csc $\{\tilde{\pi}_*(K_g;\mathbb{Z}_{p^t})\}$; will this describe a csm module structure? Again, for our prototype space D_k the answer will be "yes". It seems unlikely that $\{\mathcal{O}_A[t]\}_{t\geq 1}$ forms a csa in general (the obstacle arises in defining θ^t_{t-1}). A curious situation occurs: the images of (8Q) as t varies form a csa, even though neither the domains nor the ranges of (8Q) typically comprise a csa as t varies! Are there natural sub-dga's of $\mathcal{O}_A[t]$ larger than the image of (8Q) that together comprise a csa? A clue is provided by the reasonable expectation that if such a structure exists then $\mathcal{H}_0$ will be a morphism of csa's. If so, will $\{\tilde{\pi}_*(K_g;\mathbb{Z}_{p^t})\}$ be a csm over this csa? Lastly, is there a functorial chain complex C_A for which $\mathcal{O}_A[t]$ is $H_*(C_A;\mathbb{Z}_{p^t})$? If so, we can expect to find a small Adams-Hilton-like model for this chain complex that would permit us to compute $\mathcal{O}_A$, and we would have a powerful tool for addressing the other questions listed here as well.

If we prefer, we can regard $\pi_*(\Omega K_g;\mathbb{Z}_{p^t})$ as the graded module on which $\mathcal{O}_A$ acts functorially, rather than $\tilde{\pi}_*(K_g;\mathbb{Z}_{p^t})$. This approach has some advantages. For one thing, we are dealing with a csL instead of just a csc, so the possibility arises of additional natural relationships. Second, the Hurewicz homomorphism now takes its image in $H_*(\Omega K_g;\mathbb{Z}_{p^t})$, a considerably richer object than $H_*(K_g;\mathbb{Z}_{p^t})$, so it has the potential to reveal more information about stable homotopy operations under A. The unifying concept for all these results is the conjugation action of the total space upon the fiber, for a looped fibration.

Lemma 8.15.

(a) *Whenever (C, e) is a differential graded module over a dga (A, d), then $\mathcal{F}(C, e)$ is a csm over $\mathcal{F}(A, d)$, and $\{H_*(C, e; \mathbb{Z}_{p^t})\}$ is a csm over $\{H_*(A, d; \mathbb{Z}_{p^t})\}$.*

(b) *For any fibration $X \to Y \to Z$, the conjugation map*

$$\tilde{c}:\ \Omega Y \ltimes \Omega X \to \Omega X$$

given by $\tilde{c}(y, x) = yxy^{-1}$ induces on $\{\overline{H}_*(\Omega X;\mathbb{Z}_{p^t})\}$ the structure of a csm over the csa $\{H_*(\Omega Y;\mathbb{Z}_{p^t})\}$. This structure is natural on the category of fibrations.

(c) *On $\mathcal{M}_A$, $\{\overline{H}_*(\Omega K_g;\mathbb{Z}_{p^t})\}$ is naturally a csm over the csa $\{H_*(\Omega A;\mathbb{Z}_{p^t})\}$. We denote the csm structure homomorphism by $\tilde{\tilde{c}}_*$, to remind us of its close tie to the map $\tilde{c}$.*

Proof. Part (a) is straightforward, and (b) is an application of (a). For part (c), let (f, g) be a map pair, and let $\tilde{\tilde{c}} = \tilde{c}(\Omega f \ltimes 1) :\ \Omega A \ltimes \Omega K_g \to \Omega K_g$. Then $\tilde{\tilde{c}}_*$ is a natural csm structure homomorphism.

Proposition 8.16. *Let $t \geq 1$. For any map pair (f, g) in $\mathcal{M}_A$, this diagram commutes:*

$$\begin{array}{ccc}
\ker(\mathcal{H}_1) \otimes \pi_*(\Omega K_g;\mathbb{Z}_{p^t}) & \longrightarrow & \pi_*(\Omega K_g;\mathbb{Z}_{p^t}) \\
\Big\downarrow{\scriptstyle \mathcal{H}_0 \otimes h} & & \Big\downarrow{\scriptstyle h} \\
H_*(\Omega_A;\mathbb{Z}_{p^t}) \otimes \overline{H}_*(\Omega K_g;\mathbb{Z}_{p^t}) & \xrightarrow{\ \tilde{\tilde{c}}_*\ } & \overline{H}_*(\Omega K_g;\mathbb{Z}_{p^t}).
\end{array}$$

A formula that says the same thing is

(8S) $$\tilde{\tilde{c}}_*(\mathcal{H}_0(\chi) \otimes h(\hat{y})) = h(\chi(\hat{y})),$$

whenever $\hat{y} \in \pi_*(\Omega K_g;\mathbb{Z}_{p^t})$ and $\chi \in \mathcal{O}_A[t]$ with $\mathcal{H}_1(\chi) = 0$.

Proof. We quickly reduce to the universal case, which is $K_g = A'(q, p^t)$ or $A'(2)$ and $\hat{y} = \hat{v}_q^a$, where a superscript $(\ )^a$ denotes adjoint. We cover the case of $A'(q, p^t)$ only, since the case of $A'(2)$ is so similar.

Because $\Omega A'(q, p^t) = \Omega\Sigma A'(q-1, p^t)$, the target group for (8S) is a direct sum,

(8T) $$\overline{H}_*(\Omega A'(q, p^t);\mathbb{Z}_{p^t}) = \bigoplus_{m=1}^{\infty} \overline{H}_*((A'(q-1, p^t))^{\wedge m};\mathbb{Z}_{p^t}),$$

in which we will write an element w in terms of its components as $w = w_1 + w_2 + w_3 + \ldots$. The right–hand side of (8S) for the case we are evaluating is

$$h(\chi(\hat{v}_q^a)) = h(\chi(\hat{v}_q)^a) = h(F(\chi_q)^a),$$

call it w. Then the $m = 1$ component of w is $w_1 = h(F(\chi_q))$ because the suspension of the adjoint, composed with the evaluation map $\Sigma\Omega(\ ) \to (\ )$, is just the identity, and w_1 is measuring the homology image of this composition. Moreover, the components $w_{\geq 2}$ are zero because they are the homology images of James–Hopf invariants, which must vanish for the additive homotopy class $F(\chi_q)$. Because $\mathcal{H}_1(\chi) = 0$, we have $w = w_1 = h(F(\chi_q)) = \mathcal{H}_0(\chi) \otimes v_{q-1}$.

To evaluate the left–hand side of (8S), we make use of the diagram

$$\text{(8U)} \qquad \begin{array}{ccc} \Omega A \ltimes P^{q-1}(p^t) = A'(q-1, p^t) & \xrightarrow{1^a} & \Omega\Sigma A'(q-1, p^t) \\ {\scriptstyle 1\ltimes(\hat{v}_q)^a}\downarrow & & \downarrow{\scriptstyle =} \\ \Omega A \ltimes \Omega A'(q, p^t) & \xrightarrow{\tilde{c}} & \Omega A'(q, p^t), \end{array}$$

whose commutativity up to homotopy is left as an exercise. The left–hand side of (8S) for the case we are doing is $\tilde{\tilde{c}}_*(\mathcal{H}_0(\chi)\otimes(\hat{v}_q)^a_*(v_{q-1}))$, which by (8U) (here $\tilde{c} = \tilde{\tilde{c}}$) is $(1^a)_*(\mathcal{H}_0(\chi)\otimes v_{q-1})$. The image of $(1^a)_*$ lies entirely in the $m = 1$ component of (8T), so the left–hand side of (8S) is simply $\mathcal{H}_0(\chi) \otimes v_{q-1}$, as needed.

Corollary 8.17. *Suppose A has the property that $\mathcal{H}_0 :\ \ker(\mathcal{H}_1) \to H_*(\Omega A; \mathbf{Z}_{p^t})$ does have a section $\Psi_{[t]}$ as dga's, for each $t \geq 1$, and suppose that*

$$z \otimes \hat{x} \mapsto \Psi_{[t]}(z)(\hat{x}) \text{ for } z \in H_*(\Omega A; \mathbf{Z}_{p^t}), \hat{x} \in \pi_*(\Omega K_g; \mathbf{Z}_{p^t})$$

gives to $\{\pi_*(\Omega K_g; \mathbf{Z}_{p^t})\}$ a natural structure as a csm over $\{H_*(\Omega A; \mathbf{Z}p^t)\}$. Then the Hurewicz homomorphism

$$h :\ \{\pi_*(\Omega K_g; \mathbf{Z}_{p^t})\}_{t\geq 1} \to \{\overline{H}_*(\Omega K_g; \mathbf{Z}_{p^t})\}_{t\geq 1}$$

is a natural homomorphism of csm's over $\{H_*(\Omega A; \mathbf{Z}_{p^t})\}_{t\geq 1}$, where the csm structure on the target is given by Lemma 8.15(c).

Proof. Since h was already seen to be a homomorphism of csc's in Lemma 7.3(d), we need only to check that $h(\Psi_{[t]}(z)(\hat{x})) = \tilde{\tilde{c}}_*(z \otimes h(\hat{x}))$. But this is precisely what (8S) tells us.

Remark. By Lemma 7.3(g) h is also bracket–preserving, so it is both a csL and a csm homomorphism. In view of Definition 7.11 it is tempting to imagine that one or both of $\{\pi_*(\Omega K_g; \mathbf{Z}_{p^t})\}$ and $\{H_*(\Omega K_g; \mathbf{Z}_{p^t})\}$ might be a csL over $\{H_*(\Omega A; \mathbf{Z}_{p^t})\}$. The author suspects that there is a concept of a "csa over a Dcsa", that $\{H_*(\Omega A; \mathbf{Z}_{p^t})\}$ always has a natural diagonal structure, and that $\{H_*(\Omega K_g; \mathbf{Z}_{p^t})\}$ is always a csa over this natural Dcsa structure. We shall see in Chapter 10 that $\pi_*(\Omega K_g; \mathbf{Z}_{p^t})$ is indeed naturally a csL over $\{H_*(\Omega D_k; \mathbf{Z}_{p^t})\} = \mathcal{F}(\mathcal{U}M^k)$ for $(f, g) \in \mathcal{M}_{D_k}$, so this kind of structure can occur for at least some spaces.

We close this chapter with a full accounting of the situation for $A = P^{l+1}(p^r)$, $l \geq 2$.

Proposition 8.18. *Let $p \geq 5$, $r \geq 1$, $l \geq 2$, and put $A = P^{l+1}(p^r)$. For each $t \geq 1$, there is a section $\overline{\Psi}_{[t]}$ for $\mathcal{H}_0|_{\ker(\mathcal{H}_1)}$ as dga's. These $\overline{\Psi}_{[t]}$'s determine a functorial (on $\mathcal{M}_A$) structure for $\tilde{\pi}_*(K_g; \mathbf{Z}_{p^t})$ as a csm over $\{H_*(\Omega A; \mathbf{Z}_{p^t})\}$. When $t \leq r$, $\overline{\Psi}_{[t]}$ factors through (8Q),* *i.e., through Φ. Furthermore, for $t \le r$ there is a diagonal χ_{imt} on $\{H_*(\Omega A;\mathbb{Z}_{p^t})\}$ that has all $\chi_{imt}=0$ except for χ_{0tt}, and $\{\pi_*(\Omega K_g;\mathbb{Z}_{p^t})\}_{t\le r}$ is a csL over $\{H_*(\Omega A;\mathbb{Z}_{p^t}),\chi_{imt}\}_{t\le r}$.*

Proof. Let $(C,d)=\mathcal{A}(P^{l+1}(p^r))=\mathbb{Z}_{(p)}\langle v_l,u_{l-1}\rangle$ with $d(v_l)=p^r u_{l-1}$, and let $\underline{q}:\ \mathcal{F}(C)\to \{H_*(\Omega A;\mathbb{Z}_{p^t})\}$ be the csa surjection of (7AB). For $t\le r$, notice that $\mathcal{F}(C)_{[t]}=C=\mathbb{Z}_{(p)}\langle u_l,v_{l-1}\rangle$ and define $\Psi_{[t]}:\ C\to \mathcal{O}_A[t]$ by $\Psi_{[t]}(v_l)=\Phi((\omega_t^r)^a)$ and $\Psi_{[t]}(u_{l-1})=\Phi((\omega^r\rho^t)^a)$, where $(\ )^a$ denotes the adjoint of a homotopy class. By Lemma 8.9, it is trivial that $\mathcal{H}_1\Psi_{[t]}=0$ while $\mathcal{H}_0\Psi_{[t]}=\underline{q}_{[t]}$.

Dualizing $\Psi_{[t]}$ yields module structure homomorphisms

$$\psi_{[t]}:\ \mathcal{F}(C)_{[t]}\otimes\tilde{\pi}_*(K_g;\mathbb{Z}_{p^t})\to\tilde{\pi}_*(K_g;\mathbb{Z}_{p^t})$$

for $t\le r$ that are easily shown by direct checking to comprise a functorial (on $\mathcal{M}_A$) csm structure. By Lemma 7.10(b) these factor through a csm structure $\{\overline{\psi}_{[t]}\}$ over the csa $\{H_*(\Omega A;\mathbb{Z}_{p^t})\}$. Dualizing $\overline{\psi}_{[t]}$ gives us the factor homomorphism $\overline{\Psi}_{[t]}:\ H_*(\Omega A;\mathbb{Z}_{p^t})\to \mathcal{O}_A[t]$.

To construct $\overline{\Psi}_{[t]}$ for $t>r$ we use a similar approach. Because $E^t(\Omega A)$ is trivial for $t>r$, $N_{[t]}=\mathbb{Z}_{(p)}$, i.e. the ground ring which has no generators serves as a Bockstein sub–dga in Proposition 7.9. The hypotheses of Proposition 7.9 are vacuously satisfied. As a result

$$\psi_{[t]}:\ \mathcal{F}(C)_{[t]}\otimes\tilde{\pi}_*(K_g;\mathbb{Z}_{p^t})\to\tilde{\pi}_*(K_g;\mathbb{Z}_{p^t})$$

is a well-defined natural csm structure, for all t. Using Lemma 8.7, the composition

$$\mathcal{F}(C)_{[t]}\xrightarrow{\psi_{[t]}(-\otimes\hat{v}_a)}\tilde{\pi}_*(A'(q,p^t);\mathbb{Z}_{p^t})\xrightarrow{h}\overline{H}_*(A';\mathbb{Z}_{p^t})\approx H_*(\Omega A;\mathbb{Z}_{p^t})\oplus H_{*+1}(\Omega A;\mathbb{Z}_{p^t})$$

equals $\mathcal{H}\Psi_{[t]}$, and it is clearly a csc homomorphism. Since $\mathcal{H}\Psi_{[t]}=(\underline{q}_{[t]},0)$ for $t\le r$, the uniqueness aspect of Proposition 7.5 guarantees that $\mathcal{H}\Psi_{[t]}=(\underline{q}_{[t]},0)$ for all t. By Lemma 7.10(b), $\psi_{[t]}$ factors through a $\{H_*(\Omega A;\mathbb{Z}_{p^t})\}$-action $\overline{\psi}_{[t]}$. Its dual $\overline{\Psi}_{[t]}$ must satisfy $\mathcal{H}\overline{\Psi}_{[t]}=(1,0)$, by 7.10(a).

For the diagonal, let $\chi_{imt}=0$ if $i=1$ or $m<t$, and let χ_{0tt} be the usual multiplicative Hopf algebra diagonal $C\to C\otimes C$ for which v_l and u_{l-1} are primitive. When $\hat{x},\hat{y}\in\pi_*(\Omega K_g;\mathbb{Z}_{p^t})$, $w\in\mathcal{F}(C)_{[t]}=C$, $t\le r$, the formula (cf. (7BA))

$$\text{(8V)}\qquad \psi_{[t]}(w\otimes[\hat{x},\hat{y}])=\Sigma'_{0t}(-1)^{|w''||\hat{x}|}[\psi_{[t]}(w'\otimes\hat{x}),\psi_{[t]}(w''\otimes\hat{y})],$$

where $\chi_{0tt}(w)=\Sigma'_{0t}w'\otimes w''$, is a straightforward application of the Jacobi identity for mod p^t relative Samelson products. Formula (8V) can easily be proved directly, by induction on the length of w (i.e., length as a product of v_q's and u_{q-1}'s). Clearly $\psi_{[t]}$ can also be replaced by $\overline{\psi}_{[t]}$ in (8V).

Corollary 8.19. *Let $A=P^{l+1}(p^r)$ and $C=\mathcal{A}(A)$ and χ_{imt} be as above. Put $L=\mathbb{L}\langle v_q,u_{q-1}\rangle$, so that $(C,d)=\mathcal{U}(L,d)$ and ad : $C\otimes L\to L$ makes (L,d) into a differential (C,d)-module. Then $\{\mathcal{F}(L)\}_{1\le t\le r}$ is a csL over $\{\mathcal{F}(C)_{[t]},\chi_{imt}\}_{1\le t\le r}$, and there is a csL/csm homomorphism*

$$\{\gamma_{[t]}\}:\ \{\mathcal{F}(L)_{[t]}\}_{1\le t\le r}\to\{\pi_*(\Omega A;\mathbb{Z}_{p^t})\}_{1\le t\le r}$$

between csL's over $\{\mathcal{F}(C)_{[t]}, \chi_{imt}\}_{1\leq t\leq r}$. *If* $e : L \to C$ *denotes the inclusion, then for* $t \leq r$ *we have*

$$\text{(8W)} \qquad \Phi\gamma_{[t]} = \Psi_{[t]}\mathcal{F}(e).$$

A similar claim holds if (L, d) *is replaced by any differential Lie ideal of* (L, d).

Proof. For $t \leq r$, $\mathcal{F}(L)_{[t]} = L$, and we define $\gamma_{[t]}$ by $\gamma_{[t]}(v_l) = (\omega_l^r)^a$ and $\gamma_{[t]}(u_{l-1}) = (\omega^r\rho^t)^a$. Then (8W) is really just the definition of $\Psi_{[t]}$. The first assertion, that $\mathcal{F}(L)$ is a csL over our Dcsa, is Lemma 7.12(c)

CHAPTER 9

Statement of the Main Theorem

In Chapter 9 we at last state the main theorem of this monograph, Theorem 9.1. It asserts the existence of the spaces $\{D_k\}_{k\geq 0}$ and of diagrams like (1K) (see (9C)) for each D_k, and it outlines twelve of their major properties. The list of properties will be continued in Theorems 12.4 and 12.6. We examine in Chapter 9 what the theorem says at $k = 0$ and why it is true there. We detail how to perform the construction of D_k out of D_{k-1}. We get started on the proof of Theorem 9.1 by proving the three parts that provide descriptions of Adams–Hilton models. Theorem 9.1 is a herculean project: all of Chapters 9 through 13 will be needed in order to complete its proof.

The overall method of proof is inductive: the existence and properties of D_{k-1} are used in order to establish the existence and properties of D_k. As with many significant induction arguments, the proposition $P(k)$ for which $P(k)$ actually implies $P(k+1)$ is very subtle and complex, and it contains many more parts than one would naively even want to prove. The author tried for a long time to discover a concise proof of the existence of the D_k's, after which the properties could be spun off, one at a time. Such a proof does not appear to be in the cards. We need to include literally dozens of facts about D_k in the inductive proposition $P(k)$.

Roughly speaking, the reason for the high complexity of $P(k)$ is that we are creating from scratch a "p–primary homotopy theory under D_k." If we were only doing the case D_1, it would surprise no one that everything we know about p–primary homotopy theory — including differentials, Hurewicz homomorphisms, relative Samelson products, Jacobi identities, and so on — would be used all together. But now consider that p–primary homotopy theory is the theory of the Moore space, which is D_0. In other words, we need to know everything about D_0 before being able to build and work with D_1. When the issue is viewed this way, it seems less surprising that we need to know everything about D_{k-1} before being able to build and to work with D_k. The difficulty is, that a priori nothing is known about D_{k-1} ! Consequently our main theorem must not only assert the existence of D_k; it must be a small treatise that contains counterparts for each of the major tenets of p–primary homotopy theory.

Theorem 9.1, even when supplemented by 12.4 and 12.6, will not include all the properties that would technically need to be present as part of the inductive proposition $P(k)$. Still, most of the important ones are there. Occasionally during a proof we shall run across a property that seemed too minor or distracting to include in these principal theorems, but we still need it as part of the inductive plan. We shall introduce such properties quite casually, and we shall always feel free to assume that they hold at $k-1$ as we prove them at k. Whenever we introduce such a claim we will also have to justify why it holds at $k=0$, of course. Our words for the total inductive proposition will be "the grand inductive scheme." By these words we shall also refer to everything that is assumed to hold at $k-1$, along with whatever facts have thus far been established at k.

Let us give the reader some help sorting through the mass of information that is Theorem 9.1. Although there are twelve separate parts, labeled (i) through (xii), these are best viewed as grouped under five headings. The first of the five topics addressed by Theorem 9.1 is the existence of the spaces D_k, E_k, F_k and of the various maps among them. We have already discussed the centrality of this result to the entire monograph, in terms of its organization and motivation.

The second of the five topics in Theorem 9.1 provides the primary link between the algebra we have developed and the geometry. The dga's $\mathcal{U}M^k$ and $\mathcal{U}L^k$ are shown to serve as Adams–Hilton models for D_k and for F_k. (Technically, because $\mathcal{U}M^k$ is not a free dga it is not an Adams–Hilton model but rather an "AH quotient"; see Corollary 6.6.) This means that all the algebra we did in Chapters 2–4, and all our subsequent discussion of this algebra, suddenly becomes deeply relevant to topology.

The third topic is the ring of stable homotopy operations $\mathcal{O}_{D_k}$ under D_k. The "best case scenario" outlined in 8.14 and 8.17 comes true for the D_k's. For each $t \geq 1$, $\mathcal{O}_{D_k}[t]$ has $H_*(\Omega D_k; \mathbb{Z}_{p^t})$ as a dga retract (and we presumably understand $H_*(\Omega D_k; \mathbb{Z}_{p^t})$ because it equals $H_*(\mathcal{U}M^k; \mathbb{Z}_{p^t})$). This gets expressed in Theorem 9.1 in csa language. The notations are $\bar{\Psi}_t^k$ for the inclusion of $H_*(\Omega D_k; \mathbb{Z}_{p^t})$ into $\mathcal{O}_{D_k}[t]$, and Ψ_t^k for the more useful composite $\bar{\Psi}_t^k \underline{q}_{[t]}$ with source $\mathcal{F}(\mathcal{U}M^k)_{[t]}$. The reader who perseveres can expect to become very well acquainted with Ψ_t^k and its dual form, denoted ψ_t^k.

Recall from Chapter 8 that the elements of $\mathcal{O}_{D_k}$ act on the homotopy of the homotopy-theoretic fiber K_g, for map pairs $D_k \xrightarrow{f} Y \xrightarrow{g} Z$. Since so many spaces could potentially be K_g for some $(f,g) \in \mathcal{M}_{D_k}$, these homotopy operations could conceivably find wide application. The author suspects that the operations will ultimately prove to have greater value than the D_k's themselves.

Fourth, Theorem 9.1 finds, in effect, a copy of $H_*(\bar{L}^k; \mathbb{Z}_{p^t})$ inside $\pi_*(\Omega E_k; \mathbb{Z}_{p^t})$, for $t \leq r+k$. Again this is expressed in the language of Chapter 7. We construct a csL/csm homomorphism over $\mathcal{F}(\mathcal{U}M^k)$, from the csL $\mathcal{F}(\bar{L}^k)$ to the csL $\{\pi_*(\Omega E_k; \mathbb{Z}_{p^{r+k}})\}$. Recall (Proposition 2.7) that $H_*(\bar{L}^k; \mathbb{Z}_{p^{r+k}})$ has torsion of many different orders, all jumbled together; impressively, the csL machinery takes care of this complexity automatically. Yet our success is not unmitigated: recalling Discussion 7.7 regarding $\mathcal{F}(\bar{L}^k)$, we can be certain of our results only as high as $*$–degree $2p$. The notation for the csL/csm homomorphism, which will also appear liberally during the coming chapters, is γ_t^k.

Fifth and last, Theorem 9.1 makes certain assertions regarding a very special homotopy class. The homotopy class is special because it provides the essential link between D_k and D_{k+1}. Its existence for D_{k-1} is the very first thing we use in constructing D_k, and its existence for D_k is the very last thing we prove about D_k. Philosophically, a lot can be learned about why the overall proof of 9.1 goes the way it goes, by examining what information is required to construct this class, what properties of it we really need, and what could still be proved without it.

The proof of Theorem 9.1 is spread out across five chapters (9 through 13), but these do not correspond to the five topics. In Chapter 9 we state the theorem, prove it $k = 0$, construct D_k from D_{k-1}, and compute Adams–Hilton models for D_k and F_k. The proofs of the properties regarding AH models rely upon a number of technical facts about the relationships between the AH models of the total space and of the fiber, in a fibration over an odd–dimensional sphere. This situation is considered in detail in [A4], so a lot of the proof consists of references to [A4]. Chapter 10 focuses on the ring of operations $\mathcal{O}_{D_k}$, and we show that $\mathcal{H}_0|_{\ker(\mathcal{H}_1)}$ has a section, for D_k. We also construct the csL/csm homomorphism γ_t^k for $t < r+k$, but we hold off on defining the $(r+k)^{\text{th}}$ stage.

In order to continue the proof of 9.1, we need a certain virtual syzygy (Theorem 7.15) in $\pi_{6np^k-2}(\Omega E_k; \mathbb{Z}_{p^{r+k}})$. This turns out to be far harder than it sounds, and this is where the dgL's N^{kl} and N^{klm} enter the picture. We need another whole chapter devoted to the general theory of csL's, and to $\mathcal{F}(N^{kk})$ and $\mathcal{F}(N^{kkk})$ in particular. Chapter 11 does no topology, but Chapter 12 builds upon its algebraic results by establishing certain "Jacobi identities". In Chapter 13 we finish the proof, first by deriving the virtual syzygy from the Jacobi identities, and second by constructing the special homotopy class mentioned earlier, to prepare the way for D_{k+1}.

Notations. From Chapters 2, 3, 4, the notations $\overline{L}^k, L^k, M^k, \mathcal{U}M^k$ and the names $y_j, z'_j, x_j, a_s, b, c_s$ for their elements will be continued. Specifically, $\overline{L}^k = \mathbb{L}\langle y_j, z'_j | j \geq 2\rangle$, $L^k = \mathbb{L}\langle x_j | j \geq 1\rangle$, and M^k is a quotient of $\mathbb{L}\langle b, a_0, \ldots, a_k\rangle$. The embedding of $\overline{L}^k$ into L^k takes y_j to x_j, and the embedding of L^k into M^k takes x_{p^s} to a_s for $0 \leq s \leq k$. We denote any of the embeddings among $\overline{L}^k, L^k, M^k, \mathcal{U}\overline{L}^k, \mathcal{U}L^k, \mathcal{U}M^k$ by the symbol $\overline{\iota}$, and we supply clarification locally if it is needed.

Theorem 9.1. *Fix $n \geq 1$, $r \geq 1$, and a prime $p \geq 5$. For each $k \geq 0$ there exist spaces and maps*

$$E_k \xrightarrow{\eta_k} F_k \xrightarrow{\lambda_k} D_k \xrightarrow{\rho_k} S^{2n+1}\{p^r\} \tag{9A}$$

satisfying conditions (i) – (xii) *below.*

Description of the Spaces and Maps

(i) $D_0 = P^{2n+1}(p^r)$, *and* $\rho_0 = P^{2n+1}(p^r) \to S^{2n+1}\{p^r\}$ *is the inclusion of the* $(2n+1)$–*skeleton, given in* (1M). *For* $k \geq 1$ *there is a cofibration*

$$P^{2np^k}(p) \xrightarrow{\lambda_{k-1}\eta_{k-1}\tilde{y}_{p^k}} D_{k-1} \xrightarrow{\hat{\imath}} D_k, \tag{9B}$$

and ρ_k *extends* ρ_{k-1}, *i.e.,* $\rho_{k-1} = \rho_k\hat{\imath}$. D_k *has a CW structure with one cell in each dimension* $2np^s$ *and* $2np^s+1$, $0 \leq s \leq k$, *but no cells in any other positive dimensions. The homology of* D_k *is given by:*

$$\tilde{H}_m(D_k) = \begin{cases} \mathbb{Z}_{p^r} & \text{if } m = 2n \\ \mathbb{Z}_p & \text{if } m = 2np^s,\ 1 \leq s \leq k \\ 0 & \text{otherwise.} \end{cases}$$

The cohomology of D_k *(any coefficients) has trivial multiplication, and* ΣD_k *is a wedge of Moore spaces (i.e., all attaching maps are unstable).*

(ii) *There is a CMN diagram*

$$
\text{(9C)}\qquad
\begin{array}{ccccccc}
 & & \Omega^2 S^{2n+1} & \longrightarrow & * & \longrightarrow & \Omega S^{2n+1} \\
 & & \downarrow & & \downarrow & & \downarrow \\
 & & E_k & \xrightarrow{\lambda_k \eta_k} & D_k & \xrightarrow{\rho_k} & S^{2n+1}\{p^k\} \\
 & & \downarrow{\scriptstyle \eta_k} & & \downarrow{\scriptstyle =} & & \downarrow{\scriptstyle \bar\rho} \\
\Omega S^{2n+1} & \xrightarrow{\kappa_k} & F_k & \xrightarrow{\lambda_k} & D_k & \xrightarrow{\bar\rho\rho_k} & S^{2n+1} \\
{\scriptstyle \Omega p^r}\downarrow & & \downarrow{\scriptstyle \mu_k} & & \downarrow & & \downarrow{\scriptstyle p^r} \\
\Omega S^{2n+1} & \xrightarrow{=} & \Omega S^{2n+1} & \longrightarrow & * & \longrightarrow & S^{2n+1},
\end{array}
$$

which defines the maps κ_k and μ_k. The fact that $\rho_k \hat{i} = \rho_{k-1}$ for $k > 0$ means that there are induced maps $F_{k-1} \to F_k$ and $E_{k-1} \to E_k$ between the respective fibers. These maps will also be denoted $\hat{i}$. Then $\eta_k \hat{i} = \hat{i}\eta_{k-1} : E_{k-1} \to F_k$ and $\lambda_k \hat{i} = \hat{i}\lambda_{k-1} : F_{k-1} \to D_k$, and $\mu_k \hat{i} = \mu_{k-1}$.

Description of Adams-Hilton Models

(iii) *There is a quism $(\mathcal{U}L^k, d) \to C_*(\Omega F_k)$, i.e, $(\mathcal{U}L^k, d)$ is an Adams–Hilton model for F_k. In particular, $H^*(F_k)$ is the $\mathbb{Z}_{(p)}$–free ring R^k of Chapter 2, and the BSS for ΩF_k coincides as algebras with the BSS for $\mathcal{U}L^k$ (cf. Proposition 2.7). If we write*

$$
\mathcal{A}(\Omega S^{2n+1}) = \mathbb{Z}_{(p)}\langle a_1, a_2, \ldots \rangle, \ |a_j| = 2nj - 1,
$$

$$
d(a_j) = -\frac{1}{2}\sum_{i=1}^{j-1} (j|i)[a_i, a_{j-i}],
$$

then $\mathcal{A}(\mu_k)$ is given by

$$
\text{(9D)}\qquad \mathcal{A}(\mu_k)(x_j) = p^{(j-1)r-\alpha(j)} a_j,
$$

where $\alpha(j) = \alpha_k(j)$ is defined in (2L).

(iv) *There is a commuting diagram*

$$
\text{(9E)}\qquad
\begin{array}{ccccc}
\mathcal{A}(F_k) & \xrightarrow{\mathcal{A}(\lambda_k)} & \mathcal{A}(D_k) & \xrightarrow{\mathcal{A}(\bar\rho\rho_k)} & \mathcal{A}(S^{2n+1}) \\
{\scriptstyle =}\downarrow & & {\scriptstyle \theta^k}\downarrow & & \downarrow{\scriptstyle =} \\
\mathcal{U}(L^k, d) & \longrightarrow & \mathcal{U}(M^k, d) & \longrightarrow & (\mathbb{Z}_{(p)}[b], 0)
\end{array}
$$

in which the bottom row results from applying $\mathcal{U}$ to (3B) *and the middle vertical arrow is a surjective quism. In particular, $H_*(\Omega D_k) \approx H_*(\mathcal{U}M^k, d)$ as algebras over $\mathbb{Z}_{(p)}$, and the BSS's for ΩD_k and $\mathcal{U}M^k$ coincide as $\mathbb{Z}_p$–algebras (cf. Theorem 4.8). Furthermore, $\mathcal{A}(D_k)$ is a dgH–AH model for D_k, and θ^k is a dgH quism.*

(v) *The conjugation action $\tilde{c}_*$ of $H_*(\Omega D_k; \mathbb{Z}_{p^t})$ on $H_*(\Omega F_k; \mathbb{Z}_{p^t})$ (for the fibration that is the third row of* (9C)*) coincides with the action induced by the dgH $(\mathcal{U}M^k, d)$ on the sub–dgH $(\mathcal{U}L^k, d)$.*

Properties of the Ring $\mathcal{O}_{D_k}$

(vi) *On the category* $\mathcal{M}_{D_k}$, *there is a functorial structure for* $\{\pi_*(\Omega K_g; \mathbb{Z}_{p^t})\}_{t\geq 1}$ *as a csL over the Dcsa* $\mathcal{F}(\mathcal{U}M^k)$, *where the diagonal for* $\mathcal{F}(\mathcal{U}M^k)$ *is described in Example 7.13. We denote the module structure homomorphisms by*

$$\psi_t^k : \mathcal{F}(\mathcal{U}M^k)_{[t]} \otimes \tilde{\pi}_*(K_g; \mathbb{Z}_{p^t}) \to \tilde{\pi}_*(K_g; \mathbb{Z}_{p^t}). \tag{9F}$$

Each ψ_t^k *factors through the quotient homomorphism* $\underline{q}_{[t]}$ *of* (7AB), *and we have a functorial csm structure*

$$\overline{\psi}_t^k : H_*(\Omega D_k; \mathbb{Z}_{p^t}) \otimes \tilde{\pi}_*(K_g; \mathbb{Z}_{p^t}) \to \tilde{\pi}_*(K_g; \mathbb{Z}_{p^t}). \tag{9G}$$

Dualizing ψ_t^k *and* $\overline{\psi}_t^k$ *gives us the dga homomorphisms for each* t,

$$\overline{\Psi}_t^k : H_*(\Omega D_k; \mathbb{Z}_{p^t}) \to \mathcal{O}_{D_k}[t], \ \Psi_t^k = \overline{\Psi}_t^k \underline{q}_{[t]} : \mathcal{F}(\mathcal{U}M^k)_{[t]} \to \mathcal{O}_{D_k}[t].$$

(vii) *For* $t \geq 1$, $\mathcal{H}_1 \overline{\Psi}_t^k = 0$ *and* $\mathcal{H}_0 \overline{\Psi}_t^k$ *is the identity, i.e.,* $\overline{\Psi}_t^k$ *splits* $\mathcal{H}_0|_{\ker(\mathcal{H}_1)}$ *as dga's. In particular, Corollary 8.17 is true for* D_k.

(viii) *For* $k > 0$, Ψ_t^k *is compatible with* $\hat{\imath}$, *i.e.,*

$$\Psi_t^k \mathcal{F}(\hat{\imath}) = \mathcal{O}_{(\hat{\imath})} \Psi_t^{k-1} : \mathcal{F}(\mathcal{U}M^{k-1})_{[t]} \to \mathcal{O}_{D_k}[t], \tag{9H}$$

$$\overline{\Psi}_t^k (\Omega \hat{\imath})_* = \mathcal{O}_{(\hat{\imath})} \overline{\Psi}_t^{k-1} : H_*(\Omega D_{k-1}; \mathbb{Z}_{p^t}) \to \mathcal{O}_{D_k}[t]. \tag{9I}$$

Properties of the CSL Homomorphism

(ix) *Below* $*$*-degree* $2p+1$ *there is a homomorphism of csL's and of csm's over* $\mathcal{F}(\mathcal{U}M^k)$,

$$\{\gamma_t^k\} : \{\mathcal{F}(\overline{L}^k)_{[t]}^{*\leq 2p}\}_{1\leq t\leq r+k} \to \{\pi_*(\Omega E_k; \mathbb{Z}_{p^t})\}_{1\leq t \leq r+k}. \tag{9J}$$

For $j \in I_s$ *where* $s = \max(0, t, -r)$, *we have*

$$\underline{a}_{\eta_k}(\gamma_t^k(y_j)) = \bar{e}(y_j) = x_j, \tag{9K}$$

$$\underline{a}_{\eta_k}(\gamma_t^k(z_j')) = \bar{e}(z_j'), \tag{9L}$$

where $\bar{e}$ *is the inclusion of* $\overline{L}^k$ *into* $\mathcal{U}L^k$.

(x) *For* $1 \leq t \leq r+k-1$, γ_t^k *is compatible with* $\hat{\imath}$, *i.e.,*

$$(\Omega \hat{\imath})_{\#} \gamma_t^{k-1} = \gamma_t^k \mathcal{F}(\bar{\imath}) : \mathcal{F}(\overline{L}^{k-1})_{[t]}^{*\leq 2p} \to \pi_*(\Omega E_k; \mathbb{Z}_{p^t}), \tag{9M}$$

where $\bar{\imath} : \overline{L}^{k-1} \to \overline{L}^k$ *is the dgL inclusion.*

(xi) *For* $t \leq r+k$, $w \in \mathcal{F}(\overline{L}^k)^{*\leq 2p}_{[t]}$,

$$\text{(9N)} \qquad \Psi^k_t \mathcal{F}(\bar{e})(w) = \Phi((\Omega\lambda_k\eta_k)_\# \gamma^k_t(w)),$$

where Φ *is described in Lemma 8.9 and* $\bar{e} : \overline{L}^k \to \mathcal{U}M^k$.

Extending a Special Homotopy Class

(xii) *The homotopy class* $\gamma^k_{r+k}(y_{p^{k+1}}) : P^{2np^{k+1}-1}(p^{r+k}) \to \Omega E_k$ *has a unique extension (through* ω^{r+k+1}_{r+k}*) over* $P^{2np^{k+1}-1}(p^{r+k+1})$, *call it* $\hat{y}_{p^{k+1}}$, *that satisfies*

$$\text{(9O)} \qquad \underline{a}_{\eta_k}(\hat{y}_{p^{k+1}}) = x_{p^{k+1}}$$

and

$$\text{(9P)} \qquad \hat{y}_{p^{k+1}}\omega^{r+k+1}\rho^{r+k} = \gamma^k_{r+k}(p^{-r-k-1}d(y_{p^{k+1}})).$$

The map $\tilde{y}_{p^{k+1}}$ *that appears in* (9B) *and is used to build* D_{k+1} *out of* D_k *is the adjoint of* $\hat{y}_{p^{k+1}}\omega^{r+k+1}_1 = (\omega^{r+k}_1)^\# \gamma^k_{r+k}(y_{p^{k+1}})$.

Discussion when $k = 0$. Most parts of Theorem 9.1 when $k = 0$ were proved by Cohen–Moore–Neisendorfer in [CMN1] and [CMN2]. Indeed, their classic papers provided the road map which our proof largely follows. In many respects their terminology differs from ours. For example, they made no use of Adams–Hilton models, and they did not have the concepts of a ring of homotopy operations or a csL. Nevertheless, most parts of theorem 9.1 are present in [CMN12], even if they are phrased rather differently. In those instances where the result is not in [CMN12], it will turn out that the proof we offer for $k > 0$ will actually be valid at $k = 0$ too.

Let us run quickly through the parts of Theorem 9.1, at $k = 0$. Part (i) is trivial at $k = 0$, and (9C) serves mostly to define the spaces and maps of (9A). Cohen–Moore–Neisendorfer handled (iii) and (iv) by working with differential coalgebras that are chain equivalent to the chains on $P^{2n+1}(p^r) = D_0$, on ΩS^{2n+1}, and on F_0. Our Adams–Hilton models would be the result of applying the cobar construction to their differential coalgebras. The reasoning behind (v) will work equally well for $k > 0$ or for $k = 0$.

Parts (vi) and (vii) when $k = 0$ are essentially Proposition 8.18 when we put $A = P^{l+1}(p^r) = P^{2n+1}(p^r)$. The reader should realize up front that when $t \leq r$, the operation $\psi^k_t(b \otimes -)$ for the map pair $D_k \xrightarrow{f} Y \xrightarrow{g} Z$ in $\mathcal{M}_{D_k}$ is the relative Samelson product $[\hat{b}, -]$, where $\hat{b}$ denotes the composite

$$P^{2n+1}(p^t) \xrightarrow{\omega^r_t} P^{2n+1}(p^r) = D_0 \hookrightarrow D_k \xrightarrow{f} Y.$$

Thus $\psi^k_{r+s}(b^{p^s} \otimes -)$ is a functorial extension of $\mathrm{ad}^{p^s}(\hat{b})$. Part (viii) is vacuous when $k = 0$.

Part (x) is vacuous at $k = 0$, and part (ix) is "almost" in [CMN2]. Cohen-Moore–Neisendorfer worked with the homotopy Bockstein spectral sequence instead of with mod p^r homotopy and ${}_\pi\delta^r$. But their methods do not need any improvements to handle the latter approach equally well. When all the jargon is cleared away at $k = 0$, (ix) boils down to the assertion that there exists a dgL homomorphism

$$\gamma^0_t : \overline{L}^0 = \mathbb{L}\langle y_i, z_i | i \geq 2\rangle \to \pi_*(\Omega E_0; \mathbb{Z}_{p^t})$$

satisfying certain compatibilities with $[\omega_t^r, -]$ and $[\omega^r\rho^t, -]$. This is really what is done in [CMN2].

Formula (9N) is implicit in Corollary 8.19, so the homomorphism $\gamma_{[t]}$ of 8.19 is relevant here. Expressed in our terms, Cohen–Moore–Neisendorfer's principal motivation for constructing γ_t^0 is that it is a lift of $\gamma_{[t]}$ through $(\Omega\lambda_k\eta_k)_{\#}$. Thus (9N) at $k = 0$ merely reiterates (8W). Part (xii) was known to Joe Neisendorfer [oral communication] when $k = 0$, and its proof will not require k to be positive.

For the remainder of this chapter and Chapter 10, we shall assume that $k \geq 1$, although as we just noted, in some instances the case $k = 0$ will be covered by the proof as well.

Proof of (i) *and* (ii). The map $\tilde{y}_{p^k}$ is defined in (xii) for $k - 1$, and the attaching map $(\lambda_{k-1}\eta_{k-1})_{\#}(\tilde{y}_{p^k})$ is a mod p homotopy class in $\pi_{2np^k}(D_{k-1}; \mathbb{Z}_p)$. Let us tell the reader a little more about this attaching map. When $k = 1$, this homotopy class belongs to $\pi_{2np}(D_0; \mathbb{Z}_p) = \pi_{2np}(P^{2n+1}(p^r); \mathbb{Z}_p)$. If the canonical homotopy classes are denoted $\tilde{b} = \omega_1^r \in \pi_{2n+1}(P^{2n+1}(p^r); \mathbb{Z}_p)$ and $\tilde{a}_0 = \omega^r\rho^1 \in \pi_{2n}(P^{2n+1}(p^r); \mathbb{Z}_p)$, then $(\lambda_0\eta_0)_{\#}(\tilde{y}_p)$ will be the iterated Whitehead product, $\mathrm{ad}^{p-1}(\tilde{b})(\tilde{a}_0)$. For $k \geq 2$, there will be a homotopy class $\tilde{a}_{k-1} \in \pi_{2np^k-1}(D_{k-1}; \mathbb{Z}_p)$ – specifically, $\tilde{a}_{k-1}$ will be adjoint to $(\omega_1^{r+k-1})^{\#}\gamma_{r+k-1}^{k-1}(y_{p^{k-1}})$ – and $(\lambda_{k-1}\eta_{k-1})_{\#}(\tilde{y}_{p^k})$ will equal the iterated Whitehead product $\mathrm{ad}^{p^k-p^{k-1}}(\tilde{b})(\tilde{a}_{k-1})$. Because Whitehead products are unstable, ΣD_k is a wedge of Moore spaces.

Because the attaching map $\lambda_{k-1}\eta_{k-1}\tilde{y}_{p^k}$ in (9B) satisfies $\rho_{k-1}(\lambda_{k-1}\eta_{k-1}\tilde{y}_{p^k}) = 0$ (by second row of (9C)), there is an extension of ρ_{k-1} over D_k. *Our choice of extension is completely arbitrary.* Starting from the map ρ_k, the lower two right–hand squares of (9C) can be made to commute on the nose, if the "$*$" in the bottom row is interpreted as the cone CD_k on D_k. As a result E_k and F_k can be defined to be literally (not just up to homotopy type) the homotopy–theoretic fibers of the maps ρ_k and $\beta\rho_k$. The remaining properties listed in (i) and (ii) are now either obvious or a matter of definition.

Remark 9.2. For the record, we note that the indeterminacy or flexibility in the choice of ρ_k (given ρ_{k-1}) is given by $\pi_{2np^k+1}(S^{2n+1}\{p^r\}; \mathbb{Z}_p)$, since any element of this homotopy group could be "added" (in the sense of the coaction $D_k \to D_k \vee P^{2np^k+1}(p))$ to ρ_k without affecting the restriction to D_{k-1} being the given ρ_{k-1}.

Proof of (iii). We are concerned here with the Adams–Hilton models for the fiber and for the total space, in a fibration over S^{2n+1}. The precise relationships between these models were studied exhaustively in [A4]. The gist of [A4] is that on the Adams–Hilton model of the fiber we can expect to find a natural derivation Q of degree $+2n$. This derivation arises from the action of ΩS^{2n+1} on the fiber. When mapped into the total space, the derivation Q becomes $\mathrm{ad}(b)$, where b is a certain $2n$–dimensional element in the Adams–Hilton model of the total space.

Section 5 of [A4] deals specifically with the question of how the Adams–Hilton model of the fiber changes, when a single Moore space is attached to the total space (in such a way that the fibering over S^{2n+1} can be extended). In the language of [A4, Def. 5.4] ρ_{k-1} is "t–pinch–like" for any $t > p^{k-1}$. Furthermore, letting $\tilde{x}_{pk} \in \pi_{2np^k-1}(\Omega F_{k-1}; \mathbb{Z}_p)$ denote the adjoint of $(\eta_{k-1})_{\#}(\tilde{y}_{p^k})$, i.e.,

$$\tilde{x}_{p^k} = (\omega_1^{r+k-1})^{\#}(\eta_{k-1})_{\#}\gamma_{r+k-1}^{k-1}(y_{p^k}) = (\omega_1^{r+k})^{\#}(\eta_{k-1})_{\#}(\hat{y}_{p^k})$$

by (xii) for $k-1$, we have that $\tilde{x}_{p^k}$ is "$(r+k)$–deep" [A4, Def. 5.5] because (by (xii)) $\underline{a}_{\eta_{k-1}}(\hat{y}_{p^k}) = x_{p^k} \in \mathcal{U}L^{k-1} = \mathcal{A}(F_{k-1})$. The hypotheses of [A4, Theorem 5.6] are satisfied. The conclusion of that theorem tells us that ρ_k is t–pinch–like (any $t \geq p^k$). It also tells us that $\mathcal{A}(F_k) = \mathbb{Z}_{(p)}\langle x_1', x_2', \dots\rangle$ and that

(9Q) $$\mathcal{A}(\hat{i})(x_j) = p^{\alpha_k(j)-\alpha_{k-1}(j)}(x_j') \quad ;$$

the differential on $\mathcal{A}(F_k)$ is actually determined by the fact that $\mathcal{A}(\hat{i})$ is to be a chain map. Using Lemma 3.1 or [A4, Prop. 5.8] this is enough to conclude that $\mathcal{A}(F_k) = \mathcal{U}(L^k, d)$.

We turn our attention now to (9D). The model $\mathcal{A}(\Omega S^{2n+1})$ for ΩS^{2n+1} given in (iii) is well–known, e.g., it appears in differential coalgebra form already in [CMN1]. To prove (9D), consider the two cases. When $j < p^k$, the fact that $\mathcal{A}(\hat{i}) : \mathcal{A}(F_{k-1}) \to \mathcal{A}(F_k)$ is an isomorphism in dimensions $\leq |x_j| + 1$, together with the relation $\mu_k\hat{i} = \mu_{k-1}$, forces $\mathcal{A}(\mu_k)(x_j)$ to equal $\mathcal{A}(\mu_{k-1})(x_j)$ (we are using here that $\alpha_k(j) = \alpha_{k-1}(j)$ for $j < p^k$). When $j \geq p^k$ we will use an argument entailing induction on j.

Let ϕ denote the dga homomorphism (9D), and we are proving that $\mathcal{A}(\mu_k)(x_j) = \phi(x_j)$ under the assumption that $\mathcal{A}(\mu_k)(x_i) = \phi(x_i)$ for all $i < j$. We find that $d\mathcal{A}(\mu_k)(x_j) = \mathcal{A}(\mu_k)d(x_j) = \phi d(x_j) = d\phi(x_j)$, so $\mathcal{A}(\mu_k)(x_j) = \phi(x_j) + z$ for some cycle z. Taking mod p homology, we get $(\Omega\mu_k)_*(\bar{x}_j) = \bar{z}$, the overbars denoting homology class in $H_*(\Omega-;\mathbb{Z}_p)$. But the fact that ρ_k is p^k–pinch–like includes the fact that $\{x_i\}_{i \geq p^k}$ is an "action sequence" [A4, Def. 5.2], hence $\bar{x}_j$ is a mod p Hurewicz image. But with one exception there are no non–zero mod p Hurewicz images in $H_{2nj-1}(\Omega^2 S^{2n+1};\mathbb{Z}_p)$ for $j \geq p^k$. The exception occurs for $k = 1$ and $j = p$ and $n = 1$. But even for this exception the class $\bar{z}$ must be the mod p reduction of an integral homology class, while $H_{2np-1}(\Omega^2 S^{2n+1}) = 0$. Our conclusion is that $\bar{z} = 0$, i.e., z equals a multiple of p plus a boundary. Writing $z = pz_0 + d(z_1)$ in $\mathcal{A}(\Omega S^{2n+1})$, we see (because Adams–Hilton models are torsion–free) that z_0 is a cycle of dimension $2nj - 1$. But $H_{>2n-1}(\Omega^2 S^{2n+1})$ has exponent p, which means that pz_0 is a boundary. Thus z is a boundary, and in choosing $\mathcal{A}(\mu_k)$ we may happily replace the $\mathcal{A}(\mu_k)$–image of x_j, which was $\phi(x_j) + z$, by $\phi(x_j)$. This completes the inductive step, and the proof.

Proof of (iv). The commuting diagram (9E) is an application of [A4 ,Cor. 3.3]. For the spaces under consideration, the sequence (29) of [A4] may be recognized with the help of (iii) as our sequence (3B).

The fact that D_k has a dgH–AH model and that θ^k is a dgH quism is an application of [A4, Theorem 3.2(c)] and [A4, Lemma 3.4]. According to these results, whenever the fiber of a fibration over a sphere has a dgH–AH model that is an enveloping algebra of a dgL, say we write it as $\mathcal{U}(N,d)$, then the total space has a dgH–AH model too; and the total space's model comes with a surjective dgH quism to the enveloping algebra of a certain Lie algebra that is an extension of (N,d) by $(\mathbb{L}\langle b\rangle, 0)$. Furthermore, even if it is only known that the fiber has such a model below some dimension q, we may still draw the same conclusions for the total space, below dimension q.

Because D_k has dimension $2np^k+1$, an Adams–Hilton model can be assumed to have no generators above dimension $2np^k$, so proving that D_k has a dgH–AH model in dimensions $\leq 2np^k+1$ is the same as proving it in all dimensions. It suffices therefore to prove that F_k has a dgH–AH model below dimension $2np^k+1$. At issue is whether the composition

$$\psi_{F_k}: \quad \mathcal{A}(F_k) \xrightarrow{\mathcal{A}(\Delta_{F_k})} \mathcal{A}(F_k \times F_k) \xrightarrow{\simeq} \mathcal{A}(F_k) \otimes \mathcal{A}(F_k),$$

which we call the "topological coproduct", coincides with the "algebraic coproduct" Δ_{L^k} on $\mathcal{U}L^k$. Here $\Delta_{F_k}: F_k \to F_k \times F_k$ is the diagonal map, and $\Delta_{L^k}: \mathcal{U}L^k \to \mathcal{U}L^k \otimes \mathcal{U}L^k$ is the usual diagonal with respect to which all of L^k is primitive.

Combining (iv) for $k-1$ with Lemma 6.23(b), we find that the topological and algebraic coproducts coincide for F_{k-1}. Another way to say this is that each (algebraically primitive) generator x_j of $\mathcal{U}L^{k-1} = \mathcal{A}(F_{k-1})$ is primitive, where by "primitive" we shall always mean "primitive with respect to the topological coproduct". The fact that $\hat{\imath}: F_{k-1} \to F_k$ is an equivalence below dimension $2np^k-1$ allows us to transfer the primitivity of the generator x_j of $\mathcal{A}(F_{k-1})$ to the primitivity of $x_j \in \mathcal{A}(F_k)$ if $j < p^k$ (because $\mathcal{A}(\hat{\imath})(x_j) = x_j$ for $j < p^k$). We need only to establish the primitivity of generators below dimension $2np^k+1$, so all that remains for (iv) is to show that x_{p^k} is primitive in $\mathcal{A}(F_k)$.

When we applied [A4, Theorem 5.6] in the proof of (iii), one part of the conclusion of that theorem deals with "$(r+k-1)$–depth". In this case it is saying that there is a homotopy class $\hat{x}_{p^k}: P^{2np^k}(p^{r+k-1}) \to F_k$ for which $\mathcal{A}(\hat{x}_{p^k})(v) = x_{p^k}$. As in the argument for (9D), we know that two maps with source $\mathcal{A}(F_k)$ must differ at x_{p^k} by a cycle if they coincide below x_{p^k}, i.e., $\psi_{F_k}(x_{p^k}) = \Delta_{L^k}(x_{p^k}) + z$ for some $(2np^k-1)$–dimensional cycle z. Consider the diagram

$$\begin{array}{ccc} \mathcal{A}(P^{2np^k}(p^{r+k-1})) = \mathcal{U}\mathbb{L}\langle u,v\rangle & \xrightarrow{\mathcal{A}(\hat{x}_{p^k})} & \mathcal{A}(F_k) \\ \Big\downarrow{\scriptstyle \Delta_{\mathbb{L}\langle u,v\rangle}} & & \Big\downarrow{\scriptstyle \psi_{F_k}} \\ \mathcal{U}\mathbb{L}\langle u,v\rangle \otimes \mathcal{U}\mathbb{L}\langle u,v\rangle & \longrightarrow & \mathcal{A}(F_k) \otimes \mathcal{A}(F_k), \end{array}$$

which must commute up to dga homotopy because the algebraic and topological coproducts always coincide for a Moore space. Deduce by Lemma 6.2(b) that z may be written as $z = d(z_0) + p^{r+k-1}z_1$, where z_1 is a cycle (because z is a cycle) in $\mathcal{A}(F_k) \otimes \mathcal{A}(F_k)$. But p^{r+k-1} annihilates the $(2np^k-1)$–dimensional homology of $\mathcal{U}L^k \otimes \mathcal{U}L^k$ (see Proposition 2.7), so this z is a boundary. We can now adjust ψ_{F_k} so that $\psi_{F_k}(x_{p^k})$ equals $\Delta_{L^k}(x_{p^k})$ instead of $\Delta_{L^k}(x_{p^k}) + z$.

Remark. To summarize the pattern of the proof of (iv), we assume inductively that the topological and algebraic coproducts coincide for D_{k-1}. Then they coincide for F_{k-1}. Since F_k equals F_{k-1} below dimension $2np^k-1$, they coincide for F_k below dimension $2np^k-1$. A special argument using $(r+k-1)$–depth bumps this dimension up to $2np^k+1$. Then [A4, Lemma 3.4] says that the two coproducts coincide on D_k, up to dimension $2np^k+1$. But $\dim(D_k) = 2np^k+1$, so this means that the two coproducts coincide on D_k in all dimensions.

Proof of (v). This sort of thing is probably true in general. That is, whenever a map f has a model $\mathcal{A}(f)$ which is a dgH surjection between dgH–AH models of the source and target of f, the model for the homotopy–theoretic fiber of f is probably the HAK of $\mathcal{A}(f)$, and the conjugation action is induced by the action of the total space's model as a dgH on the HAK. However, we do not need to prove this full conjecture in order to handle this case.

For this case, it suffices to know how each generator of $\mathcal{U}M^k$ acts on $\mathcal{A}(F_k) = \mathcal{U}L^k$. We can think of M^k as being generated by b together with the images $\bar{e}(x_i)$ of all the x_i's. When it comes to conjugation, $\bar{e}(x_i)$ acts on $\mathcal{A}(F_k)$ the same way x_i does, which is (since x_i is primitive in $\mathcal{A}(F_k)$) via $\mathrm{ad}(x_i)$.

As to the action of b, because conjugation is functorial it suffices to consider a universal example. The universal example, call it $\overline{F}_0$, is the homotopy–theoretic fiber of $P^{2n+1}(p^r) \vee P^q(p^r) \xrightarrow{\tilde{\rho}\rho_0\bar{\pi}_1} S^{2n+1}$. Although conjugation was not studied in [A4], this universal example was studied at some length. When we compute how conjugation works on $\overline{F}_0$, we find that the above conjecture is true for $\overline{F}_0$. In particular, for $\overline{F}_0$, conjugation by the special element b_0 (b_0 is the unique element whose $\mathcal{A}(\tilde{\rho}\rho_0\bar{\pi}_1)$–image is the generator of $\mathcal{A}(S^{2n+1})$) coincides with $\mathrm{ad}(b_0)$. By analogy with our earlier discussion of diagonals we might describe this by saying that "the topological and the algebraic conjugations coincide" for $\overline{F}_0$. Upon mapping $\overline{F}_0$ to F_k, the topologically induced conjugation action of b is seen to coincide with the algebraically defined action $\mathrm{ad}(b)$, on any generator x_i of $\mathcal{A}(F_k)$. By using the fact that the conjugation map $\tilde{c}$ satisfies $\tilde{c}(x, yz) = \tilde{c}(x, y)\tilde{c}(x, z)$, this is enough to determine that b acts via $\mathrm{ad}(b)$ on all of $\mathcal{A}(F_k)$.

CHAPTER 10

Stable Homotopy Operations Under D_k

In Chapter 10 we continue the proof of Theorem 9.1. We construct ψ_t^k for all t and γ_t^k for $t < r + k$, and we prove all of the properties 9.1(vi)–9.1(xi) about these homomorphisms. The techniques involved include strong use of Adams–Hilton models via Theorem 6.4, as well as varied applications of the major results from Chapters 7 and 8.

An expanded outline for Chapter 10 is as follows. We begin with two useful lemmas on the location of the higher torsion in the homotopy of a Moore space or a bouquet of Moore spaces. Commencing direct work on Theorem 9.1, we construct a special homotopy class, denoted $\hat{y}^*_{p^k} \in \pi_{2np^k-1}(\Omega E_k; \mathbb{Z}_{p^{r+k-1}})$, and its $(\Omega\lambda_k\eta_k)_\#$-image, denoted $\hat{a}_k$. Using $\hat{a}_k$ we construct ψ_t^k for $t < r + k$. We then arrive at the conceptual centerpiece of Chapter 10, which is the construction of the stable homotopy operations $\Psi^k_{r+k}(c_k)$ and $\Psi^k_{r+k}(b^{p^k})$ in $\mathcal{O}_{D_k}[r+k]$. Theorem 6.4 is used repeatedly at this stage, both to perform the constructions and to verify their properties. The techniques of Chapter 7 then permit us to flesh out the skeleton generated by these two elements, so we get the csm structure homomorphisms ψ_t^k for all t.

Another application of Theorem 6.4, this time using the thin product of three spaces described in Chapter 6, establishes that $\{\pi_*(\Omega K_g; \mathbb{Z}_{p^t})\}$ is functorially a csL, not just a csm, over the Dcsa $\mathcal{F}(\mathcal{U}M^k)$. Theorem 7.15 is invoked in the proof in order to avoid a very messy computation. Lastly, we construct γ_t^k for $t < r + k$. The construction divides neatly into the range of dimensions below $2np^k - 1$, for which we build upon the properties of γ_t^{k-1}, and the range above $2np^k - 1$, for which Theorem 7.15 is the perfect tool. Fittingly, Chapter 10 comes full circle with this proof in that the special homotopy class $\hat{y}^*_{p^k}$ plays a major role.

Some philosophical comments are in order regarding parts (vi) through (xi) of Theorem 9.1. It asserts the existence of ψ_t^k and γ_t^k, but the vast number of axioms and equations they satisfy indicate that these equations are the main point. Taken as a whole, parts (vi)–(xi) present a "highly overdetermined system of equations" for the γ_t^k and ψ_t^k. The proof asserts that there is a (unique) solution. Philosophically, the principal conclusion is that tremendous consistency and elegance is built into Theorem 9.1. The astounding thing about Theorem 9.1 is not the construction, but the fact that this overdetermined system has a solution, i.e., that such a complex and coherent total structure can exist at all.

It will not be obvious why we have chosen the order that we have, for proving the various parts of Theorem 9.1. A great many proof sequences were tried, but only the method presented here managed to avoid circular logic while maintaining some semblance of a larger plan.

We continue to refer to the individual parts of Theorem 9.1 by their unprefixed small Roman numeral names, e.g., (vi) means Theorem 9.1(vi).

We begin the mathematics of Chapter 10 with two lemmas regarding the presence of higher p–torsion in the homotopy groups of a Moore space or wedge of Moore spaces. The lemmas are straightforward consequences of [N2] and their proofs are omitted. Lemma 10.1 also introduces the operations $\hat{\tau}_j$ and $\hat{\sigma}_j$, whose adjoints $\tilde{\tau}_j$ and $\tilde{\sigma}_j$ will be discussed at some length in Discussion 11.E5 and Proposition 13.24.

Lemma 10.1. *Let $l \geq 3$, $t \geq 1$, $p \geq 3$. For the Moore space $P^l(p^t)$, let $\hat{v}_l \in \pi_l(P^l(p^t); \mathbb{Z}_{p^t})$ denote the homotopy class of the identity, and let $\hat{u}_{l-1} = {}_{\pi}\delta^t(\hat{v}_l)$. Let $c(l) = p(l-1)-1$ if l is odd, let $c(l) = 2p(l-1)-1$ if l is even. Put $c = c(l)$.*

(a) $p^{t+1}\pi_*(P^l(p^t)) = 0$.

(b) $p^t\pi_q(P^l(p^t)) = 0$ for $q < c$.

(c) $p^t\pi_c(P^l(p^t)) = \mathbb{Z}_p$, i.e., $\pi_c(P^l(p^t))$ contains a single summand $\mathbb{Z}_{p^{t+1}}$.

(d) If l is odd then for each $j \geq 1$ there are homotopy classes denoted $\hat{\tau}_j(\hat{v}_l) \in \pi_{p^j(l-1)}(P^l(p^t); \mathbb{Z}_{p^{t+1}})$ and $\hat{\sigma}_j(\hat{v}_l) = {}_{\pi}\delta^{t+1}(\hat{\tau}_j(\hat{v}_l))$. The class $\hat{\tau}_j(\hat{v}_l)\omega^{t+1}$ has order p^{t+1} in $\pi_{p^j(l-1)-1}(P^l(p^t))$, and $\hat{\tau}_j(\hat{v}_l)$ satisfies

$$\hat{\tau}_j(\hat{v}_l)\omega_t^{t+1} = \tau_j(\hat{v}_l), \tag{10A}$$

where $\tau_j(\hat{v}_l)$ denotes the iterated Whitehead product $\mathrm{ad}^{p^j-1}(\hat{v}_l)(\hat{u}_{l-1})$. For any space X and homotopy class $\hat{x} \in \pi_l(X; \mathbb{Z}_{p^t})$ we set

$$\hat{\tau}_j(\hat{x}) = (\hat{x})_{\#}(\hat{\tau}_j(\hat{v}_l)) \in \pi_{p^j(l-1)}(X; \mathbb{Z}_{p^{t+1}}), \tag{10B}$$

$$\hat{\sigma}_j(\hat{x}) = (\hat{x})_{\#}(\hat{\sigma}_j(\hat{v}_l)) = {}_{\pi}\delta^{t+1}\hat{\tau}_j(\hat{x}) \in \pi_{p^j(l-1)-1}(X; \mathbb{Z}_{p^{t+1}}). \tag{10C}$$

(e) Suppose l is odd. The unique summand of $\mathbb{Z}_{p^{t+1}}$ in $\pi_c(P^l(p^t); \mathbb{Z}_{p^{t+1}})$ is generated by $\hat{\sigma}_1(\hat{v}_l)$. The unique summand of $\mathbb{Z}_{p^{t+1}}$ in $\pi_{c+1}(P^l(p^t); \mathbb{Z}_{p^{t+1}})$ is generated by $\hat{\tau}_1(\hat{v}_l)$.

(f) Suppose l is even. The unique summand of $\mathbb{Z}_{p^{t+1}}$ in $\pi_c(P^l(p^t); \mathbb{Z}_{p^{t+1}})$ is generated by $\hat{\sigma}_1([\hat{v}_l, \hat{v}_l])$. The unique summand of $\mathbb{Z}_{p^{t+1}}$ in $\pi_{c+1}(P^l(p^t); \mathbb{Z}_{p^{t+1}})$ is generated by $\hat{\tau}_1([\hat{v}_l, \hat{v}_l])$.

The next lemma locates the higher torsion in a bouquet of Moore spaces.

Notation. For $1 \leq s \leq t \leq \infty$, write $X \in \mathcal{W}_s^t$ if X has finite type and the homotopy type of $\bigvee_{i=s}^t X^i$, where each X^i is a 1–connected bouquet of mod p^i Moore spaces.

Lemma 10.2. *Let $X \in \mathcal{W}_1^\infty$ and write $X \approx \bigvee_{i=1}^\infty X^i$, where X^i is a bouquet of mod p^i Moore spaces. For each $m \geq 1$ let $a(m)$ denote the least integer (or ∞ if $X^m = *$) for which $P^{a(m)}(p^m)$ is a summand of X^m. Let $a_e(m)$, $a_o(m)$ denote the least even, odd* *integers for which $P^{a_e(m)}(p^m)$, $P^{a_o(m)}(p^m)$ is a summand of X^m. Obviously, $a(m) = \min(a_e(m), a_o(m))$. Let*

$$b(m) = \min(p(a_o(m) - 1) - 1,\ 2p(a_e(m) - 1) - 1).$$

Let $a_+(m) = inf\{a(i) | i \geq m\}$ and let $X^{\leq m}$ denote $\bigvee_{i=1}^{m} X^i$.

(a) $p^{m+1}\pi_*(X^{\leq m}) = 0$.

(b) $p^m\pi_q(X^{\leq m}) = 0$ for $q < b(m)$.

(c) $p^m\pi_{b(m)}(X^{\leq m})$ is a $\mathbb{Z}_p$ vector space whose dimension equals b_o if $b(m) \equiv 3(\mod 4)$, $b_o + b_e$ if $b(m) \equiv 1(\mod 4)$. Here b_o (resp. b_e) denotes the number of Moore space summands of dimension $\frac{1}{p}(b(m)+1)$ (resp. $\frac{1}{2p}(b(m)+1)+1$) in X^m.

(d) For $j \leq \min(a_+(m+1) + 2p - 6,\ b(m) - 2)$, the Hurewicz homomorphism

$$h:\ p^m\pi_j(\Omega X) \to p^m H_j(\Omega X)$$

is one–to–one.

Construction of $\hat{y}^*_{p^k}$ and $\hat{a}_k$

Our first task is to construct the special homotopy class $\hat{y}^*_{p^k}$ in $\pi_{2np^k-1}(\Omega E_k; \mathbb{Z}_{p^{r+k-1}})$. The importance of $\hat{y}^*_{p^k}$ would be difficult to exaggerate: it is the source from which springs all the remaining work in Chapter 10. From (xii) at $k-1$, we have a map $\hat{y}_{p^k} : P^{2np^k-1}(p^{r+k}) \to \Omega E_{k-1}$ for which

(10D) $$\hat{y}_{p^k}\omega^{r+k}_{r+k-1} = \gamma^{k-1}_{r+k-1}(y_{p^k}).$$

Consider the following diagram, in which the top row is a cofibration and the lower three rows form a CMN diagram:

(10E)
$$\begin{array}{ccccc}
P^{2np^k}(p) & \xrightarrow{\omega_1^{r+k}} & P^{2np^k}(p^{r+k}) & \xrightarrow{\rho_{r+k}^{r+k-1}} & P^{2np^k}(p^{r+k-1}) \\
\downarrow & & \downarrow{\scriptstyle (\hat{y}_{p^k})^{\bar{a}}} & & \vdots{\scriptstyle (\hat{y}^*_{p^k})^{\bar{a}}} \\
P^{2np^k}(p) \cup (\text{higher cells}) & \longrightarrow & E_{k-1} & \xrightarrow{\hat{i}} & E_k \\
=\downarrow & & {\scriptstyle \lambda_{k-1}\eta_{k-1}}\downarrow & & \downarrow{\scriptstyle \lambda_k\eta_k} \\
P^{2np^k}(p) \cup (\text{higher cells}) & \longrightarrow & D_{k-1} & \xrightarrow{\hat{i}} & D_k \\
\downarrow & & \downarrow{\scriptstyle \rho_{k-1}} & & \downarrow{\scriptstyle \rho_k} \\
* & \longrightarrow & S^{2n+1}\{p^r\} & \xrightarrow{=} & S^{2n+1}\{p^r\}.
\end{array}$$

In (10E), the notation $(\ )^{\bar{a}}$ means the coadjoint of a map into a loop space. The top left square of (10E) commutes on the nose, and as a result there is a canonical map $\hat{y}^*_{p^k}$ for which

(10F) $$(\Omega\hat{i})\hat{y}_{p^k} = \hat{y}^*_{p^k}\rho^{r+k-1}_{r+k}.$$

Notice immediately that

$$(10G)\qquad (\Omega\hat{i})_{\#}\gamma_{r+k-1}^{k-1}(y_{p^k}) = (\Omega\hat{i})_{\#}\hat{y}_{p^k}\omega_{r+k-1}^{r+k} = (p)\hat{y}_{p^k}^*$$

by Lemma 6.1(e); (10G) is an extremely important equation. Furthermore,

$$(10H)\qquad \begin{aligned} {}_{\pi}\delta^{r+k-1}(\hat{y}_{p^k}^*) &= \hat{y}_{p^k}^*\omega^{r+k-1}\rho^{r+k-1} = (\Omega\hat{i})\hat{y}_{p^k}\omega^{r+k}\rho^{r+k-1} \\ &= (\Omega\hat{i})_{\#}\gamma_{r+k-1}^{k-1}(p^{-r-k}d(y_{p^k}))\end{aligned}$$

by (9P) at $k-1$. We note in addition that

$$(10I)\qquad \underline{a}_{\eta_k}(\hat{y}_{p^k}^*) = x_{p^k};$$

this is in a sense the definition of $x_{p^k} \in \mathcal{A}(F_k)$ in [A4, Theorem 5.6]. Lastly, we define

$$(10J)\qquad \hat{a}_k = (\Omega\lambda_k\eta_k)_{\#}(\hat{y}_{p^k}^*) \in \pi_{2np^k-1}(\Omega D_k; \mathbb{Z}_{p^{r+k-1}}).$$

Construction of ψ_t^k for $t < r+k$

Fortified with the elements $\hat{y}_{p^k}^*$ and $\hat{a}_k$, we commence the construction of ψ_t^k for $t < r+k$. The key step is the definition of $\Psi_t^k(c_k)$, since everything else is determined by (9H) and the csm axioms. Once $\Psi_t^k(c_k)$ is defined, we obtain a natural csm structure using Proposition 7.9. Part (viii) will be implicit in our definition, and (vii) will be an easy corollary.

Let $0 \le s \le k$. Recall from Chapter 4 that $\mathcal{F}(\mathcal{U}M^k)_{[r+s]}$ contains the sub-dga

$$R_s^k = \mathbb{Z}_{(p)}\langle b^{p^s}, c_s, c_{s+1}, \ldots, c_k\rangle / \langle \epsilon(b^{p^m}, c_m) - pc_{m+1} \text{ for } s \le m < k\rangle,$$

where $\epsilon(b^{p^m}, c_m) = (b^{p^m})^{p-1}c_m + (b^{p^m})^{p-2}c_m(b^{p^m}) + \ldots + c_m(b^{p^m})^{p-1}$ as in Lemma 4.1. We begin by defining $\Psi_t^k\colon R_{t-r}^k \to \mathcal{O}_{D_k}[t]$, where we interpret $t-r$ as zero (we continue this convention throughout the proof) if $t < r$. There are four steps involved in defining Ψ_t^k. We must declare what $\Psi_t^k(b^{p^{t-r}})$ and $\Psi_t^k(c_m)$ for $t-r \le m \le k$ are to be, we must extend to a multiplicative homomorphism, we must show that the Ψ_t^k-image of the defining relators for R_{t-r}^k is zero, and we must verify on generators that Ψ_t^k is a chain map.

We are considering for now only the case $t < r+k$, so we may define $\Psi_t^k(b^{p^{t-r}})$ to be $\mathcal{O}_{(\hat{i})}\Psi_t^{k-1}(b^{p^{t-r}})$ and $\Psi_t^k(c_m)$ to be $\mathcal{O}_{(\hat{i})}\Psi_t^{k-1}(c_m)$ when $t-r \le m < k$. Then the relations

$$\Psi_t^k(\epsilon(b^{p^m}, c_m) - pc_{m+1}) = \mathcal{O}_{(\hat{i})}\Psi_t^{k-1}(\epsilon(b^{p^m}, c_m) - pc_{m+1}) = 0$$

hold true, for $t-r \le m < k-1$, and the differentials commute with Ψ_t^k on $b^{p^{t-r}}$ and on $c_{t-r}, \ldots, c_{k-1}$. The next step is to choose $\Psi_t^k(c_k)$.

Recall from Proposition 4.3(c) that

$$(10K)\qquad c_k = a_k - \sum_{m=0}^{k-1}\sum_{i=0}^{p^m(p-1)-1}(p^{-q}\lambda_i^{k,m})(p^q b^i c_m b^{p^k-p^m-i}),$$

where in the summation q denotes the integer that satisfies $p^q \gcd(p^m, i) = p^{k-1}$, and the $\{p^{-q}\lambda_i^{k,m}\}$ are all integers. Each summand on the right-hand side of (10K) belongs to

$\mathcal{F}(\mathcal{U}M^k)_{[r+k-1]}$, and all right–hand terms except a_k belong to $\mathcal{F}(\mathcal{U}M^{k-1})_{[r+k-1]}$. Define $\Psi_t^k(c_k)$ by

$$\text{(10L)} \qquad \Psi_t^k(c_k) = \Phi(\hat{a}_k\omega_t^{r+k-1}) - \sum_{m=0}^{k-1}\sum_{i=0}^{p^m(p-1)-1}(p^{-q}\lambda_i^{k,m})\mathcal{O}_{(\hat{i})}\Psi_t^{k-1}(p^q b^i c_m b^{p^k-p^m-i})$$

The notation $\Phi : \pi_*(\Omega D_k; \mathbb{Z}_{p^t}) \to \mathcal{O}_{D_k}[t]$ comes from Lemma 8.9.

For Ψ_t^k to be well–defined on the non–free algebra R_{t-r}^k, it must vanish on all defining relators. This has been established, except for the equation

$$\Psi_t^k(pc_k) = \Psi_t^k(\epsilon(b^{p^{k-1}}, c_{k-1})),$$

which we prove next. Using (10J) and (10G),

$$p\hat{a}_k = (\Omega\lambda_k\eta_k)_\#(\Omega\hat{i})_\#\gamma_{r+k-1}^{k-1}(y_{p^k}) = (\Omega\hat{i})_\#(\Omega\lambda_{k-1}\eta_{k-1})_\#\gamma_{r+k-1}^{k-1}(y_{p^k}),$$

hence $p\Phi(\hat{a}_k\omega_t^{r+k-1}) = \mathcal{O}_{(\hat{i})}\Phi((\Omega\lambda_{k-1}\eta_{k-1})_\#\gamma_t^{k-1}(y_{p^k}))$, where we have used the csL homomorphism axiom $\theta_t^{r+k-1}\gamma_{r+k-1}^{k-1} = \gamma_t^{k-1}\theta_t^{r+k-1}$ to replace $\gamma_{r+k-1}^{k-1}(y_{p^k})\omega_t^{r+k-1}$ by $\gamma_t^{k-1}(y_{p^k})$. Since the $\bar{e}$–image of $y_{p^k} \in \overline{L}^k$ in $\mathcal{U}M^k$ is $\mathrm{ad}(b^{p^k-p^{k-1}}(a_{k-1}))$, (xi) at $k-1$ tells us that

$$\text{(10M)} \qquad p\Phi(\hat{a}_k\omega_t^{r+k-1}) = \mathcal{O}_{(\hat{i})}\Psi_t^{k-1}(\mathrm{ad}(b^{p^k-p^{k-1}}(a_{k-1})).$$

By (4H), the right–hand side of (10M) may be replaced by the $(\mathcal{O}_{(\hat{i})}\Psi_t^{k-1})$ image of

$$\epsilon(b^{p^{k-1}}, c_{k-1}) + (p)\sum_{m=0}^{k-1}\sum_{i=0}^{p^m(p-1)-1}(p^{-q}\lambda_i^{k,m})(p^q b^i c_m b^{p^k-p^m-i}).$$

Multiply (10L) through by p and make the indicated substitutions. The result is the desired equation,

$$p\Psi_t^k(c_k) = \mathcal{O}_{(\hat{i})}\Psi_t^{k-1}(\epsilon(b^{p^{k-1}}, c_{k-1})) = \Psi_t^k(\epsilon(b^{p^{k-1}}, c_{k-1})).$$

We leave it as an exercise to show that $\Psi_t^k(c_k)$ is a cycle, giving only the hint that (9N) is applied at $k-1$, with $w = p^{-(r+k)}d(y_{p^k})$.

So far, we have defined $\Psi_t^k : R_{t-r}^k \to \mathcal{O}_{D_k}[t]$. We dualize this to obtain module structure homomorphisms

$$\psi_t^k : R_{t-r}^k \otimes \tilde{\pi}_*(K_g; \mathbb{Z}_{p^t}) \to \tilde{\pi}_*(K_g; \mathbb{Z}_{p^t}).$$

The hypotheses of Proposition 7.9 are straightforward to verify in an induction on t, with $N_{[t]} = R_{t-r}^k$ (which is a Bockstein sub–dga by Lemma 7.4 and Theorem 4.8), and we omit this. The conclusion is that ψ_t^k extends to

$$\psi_t^k : \mathcal{F}(\mathcal{U}M^k)_{[t]} \otimes \tilde{\pi}_*(K_g; \mathbb{Z}_{p^t}) \to \tilde{\pi}_*(K_g; \mathbb{Z}_{p^t}).$$

Dualizing again, we obtain our extension of Ψ_t^k over all of $\mathcal{F}(\mathcal{U}M^k)_{[t]}$. That ψ_t^k and Ψ_t^k factor through $H_*(\mathcal{U}M^k; \mathbb{Z}_{p^t})$ is an application of Corollary 7.10(b).

This completes the construction of Ψ_t^k and $\overline{\Psi}_t^k$ for $t < r+k$. Part (viii) is implicit in our definition, and a technically precise proof goes as follows. ψ_t^{k-1} and $\psi_t^k \circ (\mathcal{F}(\bar{i}) \otimes 1)$ both define structures for $\{\tilde{\pi}_*(K_g; \mathbb{Z}_{p^t})\}_{t<r+k}$ as a csm over $\mathcal{F}(\mathcal{U}M^{k-1})$ for mapping pairs $(f,g) \in \mathcal{M}_{D_{k-1}}$, where $\bar{i} : \mathcal{U}M^{k-1} \to \mathcal{U}M^k$ is the embedding that takes b to b, c_s to c_s. Also, by definition of ψ_t^k, they concur on the Bockstein sub-dga $N_{[t]} = R_{t-r}^{k-1}$. By the uniqueness aspect of Proposition 7.9, the coincide. In other words, we use the uniqueness aspect of Proposition 7.9 to extend our knowledge that (9H) holds on R_{t-r}^{k-1} to having it on all of $\mathcal{F}(\mathcal{U}M^{k-1})_{[t]}$.

Consider (vii). To compute $\mathcal{H}_0\Psi_t^k$, write $\mathcal{H}_0\Psi_t^k(c_m)$ for $t-r \le m < k$ as

$$\mathcal{H}_0\mathcal{O}_{(\hat{i})}\Psi_t^{k-1}(c_m) = (\Omega\hat{i})_*\mathcal{H}_0\Psi_t^{k-1}(c_m) = (\Omega\hat{i})_*(c_m) = c_m,$$

where these equations take place in $H_*(\Omega D_k; \mathbb{Z}_{p^t}) = H_*(\mathcal{U}M^k; \mathbb{Z}_{p^t})$. Likewise, we have $\mathcal{H}_0\Psi_t^k(b^{p^{t-r}}) = b^{p^{t-r}}$. For the same reason, $\mathcal{H}_1\Psi_t^k$ vanishes on these generators. Next, $\Psi_t^k(c_k) \in \ker(\mathcal{H}_1)$ because it is built out of a piece in $\operatorname{im}(\Phi)$ and a piece in $\operatorname{im}(\mathcal{O}_{(\hat{i})}\Psi_t^{k-1})$, and both of these lie in $\ker(\mathcal{H}_1)$. Now $\mathcal{H}_0\Psi_t^k(c_k)$ can be computed directly from (10L), using Lemma 8.9 and (vii) at $k-1$. Not surprisingly, the answer is the right-hand side of (4I), which (as (4I) tells us) is c_k. It suffices to establish (vii) on the generators of a Bockstein sub-dga $N_{[t]}$, as we have now done, because Lemma 8.7 extends the formulas to all of $N_{[t]}$, and then the fact that $\mathcal{H}$ is a csc (not a priori a csa) homomorphism implies by the uniqueness aspect of Proposition 7.5 for csc's (view $N_{[t]}$ as merely a Bockstein subcomplex) that $\mathcal{H}\Psi_t^k = (1,0)$ on all of $\mathcal{F}(\mathcal{U}M^k)_{[t]}$.

We have finished proving all of (vi)–(viii) for $t < r+k$, except that we have not said why $\{\pi_*(\Omega K_g; \mathbb{Z}_{p^t})\}$ is a csL over $\mathcal{F}(\mathcal{U}M^k)$ as a Dcsa. This is trivial, since by Example 7.13 we need only to verify (7BC) and (7BD). For (7BC), we are verifying

$$\begin{aligned}&\psi_t^k(b^{p^s} \otimes [\hat{x},\hat{y}]) = [\psi_t^k(b^{p^s} \otimes \hat{x}), \hat{y}] + [\hat{x} \otimes \psi_t^k(b^{p^s} \otimes \hat{y})]\\ \text{(10N)}\qquad &+ \sum_{j=1}^{p^s-1}((p^s|j)p^{m-t})\zeta_t^m([\psi_m^k(b^j \otimes \theta_m^t(\hat{x})), \psi_m^k(b^{p^s-j} \otimes \theta_m^t(\hat{y}))]),\end{aligned}$$

where $s = \max(0, t-r)$ and $p^m = \gcd(p^r j, p^t)$, for $\hat{x}, \hat{y} \in \pi_*(\Omega K_g; \mathbb{Z}_{p^t})$ with $(f,g) \in \mathcal{M}_{D_k}$. Let (10N′) be the equation obtained by replacing all superscripts of k in (10N) by $k-1$, and where we think of b^{p^s} and b^j as elements of $\mathcal{F}(\mathcal{U}M^{k-1})_{[t]}$ and of $\mathcal{F}(\mathcal{U}M^{k-1})_{[m]}$. Then (10N′) is true because it is part of (vi) at $k-1$. By (9H), (10N) follows at once from (10N′). Formula (7BD) for Ψ_t^k likewise follows from (9H) for a_q when $q < k$, and for a_k it is the lovely fact that $\Psi_t^k(a_k) = \Phi(\hat{a}_k w_t^{r+k-1})$, together with the fact that the Φ–image of any mod p^t homotopy class is derivation.

Construction of ψ_{r+k}^k

We move on now to the segment that we have called the conceptual centerpiece of Chapter 10, the construction and properties of ψ_{r+k}^k. In an important way, this construction must differ from all the work we have done so far toward Theorem 9.1. Until now, we have been shuffling and comparing mod p^t homotopy classes, $t < r+k$, and using mod p^t classes to build new mod p^t classes. For ψ_{r+k}^k, however, we are required to concoct a mod p^{r+k} class where none has existed previously. Once one mod p^{r+k} class has been concocted (actually, we will build two), it can be transferred, amplified, and iterated so as to satisfy all the requirements of Theorem 9.1. The tool that will facilitate the concoction process is Theorem 6.4.

Let us look ahead concerning our intended application of Theorem 6.4. In reviewing the numerous hypotheses that go into Theorem 6.4, the most difficult one to satisfy is the exactness of (6H) at $p^t\pi_{j+1}(X)$; it is easier to see what the problem is by examining the closely related condition that (6K) be bijective. Suppose the $\bmod p^{r+k}$ homotopy class we are constructing has dimension q. Since the p–local Hurewicz theorem is valid only in a small range of dimensions starting at the connectivity of X, h would typically be injective for X only if X is at least $q-(2p-4)$–connected. In other words, the q–dimensional homotopy class we are defining would have to live in some space X whose connectivity is nearly q itself. Fortunately, Theorem 6.4 demands a weaker condition on h than true injectivity (cf. (6K)). By Lemma 10.2(d), if X is a wedge of Moore spaces, it suffices that there be no $\bmod p^{r+k}$ or higher Moore space summands of X, below dimension $q-(2p-4)$ or so. (We also have to be careful about where the first $\bmod p^{r+k-1}$ Moore space occurs.) If just one idea had to be singled out as the key idea in the construction of ψ^k_{r+k}, a good candidate would be the idea of carefully selecting the space X having the right connectivity properties to serve as the "home" where the first $\bmod p^{r+k}$ homotopy classes are defined.

Our goal is to construct Ψ^k_{r+k} that satisfies (vi). Since Proposition 7.9 is obviously going to be invoked at some point, it probably suffices to define Ψ^k_{r+k} on a suitable sub–dga $N_{[r+k]}$, and the obvious candidate for $N_{[r+k]}$ is $R^k_k = \mathbb{Z}_{(p)}\langle b^{p^k}, c_k\rangle$. So we are seeking to construct the operations $\Psi^k_{r+k}(c_k)$ and $\Psi^k_{r+k}(b^{p^k})$ in $\mathcal{O}_{D_k}[r+k]$. By Chapter 8, such an operation can be identified with a homotopy class (actually, with a "stable" list of homotopy classes) in the universal object(s), which is $D'_k(q+1, p^{r+k})$. Because $|c_k| = 2np^k - 1$ and $|b^{p^k}| = 2np^k$, we are therefore looking for homotopy classes $(\tilde{u}_{q-1})^{\overline{a}} \in \pi_{2np^k+q}(D'_k(q+1,p^{r+k}); \mathbb{Z}_{p^{r+k}})$ and $(\tilde{v}_q)^{\overline{a}} \in \pi_{2np^k+q+1}(D'_k(q+1,p^{r+k}); \mathbb{Z}_{p^{r+k}})$.

Let us digress to establish an important and useful property of D'_k, namely, that it is a wedge of Moore spaces. We will actually prove that $\Sigma^2\Omega D_k \in \mathcal{W}^{r+k}_r$ by providing a wedge decomposition for $D'_k(2) = (\Omega D_k) \ltimes S^2 = S^2 \vee \Sigma^2\Omega D_k$. Then $D'_k = (\Omega D_k)_+ \wedge P^{q+1}(p^{r+k})$ belongs to $\mathcal{W}^{r+k}_r$ because it is the cofiber of $(\Omega D_k)_+ \wedge S^q \xrightarrow{1\wedge p^{r+k}} (\Omega D_k)_+ \wedge S^q$.

To prove that $\Sigma^2\Omega D_k \in \mathcal{W}^{r+k}_r$, let R^k_{k+1} denote the trivial graded algebra $\mathbb{Z}_{(p)}$. For $0 \le s \le k$ let $\overline{R}^k_s$ denote a complementary summand to R^k_{s+1}, as a chain subcomplex of R^k_s, i.e., $R^k_s = R^k_{s+1} \oplus \overline{R}^k_s$ as differential graded $\mathbb{Z}_{(p)}$–modules. Then $\overline{R}^k_s$ is $\mathbb{Z}_{(p)}$–free and δ^{r+s}–acyclic. Notice that $\mathcal{U}M^k = R^k_0 = \overline{R}^k_0 \oplus \ldots \oplus \overline{R}^k_k \oplus \mathbb{Z}_{(p)}$, so

$$\overline{H}_*(\Omega D_k) = \overline{H}_*(\mathcal{U}M^k) = H_*(\overline{R}^k_0) \oplus \ldots \oplus H_*(\overline{R}^k_k).$$

Let $\{v_{sj}, u_{sj}\}$ denote a $\mathbb{Z}_{(p)}$–basis for $\overline{R}^k_s$ that satisfies $\delta^{r+s}(v_{sj}) = u_{sj}$. Each v_{sj} for $s < k$ determines a map

$$\psi^k_{r+s}(v_{sj} \otimes \hat{v}):\ P^{2+|v_{sj}|}(p^{r+s}) \to D'_k(2) = S^2 \vee \Sigma^2\Omega D_k,$$

whose image in integral homology is the copy of $\mathbb{Z}_{p^{r+s}}$ generated by the suspension $\Sigma^2 u_{sj}$ of the cycle $u_{sj} \in H_*(\overline{R}^k_s) \subseteq \overline{H}_*(\Omega D_k)$. (Here ψ^k_{r+s} is being evaluated for the map pair $D_k \to D_k \vee S^2 \to D_k$). For each element v_{kj} we can compute that $\psi^k_{r+k-1}(v_{kj} \otimes$

$\hat{v})$ has Hurewicz image $\Sigma^2 v_{kj}$ in $\overline{H}_*(\Sigma^2\Omega D_k \vee S^2; \mathbb{Z}_{p^{r+k-1}})$, and we can compute that ${}_\pi\delta^{r+k-1}\psi^k_{r+k-1}(v_{kj}\otimes\hat{v}) = (p)\psi^k_{r+k-1}(u_{kj}\otimes\hat{v})$ (there is no ${}_\pi\delta^{r+k-1}(\hat{v})$ term because $|\hat{v}| = 2$ here and its homotopy Bockstein vanishes). It follows that $\psi^k_{r+k-1}(v_{kj}\otimes\hat{v})$ extends (through w^{r+k}_{r+k-1}) to a map $\tilde{v}_{kj} : P^{2+|v_{kj}|}(p^{r+k}) \to D'_k(2)$ whose Hurewicz image, now in $\overline{H}_*(\Sigma^2\Omega D_k \vee S^2; \mathbb{Z}_{p^{r+k}})$, is $\Sigma^2 v_{kj}$. We wedge together all the $\psi^k_{r+s}(v_{sj}\otimes\hat{v})$ for $s < k$ and all the $\tilde{v}_{kj}$ to realize a homology equivalence from a wedge of Moore spaces to $\Sigma^2\Omega D_k$.

With this digression completed, we return to the business of hunting for the homotopy classes $(\tilde{u}_{q-1})^{\overline{a}}$ and $(\tilde{v}_q)^{\overline{a}}$ mentioned above. They exist in $\pi_*(D'_k; \mathbb{Z}_{p^{r+k}})$, but it turns out that D'_k is not the best place to find them because it is not sufficiently connected. Specifically, the summand $P^{q+1}(p^{r+k})$ in $D'_k = D'_k(q+1, p^{r+k})$ (for which the inclusion map is denoted $\hat{v}_{q+1}$) occurs in a low dimension that utterly ruins the needed condition (6K). We remove this troublesome summand by transferring our attention to the space $D''_k = D''_k(q+1, p^{r+k})$, which is defined as the homotopy–theoretic fiber of the composition $D'_k(q+1, p^{r+k}) \xrightarrow{d'_k} D_k \vee P^{q+1}(p^{r+k}) \xrightarrow{\bar{\pi}_2} P^{q+1}(p^{r+k})$. Here d'_k denotes the fiber inclusion map from D'_k to $D_k \vee P^{q+1}(p^{r+k})$. Several fibrations of interest occur in the CMN diagram

$$
\text{(10O)}\qquad
\begin{array}{ccccc}
\Omega P^{q+1}(p^{r+k}) & \longrightarrow & * & \longrightarrow & P^{q+1}(p^{r+k}) \\
\downarrow & & \downarrow & & \downarrow \\
D''_k & \xrightarrow{d'_k d''_k} & D_k \vee P^{q+1}(p^{r+k}) & \longrightarrow & D_k \times P^{q+1}(p^{r+k}) \\
\downarrow{\scriptstyle d''_k} & & \downarrow{\scriptstyle =} & & \downarrow{\scriptstyle \bar{\pi}_1} \\
D'_k & \xrightarrow{d'_k} & D_k \vee P^{q+1}(p^{r+k}) & \xrightarrow{\bar{\pi}_1} & D_k, \\
\downarrow{\scriptstyle \bar{\pi}_2 d'_k} & & & & \\
P^{q+1}(p^{r+k}) & & & &
\end{array}
$$

which serves to define d''_k.

Because D''_k is the fiber in the second row of (10O), we have

$$D''_k = \Sigma(\Omega D_k) \wedge (\Omega P^{q+1}(p^{r+k})) \in \mathcal{W}^{r+k}_r.$$

An homology calculation for this smash product shows that the first mod p^{r+k-1} Moore space summand of D''_k is a $P^{2np^{k-1}+q}(p^{r+k-1})$ while the first mod p^{r+k} summand is a $P^{2np^k+q}(p^{r+k})$. By Lemma 10.2(d), $h : p^{r+k-1}\pi_*(\Omega D''_k) \to H_*(\Omega D''_k)$ is an isomorphism in dimensions $\le 2np^k + q$, which includes the dimensions where $\tilde{u}_{j-1}$ and $\tilde{v}_j$ will occur.

We will be using the middle row of (10O) as our fibration (6G) in Theorem 6.4. Let us check the hypotheses for 6.4, as interpreted by Lemmas 6.5 and 6.6, one at a time. That the fibration splits when looped once is discussed in Chapter 6, and we have already seen why conditions 6.5(ii) and 6.5(iii) are true for D''_k. Verifying 6.5(iv) is an easy homology calculation (use $m = 3$ in 6.5(iv), but remember that the dimension "q" of 6.5 is now called $2np^k + q$ or $2np^k + q - 1$, and the "t" of 6.5 is $r + k - 1$). What shall we use for $\hat{y}$ and $\hat{z}$ in 6.4? We want them to be preimages in D''_k of certain homotopy classes in D'_k. Fortunately $(d''_k)_\#$ is monomorphic, because the map $\bar{\pi}_2 d'_k$ of (10O) has a section (namely $\hat{v}_{q+1}$). We make it part of the grand inductive scheme that

$$\text{(10P)}\qquad \psi^k_t(w\otimes\hat{v}_{q+1}) \in \operatorname{im}(d''_k)_\# \text{ whenever } |w| > 0,\ w \in \mathcal{F}(\mathcal{U}M^k)_{[t]}.$$

(This is true when $k = 0$ and $t \le r$ because $\Psi^0_t(w) \in \mathrm{im}(\Phi)$ and any relative Samelson product of $\hat{v}_{q+1}$ with something in D_k lies in $\mathrm{im}(d''_k)_\#$. When $k = 0$ and $t > r$ it is true because $\mathrm{im}(d''_k)_\#$ is closed under $\zeta^r_t = (\rho^r_t)^\#$ and $d_{[t]}\zeta^r_t = (\omega^r\rho^t)^\#$.) We are thus assuming that (10P) holds at $k-1$, and it is easily seen to hold for ψ^k_t with $t < r+k$, according to the definition we gave for ψ^k_t.

Write $\hat{v}$ for $\hat{v}_{q+1} \in \pi_{q+1}(D'_k(q+1, p^{r+k}); \mathbb{Z}_{p^{r+k}})$ and put $\hat{u} = \hat{u}_q = {}_\pi\delta^{r+k}(\hat{v})$. For our $\hat{y}$ and $\hat{z}$ in Theorem 6.4 we will take

$$\hat{y} = (d''_k)^{-1}_\# \psi^k_{r+k-1}(c_k \otimes \hat{v}\omega^{r+k}_{r+k-1})$$

and

$$\hat{z} = -(d''_k)^{-1}_\# \psi^k_{r+k-1}(c_k \otimes \hat{u}\omega^{r+k}_{r+k-1}).$$

Also, ${}_\pi\delta^{r+k-1}(\hat{y}) = p\hat{z}$ and ${}_\pi\delta^{r+k-1}(\hat{z}) = 0$ because these equations hold for their $(d''_k)_\#$-images. Theorem 6.4 also requires us to check certain hypotheses regarding $y = \underline{a}_{d'_k d''_k}(\hat{y})$ and $z = \underline{a}_{d'_k d''_k}(\hat{z})$. Using Corollary 8.17 we find that the mod p^{r+k-1} homology classes of y and of z are

$$\bar{y} = (\underline{a}_{d'_k})_* \psi^k_{r+k-1}(c_k \otimes \hat{v}\omega^{r+k}_{r+k-1}) = \tilde{c}_*(c_k \otimes h(\hat{v}\omega^{r+k}_{r+k-1})) = \mathrm{ad}(c_k)(v_q),$$
$$\bar{z} = (\underline{a}_{d'_k})_* \psi^k_{r+k-1}(c_k \otimes \hat{u}\omega^{r+k}_{r+k-1}) = -\tilde{c}_*(c_k \otimes h(\hat{u}\omega^{r+k}_{r+k-1}) = -\mathrm{ad}(c_k)(u_{q-1}),$$

where v_q and u_{q-1} denote the generators of $\mathcal{A}(P^{q+1}(p^{r+k}))$ viewed as a summand of the model $\mathcal{U}M^k \amalg \mathcal{A}(P^{q+1}(p^{r+k}))$ for $D_k \vee P^{q+1}(p^{r+k})$. We are free to use $y = \mathrm{ad}(c_k)(v_q)$ and $z = -\mathrm{ad}(c_k)(u_{q-1})$. Sure enough, $d(y) = p^{r+k}z$ and $\mathcal{A}(g)(y) = 0$, since our model for $g : D_k \vee P^{q+1}(p^{r+k}) \to D_k \times P^{q+1}(p^{r+k})$ is the projection $\mathcal{U}M^k \amalg \mathbb{Z}_{(p)}\langle u_{q-1}, v_q\rangle \to \mathcal{U}M^k \otimes \mathbb{Z}_{(p)}\langle u_{q-1}, v_q\rangle$. We have confirmed all the hypotheses of Theorem 6.4.

The conclusion of Theorem 6.4 is the existence of a mod p^{r+k} homotopy class, whose $(d''_k)_\#$-image we shall denote by $\psi^k_{r+k}(c_k \otimes \hat{v}_{q+1})$. Theorem 6.4 tells us that

$$\psi^k_{r+k}(c_k \otimes \hat{v}_{q+1})\omega^{r+k}_{r+k-1} = \psi^k_{r+k}(c_k \otimes \hat{v}_{q+1}\omega^{r+k}_{r+k-1}),$$

which, with the help of naturality, is going to be the formula (7AP) that we need in order to apply Proposition 7.9. We have also

$$\text{(10Q)} \qquad \underline{a}_{d'_k}\psi^k_{r+k}(c_k \otimes \hat{v}_{q+1}) = y = \mathrm{ad}(c_k)(v_q)$$

and

$$({}_\pi\delta^{r+k}\psi^k_{r+k}(c_k \otimes \hat{v}_{q+1}))\omega^{r+k}_{r+k-1} = (d''_k)_\#(\hat{z}).$$

For $(f, g) \in \mathcal{M}_{D_k}$ and $\hat{x} \in \tilde{\pi}_{q+1}(K_g; \mathbb{Z}_{p^{r+k}})$, define $\psi^k_{r+k}(c_k \otimes \hat{x})$ to be $\mathfrak{f}(\hat{x})_\#\psi^k_{r+k}(c_k \otimes \hat{v}_{q+1})$. In particular, $\psi^k_{r+k}(c_k \otimes \hat{u}_q) = \mathfrak{g}(\omega^{r+k}\rho^{r+k})_\#\psi^k_{r+k}(c_k \otimes \hat{v}_q)$. The proof that ${}_\pi\delta^{r+k}\psi^k_{r+k}(c_k \otimes \hat{v}_{q+1}) = -\psi^k_{r+k}(c_k \otimes \hat{u}_q)$, which is the universal case for (7AO) when $w = c_k$, is rather interesting. We are proving the equation

$$\text{(10R)} \qquad {}_\pi\delta^{r+k}(\psi^k_{r+k}(c_k \otimes \hat{v}_{q+1})) = -\mathfrak{g}(\omega^{r+k}\rho^{r+k})_\#\psi^k_{r+k}(c_k \otimes \hat{v}_q).$$

The way to prove (10R), which is an equation between mod p^{r+k} homotopy classes, is to exploit the uniqueness aspect of Theorem 6.4. Both of these classes have the same

$(\omega^{r+k}_{r+k-1})^{\#}$–image, namely $(d''_k)_{\#}(\hat{z})$, and both have the same $(\underline{a}_{d'_k})$–image, namely z (for $\underline{a}_{d'_k}$ applied to the right–hand side of (10R), the expression we need to compute is $-\mathcal{A}(1 \vee \omega^{r+k}\rho^{r+k})(\mathrm{ad}(c_k)(v_{q-1}))$ because $d'_k\mathfrak{g}(\omega^{r+k}\rho^{r+k}) = (1 \vee \omega^{r+k}\rho^{r+k})d'_k$, and the answer is $-\mathrm{ad}(c_k)(u_{q-1}) = z$). Theorem 6.4 does apply, with the "$\hat{y}$" of 6.4 now being $\hat{z}$ and the "$\hat{z}$" being zero. Since both sides of (10R) satisfy the conclusion of Theorem 6.4, they must be equal. Formula (10R) leads easily to the desired (7AO). The proof of (7AQ) is similar, so we omit it.

In a similar fashion we define $\psi^k_{r+k}(b^{p^k} \otimes \hat{v}_{q+1})$. Having illustrated the technique in complete detail once, we omit the construction entirely, mentioning only that $\hat{y} = (d''_k)^{-1}_{\#}\psi^k_{r+k-1}(b^{p^k} \otimes \hat{v}\omega^{r+k}_{r+k-1})$ and $y = \mathrm{ad}(b^{p^k})(v_q)$ and

$$\hat{z} = (d''_k)^{-1}_{\#}(-\psi^k_{r+k-1}(c_k \otimes \hat{v}\omega^{r+k}_{r+k-1}) + \psi^k_{r+k-1}(b^{p^k} \otimes \hat{u}\omega^{r+k}_{r+k-1})).$$

Once we have $\psi^k_{r+k}(b^{p^k} \otimes \hat{v})$, we define $\psi^k_{r+k}(b^{p^k} \otimes \hat{x})$ to be $\mathfrak{f}(\hat{x})_{\#}\psi^k_{r+k}(b^{p^k} \otimes \hat{v})$. Mimicking our method for $w = c_k$, we check (7AO)–(7AQ) for $w = b^{p^k}$, a check which again we omit as it offers no new insights.

The homotopy operations under D_k, $\psi^k_{r+k}(c_k \otimes -)$ and $\psi^k_{r+k}(b^{p^k} \otimes -)$ are additive. The notation may obscure the fact that their additivity needs to be proved! Letting w stand for either b^{p^k} or c_k, we can verify the additivity of $\psi^k_{r+k}(w \otimes -)$ by looking at a universal example that allows for the addition of two homotopy classes. The right fibration turns out to be

$$D''_k \to D_k \vee P^{q+1}(p^{r+k}) \vee P^{q+1}(p^{r+k}) \longrightarrow D_k \times (P^{q+1}(p^{r+k}) \vee P^{q+1}(p^{r+k})).$$

The uniqueness aspect of Theorem 6.4, when applied to this fibration, can be used to establish the equation

$$\psi^k_{r+k}(w \otimes (v^{(1)}_{q+1} + v^{(2)}_{q+1})) = \psi^k_{r+k}(w \otimes v^{(1)}_{q+1}) + \psi^k_{r+k}(w \otimes v^{(2)}_{q+1}),$$

where w is either b^{p^k} or c_k.

We leave the details to the reader. All the hypotheses of Proposition 7.9 are fulfilled, with $N_{[r+k]} = R^k_k = \mathbb{Z}_{(p)}\langle b^{p^k}, c_k\rangle$. Proposition 7.9 shows that ψ^k_{r+k} extends uniquely over $\mathcal{F}(\mathcal{U}M^k)_{[r+k]}$, and that $\{\psi^k_t\}_{1 \le t \le r+k}$ is a natural csm structure on $\tilde{\pi}_*(K_g; \mathbb{Z}_{p^{r+k}})$.

Dualizing ψ^k_{r+k} will give us Ψ^k_{r+k}, but we have to check tht the element lists $\{\psi^k_{r+k}(c_k \otimes \hat{v}_{q+1})\}_{q \ge 1}$ and $\{\psi^k_{r+k}(b^{p^k} \otimes \hat{v}_{q+1})\}_{q \ge 1}$ are stable, in the sense of Chapter 8. Again, this is proved via Theorem 6.4, capitalizing on the uniqueness aspect to verify that $\Sigma\psi^k_{r+k}(c_k \otimes \hat{v}_q)$ and $\psi^k_{r+k}(c_k \otimes \hat{v}_{q+1})$ must coincide, and likewise for $\Sigma\psi^k_{r+k}(b^{p^k} \otimes \hat{v}_q)$ and $\psi^k_{r+k}(b^{p^k} \otimes \hat{v}_{q+1})$.

The idea behind the stability proof is that the construction in Theorem 6.4 is compatible with suspension, in this case. The way the proof of 6.4 goes, is that a number of homotopy classes in X are chosen, added and compared in the course of the proof. As we perform the construction in 6.4 to build, say $\psi^k_{r+k}(c_k \otimes \hat{v}_q)$, we keep track of the intermediate elements (denoted $\hat{w}$, $\hat{z}_0$, $\hat{z}_1$, etc.). It turns out that we can then perform the construction of $\psi^k_{r+k}(c_k \otimes \hat{v}_{q+1})$ by using precisely these elements' suspensions as we go along (to be exact, we use $(d''_k)^{-1}_{\#}\Sigma(d''_k)_{\#}(\hat{w})$, $(d''_k)^{-1}_{\#}\Sigma(d''_k)_{\#}(\hat{z}_0)$, etc.). The net result is that $\Sigma\psi^k_{r+k}(c_k \otimes \hat{v}_q)$ is one possible outcome of the construction procedure. Since uniqueness means that there is only one possible outcome, however, $\Sigma\psi^k_{r+k}(c_k \otimes \hat{v}_q)$ must equal $\psi^k_{r+k}(c_k \otimes \hat{v}_{q+1})$.

We have now shown that the dual Ψ^k_{r+k} to ψ^k_{r+k} is a dga homomorphism from $\mathcal{F}(\mathcal{U}M^k)_{[r+k]}$ to $\mathcal{O}_{D_k}[r+k]$. By Lemma 7.10(b) ψ^k_{r+k} induces $\overline{\psi}^k_{r+k}: H_*(\Omega D_k; \mathbf{Z}_{p^{r+k}}) \otimes \tilde{\pi}_*(K_g; \mathbf{Z}_{r+k}) \to \tilde{\pi}_*(K_g; \mathbf{Z}_{p^{r+k}})$, and Ψ^k_{r+k} factors through the dual $\overline{\Psi}^k_{r+k}$ of $\overline{\psi}^k_{r+k}$.

The fact that $\psi^k_{r+k}(c_k \otimes \hat{v}_{q+1}) = \Sigma\psi^k_{r+k}(c_k \otimes \hat{v}_q)$ enables us to compute $\mathcal{H}\overline{\Psi}^k_{r+k}$. With the help of (8U), we find that

$$\underline{a}_{d'_k}(\Sigma\overline{\psi}^k_{r+k}(w \otimes \hat{v}_q)) = \mathrm{ad}(\mathcal{H}_0\overline{\Psi}^k_{r+k}(w))(v_q) + \mathrm{ad}(\mathcal{H}_1\overline{\Psi}^k_{r+k}(w))(u_{q-1}).$$

Comparing this with (10Q), the only possibility is $\mathcal{H}_0\overline{\Psi}^k_{r+k}(c_k) = c_k$, $\mathcal{H}_1\overline{\Psi}^k_{r+k}(c_k) = 0$. Likewise, $\mathcal{H}\overline{\Psi}^k_{r+k}(b^{p^k}) = (b^{p^k}, 0)$. By Lemma 8.7, $\mathcal{H}\Psi^k_{r+k} = (\underline{q}_{[r+k]}, 0)$ on all of $R^k_k = \mathbf{Z}_{(p)}\langle b^{p^k}, c_k\rangle$. A trivial application (i.e., $N_{[t]} = 0$) of Proposition 7.9 shows that ψ^k_t is defined and $\mathcal{H}\Psi^k_t = (\underline{q}, 0)$ for all $t > r+k$. In a similar vein, (viii) is true trivially for $t \geq r+k$. We have proved all of (vi)–(viii) for all t, except for the assertion that $\{\pi_*(\Omega K_g; \mathbf{Z}_{p^t})\}$ is a csL over the Dcsa $\mathcal{F}(\mathcal{U}M^k)$.

The First Jacobi Identity for D_k

We turn next to the proof that $\{\pi_*(\Omega K_g; \mathbf{Z}_{p^t})\}$ is a csL over $\mathcal{F}(\mathcal{U}M^k)$. We have already established this for $t < r+k$. As we saw in Example 7.13, it suffices to prove the single identity (the brackets denote Whitehead products):

$$\begin{aligned}\psi^k_{r+k}(b^{p^k} \otimes [\hat{x}, \hat{y}]) &= [\psi^k_{r+k}(b^{p^k} \otimes \hat{x}), \hat{y}] + [\hat{x}, \psi^k_{r+k}(b^{p^k} \otimes \hat{y})] \\ &+ \sum_{j=1}^{p^k-1} ((p^k|j)p^{m-k})\zeta^{r+m}_{r+k}([\psi^k_{r+m}(b^j \otimes \theta^{r+k}_{r+m}(\hat{x})), \psi^k_{r+m}(b^{p^k-j} \otimes \theta^{r+k}_{r+m}(\hat{y}))]),\end{aligned} \tag{10S}$$

where $p^m = \gcd(j, p^k)$, for $\hat{x}, \hat{y} \in \tilde{\pi}_*(K_g; \mathbf{Z}_{p^{r+k}})$, for any $(f,g) \in \mathcal{M}_{D_k}$. For reasons that will become clear later, we refer to (10S) as the "first Jacobi identity for D_k".

We will write q and q' for $|\hat{x}| - 1$ and $|\hat{y}| - 1$ in (10S). For conciseness, in this proof we write P^l for $P^l(p^{r+k})$ if $l \geq 3$. If $l = 2$ we let P^2 denote S^2. The reason for this is that (10S) is a formula about $\tilde{\pi}_*(\ ; \mathbf{Z}_{p^{r+k}})$, so the universal source object in dimension 2 is S^2, not $P^2(p^{r+k})$.

It suffices to prove (10S) in the universal case, which is the map pair $D_k \to D_k \vee P^{q+1} \vee P^{q'+1} \xrightarrow{\bar{\pi}_1} D_k$. However, the resulting fiber $K_{\bar{\pi}_1}$ is insufficiently connected for Theorem 6.4 to be usable, so we look to a different fibration. Recall the homotopy limit $\mathcal{R}(D_k, P^{q+1}, P^{q'+1})$ of Chapter 6, and consider the diagram

$$\begin{array}{ccccc} G = \mathcal{G}(D_k, P^{q+1}, P^{q'+1}) & \xrightarrow{\nu} & D_k \vee P^{q+1} \vee P^{q'+1} & \xrightarrow{\tilde{\lambda}} & \mathcal{R}(D_k, P^{q+1}, P^{q'+1}) \\ \downarrow & & \downarrow = & & \downarrow \\ K_{\bar{\pi}_1} & \xrightarrow{e'_k} & D_k \vee P^{q+1} \vee P^{q'+1} & \xrightarrow{\bar{\pi}_1} & D_k. \end{array} \tag{10T}$$

Because both rows of (10T) are fibrations that split when looped oncé, we know that $\nu_\#$ and $(e'_k)_\#$ are monomorphisms.

Let $\hat{v}_{q+1} : P^{q+1} \to D_k \vee P^{q+1} \vee P^{q+1}$ and $\hat{v}'_{q'+1} : P^{q'+1} \to D_k \vee P^{q+1} \vee P^{q'+1}$ denote the obvious inclusions, and let $\hat{y}_0$ denote the difference beween the two sides of (10S) when $\hat{x} = \hat{v}_{q+1}$ and $\hat{y} = \hat{v}'_{q'+1}$. Then $\hat{y}_0 \in \tilde{\pi}_{2np^k+q+q'+1}(D_k \vee P^{q+1} \vee P^{q'+1}; \mathbb{Z}_{p^{r+k}})$ and $(\bar{\pi}_1)_\#(\hat{y}_0) = 0$ in (10T), so $(e'_k)^{-1}_\#(\hat{y}_0)$ is well–defined. The universal case for (10S) is the assertion that $(e'_k)^{-1}_\#(\hat{y}_0) = 0$, which (because $(e'_k)_\#$ is one–to–one) is equivalent to the claim that $\hat{y}_0 = 0$.

In order to prove that $\hat{y}_0 = 0$, we will apply Corollary 6.7 to the top row of (10T). The "t" of 6.7 will be $r + k - 1$ and the "q" will be $|\hat{y}_0| - 1 = 2np^k + q + q'$. We already noted that the top row of (10T) splits when looped once, by Lemma 6.24(b). Also, $G \in \mathcal{W}_r^{r+k}$, so 6.5(iii) holds. Let us check 6.5(ii), with the help of Lemma 10.2(d). Referring to the notation $a(\ )$ used in Lemma 10.2, we find that $a(r + k - 1) = 2np^{k-1} + q + q' - 1$ and $a(r + k) = 2np^k + q + q' - 1$ if $q + 1 \geq 3$ and $q' + 1 \geq 3$. A calculation based upon Lemma 10.2(d) shows us that Condition 6.5(ii) will hold if both $2p \geq 7$ and $(p-1)(q+q'-2) \geq 3$, which are always true for $p \geq 5$. If $q + 1$ or $q' + 1$ is 2, the calculation is slightly different because we would be using S^2 in place of P^{q+1} or $P^{q'+1}$, and S^2 is more connected than P^2. The actual formula in all cases is $a(r + s) = 2np^s + \max(2, q) + \max(2, q') - 1$, $0 \leq s \leq k$. Again, Lemma 10.2(d) assures us that 6.5(ii) is true for this fibration.

In the hypotheses of Corollary 6.7 we also have four criteria that $\hat{y}_0$ must satisfy. We know $\tilde{\lambda}_\#(\hat{y}_0) = 0$ because the projections of $\hat{y}_0$ to each of the three subspaces $D_k \vee P^{q+1}, D_k \vee P^{q'+1}, P^{q+1} \vee P^{q'+1}$, are null–homotopic, whence $\hat{y}_0$ becomes null in their homotopy limit. A direct calculation of $h((\hat{y}_0)^a)$ can be performed in the dga $\mathcal{A}(D_k \vee P^{q+1} \vee P^{q'+1})$, which we may replace by $\mathcal{U}L$, where $L = M^k \amalg \mathbb{L}\langle v_q, u_{q-1}, v'_{q'}, u'_{q'-1}\rangle$. The way to perform the calculation is to start with the difference between the two sides of (10S) and then make a series of substitutions that convert homotopy to loop space homotopy. Brakcets in (10S) become (graded) commutators, $\hat{x}$ and $\hat{y}$ become v_q and v'_q respectively, and $\psi^k_{r+m}(b^j \otimes -)$ becomes $\mathrm{ad}(b^j)$ by Corollary 8.17. θ^{r+k}_{r+m} becomes the identity and ζ^{r+m}_{r+k} becomes multiplication by p^{k-m}. After making these substitutions we end up with

$$h((\hat{y}_0)^a) = \mathrm{ad}(b^{p^k})[v_q, v'_{q'}] - \sum_{j=0}^{p^k}(p^k|j)[\mathrm{ad}(b^j)(v_q), \mathrm{ad}(b^{p^k-j})(v'_{q'})] = 0.$$

Here we are making use of the well–known identity,

$$\mathrm{ad}(b^l)[x, y] = \sum_{j=0}^{l}(l|j)[\mathrm{ad}(b^j)(x), \mathrm{ad}(b^{l-j})(y)],$$

valid in any Lie algebra, which is easily established via induction on l.

We have verified all the hypothesis of Corollary 6.7 save two, for the element $\hat{y}_0 \in \tilde{\pi}_*(D_k \vee P^{q+1} \vee P^{q'+1}; \mathbb{Z}_{p^{r+k}})$, so (10S) will be proved as soon as we show that $\theta^{r+k}_{r+k-1}(\hat{y}_0) = 0$ and $\theta^{r+k}_{r+k-1}({}_\pi\delta^{r+k}(\hat{y}_0)) = 0$.

We can prove these two equations using a clever trick rather than an involved computation, since they are in a sense formalities. We begin by observing, because (vi) holds up to $r + k - 1$, that $\{\tilde{\pi}_*(D_k \vee P^{q+1}; \mathbb{Z}_{p^t})\}_{1 \leq t \leq r+k-1}$ is a csL over $\mathcal{F}(\mathcal{U}M^k)$. (The map pair is $D_k \to D_k \vee P^{q+1} \vee P^{q'+1} \to *$.) Another csL over $\mathcal{F}(\mathcal{U}M^k)$ is $\mathcal{F}(K)$, where K is defined to be the kernel in the split short exact sequnce of dgL's,

(10U) $$0 \to K \xrightarrow{\bar{e}} L = M^k \amalg \mathbb{L}\langle u_{q-1}, v_q, u'_{q'-1}, v'_{q'}\rangle \to M^k \to 0.$$

We know that K is a $\mathcal{U}M^k$–module, and in fact by Lemma 2.4 $K = \mathcal{L}(\mathcal{U}M^k \otimes V)$, where $\mathcal{L}(V) = \mathbf{L}\langle u_{q-1}, v_q, u'_{q'-1}, v'_{q'}\rangle$. That is, we define V to be the $\mathbf{Z}_{(p)}$–free chain complex with the four generators $\{u_{q-1}, v_q, u'_{q'-1}, v'_{q'}\}$ and differential given by $d(v_q) = p^{r+k}u_{q-1}$, $d(v'_{q'}) = p^{r+k}u'_{q-1}$. Then $K = \mathcal{L}(\mathcal{U}M^k \otimes V, d)$ as a dgL. By Lemma 7.12(c), $\mathcal{F}(K)$ is a csL over $\mathcal{F}(\mathcal{U}(M^k \amalg \mathcal{L}(V)))$, hence a csL over $\mathcal{F}(\mathcal{U}M^k)$.

We are going to construct a csL homomorphism out of $\mathcal{F}(K)$, and for this purpose we want to be able to describe $\mathcal{L}(R^k_{t-r} \otimes V)$ as a Bockstein sub–dgL of $\mathcal{F}(K)_{[t]}$. We run up against the issue, which has already been lengthily discussed, that the $\tau_j(y)$'s and $\sigma_j(y)$'s are going to cause a problem for (7AC). We make "the usual compromise", which is to impose another grading, call it "\$–degree", on $R^k_s \otimes V$, and then exclude \$–degrees that are too high. Specifically, let all of V have \$–degree 1, and let all of M^k have \$–degree zero. Then K lies entirely in positive \$–degrees, and $\mathcal{L}(R^k_{t-r} \otimes V)$ is a Bockstein sub–dgL of $\mathcal{F}(K)_{[t]}$ in \$–degree below p.

We are going to construct a csL/csm homomorphism $g = \{g_{[t]}\}_{1\le t\le r+k-1}$ that fills in the top arrow of

(10V)
$$\begin{array}{ccc} \mathcal{F}(K)^{\$<p} & \overset{g}{\dashrightarrow} & \{\pi_*(\Omega(D_k \vee P^{q+1} \vee P^{q'+1}); \mathbf{Z}_{p^t})\} \\ {\scriptstyle \underline{q}}\downarrow & \overset{\bar{g}}{\nearrow} & \downarrow{\scriptstyle h} \\ \{H_*(K; \mathbf{Z}_{p^t})\} & \xrightarrow{(\bar{e})_*} & \{H_*(\mathcal{U}L; \mathbf{Z}_{p^t})\}, \end{array}$$

where $\bar{e}$ is our generic name for the dgL inclusion that appears in (10U).

In order to construct g, we will use Theorem 7.15. (Note: because the target csL of g is not yet known to be a csL over $\mathcal{F}(\mathcal{U}M^k)$ at the $(r+k)^{\text{th}}$ stage, we can only get $g_{[t]}$ for $t < r+k$ this way.) When constructing the tth stage, $1 \le t \le r+k-1$, the $P_{[t]}$ of 7.15 will be R^k_{t-r} and the $N_{[t]}$ will be $\mathcal{L}(R^k_{t-r} \otimes V)$. The $\mathbf{Z}_p$–algebra P of 7.15 is $R^k_{t-r} \otimes \mathbf{Z}_p$, and the P–module N is $P \otimes (V \otimes \mathbf{Z}_p)$. As a left P module, N is free, so there are no syzygies to be concerned with. Furthermore, we are not dividing out by any virtual relators. Conditions (iii) and (iv) of Theorem 7.15 are vacuously satisfied. The lift to $R^k_{t-r} \otimes V$ of the generator set for N as a P–module is precisely the four–element set $\{u_{q-1}, v_q, u'_{q'-1}, v'_{q'}\}$, so we need only to declare where this set is to be sent, and to check 7.15(i) and 7.15(ii).

Where to send these four generators is obvious, given that we want (10V) to commute. Send $v_q, u_{q-1}, v'_{q'}, u'_{q'-1}$ respectively to $(\omega^{r+k}_t)^\#(\hat{v}_{q+1})^a$, $(\omega^{r+k}_t)^\#({}_\pi\delta^{r+k}(\hat{v}_{q+1}))^a$, $(\omega^{r+k}_t)^\#(\hat{v}'_{q'+1})^a$, $(\omega^{r+k}_t)^\#({}_\pi\delta^{r+k}(\hat{v}'_{q'+1}))^a$. Now (10V) commutes on these four generators, for any $t < r+k$. Conditions 7.15(i) and 7.15(ii) are easily checked, so Theorem 7.15 gives us a csL/csm homomorphism $g = \{g_{[t]}\}_{1\le t\le r+k-1}$ filling in the top arrow of (10V).

We claim that (10V) commutes on all of $\{\mathcal{F}(K)^{\$<p}_{[t]}\}_{1\le t\le r+k-1}$. To see this, notice that the three solid arrows of (10V) are all bracket–preserving csm homomorphisms, so hg and $(\bar{e})_*\underline{q}$ are two bracket–preserving csm homomorphisms that coincide on $\{u_{q-1}, v_q, u'_{q'-1}, v'_{q'}\}$. By the uniqueness aspect of Theorem 7.15, we have $hg = (\bar{e})_*\underline{q}$. Next, by Lemma 7.10(a), g factors through $\underline{q}$, so the diagonal arrow of (10V) exists, call it $\bar{g} = \{\bar{g}_{[t]}\}_{1\le t\le r+k-1}$. Of course, all of this is valid only in \$–degree below p, which is fine because for formula (10S) we will only need \$–degree 2. Recall that our entire purpose for constructing g and $\bar{g}$ is to prove that $\theta^{r+k}_{r+k-1}(\hat{y}_0)^a = 0$ and $\theta^{r+k}_{r+k-1}({}_\pi\delta^{r+k}(\hat{y}_0))^a = 0$.

We are nearly done. Because $(\bar{e})_*$ is one–to–one in \$–degrees below p, we must have

$$0 = \ker(h) \cap \operatorname{im}(\bar{g}_{[t]}) = \ker(h) \cap \operatorname{im}(g_{[t]})$$

for $1 \leq t \leq r+k-1$. But $(\hat{y}_0)^a$ was already shown to belong to $\ker(h)$, hence $h(\theta^{r+k}_{r+k-1}(\hat{y}_0)^a) = 0$ and $h(\theta^{r+k}_{r+k-1}({}_\pi\delta^{r+k}(\hat{y}))^a) = 0$. At the same time, $\theta^{r+k}_{r+k-1}(\hat{y}_0)^a$ belongs to $\operatorname{im}(g_{[r+k-1]})$. To see this, we let $\hat{x} = \hat{v}_{q+1}$ and $\hat{y} = \hat{v}'_{q'+1}$ in (10S); then apply θ^{r+k}_{r+k-1} to both sides of it and simplify as much as possible using the csL and csm axioms. The resulting expressions are built entirely upon $\theta^{r+k}_{r+k-1}(\hat{v}_{q+1})$ and $\theta^{r+k}_{r+k-1}(\hat{v}'_{q'+1})$, which are the $g_{[r+k-1]}$–images of v_q and v'_q. Deduce that $\theta^{r+k}_{r+k-1}(\hat{y}_0)^a \in \operatorname{im}(g_{[r+k-1]})$, whence $\theta^{r+k}_{r+k-1}(\hat{y}_0)^a = 0$. Similar reasoning shows that $\theta^{r+k}_{r+k-1}({}_\pi\delta^{r+k}(\hat{y}_0))^a = 0$. This completes the proof that $\{\pi_*(\Omega K_g; \mathbb{Z}_{p^t})\}_{1 \leq t \leq r+k}$ is a csL over $\mathcal{F}(\mathcal{U}M^k)$.

Remark 10.3. The above proof and the one before it illuminate a deep principle that pervades all of our work toward establishing identities among homotopy classes. The question that naturally arises is, "What is the universal example?" for a formula. Often, the true universal example is not highly enough connected to serve as a place where the formula can be proved. The better question seems to be, "What is the universal space where both sides of the formula have some kind of interpretation?" We found D''_k preferable to D'_k when we were defining ψ^k_{r+k}, and we found G preferable to the true universal example $K_{\hat{\pi}_1}$ of (10T) for the first Jacobi identity. The drawback of the true universal example is that it is not universal enough, in the sense that it is not really the initial object in some category consisting of all settings where the formula makes sense. The applicability of this principle in the case of ordinary p–primary relative Samelson products (the case $k = 0$, for us) was pointed out to the author by Joe Neisendorfer. We remark that the proof given above can indeed be adapted to give a simple proof of the classical Jacobi identity for mod p^r relative Samelson products, when $k = 0$.

Construction of γ^k_t for $t < r + k$

Our last task in Chapter 10 is the construction of γ^k_t satisfying (ix)–(xi), for $t < r + k$. The construction is performed in three stages. First, in dimensions below $2np^k - 1$, Formula (9M) effectively determines γ^k_t. Second, we make an individual definition at the special element y_{p^k}. Third, the fact that both source and target in (9J) are csL's over $\mathcal{F}(\mathcal{U}M^k)$ means that we may invoke Theorem 7.15 to construct γ^k_t in dimensions above $2np^k - 1$.

Since (x) utilizes the dgL inclusion $\bar{i} : \bar{L}^{k-1} \to \bar{L}^k$, we begin by recalling some of its properties. Recall that $\bar{e} : \bar{L}^{k-1} \to L^{k-1}$ has $\bar{e}(y_i) = x_i$, and likewise for the embedding of $\bar{L}^k$ into L^k. Recall from Lemma 3.1 that $\bar{i} : L^{k-1} \to L^k$ is defined by $\bar{i}(x_j) = p^{\alpha_k(j) - \alpha_{k-1}(j)}(x_j)$, hence $\bar{i}(y_j) = p^{\alpha_k(j)-\alpha_{k-1}(j)} y_j$, i.e.,

$$\bar{i}(y_j) = \begin{cases} y_j & \text{if } j < p^k \\ p y_j & \text{if } j \geq p^k. \end{cases}$$

Similarly,

$$\bar{i}(z'_j) = \begin{cases} z'_j + z''_j & \text{if } j \leq p^k, \\ p(z'_j + z''_j) & \text{if } j > p^k, \end{cases}$$

where $z''_j = 0$ if $j \notin I_{k-1}$, $z''_j \in J^k_{k-1}$ (hence z''_j is decomposable) if $j \in I_{k-1} - I_k$, and $z''_j \in J^k_k$ if $j \in I_k$. (The notation J^k_s is defined just before Lemma 4.6.)

The important thing about these formulas is that $\bar{i}$ becomes an isomorphism viewed as a homomorphism from $\bar{\bar{L}}^{k-1}$ to $\bar{\bar{L}}^{k}$, where

$$\bar{\bar{L}}^{k-1} = \mathbb{L}\langle y_2, y_3, \dots, y_{p^k-1}, z'_2, z'_3, \dots, z'_{p^k}\rangle$$

and

$$\bar{\bar{L}}^{k} = \mathbb{L}\langle y_2, y_3, \dots, y_{p^k-1}, z'_2, z'_3, \dots, z'_{p^k}\rangle.$$

Here $\bar{\bar{L}}^{k}$ is a sub-dgL of $\bar{L}^k$, and $\bar{i}(\bar{\bar{L}}^{k-1}) \subseteq \bar{\bar{L}}^{k}$, and by Lemma 2.5 $\bar{i} : \bar{\bar{L}}^{k-1} \to \bar{\bar{L}}^{k}$ is an isomorphism. Clearly $\bar{i}$ preserves $*$-degree too, so $\mathcal{F}(\bar{i}) : \mathcal{F}(\bar{\bar{L}}^{k-1})^{*\leq 2p} \to \mathcal{F}(\bar{\bar{L}}^{k})^{*\leq 2p}$ is an isomorphism.

Define γ_t^k on $\mathcal{F}(\bar{\bar{L}}^{k})^{*\leq 2p}_{[t]}$, $1 \leq t \leq r+k-1$, by

$$\gamma_t^k = (\Omega\hat{i})_{\#}\gamma_t^{k-1}\mathcal{F}(\bar{i})^{-1} \tag{10W}$$

On $\mathcal{F}(\bar{\bar{L}}^{k})^{*\leq 2p}$, (ix) and (x) can be recovered easily from the corresponding facts at $k-1$, and the same is true for (xi) on $\mathcal{F}(\bar{\bar{L}}^{k})^{*\leq 2p}$. We leave these checks as exercises, remarking only that (9H) is also needed to get (9N), and that (9K) and (9L) also utilize (9P) (the x'_j of (9P) are now being called x_j, so (9Q) says $\mathcal{A}(\hat{i})(x_j) = \bar{i}(x_j)$).

Our next step is to define γ_t^k on a sub–csL containing y_{p^k}. Write $\widetilde{L}^k$ for $\mathbb{L}\langle y_j, z'_j | j \leq p^k\rangle$ and $\widetilde{L}^k_s$ for $\mathbb{L} < \langle y_j, z'_j | j \leq p^k, j \in I_s\rangle$; these are sub-dgL's of $\bar{L}^k$. In $*$-degree below $2p+1$, $\widetilde{L}^k_{t-r}$ is a Bockstein sub-dgL of $\mathcal{F}(\widetilde{L}^k)_{[t]}$; we extend our previous convention for $\bar{L}^k_{t-r}$, so that $\widetilde{L}^k_{t-r}$ is interpreted as $\widetilde{L}^k_0$ if $t < r$. By Proposition 7.5, in order to extend

$$\gamma_t^k : \mathcal{F}(\bar{\bar{L}}^{k})_{[t]} \to \pi_*(\Omega E_k; \mathbb{Z}_{p^t})$$

over $\mathcal{F}(\widetilde{L}^k)_{[t]}$, we need only to say where the single new generator y_{p^k} should go, and to check (7AG) and (7AH).

Recall the homomotopy class $\hat{y}^*_{p^k}$ defined via (10E), which satisfies Formulas (10FGHI). Define $\gamma_t^k(y_{p^k})$ by

$$\gamma_t^k(y_{p^k}) = (\omega_t^{r+k-1})^{\#}(\hat{y}^*_{p^k}).$$

Then (7AG) obviously holds, and

$$\begin{aligned} {}_{\pi}\delta^t\gamma_t^k(y_{p^k}) &= p^{r+k-1-t}(\omega_t^{r+k-1})^{\#}\ {}_{\pi}\delta^{r+k-1}(\hat{y}^*_{p^k}) = (\Omega\hat{i})_{\#}\gamma_t^{k-1}(p^{-t-1}d(y_{p^k})) \\ &= (\Omega\hat{i})_{\#}\gamma_t^{k-1}(\bar{i})^{-1}(p^{-t}d(y_{p^k})) = \gamma_t^k(p^{-t}d(y_{p^k})) \end{aligned}$$

by (10H) and (10W), so (7AH) holds for y_{p^k}. By Proposition 7.5, this suffices to extend $\{\gamma_t^k\}$ to a unique csL homomorphism

$$\{\gamma_t^k\}_{1\leq t\leq r+k-1} : \mathcal{F}(\widetilde{L}^k)^{*\leq 2p} \to \{\pi_*(\Omega E_k; \mathbb{Z}_{p^t})\}.$$

Let us check (ix)–(xi) for this extended $\{\gamma_t^k\}$. Property (9K) at y_{p^k} merely reiterates (10I). For (9M) at y_{p^k} we have

$$\begin{aligned} (\Omega\hat{i})_{\#}\gamma_t^{k-1}(y_{p^k}) &= (\Omega\hat{i})_{\#}\hat{y}_{p^k}(\omega^{r+k}_{r+k-1})(\omega_t^{r+k-1}) = (p)\hat{y}^*_{p^k}\omega_t^{r+k-1} \\ &= (p)\gamma_t^k(y_{p^k}) = \gamma_t^k(\hat{i}(y_{p^k})) \end{aligned}$$

by (10F). The uniqueness aspect of Proposition 7.5 permits us to compare the two csL homomorphisms $(\Omega\hat{i})_{\#} \circ \{\gamma_t^{k-1}\}$ and $\{\gamma_t^k\} \circ \mathcal{F}(\bar{i})$. Since they coincide on the generators of our Bockstein sub–dgL $\widetilde{L}_{t-r}^{k-1}$, they coincide on all of $\mathcal{F}(\widetilde{L}^{k-1})$. I.e., (9M) is true of all $\mathcal{F}(\widetilde{L}^k)_{[t]}^{*\leq 2p}$.

We prove next that (9N) holds, for any $w \in \mathcal{F}(\widetilde{L}^k)_{[t]}^{*\leq 2p}$. Put

$$Q_t' = \{w \in \mathcal{F}(\widetilde{L}^k)_{[t]}^{*\leq 2p} \mid \text{(9N) is true for } w\}.$$

It is easily shown that Q_t' is a sub–dgL of $\mathcal{F}(\widetilde{L}^k)_{[t]}^{*\leq 2p}$, for $1 \leq t \leq r+k-1$. Furthermore, $\zeta_t^{t-1}(Q_{t-1}') \subseteq Q_t'$. Since we already checked that $\mathcal{F}(\bar{\bar{L}}^k)_{[t]}^{*\leq 2p} \subseteq Q_t'$, we need only to see that $y_{p^k} \in Q_t'$, and then Q_t' will equal all of $\mathcal{F}(\widetilde{L}^k)_{[t]}^{*\leq 2p}$.

That $y_{p^k} \in Q_t'$ is quite true, but the proof feels indirect. The actual definition of $\Psi_t^k\mathcal{F}(\bar{e})(y_{p^k})$ has to be obtained by writing $a_k = \bar{e}(y_{p^k})$ as an element of $\mathcal{F}(\mathcal{U}M^k)_{[t]} = R_{t-r}^k + \operatorname{im}(\zeta_t^{t-r}) + \operatorname{im}(d_{[t]}\zeta_t^{t-1})$ and then adding the three Ψ_t^k–images. To do this, we must solve (10K) for a_k, then apply Ψ_t^k to both sides and simplify using (10L). The double summation cancels out, and we are left with the simple equation

$$\Psi_t^k(a_k) = \Phi(\hat{a}_k\omega_t^{r+k-1}), \tag{10X}$$

which is another form of (9N) for $w = y_{p^k}$. In some sense, (10L) was chosen deliberately so that (9N) would be true; but because a_k does not belong to R_s^k if $s > 0$, (9N) could not be used in place of the definition (10L). This completes the second stage of the construction of γ_t^k.

We turn now to the construction of γ_t^k above dimension $2np^k - 1$, which will utilize Theorem 7.15. Let us see what are $N_{[t]}$, $P_{[t]}$, etc. for Theorem 7.15 in this case. Since we are limiting ourselves to $*$–degree at or below $2p$, we may use

$$\bar{L}_{t-r}^k = \mathbb{L}\langle y_j, z_j' \mid j \in I_{t-r}\rangle$$

as our Bockstein sub–dgL, where as usual we interpret $t-r$ as 0 if $t < r$. The submodule V of Theorem 7.15 is the $\mathbb{Z}_{(p)}$–span of $\{y_j, z_j' | j \in I_{t-r}\}$, and $P_{[t]}$ is R_{t-r}^k. We observed in Example 7.16 that there are some virtual relators for $\mathcal{F}(\bar{L}^k)$ as a csm over $\mathcal{F}(\mathcal{U}M^k)$. At the tth stage these are (we put $s = \max(0, t-r)$)

$$b^{p^s}c_s - c_s b^{p^s},\ (c_s)^2, \text{ and } c_{m+1} - c_m b^{p^{m+1}-p^m} \text{ for } s \leq m \leq k-1.$$

The quotient $\mathbb{Z}_p$–algebra P is $\mathbb{Z}_p[b^{p^s}] \otimes \Lambda(c_s)$ as in Example 7.16, and the generators and syzygies for $N = \mathbb{Z}_p \otimes V$ are given by (7CD) and (7CE).

Let us check condition 7.15(iv), i.e., these virtual relators should be virtual relators for $\{\pi_*(\Omega E_k; \mathbb{Z}_{p^t})\}$ as a csL over $\mathcal{F}(\mathcal{U}M^k)$. The way in which $\{\pi_*(\Omega E_k; \mathbb{Z}_{p^t})\}$ is a csL over $\mathcal{F}(\mathcal{U}M^k)$ comes from the proved property (vi), with $D_k \xrightarrow{=} D_k \xrightarrow{\rho_k} S^{2n+1}\{p^r\}$ as the map pair. Following (7CA), the expression that must hold on all of $\pi_*(\Omega E_k; \mathbb{Z}_{p^t})$ in order to fulfill 7.15(iv) is (again, $s = \max(0, t-r)$)

$$\begin{aligned}\psi_t^k((b^{p^s}c_s - c_s b^{p^s}) \otimes -) &= [\gamma_t^k(y_{2p^s}), -] \\ &+ \sum_{i=2}^{2p^s-1} (\mu_i p^{m-t})\zeta_t^m([\gamma_m^k(y_i), \psi_m^k(b^{2p^s-i} \otimes \theta_m^t(-))]),\end{aligned} \tag{10Y}$$

where m is the greatest integer for which $m \le t$ and $i \in I_{m-r}$. Since our sum runs only up to $i = 2p^s - 1$, and $s = \max(0, t-r) \le k-1$, each $\gamma^k_m(y_i)$ is already known to be defined and to satisfy (9N). Also, for any $\hat{w} \in \pi_*(\Omega E_k; \mathbb{Z}_{p^t})$, we have

$$[\hat{w}, -] = \Phi((\Omega\lambda_k\eta_k)_\#(\hat{w}));$$

this is a standard fact about relative and absolute Samelson products, which we apply to the fibration

$$\Omega E_k \xrightarrow{\Omega\lambda_k\eta_k} \Omega D_k \xrightarrow{\Omega\rho_k} \Omega S^{2k+1}\{p^r\}.$$

The right–hand side of (10Y) becomes

$$\Psi^k_t(x_{2p^s} + \sum_{i=2}^{2p^s-1} \mu_i x_i b^{2p^s-i}),$$

which is $\Psi^k_t(b^{p^s}c_s - c_s b^{p^s})$ by Lemma 4.4(c).

In a similar fashion, $c_{m+1} - c_m b^{p^{m+1}-p^m}$ is shown to be a virtual relator for $\pi_*(\Omega E_k; \mathbb{Z}_{p^t})$ as a csL over $\mathcal{U}M^k$. For $(c_s)^2$, we may use (10Y) and compute the differential of both sides. This completes the verification of condition 7.15(iv).

All the hard work for the construction of γ^k_t is now behind us. We need to define γ^k_t only in dimensions $2np^k$ and higher. As (7CD) and (7CE) show, there are no generators and no syzygies in dimensions $\ge 2np^k$ when $t \le r+k-1$, since the last ones occur at dimension $2np^k - 1$. Conditions 7.15(i,ii,iii) are vacuously satisfied. Deduce that there is a unique extension of γ^k_t over all of $\mathcal{F}(\bar{L}^k)^{*\le 2p}_{[t]}$.

It remains only to check (ix)–(xi) in dimensions above $2np^k - 1$. For (x), we may use the uniqueness aspect of Theorem 7.15 to show that $(\Omega\hat{i})_\# \circ \{\gamma^{k-1}_t\} = \{\gamma^k_t\} \circ \mathcal{F}(\bar{i})$. We already know this below dimension $2np^k$. Recycling our previous application of Theorem 7.15, we already know that 7.15(iv) holds for the virtual relators. Again 7.15(i,ii,iii) are true vacuously in dimensions $\ge 2np^k$. So (x) is true in all dimensions.

Formulas (9K) and (9L) are actually special cases of the commutativity of the diagram

$$\text{(10Z)} \qquad \begin{array}{ccc} \mathcal{F}(\bar{L}^k)^{*\le 2p+1} & \xrightarrow{\gamma^k_t} & \pi_*(\Omega E_k; \mathbb{Z}_{p^t}) \\ \downarrow{\scriptstyle q} & & \downarrow{\scriptstyle (\Omega\eta_k)_* h} \\ H_*(\bar{L}^k; \mathbb{Z}_{p^t}) & \xrightarrow{(\bar{\epsilon})_*} & H_*(\mathcal{U}L^k; \mathbb{Z}_{p^t}). \end{array}$$

To show that (10Z) commutes, we can use the ideas of Theorem 7.15 even though that theorem itself does not apply because we have never discussed how or whether $\{H_*(\mathcal{U}L^k; \mathbb{Z}_{p^t})\}$ is a csL (or csa?) over $\mathcal{F}(\mathcal{U}M^k)$. Nevertheless, all four arrows of (10Z) are certainly csL homomorphisms. All four are also homomorphisms of csm's over $\mathcal{F}(\mathcal{U}M^k)$. Given that the two ways of going around (10Z) concur below dimension $2np^k$, and all csL/csm generators of $\mathcal{F}(\bar{L}^k)$ (i.e., all the $\{x_i\}_{i\in I}$ in the notation of 7.15) occur below dimension $2np^k$, the two ways must coincide in all dimensions.

There is a subtlety in the above proof that is worth noting. There are actually two structures for $\{H_*(\mathcal{U}L^k; \mathbb{Z}_{p^t})\} = \{H_*(\Omega F_k; \mathbb{Z}_{p^t})\}$ as a csm over $\mathcal{F}(\mathcal{U}M^k)$. The first is "topological", arising via Lemma 8.15(c), and $(\Omega\eta_k)_* h$ is a homomorphism for this structure.

The second is "algebraic", arising because a dgH acts via conjugation on any differential Hopf ideal, and $(\bar{e})_*$ is a homomorphism for that structure. Part (v) shows that these two csm structures are identical. Indeed, this argument for the commutativity of (10Z) is the entire reason for including (v) in Theorem 9.1 at all.

Lastly, let us finish off (xi). Like before, we put

$$Q_t = \{w \in \mathcal{F}(\bar{L}^k)^{*\leq 2p}_{[t]} \mid \text{(9N) is true for } w\}.$$

Like before, Q_t is a dgL and $\zeta_t^{t-1}(Q_{t-1}) \subseteq Q_t$, and Q_t does equal $\mathcal{F}(\bar{L}^k)^{*\leq 2p}_{[t]}$ below dimension $2np^k$. Furthermore, Q_t is closed under $\psi_{[t]}(b^{p^s} \otimes -)$ where $s = \max(0, t-r)$; this is a delicate but straightforward exercise in using the first Jacobi identity. Since $y_{p^k} \in Q_t$, we get $y_j \in Q_t$ for all $j > p^k$, $j \in I_s$, by induction on j. To get the z'_j to belong to Q_t, where $j > p^k$ and $j \in I_s$, consider two cases. If $t \leq r$, the fact that $\psi_{[t]}(a_0 \otimes y_j) = z_{j+1}$ (this was the original definition of z_{j+1}) together with $\Psi^k_t(a_0) = \Phi(\hat{a}_0)$ ($\hat{a}_0$ is the composition $P^{2n}(p^t) \xrightarrow{(-1)} P^{2n}(p^t) \xrightarrow{\omega^r \rho^t} P^{2n+1}(p^r) = D_0 \hookrightarrow D_k$) yields $z_{j+1} \in Q_t$. If $t > r$, combine $y_{p^k+p^s}$ belonging to Q_t with Q_t being closed under $\delta^t = p^{-t}d$ to get $z'_{p^k+p^s} \in Q_t$, and then the closure of Q_t under $\psi_{[t]}(b^{p^s} \otimes -)$ gives us $z'_j \in Q_t$ for all $j \in I_s$, $j > p^k$, by induction on j. Thus $Q_t = \mathcal{F}(\bar{L}^k)^{*\leq 2p}_{[t]}$, as needed for (xi).

Summary 10.4. To close Chapter 10, we summarize where we are in the overall proof of Theorem 9.1. Parts (i)–(viii) of Theorem 9.1 are completely proved, within the framework of the grand inductive scheme. Parts (ix), (x), (xi) are true for $t < r + k$. But (xii) remains a distant dream, since γ^k_{r+k} has not yet even been defined.

CHAPTER 11

General Theory of CSL's

Chapter 11 begins with a long discussion of what remains to be done in order to complete the proof of Theorem 9.1. We outline the overall plan, along with some motivation for it. We describe four problems or "snags" that arise in carrying out the plan, and we indicate how we intend to get around them.

The mathematical content of Chapter 11 is divided into four units. Each of them advances the general theory of CSL's in some direction, but we keep an eye on our intended applications as well. We first introduce the important operation ξ_j on csL's and explore its properties. We define the functor $\mathcal{F}^0\mathcal{L}$ from free chain complexes (V, d) to csL's, and we offer $\mathcal{F}^0\mathcal{L}(V)$ as a substitute for $\mathcal{F}(\mathcal{L}(V, d))$. The new csL $\mathcal{F}^0\mathcal{L}(V)$ has Bockstein sub-dgL's that are free as Lie algebras, an appealing feature that greatly facilitates the construction of csL homomorphisms.

The next unit examines $\mathcal{F}^0\mathcal{L}(V)$ as a csL over a Dcsa. We explain, state, prove, and apply a general purpose result called the Virtual Structure Lemma. The result permits us to download a virtual structure from $\mathcal{FL}(V, d)$ to $\mathcal{F}^0\mathcal{L}(V)$, an essential step as we prepare to apply Theorem 7.15.

One hypothesis of Theorem 7.15 is that the source csL should be proper, but as it turns out $\mathcal{F}^0\mathcal{L}(V)$ is not proper (Axiom 7.1(h) fails). We proceed to modify $\mathcal{F}^0\mathcal{L}(V)$ so as to fix this deficiency. The modification process involves a careful analysis of precisely how properness fails; we compute a "properness obstruction group." We can replace $\mathcal{F}^0\mathcal{L}(V)$ by a central extension of it that does satisfy 7.1(h). When we are finished we discover that we have built the universal example of a csL containing the csc $\mathcal{F}(V)$.

Our final unit looks at the homotopy csL. Even after the major surgery that we have performed on the source csL, we may still need to modify the target a little bit, for the sake of the connectedness hypothesis in Theorem 6.4. Nearly half of this unit consists of philosophical discussion of the mathematical issues raised in Chapter 11, along with some very broadly sketched new ideas.

Other than the introductory motivational material and one reference at the very end to Φ, none of Chapter 11 depends upon Chapters 8–10, and this chapter could logically have been placed immediately after Chapter 7.

The Plan and the Four Snags

Discussion 11.A1. As Summary 10.4 indicates, what remains to be done for Theorem 9.1 is the construction and properties of γ^k_{r+k}. Why does this piece of the proof require three whole chapters? Let us explain the overall plan, emphasizing the major snags that we encounter.

Since γ^k_{r+k} is to be a csL/csm homomorphism, the tool of choice for its construction is Theorem 7.15. Recall what Example 7.16 tells us, for $t = r + k$. Our Bockstein sub-dga, the "$P_{[t]}$" of Theorem 7.15, is $R^k_k = \mathbf{Z}_{(p)}\langle b^{p^k}, c_k\rangle$, a free algebra on two generators. There will be two virtual relators, namely $[b^{p^k}, c_k]$ and $(c_k)^2$, just like we had in the construction of γ^k_t for $t < r + k$. Getting these two elements to be virtual relators for $\pi_*(\Omega E_k, \mathbf{Z}_{p_r+k})$ poses no problem. So the quotient $\mathbf{Z}_p$-algebra, the "P" of Theorem 7.15, is $\mathbf{Z}_p[b^{p^k}] \otimes \Lambda(c_k)$.

Example 7.16 shows that there will be two generators in our resolution for N as a P-module, namely, the mod p reductions of y_{2p^k} and z'_{2p^k}, and there will be two minimal syzygies. The generators y_{2p^k} and z'_{2p^k} will have to be sent by γ^k_{r+k} to mod p^{r+k} homotopy classes in $\pi_*(\Omega E_k; \mathbf{Z}_{p^{r+k}})$, of dimension $4np^k - 1$ and $4np^k - 2$, respectively. These will be the first mod p^{r+k} homotopy classes in ΩE_k, and it turns out that Theorem 6.4 can be used to construct them. We denote these homotopy classes as $\hat{y}_{2p^k}$ and $\hat{z}'_{2p^k}$. It remains only to show that the two virtual syzygy equations are true in $\pi_*(\Omega E_k; \mathbf{Z}_{p^r+k})$, and Theorem 7.15 will take care of the rest.

Believe it or not, these two innocent-looking virtual syzygy equations turn out to be extremely deep. Here is why. Quoting 7.16, the two syzygies (when reduced to N) are "$b^{p^k} \cdot z'_{2p^k} = c_k \cdot y_{2p^k}$" and "$c_k \cdot z'_{2p^k} = 0$." The second equation will follow from the first by differentiating both sides ($p \neq 3$ is used here!), so we really only need to be concerned with the first equation. It is an equation that lives in dimension $6np^k - 2$, so the corresponding equation in $\pi_*(\Omega E_k; \mathbf{Z}_{p^{r+k}})$ (i.e. the equation that serves as (7BL)) lives in dimension $6np^k - 2$. Thus we must compare two mod p^{r+k} homotopy classes in dimension $6np^k - 2$ and prove that they are equal. Specifically, we want to know that $\psi^k_{r+k}(c_k \otimes \hat{y}_{2p^k}) \equiv \psi^k_{r+k}(b^{p^k} \otimes \hat{z}'_{2p^k})$, modulo stuff in $\mathrm{im}(\zeta^{r+k-1}_{r+k}) + \mathrm{im}({}_\pi\delta^{r+k}\zeta^{r+k-1}_{r+k})$.

This congruence is far harder than it sounds. The only mod p^{r+k} homotopy classes we know about in $\pi_*(\Omega E_k; \mathbf{Z}_{p^{r+k}})$ are the images $\hat{y}_{2p^k}$ and $\hat{z}'_{2p^k}$ of the generators y_{2p^k} and z'_{2p^k} of $\mathcal{F}(\overline{L}^k)_{[r+k]}$, which occur in dimensions $4np^k - 1$ and $4np^k - 2$, and the classes that result from operating on these two via $\Psi^k_{r+k}(b^{p^k})$ or $\Psi^k_{r+k}(c_k)$, which take us to dimensions around $6np^k$. There are simply not enough homotopy classes around to use, for showing that the two classes of interest coincide.

When this happens, the approach we would expect to adopt is to employ the uniqueness aspect of Theorem 6.4 in order to get the two homotopy classes to be equal. The snag is, that the fibration $E_k \longrightarrow D_k \longrightarrow S^{2n+1}\{p^r\}$ may not be used because the first p^{r+k}-torsion in $\pi_*(E_k)$ occurs around dimension $4np^k$. In view of the exactness hypothesis on (6H) in Theorem 6.4, this dimension is too low for the fibration to give us any information at the desired dimension of $6np^k - 2$. Reflecting on Remark 10.3, it appears that the space E_k is "not universal enough" as a place for proving the virtual syzygy equation that we need. But what could be more universal than E_k, for an equation that talks explicitly about homotopy classes in E_k?

Consider the inclusion $D_k \vee D_k \longrightarrow D_k \times D_k$, whose homotopy theoretic fiber we denote by D_{kk}. Its algebraic analog is the kernel of the dgL surjection $M^k \amalg M^k \longrightarrow M^k \oplus M^k$, a dgL that we denoted in Chapter 5 as N^{kk}. As Lemma 5.14 shows, the elements y_{2p^k} and z'_{2p^k} are images of elements in N^{kk} that we denoted $g(b_k \otimes c_k)$ and $(\frac{1}{2})g(c_k \otimes c_k)$. Thus D_{kk} is "more universal" than E_k in the sense that the homotopy classes of interest, and the operations $\Psi^k_{r+k}(b^{p^k})$ and $\Psi^k_{r+k}(c_k)$, can be pulled back to it. Our plan becomes to try to prove a pulled back version of the virtual syzygy equation, in $\pi_*(\Omega D_{kk}; \mathbb{Z}_{p^{r+k}})$. Then push the proved equation forward, using the induced map on fibers in

$$\text{(11AA)}\qquad \begin{array}{ccccc} D_{kk} & \longrightarrow & D_k \vee D_k & \longrightarrow & D_k \times D_k \\ \nabla_k \big\downarrow & & \big\downarrow \nabla & & \big\downarrow \rho_k \cdot \rho_k \\ E_k & \longrightarrow & D_k & \xrightarrow{\rho_k} & S^{2n+1}\{p^r\}. \end{array}$$

In (11AA) the right-most vertical arrow is defined using the H-space multiplication on $S^{2n+1}\{p^r\}$, and ∇ is the folding map.

It appears at first that we have merely exchanged one impossible dream for another, since $p^{r+k-1}\pi_*(\Omega D_{kk}; \mathbb{Z}_{p^{r+k}})$ also begins in dimension $4np^k - 2$, far too low to help us with dimension $6np^k - 2$. But we are on the right track. The key to this puzzle turns out to be adding on yet another D_k. The answer is the fibration

$$\text{(11AB)}\qquad G = \mathcal{G}(D_k, D_k, D_k) \longrightarrow D_k \vee D_k \vee D_k \longrightarrow R = \mathcal{R}(D_k, D_k, D_k),$$

where the base is the homotopy limit of three D_k's described in Chapter 6. The fiber G in (11AB) happens to be a wedge of Moore spaces with the correct connectivity properties, so Theorem 6.4 can tell us something about $\pi_*(\Omega G; \mathbb{Z}_{p^{r+k}})$ in dimension $6np^k - 2$. Indeed, there is a "third Jacobi identity" that becomes the desired syzygy equation in E_k when all three D_k's are folded together. Further discussion of the precise equations in G is best left until Chapter 12.

Having found a plan for circumventing our first snag, we attempt to implement it. We quickly encounter a second major snag. One of our first steps is the construction of a csL/csm homomorphism from $\mathcal{F}(N^{kk})$ to $\{\pi_*(\Omega D_{kk}; \mathbb{Z}_{p^t})\}$ that splits the Hurewicz homomorphism (see Theorem 12.4 for a precise statement). The snag that arises occurs in connection with the problems described in Discussion 7.7. In N^{kk}, unlike $\overline{L}^k$, there are even-dimensional elements y of $*$-degree zero, so the problem elements $\tilde{\tau}_j(y)$ and $\tilde{\sigma}_j(y)$ occur in $*$-degrees 1 and 2. Earlier we threw out the high $*$-degrees where $\tilde{\tau}_j$'s and $\tilde{\sigma}_j$'s occur, but we clearly cannot afford to throw away $*$-degrees 1 and 2. Nor does there exist another grading (like the "\$-degree" used for the first Jacobi identity) for which the problem disappears, since this alternative grading would still have to be compatible with brackets, with the differential, and with both $\mathcal{F}(\mathcal{U}M^k)$-actions. So we are obliged to grab the bull by the horns, i.e., we must find a different compromise from the compromise of evasion adopted in 7.7, one which will permit all or almost all of $\mathcal{F}(N^{kk})$ to be mapped into $\{\pi_*(\Omega D_{kk}; \mathbb{Z}_{p^{r+k}})\}$.

The first thing we do is to find a good sub-csL of $\mathcal{F}(N^{kk})$. Heuristically speaking, it is the sub-csL obtained by throwing out all of the unwanted $\tilde{\tau}_j(y)$'s and $\tilde{\sigma}_j(y)$'s. We have to work rather hard to show, first, that the resulting object is a well-defined functorial csL, and second, that it remains a csL over the Dcsa $\mathcal{F}(\mathcal{U}M^k \amalg \mathcal{U}M^k)$. We are assisted tremendously in this task by the good fortune that N^{kk} (and N^{kkk} too) has a linear differential. The theory would probably be far more complex if we could not build upon certain results in [CMN1], which apply only to dgL's with linear differentials.

After making these adaptations to $\mathcal{F}(N^{kk})$ we encounter our third major snag. Having removed a piece from $\mathcal{F}(N^{kk})$, the remaining csL is no longer proper (Axiom 7.1(h) now fails). Properness of the source csL is an essential hypothesis in both Proposition 7.5 and Theorem 7.15. To circumvent this snag we alter our csL yet again, this time by restoring the removed piece, but putting it back in slightly differently. The result is a csL that is not proper but it is "semi-proper", which is good enough. This csL turns out to be a certain kind of universal object in the category CSL; the author considers this find to be one of our most beautiful serendipitous discoveries.

It looks as though our application of Theorem 7.15 is finally ready to roll, but we first need to use Theorem 6.4. We use it both to choose suitable images (the $\tilde{x}_i$'s of Theorem 7.15) for the virtual generators of $\mathcal{F}(N^{kk})$, and we need it to confirm that the virtual syzygies for $\mathcal{F}(N^{kkk})$ hold, in the target csL. We run into our fourth snag at this juncture. One of the hypotheses of Theorem 6.4, namely the injectivity of $(\Omega f)_* h$ in (6H), just barely fails in the dimensions we need. Fortunately it fails in a "controlled" way, and we are able to tip-toe around the non-zero kernel of $(\Omega f)_* h$. One trick is to divide out the homotopy csL by a small csL ideal. Another is to consider some alterations of the hypotheses of Theorem 6.4.

To summarize, the grand inductive scheme encounters a high hurdle at the point of proving the virtual syzygy equation needed for γ^k_{r+k}. This equation is much deeper than any of the results in Chapters 9 or 10, since it depends upon a higher Jacobi identity that can only be proved by looking in $\mathcal{G}(D_k, D_k, D_k)$. A number of snags present themselves along the path to proving the equation, and only a combination of hard work and good luck enables us to overcome them. The hard work includes a few ad hoc patches that feel clumsy, but much of it consists of results in the theory of csL's that seem like natural developments anyhow.

Properties of τ_j, σ_j, ξ_j

We begin the mathematical content of Chapter 11 with a unit on general csL theory. We begin by writing down some identities that hold in any dgL. We apply these identities, to see what they tell us about the operations $\tilde{\tau}_j$, $\tilde{\sigma}_j$, and ξ_j. We define csL ideals, and we find some very small central csL ideals. We finish the unit by examining the sub-csL $\mathcal{F}^0\mathcal{L}(V,d)$ of $\mathcal{FL}(V,d)$ defined by "throwing away" the $\tilde{\tau}_j$'s and $\tilde{\sigma}_j$'s from $\mathcal{FL}(V,d)$.

We first recall three notations from Chapter 5.

Notation. (cf. (5H-J)) In any dgL (L,d) over $\mathbb{Z}_{(p)}$, let $\tau_j(y) = 0$ if $|y|$ is odd, and for $y_j \in L_{\text{even}}$ put

(11BA) $$\tau_j(y) = \mathrm{ad}^{p^j-1}(y)d(y).$$

Similarly, if $|y|$ is odd put $\sigma_j(y) = 0$ but if $|y|$ is even let

(11BB) $$\sigma_j(y) = \sum_{i=1}^{p^j-1} (2p)^{-1}(p^j|i)[\mathrm{ad}^{i-1}(y)d(y), \mathrm{ad}^{p^j-i-1}(y)d(y)].$$

Then

(11BC) $$d(\tau_j(y)) = (p)\sigma_j(y)\ ,\ d(\sigma_j(y)) = 0.$$

In any dga, let $e_j(y) = 0$ if $|y|$ is odd, and if $|y|$ is even put

(11BD) $$e_j(y) = \sum_{i=0}^{p^j-1} (p^{-1})(1-(-1)^i(p^j-1|i))y^{p^j-i-1}d(y)y^i.$$

In $\mathcal{U}(L,d)$ we have the two equations for $|y|$ even,

(11BE) $$d(y^{p^j}) = \tau_j(y) + (p)e_j(y)$$

and

(11BF) $$d(e_j(y)) = -\sigma_j(y)$$

In Lemma 11.B1 we refer to "binary m-fold brackets." By a ***binary m-fold bracket*** we mean a linear combination of m-fold brackets, where the entries in the brackets come from the set $\{x, y, d(x), d(y)\}$. Equivalently, it is any element of homogeneous bracket length m, in the free dgL $\mathbf{L}\langle x, y, d(x), d(y)\rangle$. We think of a binary m-fold bracket as a binary operation the elements of a dgL, and we denote it as a function of the two indeterminates x and y.

Lemma 11.B1. *In any dgL (L,d), the elements $\tau_j(y)$ and $\sigma_j(y)$ have the following properties, for any $j \geq 1$.*

(a) $\tau_j(cy) = c^{p^j}\tau_j(y)$ *and* $\sigma_j(cy) = c^{p^j}\sigma_j(y)$, *for* $c \in \mathbb{Z}_{(p)}$.

(b) *There exist binary p^j-fold brackets $\alpha_{0j}(x,y)$ and $\alpha_{1j}(x,y)$ such that*

(11BG) $$\tau_j(x+y) = \tau_j(x) + \tau_j(y) + d(\alpha_{0j}(x,y)) + (p)\alpha_{1j}(x,y),$$

(11BH) $$\sigma_j(x+y) = \sigma_j(x) + \sigma_j(y) + d(\alpha_{1j}(x,y)).$$

$\alpha_{0j}(x,y)$ and $\alpha_{1j}(x,y)$ are "universal", i.e. *they are independent of the dgL (L,d).*

(c) *For any $z \in L$ we have*

(11BI) $$[\tau_j(y), z] = d(\mathrm{ad}(y^{p^j})(z)) - \mathrm{ad}(y^{p^j})d(z) - (p)\,\mathrm{ad}(e_j(y))(z);$$

(11BJ) $$[\sigma_j(y), z] = -d(\mathrm{ad}(e_j(y))(z)) - \mathrm{ad}(e_j(y))d(z).$$

Proof. Straightforward.

One of the most important operations of csL theory is defined next. Most of Chapter 11 deals in one way or another with ξ_j.

Definition/Notation 11.B2. In any csL $\{M_{[t]}, d_{[t]}\}$, define $\xi_j : M_{[s]} \longrightarrow M_{[s+1]}$ by

(11BK) $$\xi_j(y) = d_{[s+1]}\zeta^s_{s+1}\tau_j(y) - \zeta^s_{s+1}\sigma_j(y).$$

Hopefully it is clear that the "d" used in (11BA) and (11BB) is now understood to be $d_{[s]}$.

Lemma 11.B3. *In any csL* $\{M_{[t]}, d_{[t]}\}$,

(a) $\theta_s^{s+1}(\xi_j(y)) = 0$, *hence* $(p)\xi_j(y) = 0$. *Similarly,* $\theta_s^{s+1}d_{[s+1]}\xi_j(y) = 0$.

(b) $\zeta_t^{s+1}\xi_j : M_{[s]} \longrightarrow M_{[t]}$ *and* $d_{[t]}\zeta_t^{s+1}\xi_j : M_{[s]} \longrightarrow M_{[t]}$ *are* $\mathbb{Z}_{(p)}$*-homomorphisms whose kernels contain* $pM_{[s]}$, *for* $s < t$.

(c) $\mathrm{Im}(\zeta_t^{s+1}\xi_j) + \mathrm{im}(d_{[t]}\zeta_t^{s+1}\xi_j)$ *lies in the center of the Lie algebra* $M_{[t]}$.

Proof. (a) $\theta_s^{s+1}\xi_j(y) = d_{[s]}\tau_j(y) - (p)\sigma_j(y) = 0$, by (11BC). Apply ζ_{s+1}^s to get $(p)\xi_j(y) = 0$. Easily, $\theta_s^{s+1}d_{[s+1]}\xi_j(y) = -d_{[s]}\sigma_j(y) = 0$.

(b) It clearly suffices to show this for ξ_j itself. We have $\xi_j(py) = p^{p^j}\xi_j(y) = 0$. That $\xi_j(cy) = c\xi_j(y)$ is a consequence of 11B1(a) and Fermat's little theorem. Finally, by 11.B1(b),

$$\xi_j(x+y) - \xi_j(x) - \xi_j(y) = d_{[s+1]}\zeta_{s+1}^s d_{[s]}\alpha_{0j}(x,y)$$
$$+(p)d_{[s+1]}\zeta_{s+1}^s\alpha_{1j}(x,y) - \zeta_{s+1}^s d_{[s]}\alpha_{1j}(x,y) = 0,$$

because

$$\zeta_{s+1}^s d_{[s]} = \zeta_{s+1}^s\theta_s^{s+1}d_{[s+1]}\zeta_{s+1}^s = (p)d_{[s+1]}\zeta_{s+1}^s. \tag{11BL}$$

(c) Let $Z(\ )$ denote the center of a dgL. It is easy to check that $Z(M_{[s]})$ is closed under $d_{[s]}$, and that $\zeta_{s+1}^s(Z(M_{[s]})) \subseteq Z(M_{[s+1]})$. So it suffices to see that $[\xi_j(y), z] = 0$ for any $y \in M_{[s]}$ and any $z \in M_{[s+1]}$. The calculation is straightforward, using (11BL) and its twin formula

$$d_{[s]}\theta_s^{s+1} = (p)\theta_s^{s+1}d_{[s]}, \tag{11BM}$$

but we provide it here as an illustration of the techniques.

Skipping some steps and omitting all superscripts and subscripts, the computation is (remember "e" is e_j):

$$\begin{aligned}
[\xi_j(y), z] &= [d\zeta\tau(y), z] - [\zeta\sigma(y), z] \\
&= d\zeta[\tau(y), \theta z] + \zeta[\tau(y), \theta dz] - \zeta[\sigma(y), \theta z] \\
&= \ (11BI)\ d\zeta[d(\mathrm{ad}(y^{p^j})\theta z) - \mathrm{ad}(y^{p^j})d\theta z - (p)\,\mathrm{ad}(e(y))\theta z] \\
&\quad + \zeta[d(\mathrm{ad}(y^{p^j})\theta dz - \mathrm{ad}(y^{p^j})d\theta dz - (p)\,\mathrm{ad}(e(y))\theta dz] - \zeta[\sigma(y), \theta z] \\
&= -(p)d\zeta\,\mathrm{ad}(e(y))\theta z - (p)\zeta\,\mathrm{ad}(e(y))\theta dz - \zeta[\sigma(y), \theta z] \\
&= -\zeta d(\mathrm{ad}(e(y))\theta z) - \zeta\,\mathrm{ad}(e(y))d\theta z - \zeta\,\mathrm{ad}(\sigma(y))\theta z = \zeta(0) = 0.
\end{aligned}$$

Notation. When (L,d) is a $\mathbb{Z}_{(p)}$-free dgL, the csL $\mathcal{F}(L,d)$ has the special property that each θ_t^{t+1} is monomorphic. Recall from Chapter 5 the elements $\tilde{\tau}_j(y)$ and $\tilde{\sigma}_j(y)$. Now that we have the language of csL's to help us say exactly what we mean, we can define $\tilde{\tau}_j : \mathcal{F}(L)_{[s]} \longrightarrow \mathcal{F}(L)_{[s+1]}$ by $\tilde{\tau}_j(y) = 0$ if $|y|$ is odd,

$$\tilde{\tau}_j(y) = (\theta_s^{s+1})^{-1}(\tau_j(y)) \tag{11BN}$$

if $|y|$ is even. Similarly, $\tilde{\sigma}_j : \mathcal{F}(L)_{[s]} \longrightarrow \mathcal{F}(L)_{[s+1]}$ is given by $\tilde{\sigma}_j(y) = 0$ if $|y|$ is odd, and if $|y|$ is even then

(11BO) $$\tilde{\sigma}_j(y) = (\theta_s^{s+1})^{-1}(\sigma_j(y)).$$

Discussion 11B4. There are two things "wrong" with the csL $\mathcal{F}(L)$, from a practical standpoint. By this we mean that it deviates in two respects from being a "universal" csL, and both respects make it harder to build csL homomorphisms out of $\mathcal{F}(L)$. We have just alluded to the first difficulty. $\mathcal{F}(L)$ contains some elements, namely the $\tilde{\tau}_j$'s and $\tilde{\sigma}_j$'s, whose existence is not mandated by the csL axioms. We have already seen (e.g. Discussions 7.7 and 7.17) how the presence of these "superfluous" elements can complicate our lives.

Second, there is an identity that holds in $\mathcal{F}(L)$ that is not a generic identity deducible strictly from the csL axioms. For (L,d) a $\mathbb{Z}_{(p)}$-free dgL, consider the fact that the homomorphism $p\theta_s^{s+1}$ is one-to-one. Since

$$\begin{aligned} p\theta_s^{s+1}(d_{[s+1]}\tilde{\tau}_j(y) - \tilde{\sigma}_j(y)) &= \theta_s^{s+1}d_{[s+1]}\zeta_{s+1}^s(\theta_s^{s+1}\tilde{\tau}_j(y) - (p)\sigma_j(y)) \\ &= d_{[s]}\tau_j(y) - (p)\sigma_j(y) = 0, \end{aligned}$$

we have

(11BP) $$d_{[s+1]}\tilde{\tau}_j(y) = \tilde{\sigma}_j(y)$$

in $\mathcal{F}(L,d)$. This formula has as a consequence

(11BQ) $$\theta_s^{s+1}d_{[s+1]}\tilde{\tau}_j(y) = \sigma_j(y),$$

which in turn implies

$$d_{[s+1]}\zeta_{s+1}^s\tau_j(y) = d_{[s+1]}\zeta_{s+1}^s\theta_s^{s+1}\tilde{\tau}_j(y) = \zeta_{s+1}^s\theta_s^{s+1}d_{[s+1]}\tilde{\tau}_j(y) = \zeta_{s+1}^s\sigma_j(y),$$

i.e.,

(11BR) $$\xi_j(y) = 0 \text{ for any } y \in \mathcal{F}(L,d).$$

However, (11BR) is not an identity that holds in every csL.

These two "problems" with $\mathcal{F}(L,d)$, i.e., that it has "extra" elements and satisfies a non-generic identity, are the bottom-line reason why much of Chapter 11 became necessary in the first place.

In our next lemma, we exploit the injectivity of θ_s^{s+1} on $\mathcal{F}(L,d)$ to establish some more identities concerning the $\tilde{\tau}_j$'s and $\tilde{\sigma}_j$'s. All three parts are corollaries of Lemma 11.B1, and we leave the proofs as exercises.

Lemma 11.B5. *Let (L,d) be a $\mathbb{Z}_{(p)}$-free dgL. Let $x, y \in \mathcal{F}(L,d)_{[s]}, z \in \mathcal{F}(L,d)_{[s+1]}$.*

(a) $$\tilde{\tau}_j(cy) = c^{p^j}\tilde{\tau}_j(y), \quad \tilde{\sigma}_j(cy) = c^{p^j}\tilde{\sigma}_j(y), \text{ for } c \in \mathbb{Z}_{(p)}.$$

(b) $$\begin{aligned} &\tilde{\tau}_j(x+y) = \tilde{\tau}_j(x) + \tilde{\tau}_j(y) + d_{[s+1]}\zeta_{s+1}^s(\alpha_{0j}(x,y)) + \zeta_{s+1}^s(\alpha_{1j}(x,y)); \\ &\tilde{\sigma}_j(x+y) = \tilde{\sigma}_j(x) + \tilde{\sigma}_j(y) + d_{[s+1]}\zeta_{s+1}^s(\alpha_{1j}(x,y)). \end{aligned}$$

(c) $$\begin{aligned} [\tilde{\tau}_j(y), z] &= d_{[s+1]}\zeta_{s+1}^s(\mathrm{ad}(y^{p^j})\theta_s^{s+1}(z)) - \zeta_{s+1}^s\,\mathrm{ad}(y^{p^j})\theta_s^{s+1}d_{[s+1]}(z) \\ &\quad - \zeta_{s+1}^s\,\mathrm{ad}(e_j(y))\theta_s^{s+1}(z); \\ [\tilde{\sigma}_j(y), z] &= -d_{[s+1]}\zeta_{s+1}^s(\mathrm{ad}(e_j(y))(z)) - \zeta_{s+1}^s\,\mathrm{ad}(e_j(y))\theta_s^{s+1}d_{[s+1]}(z). \end{aligned}$$

Definition. Let $\{M_{[t]}, d_{[t]}\}$ be a csL. For each $t \geq 1$, let $I_{[t]}$ be a differential Lie ideal of $M_{[t]}$. We call $\{I_{[t]}\}$ a ***csL ideal*** if and only if $\theta_t^{t+1}(I_{[t+1]}) \subseteq I_{[t]}$ and $\zeta_{t+1}^t(I_{[t]}) \subseteq I_{[t+1]}$, for each $t \geq 1$.

Lemma 11.B6. *Let $\{M_{[t]}, d_{[t]}\}$ be a csL, and let $I_{[t]} \subseteq M_{[t]}$ be subsets. $\{I_{[t]}\}$ is a csL ideal if and only if there is a csL homomorphism $\{f_{[t]}\} : \{M_{[t]}\} \longrightarrow \{M'_{[t]}\}$ such that $I_{[t]} = \ker(f_{[t]})$ for all t. Furthermore, a csL ideal is always a sub-csL.*

Proof. Obvious. When $\{I_{[t]}\}$ is a csL ideal, $\{M_{[t]}/I_{[t]}, d_{[t]}\}$ is a quotient csL.

Lemma 11.B7. *Let $\{M_{[t]}, d_{[t]}\}$ denote any csL. Let $s \geq 1$, let $j \geq 1$, and let y be an even-dimensional element of $M_{[s]}$. Let $\langle \xi_j(y) \rangle_{[t]} = 0$ if $t \leq s$ and put*

$$\langle \xi_j(y) \rangle_{[t]} = \operatorname{Span}(\zeta_t^{s+1}\xi_j(y), d_{[t]}\zeta_t^{s+1}\xi_j(y))$$

if $t > s$. Then $\langle \xi_j(y) \rangle = \{\langle \xi_j(y) \rangle_{[t]}\}$ is csL ideal of $\{M_{[t]}, d_{[t]}\}$.

Proof. Clear, from 11.B3(a), (11BL), and 11.B3(c).

Notation. Given a csL $\{M_{[t]}, d_{[t]}\}$, let $S = \{(s_\alpha, j_\alpha, y_\alpha)\}$ denote any indexed collection of triples, where $s_\alpha \geq 1, j_\alpha \geq 1$, $y_\alpha \in M_{[s_\alpha]}$. We let $\langle \xi_*(S) \rangle = \{\langle \xi_*(S) \rangle_{[t]}\}$ denote

$$\langle \xi_*(S) \rangle = \sum_\alpha \langle \xi_{j_\alpha}(y_\alpha) \rangle. \tag{11BS}$$

Since the sum of csL ideals is clearly a csL ideal, we have at once that $\langle \xi_*(S) \rangle$ is a csL ideal, for any such collection S. If $\{M_{[t]}\} = \mathcal{F}(L, d)$ for a $\mathbb{Z}_{(p)}$-free dgL (L, d), then of course any $\langle \xi_*(S) \rangle$ is zero, by (11BR).

Note. The next lemma has as a hypothesis that $p^t M_{[t]} = 0$ for all $t \geq 1$. This condition is satisfied by a homotopy or homology csL, and it is equivalent to the easily checked condition that $pM_{[1]} = 0$.

Lemma 11.B8. *Suppose the csL $\{M_{[t]}, d_{[t]}\}$ satisfies $p^t M_{[t]} = 0$ for all $t \geq 1$.*

(a) *Whenever $t > s \geq 1, j \geq 1$, the functions*

$$p^{s-1}\zeta_t^s \tau_j : M_{[s]} \longrightarrow M_{[t]}$$

and

$$p^{s-1} d_{[t]}\zeta_t^s \tau_j : M_{[s]} \longrightarrow M_{[t]}$$

are homomorphisms whose images lie in the center of $M_{[t]}$.

(b) *Let $y \in M_{[s]}$ be even-dimensional. Let $\langle p^{s-1}\zeta_{s+1}^s \tau_j(y) \rangle_{[t]}$ denote 0 if $t \leq s$, and if $t > s$ put*

$$\langle p^{s-1}\zeta_{s+1}^s \tau_j(y) \rangle_{[t]} = \operatorname{Span}(p^{s-1}\zeta_t^s \tau_j(y), p^{s-1} d_{[t]}\zeta_t^s \tau_j(y)). \tag{11BT}$$

Then $\langle p^{s-1}\zeta_{s+1}^s \tau_j(y) \rangle = \{\langle p^{s-1}\zeta_{s+1}^s \tau_j(y) \rangle_{[t]}\}$ is a csL ideal of $\{M_{[t]}\}$.

Proof. Write $\mu(\ )$ for $p^{s-1}\zeta_{s+1}^{s}\tau_j(\ )$. (a) If $c \in \mathbb{Z}_{(p)}$, then $\mu(cy) = c^{p^j}\mu(y)$. By Fermat's little theorem we can write $c^{p^j} = c + pc'$ for some $c' \in \mathbb{Z}_{(p)}$. Then $\mu(cy) = c\mu(y)$ $+c'\zeta_{s+1}^{s}(p^s\tau_j(y)) = c\mu(y)$, because $p^s\tau_j(y) \in p^s M_{[s]} = 0$. Using (11BG) and (11BL),

$$\mu(x+y) - \mu(x) - \mu(y) = \zeta_{s+1}^{s}(p^s\alpha_{1j}(x,y)) + p^{s-1}\zeta_{s+1}^{s}d_{[s]}\alpha_{0j}(x,y) = 0.$$

It remains only to see that $p^{s-1}\zeta_{s+1}^{s}\tau_j(y)$ belongs to the center of $M_{[s+1]}$. For any $z \in M_{[s+1]}$, using (11BI),

$$[\mu(y), z] = p^{s-1}\zeta_{s+1}^{s}[\tau_j(y), \theta_s^{s+1}(z)] = p^{s-1}\zeta_{s+1}^{s}d_{[s]}(\mathrm{ad}(y^{p^j})\theta_s^{s+1}(z))$$
$$-p^{s-1}\zeta_{s+1}^{s}\,\mathrm{ad}(y^{p^j})d_{[s]}\theta_s^{s+1}(z) - (0) = 0,$$

by (11BL) and (11BM).

(b) Easily, $\theta_s^{s+1}\mu(y) = p^s\tau_j(y) = 0$ and $\theta_s^{s+1}d_{[s+1]}\mu(y) = d_{[s]}(p^{s-1}\tau_j(y)) = p^s\sigma_j(y) = 0$. It is now trivial that $\langle p^{s-1}\zeta_{s+1}^{s}\tau_j(y)\rangle$ is closed under ζ's, θ's, d's, and (by (a)) brackets.

With these general purpose preliminaries behind us, let us turn to the first of the issues raised in Discussion 11.B4, namely that the $\tilde{\tau}_j$'s and $\tilde{\sigma}_j$'s that are present in $\mathcal{F}(L,d)$ are not forced to be there by the csL axioms by themselves. These "superfluous" elements also cause a lot of trouble. Specifically, they frequently prevent the "natural" choice for a Bockstein sub-dgL from satisfying (7AC) (cf. Discussion 7.17). We have had enough of these trouble-makers and we wish to be rid of them. This can be done, by carefully selecting a sub-csL of $\mathcal{F}(L,d)$. To simplify the exposition and because both our intended applications will have this property, we develop the theory only for the case where (L,d) is free and has a linear differential.

Definition/Notation 11.B9. Whenever we employ the notation (V,d), it will stand for a positively graded $\mathbb{Z}_{(p)}$-free chain complex satisfying $d(V) \subseteq pV$. Given a (V,d), define $\mathcal{F}^0\mathcal{L}(V,d) = \{\mathcal{F}^0\mathcal{L}(V)_{[t]}\}$ to be the sub-csL of $\mathcal{F}\mathcal{L}(V,d)$ generated by $\mathcal{F}(V)$ I.e., it is the intersection of all sub-csL's of $\mathcal{F}\mathcal{L}(V,d)$ such that the tth stage contains $\mathcal{F}(V)_{[t]}$. (Recall from Chapter 7 that $\mathcal{F}(V)_{[t]}$ consists of those $x \in V$ for which p^t divides $d(V)$.)

Lemma 11.B10. *Let (V,d) be a $\mathbb{Z}_{(p)}$-free chain complex as above, and let*

(11BU) $$V = V^{(1)} \oplus V^{(2)} \oplus \cdots \oplus V^{(j)} \oplus \cdots \oplus V^{(\infty)}$$

be a decomposition into differential submodules as in (5M), where $V^{(j)}$ is acyclic with respect to $p^{-j}d$.

(a) *$\mathcal{F}^0\mathcal{L}(V)$ satisfies axiom 7.1(g) and (11BR), since $\mathcal{F}\mathcal{L}(V)$ does.*

(b) *Viewed as a subset of $\mathcal{L}(V)$, we have*

$$\mathcal{F}^0\mathcal{L}(V)_{[t]} = \mathcal{L}(V^{(\geq t)}) + \sum_{s=1}^{t-1} p^{t-s}\mathcal{L}(V^{(\geq s)}) + \sum_{s=1}^{t-1} p^{-s}d(\mathcal{L}(V^{(\geq s)})).$$

(c) *Another way to express part (b) is*

$$\mathcal{F}^0\mathcal{L}(V)_{[t]} = \mathcal{L}(V^{(\geq t)}) + \sum_{s=1}^{t-1} \zeta_t^s\mathcal{L}(V^{(\geq s)}) + \sum_{s=1}^{t-1} d_{[t]}\zeta_t^s\mathcal{L}(V^{(\geq s)}).$$

(d) *For each $t \geq 1$, $\mathcal{L}(V^{(\geq t)})$ is a Bockstein sub-dgL of $\mathcal{F}^0\mathcal{L}(V)_{[t]}$.*

Proof. Straightforward.

Remark. The decomposition (11BU) is not canonical, but Definition 11.B9 does not depend upon the choice, so $\mathcal{F}^0\mathcal{L}$ is a functor. Unfortunately, it is not a functor of $\mathcal{L}(V,d)$ (i.e. it does depend upon the choice of generating submodule V) when $\mathcal{L}(V)$ has a linear differential.

What does $\mathcal{FL}(V,d)$ have that $\mathcal{F}^0\mathcal{L}(V,d)$ does not have? We have already answered this informally: it has $\tilde{\tau}_j$'s and $\tilde{\sigma}_j$'s. To make this precise, we need to recall the modules $W^+(V^{(\geq m)}) = \bigoplus_{\alpha\in I^{(m)}} \mathbb{Z}_{(p)}(y_\alpha, x_\alpha)$ and

$$\widetilde{W}^m(V^{(\geq m)}) = \bigoplus_{j=1}^{\infty} \bigoplus_{\alpha\in I^{(m)}} \mathbb{Z}_{(p)}(\tilde{\tau}_j(y_\alpha), \tilde{\sigma}_j(y_\alpha))$$

from Lemma 5.6, where $I^{(m)}$ denotes a suitable indexing set.

The notations $\tilde{\tau}_j(y_\alpha)$ and $\tilde{\sigma}_j(y_\alpha)$ of Chapter 5 are compatible with the way we defined $\tilde{\tau}_j$ and $\tilde{\sigma}_j$ in (11BN) and (11BO). Specifically, the pairs $\{y_\alpha, x_\alpha\}_{\alpha\in I^{(m)}}$ can be viewed as belonging to a $\mathbb{Z}_{(p)}$-basis for $\mathcal{FL}(V)_{[m]}$; another way to write them would be $(y_\alpha, d_{[m]}(y_\alpha))$. The elements denoted $\tilde{\tau}_j(y_\alpha)$ and $\tilde{\sigma}_j(y_\alpha)$ in Chapter 5 have boundaries that are divisible by p^{m+1}, hence they make sense as elements of $\mathcal{FL}(V)_{[m+1]}$. When viewed as elements of $\mathcal{FL}(V)_{[m+1]}$ they are precisely the elements $\tilde{\tau}_j(y_\alpha)$ and $\tilde{\sigma}_j(y_\alpha)$ described by (11BN) and (11BO). We will henceforth use the notations $\tilde{\tau}_j$ and $\tilde{\sigma}_j$ for whichever of these interpretations is convenient, without further explanation.

Let $\widetilde{W}^{(t)} = \widetilde{W}^{(t)}(V)$ be the differential submodule of $\mathcal{FL}(V)_{[t]}$ generated by $\tilde{\tau}_j$'s and $\tilde{\sigma}_j$'s. To be precise, let $I^{(m)}$ denote the indexing set as above, and in $\mathcal{L}(V)$ put

$$\widetilde{W}^{(t)} = \bigoplus_{s=1}^{t-1}\bigoplus_{j=1}^{\infty} \bigoplus_{\alpha\in I^{(s)}} \mathbb{Z}_{(p)}(p^{t-s-1}\tilde{\tau}_j(y_\alpha), \tilde{\sigma}_j(y_\alpha)).$$

Here $p^{t-s-1}\tilde{\tau}_j(y_\alpha) \in \mathcal{L}(V)$ can also be thought of as $\zeta_t^{s+1}\tilde{\tau}_j(y_\alpha) \in \mathcal{FL}(V)_{[t]}$, and $\tilde{\sigma}_j(y_\alpha) \in \mathcal{L}(V)$ can be thought of as $(\theta_{s+1}^t)^{-1}(\tilde{\sigma}_j(y_\alpha)) \in \mathcal{FL}(V)_{[t]}$.

Lemma 11.B11. *$(p)\widetilde{W}^{(t)} \subseteq \mathcal{F}^0\mathcal{L}(V)_{[t]}$, and the induced homomorphism on cokernels in the diagram of short exact sequences,*

$$\begin{array}{ccccccccc}
0 & \longrightarrow & p\widetilde{W}^{(t)} & \longrightarrow & \widetilde{W}^{(t)} & \longrightarrow & \widetilde{W}^{(t)}/p\widetilde{W}^{(t)} & \longrightarrow & 0 \\
 & & \downarrow & & \downarrow & & \downarrow\approx & & \\
0 & \longrightarrow & \mathcal{F}^0\mathcal{L}(V)_{[t]} & \longrightarrow & \mathcal{FL}(V)_{[t]} & \longrightarrow & \mathcal{FL}(V)_{[t]}/\mathcal{F}^0\mathcal{L}(V)_{[t]} & \longrightarrow & 0,
\end{array}$$

is an isomorphism.

Proof. To see that $p\widetilde{W}^{(t)} \subseteq \mathcal{F}^0\mathcal{L}(V)_{[t]}$, observe that $(p)\zeta_t^{s+1}\tilde{\tau}_j(y) = \zeta_t^s\tau_j(y)$, and that $(p)(\theta_t^{s+1})^{-1}\tilde{\sigma}_j(y) =$

$(\theta_t^{s+1})^{-1}\zeta_{s+1}^s\ \sigma_j(y) = d_{[t]}\zeta_t^s\tau_j(y)$, by (11BR). The induced homomorphism on cokernels will be epimorphic if and only if

$$\mathcal{F}^0\mathcal{L}(V)_{[t]} + \widetilde{W}^{(t)} = \mathcal{F}\mathcal{L}(V)_{[t]} \tag{11BV}$$

as $\mathbf{Z}_{(p)}$-modules, and it will be monomorphic if and only if

$$\mathcal{F}^0\mathcal{L}(V)_{[t]} \cap \widetilde{W}^{(t)} = p\widetilde{W}^{(t)}. \tag{11BW}$$

To prove (11BV) and (11BW), we need another notation. If $U^{(m)}$ is a $\mathbf{Z}_{(p)}$-free chain complex that is acyclic with respect to $p^{-m}d$, we can write $U^{(m)} = \mathbf{Z}_{(p)}(z_\beta, p^{-m}d(z_\beta))$ as β runs through some indexing set. Define

$$\tilde{\zeta}_t^m(U^{(m)}) = \mathbf{Z}_{(p)}(p^{t-m}z_\beta, p^{-m}d(z_\beta)),$$

which is acyclic with respect to $p^{-t}d$. Notice that

$$\widetilde{W}^{(t)} = \bigoplus_{s=1}^{t-1} \tilde{\zeta}_t^{s+1}(\widetilde{W}^s(V^{(\geq s)})).$$

If $U = U^{(1)} \oplus U^{(2)} \oplus \cdots \oplus U^{(\infty)}$ is the decomposition (like (5M)) of a $\mathbf{Z}_{(p)}$-free chain complex (U, d) having $d(U) \subseteq pU$, then

$$\mathcal{F}(U)_{[t]} = U^{(\geq t)} \oplus \left(\bigoplus_{s=1}^{t-1} \tilde{\zeta}_t^s(U^{(s)}) \right). \tag{11BX}$$

We apply (11BX) to the decomposition (5Q) for $\mathcal{L}(V)$ to obtain

$$\begin{aligned} \mathcal{F}\mathcal{L}(V)_{[t]} = \mathcal{L}(V^{(\infty)}) &\oplus \left(\bigoplus_{m=t}^{\infty} L^{(m,m)} \right) \oplus \left(\bigoplus_{s=1}^{t-1} \tilde{\zeta}_t^s(L^{(s,s)}) \right) \\ &\oplus \left(\bigoplus_{m=t}^{\infty} W^m(V^{(\geq m)}) \right) \oplus \widetilde{W}^{(t)}. \end{aligned} \tag{11BY}$$

Furthermore,

$$\begin{aligned} \mathcal{F}^0\mathcal{L}(V)_{[t]} = \mathcal{L}(V^{(\infty)}) &\oplus \left(\bigoplus_{m=t}^{\infty} L^{(m,m)} \right) \oplus \left(\bigoplus_{s=1}^{t-1} \tilde{\zeta}_t^s(L^{(s,s)}) \right) \\ &\oplus \left(\bigoplus_{m=t}^{\infty} W^m(V^{(\geq m)}) \right) \oplus (p\widetilde{W}^{(t)}), \end{aligned} \tag{11BZ}$$

since both $\subseteq$ and $\supseteq$ in (11BZ) are straightforward to verify. Now both (11BV) and (11BW) follow immediately.

Lemma 11.B12. *For $t = 1$, $\mathcal{F}\mathcal{L}(V)_{[1]} = \mathcal{F}^0\mathcal{L}(V)_{[1]}$. For $t \geq 2$, $[\mathcal{F}\mathcal{L}(V)_{[t]}, \mathcal{F}\mathcal{L}(V)_{[t]}] \subseteq \mathcal{L}^{\geq 2}(V^{(\geq t)}) + \zeta_t^{t-1}(\mathcal{F}^0\mathcal{L}(V)_{[t-1]}) + d_{[t]}\zeta_t^{t-1}(\mathcal{F}^0\mathcal{L}(V)_{[t-1]})$.*

Proof. When $t = 1$, $\widetilde{W}^{(1)} = 0$. For $t \geq 2$, it is an easy corollary of (11BV), 11.B5(c), 11.B10(c), and the csL axioms.

Corollary 11.B13. *Suppose $\mathcal{FL}(V,d)$ is a csL over a Dcsa, and let F denote a formal linear expression. For any $x \in \mathcal{FL}(V)_{[t]}$,*

$$F(x) \in \mathcal{L}^{\geq 2}(V^{(\geq t)}) + \zeta_t^{t-1}(\mathcal{FL}(V)_{[t-1]}) + d_{[t]}\zeta_t^{t-1}(\mathcal{FL}(V)_{[t-1]}).$$

Proof. Clear, from (7BF) and (7BG) and 11.B12.

Corollary 11.B13 can be strengthened if we make a mild further assumption about the fle F. Call an fle ***positive*** if it is a linear combination of expressions (7BF) and (7BG) in which every $|y_t| > 0$ and every $|y_m| > 0$. (Recall that ordinarily in an fle only the $|y_t|$ are required to be positive.)

Corollary 11.B14. *Suppose $\mathcal{FL}(V)$ is a csL over a Dcsa, and let F be a positive fle. For any $x \in \mathcal{FL}(V)_{[t]}$,*

$$F(x) \in \mathcal{L}^{\geq 2}(V^{(\geq t)}) + \zeta_t^{t-1}(\mathcal{F}^0\mathcal{L}(V)_{[t-1]}) + d_{[t]}\zeta_t^{t-1}(\mathcal{F}^0\mathcal{L}(V)_{[t-1]}).$$

Lemma 11.B15. *As always $(\ )_q$ denotes the dimension q component of a graded object.*

(a) $(\mathcal{F}^0\mathcal{L}(V))_q = (\mathcal{FL}(V))_q$ *if $q \not\equiv -1$ or -2, modulo $2p$.*

(b) $d_{[t]} : (\mathcal{FL}(V)/\mathcal{F}^0\mathcal{L}(V))_q \longrightarrow (\mathcal{FL}(V)/\mathcal{F}^0\mathcal{L}(V))_{q-1}$ *is an isomorphism if $q \equiv -1$ modulo $2p$.*

(c) $\zeta_{t+1}^t(\mathcal{FL}(V)_{[t]})_q \subseteq (\mathcal{F}^0\mathcal{L}(V)_{[t]})_q$ *if $q \equiv -2$ modulo $2p$.*

Proof. Use (11BY) and (11BZ) to see that

$$\mathcal{FL}(V)_{[t]}/\mathcal{F}^0\mathcal{L}(V)_{[t]} = \widetilde{W}^{(t)} \otimes \mathbb{Z}_p,$$

which is the $\mathbb{Z}_p$-vector space spanned by $\{\zeta_t^s\tilde{\tau}_j(y), d_{[t]}\zeta_t^s\tilde{\tau}_j(y)\}$ as (s,j,y) runs through a certain set of triples determined by V. Parts (a) and (b) follow at once, since any $\tilde{\tau}_j(y)$ has $|\tilde{\tau}_j(y)| \equiv -1$ modulo $2p$. When $q \equiv -2$ modulo $2p$, use (b) to write an arbitrary $z \in (\mathcal{FL}(V)_{[t]})_q$ as $z = d_{[t]}(x) + z'$, where $z' \in \mathcal{F}^0\mathcal{L}(V)_{[t]}$ and $x \in \mathcal{FL}(V)_{[t]}$. Then $px \in \mathcal{F}^0\mathcal{L}(V)_{[t]}$, and

$$\zeta_{t+1}^t(z) = \zeta_{t+1}^t(z') + (p)d_{[t+1]}\zeta_{t+1}^t(x) = \zeta_{t+1}^t(z') + d_{[t+1]}\zeta_{t+1}^t(px),$$

which belongs to $\mathcal{F}^0\mathcal{L}(V)_{[t]}$.

The Virtual Structure Lemma

We saw in Lemma 11.B10 that $\mathcal{F}^0\mathcal{L}(V,d)$ is a csL having the property that the free dgL $\mathcal{L}(V^{(\geq t)})$ is a true Bockstein sub-dgL of $\mathcal{F}^0\mathcal{L}(V,d)_{[t]}$. This property can greatly streamline applications of Theorem 7.5, as Discussion 7.7 indicates. Since we are frequently interested in csL's that carry the additional structure of being a csL over a Dcsa, we can wonder whether $\mathcal{F}^0\mathcal{L}(V,d)$ inherits such a structure when $\mathcal{FL}(V,d)$ has it. The "Virtual Structure lemma" gives sufficient conditions for an affirmative answer.

In this unit of Chapter 11, we begin by listing some identities that hold for csL's over $\mathcal{F}(\mathcal{U}M^k)$. The Dcsa $\mathcal{F}(\mathcal{U}M^k)$ is particularly congenial to work over, because its entire diagonal structure is encoded in the single formula (7BC). We obtain analogs for the $\mathcal{F}(\mathcal{U}M^k)$-action, of several of the lemmas from the previous unit. We then explain, state, and prove the Virtual Structure lemma. We finish by applying the lemma to the csL's $\mathcal{F}^0(N^{kl})$ and $\mathcal{F}^0(N^{klm})$, where N^{kl} and N^{klm} are described in Chapter 5.

Our first result in this unit is very similar to Lemma 11.B1. In Lemma 11.B1(c) we looked at the effect on τ_j's and σ_j's of bracketing with something. Now let us suppose that we are in a csL over $\mathcal{F}(\mathcal{U}M^k)$, and we are interested in knowing the effect of the action on τ_j's and σ_j's. This action is deducible entirely from knowing what the b_s's and the a_m's do. Since the a_q's are primitive, $\psi_{[t]}(a_m \otimes -)$ behaves just like bracketing with something. We therefore concentrate on $\psi_{[t]}(b_s \otimes \tau_j(\ ))$ and $\psi_{[t]}(b_s \otimes \sigma_j(\ ))$.

Given $t \geq 1$, define a *q-fold bracket expression over* $\mathcal{U}M^k$ to be any linear combination of ζ_t^m-images of q-fold iterated brackets, where each entry in each iterated bracket has the form $\psi_{[m]}(w \otimes \theta_m^t(x))$ or $\psi_{[t]}(w \otimes \theta_m^t d_{[t]}(x))$ for some $w \in \mathcal{F}(\mathcal{U}M^k)_{[m]}$, $m \leq t$. We think of a q-fold bracket expression as a unary operation on the variable x. It has an obvious interpretation as an operation in any csL over $\mathcal{F}(\mathcal{U}M^k)$.

Lemma 11.C1. *Let $k \geq 0, j \geq 1$. Let t be in the range $r \leq t < r+k$, and let $s = t - r$. There exist p^j-fold bracket expressions over $\mathcal{U}M^k$, denoted $\beta_{0j}(x), \beta_{1j}(x), \gamma_{1j}(x), \gamma_{2j}(x)$ (the first subscript indicates $*$-degree), with the following property. Let $\{M_{[q]}, d_{[q]}\}$ be a csL over $\mathcal{F}(\mathcal{U}M^k)$; denote the action by $\psi_{[q]}$, and let $y \in M_{[t]}$. Then*

(11CA) $$\psi_{[t]}((b_s)^p \otimes \tau_j(y)) = d_{[t]}\beta_{0j}(y) + (p)\beta_{1j}(y);$$

(11CB) $$\psi_{[t]}((b_s)^p \otimes \sigma_j(y)) = d_{[t]}\gamma_{1j}(y) + (p)\gamma_{2j}(y).$$

Proof. A long but straightforward calculation, which we omit. Another calculation, like that of the proof of 11.B3(c), shows

Lemma 11.C2. *Let $k \geq 0, j \geq 1, r \leq t < r+k, s = t-r$. Let $\{M_{[q]}, d_{[q]}\}$ be a csL over $\mathcal{F}(\mathcal{U}M^k)$, and let $y \in M_{[t]}$. Then*

$$\psi_{[t+1]}(b_{s+1} \otimes \xi_j(y)) = 0;$$
$$\psi_{[t+1]}(b_{s+1} \otimes d_{[t+1]}\xi_j(y)) = 0;$$

where $\xi_j(y)$ is given by (11BK).

Just as Lemma 11.B5 is a corollary of Lemma 11.B1, the next lemma is a corollary of Lemma 11.C1, derived using the injectivity of θ_t^{t+1} on $\mathcal{FL}(V,d)$.

Lemma 11.C3. *Let $k \geq 0, j \geq 1, r \leq t < r+k, s = t-r$. Let $\{M_{[q]}, d_{[q]}\}$ be a csL over $\mathcal{F}(\mathcal{U}M^k)$. Suppose that $\{M_{[q]}, d_{[q]}\}$ equals $\mathcal{F}(L,d)$ for some $\mathbb{Z}_{(p)}$-free dgL (L,d). Then*

(11CC) $$\psi_{[t+1]}(b_{s+1} \otimes \tilde{\tau}_j(y)) = d_{[t+1]}\zeta_{t+1}^t\beta_{0j}(y) + \zeta_{t+1}^t\beta_{1j}(y);$$

(11CD) $$\psi_{[t+1]}(b_{s+1} \otimes \tilde{\sigma}_j(y)) = d_{[t+1]}\zeta_{t+1}^t\gamma_{1j}(y) + \zeta_{t+1}^t\gamma_{2j}(y);$$

where $\beta_{0j}, \beta_{1j}, \gamma_{1j}, \gamma_{2j}$ are the bracket expressions of Lemma 11.C1.

Let (L,e) be a dgL which is also a differential module over a dga (A,d). Let I be a differential Lie ideal of L. The annihilator of I in A is a two-sided differential ideal of A. If $A = \mathcal{F}(\mathcal{U}M^k)_{[t]}$ and $\{L_{[q]}, e_{[q]}\}$ is a csL over $\mathcal{F}(\mathcal{U}M^k)$, then the annihilator of $L_{[t]}$ equals all of $\mathcal{F}(\mathcal{U}M_+^k)_{[t]}$ if and only if b_{t-r} (where $t-r$ means 0 if $t < r$) and each a_m for $t - r < m \leq k$ annihilates $L_{[t]}$.

Lemma 11.C4. *Let $\{M_{[q]}, d_{[q]}\}$ be any csL over $\mathcal{F}(\mathcal{U}M^k)$.*

(a) *For any $j \geq 1$ and any $y \in M_{[s]}$, the csL ideal $\langle \xi_j(y) \rangle$ is annihilated by $\mathcal{F}(\mathcal{U}M^k_+)$.*

(b) *Recall the notation (11BS). For any set $S = \{(s_\alpha, j_\alpha, y_\alpha)\}$ of triples, $\langle \xi_*(S) \rangle$ is annihilated by $\mathcal{F}(\mathcal{U}M^k)$, and*

$$\{M_{[t]}/\langle \xi_*(S) \rangle_{[t]}, d_{[t]}\}$$

is a csL over $\mathcal{F}(\mathcal{U}M^k)$.

Proof. By the paragraph preceding the lemma, for $y \in M_{[s]}$ it suffices that b_{s-r+1} and a_m for $s + 1 < m \leq k$ should annihilate $\xi_j(y)$ and $d_{[s+1]}\xi_j(y)$. When $s \geq r$, for b_{s-r+1}, this is precisely what Lemma 11.C2 says. For a_m, and for b_0 when $s < r$, the diagonal structure of the Dcsa $\mathcal{F}(\mathcal{U}M^k)$ says that a_m and (if $s < r$) b_0 are "primitive", i.e., $\psi_{[s+1]}(a_m \otimes —)$ is a derivation on $M_{[s+1]}$ and likewise for b_0 (see (7BD)). A careful review of the proof of Lemma 11.B3(c) reveals that the same argument works to get $\xi_j(y)$ to be annihilated by any primitive, in particular, by both a_m and $d_{[s+1]}(a_m)$. Then $\langle \xi_j(y) \rangle$ is annihilated by a_m and (if $s < r$) b_0, as needed.

Remark. We admit that we are omitting a number of moderately significant details here, e.g. what happens when $s \geq r + k$. They are not difficult, and we leave them to the reader to fill in.

Using Lemmas 11.C1 and 11B8(a), we prove similarly

Lemma 11.C5. *Let $\{M_{[t]}, d_{[t]}\}$ be any csL over $\mathcal{F}(\mathcal{U}M^k)$. For any $j \geq 1$ and any element $y \in M_{[s]}$, the csL ideal $\langle p^{s-1}\zeta^s_{s+1}\tau_j(y) \rangle$ is annihilated by $\mathcal{F}(\mathcal{U}M^k_+)$. For any set of triples $S = \{(s_\alpha, j_\alpha, y_\alpha)\}$ with $y_\alpha \in M_{[s_\alpha]}$ and $j_\alpha \geq 1$, the quotient csL*

$$\{M_{[t]}/(\sum_\alpha \langle p^{s_\alpha - 1}\zeta^{s_\alpha}_{s_\alpha+1}\tau_j(y_\alpha) \rangle_{[t]}), \quad d_{[t]}\}$$

is a csL over $\mathcal{F}(\mathcal{U}M^k)$.

Notation. We denote by $\langle p^*\tau_*(S) \rangle$ the csL ideal

$$\sum_\alpha \langle p^{s_\alpha - 1}\zeta^{s_\alpha}_{s_\alpha+1}\tau_{j_\alpha}(y_\alpha) \rangle$$

associated to the set of triples $S = \{(s_\alpha, j_\alpha, y_\alpha)\}$ in Lemma 11.C5.

Remark 11.C6. We are about to assert that Lemmas 11.C4 and 11.C5 are true also for csL's over $\mathcal{F}(\mathcal{U}M^k \amalg \mathcal{U}M^l)$, so let us review carefully what it means to be a csL over $\mathcal{F}(\mathcal{U}M^k \amalg \mathcal{U}M^l)$. If $\{L_{[q]}, d_{[q]}\}$ is a csL over $\mathcal{F}(\mathcal{U}M^k \amalg \mathcal{U}M^l)$, then it obviously has simultaneous structures as a csL over $\mathcal{F}(\mathcal{U}M^k)$ and as a csL over $\mathcal{F}(\mathcal{U}M^l)$. A partial converse also holds. If $\{L_{[q]}, d_{[q]}\}$ is a csm over $\mathcal{F}(\mathcal{U}M^k \amalg \mathcal{U}M^l)$, and it is simultaneously both a csL over $\mathcal{F}(\mathcal{U}M^k)$ and (with the same bracket) a csL over $\mathcal{F}(\mathcal{U}M^l)$, then it is a csL over

$\mathcal{F}(\mathcal{U}M^k \amalg \mathcal{U}M^l)$. Just as a csL that is a csm over $\mathcal{F}(\mathcal{U}M^k)$ can be certified a csL over $\mathcal{F}(\mathcal{U}M^k)$ by checking the single formula (7BC) and the primitivity of the a_m's, a csL that is a csm over $\mathcal{F}(\mathcal{U}M^k \amalg \mathcal{U}M^l)$ can be certified to be a csL over $\mathcal{F}(\mathcal{U}M^k \amalg \mathcal{U}M^l)$ merely by checking the primitivity of the $a_m^{(i)}$'s and (7BC) for $b_{t-r}^{(i)}$, for both $i = 1$ and $i = 2$.

In particular, $\mathcal{F}((\mathcal{U}M^k \amalg \mathcal{U}M^l)_+)_{[t]}$ will annihilate a differential Lie ideal of $M_{[t]}$ if and only if the appropriate elements of the set $\{b_{t-r}^{(1)}, b_{t-r}^{(2)}, a_m^{(1)}, a_m^{(2)}, b_0^{(1)}, b_0^{(2)}\}$ all annihilate it. An element is considered "appropriate" here according to whether t or m lies in the proper range.

Comparable remarks clearly apply to csL's over $\mathcal{F}(\mathcal{U}M^k \amalg \mathcal{U}M^l \amalg \mathcal{U}M^m)$. We conclude

Corollary 11.C7. *Lemmas 11.C4 and 11.C5 are also true, for csL's over $\mathcal{F}(\mathcal{U}M^k \amalg \mathcal{U}M^l)$ or over $\mathcal{F}(\mathcal{U}M^k \amalg \mathcal{U}M^l \amalg \mathcal{U}M^m)$ instead of over $\mathcal{F}(\mathcal{U}M^k)$.*

We have arrived at the Virtual Structure Lemma. We will first spend a while explaining the context to which the lemma applies. Suppose

$$0 \longrightarrow N \longrightarrow M \longrightarrow M/N \longrightarrow 0 \tag{11CE}$$

is a short exact sequence of $\mathbb{Z}_{(p)}$-free dgL's. We saw in Lemma 7.12(c) that $\mathcal{F}(N)$ is a csL over $\mathcal{F}(\mathcal{U}M)$. Now suppose further that (N, d) happens to be a free dgL with a linear differential, and that $d(N) \subseteq pN$. We choose an indecomposable differential submodule $(V, d) \subseteq (N, d)$, i.e., $d(V) \subseteq pV$ and the composition $V \longrightarrow N \longrightarrow N/([N, N] + (N_{\mathrm{odd}})^2)$ is an isomorphism. We identify (N, d) with $\mathcal{L}(V, d)$. Since $\mathcal{F}\mathcal{L}(V, d) = \mathcal{F}(N)$ is a csL over $\mathcal{F}(\mathcal{U}M)$, and the sub-csL $\mathcal{F}^0\mathcal{L}(V, d)$ is defined, we ask: under what conditions is $\mathcal{F}^0\mathcal{L}(V, d)$ a csL over $\mathcal{F}(\mathcal{U}M)$? Equivalently, when is $\mathcal{F}^0\mathcal{L}(V, d)$ closed under the action of $\mathcal{F}(\mathcal{U}M)$, or, when do we have

$$\psi_{[t]}(\mathcal{F}(\mathcal{U}M^k)_{[t]} \otimes \mathcal{F}^0\mathcal{L}(V)_{[t]}) \subseteq \mathcal{F}^0\mathcal{L}(V)_{[t]}$$

for all t?

We answer this question by embedding it in the context of a broader question regarding virtual structures for csL's over Dcsa's (see Chapter 7). Since $V = \mathcal{L}(V)/\mathcal{L}^{\geq 2}(V)$, and $\mathcal{L}^{\geq 2}(V)$ is closed under the action of $\mathcal{U}M$, V is also a $\mathcal{U}M$-module. Consequently the csc $\mathcal{F}(V)$ is a csm over $\mathcal{F}(\mathcal{U}M)$. Define $\overline{V}_{[t]}$ by

$$\overline{V}_{[t]} = \mathcal{F}(V)_{[t]}/(\operatorname{im} \zeta_t^{t-1} + \operatorname{im}(d_{[t]}\zeta_t^{t-1})).$$

If

$$V = V^{(1)} \oplus V^{(2)} \oplus \cdots \oplus V^{(\infty)} \tag{11CF}$$

is any chain complex decomposition like (11BU), notice that we have an isomorphism of $\mathbb{Z}_p$-vector spaces,

$$V^{(\geq t)} \otimes \mathbb{Z}_p \approx \overline{V}_{[t]}.$$

Since $\operatorname{im}(\zeta_t^{t-1}) + \operatorname{im}(d_{[t]}\zeta_t^{t-1})$ is closed under the $\mathcal{F}(\mathcal{U}M)_{[t]}$-action we see that $\overline{V}_{[t]}$ is a left module over $\mathcal{F}(\mathcal{U}M)_{[t]}$.

Given this situation, suppose r_α is a virtual relator (henceforth VR) in $\mathcal{F}(\mathcal{U}M)_{[t]}$ for the module $\mathcal{F}\mathcal{L}(V)_{[t]}$. Since any formal linear expression evaluated at any $x \in \mathcal{F}\mathcal{L}(V)_{[t]}$ lands in $\mathcal{L}^{\geq 2}(V^{(\geq t)}) + \operatorname{im}(\zeta_t^{t-1}) + \operatorname{im}(d_{[t]}\zeta_t^{t-1})$ by Corollary 11.B13, we see that r_α annihilates the module $\overline{V}_{[t]}$. If $\overline{P}_{[t]}$ denotes the quotient of $\mathcal{F}(\mathcal{U}M)_{[t]}$ by the ideal of all VR's, then the $\mathbf{Z}_p$-vector space $\overline{V}_{[t]}$ is a left module over the $\mathbf{Z}_p$-algebra $\overline{P}_{[t]}$. We are tempted to write down a minimal set of generators and syzygies for $\overline{V}_{[t]}$ as a $\overline{P}_{[t]}$-module, and then to try to lift them to $\mathcal{F}^0\mathcal{L}(V)_{[t]}$.

Let us give in to this temptation. Fix $t \geq 1$. Let $\{\bar{x}_i\}_{i\in I}$ be a minimal set of generators for $\overline{V}_{[t]}$ as a $\overline{P}_{[t]}$-module, and let $\{\bar{y}_j\}_{j\in J}$ be a minimal set of syszgies, i.e., $\{\overline{\delta}^{(1)}(\bar{y}_j)\}_{j\in J}$ generate the $\overline{P}_{[t]}$-module $\ker(\bar{\delta}^{(0)})$ in the resolution

$$\text{(11CG)} \qquad 0 \longleftarrow \overline{V}_{[t]} \overset{\bar{\delta}^{(0)}}{\longleftarrow} \bigoplus_{i\in I} \overline{P}_{[t]}x_i \overset{\bar{\delta}^{(1)}}{\longleftarrow} \bigoplus_{j\in J} \overline{P}_{[t]}\bar{y}_j \longleftarrow \cdots.$$

In Chapter 7 we lifted the $\{\bar{x}_i\}$ to arbitrary $x_i \in \mathcal{FL}(V)$. When $\bar{y}_j$ was written in the form

$$\bar{y}_j = \sum_{i\in I} \bar{b}_{ij}(\bar{x}_i)$$

with $\bar{b}_{ij} \in \overline{P}_{[t]}$, we lifted the $\bar{b}_{ij}$ to $b_{ij} \in \mathcal{F}(\mathcal{U}M)_{[t]}$, and we considered the expressions

$$\text{(11CH)} \qquad y_j = \sum_{i\in I} \psi_{[t]}(b_{ij} \otimes x_i) \in \mathcal{F}(N)_{[t]} = \mathcal{FL}(V)_{[t]}.$$

Since we know that y_j projects to zero in $\overline{V}_{[t]}$, we see that we may write y_j in the form

$$\text{(11CI)} \qquad y_j = \zeta_t^{t-1}(z_j^1) + d_{[t]}\zeta_t^{t-1}(z_j^2) + z_j^0,$$

where $z_j^0 \in \mathcal{F}(\mathcal{L}^{\geq 2}(V))_{[t]}$. (Here we are exploiting the decomposition $\mathcal{FL}(V) = \mathcal{F}(V) \oplus \mathcal{F}(\mathcal{L}^{\geq 2}(V))$ as csc's, arising from the splitting $\mathcal{L}(V) = V \oplus \mathcal{L}^{\geq 2}(V)$ as chain complexes.) If $\mathcal{L}(V^{(\geq t)})$ were a Bockstein sub-dgL of $\mathcal{FL}(V)_{[t]}$, then we could assume, by altering z_j^1 and z_j^2 if necessary, that $z_j^0 \in \mathcal{L}^{\geq 2}(V^{(\geq t)})$, and we would have exactly the situation described in Chapter 7 in connection with (7BJ).

Now $\mathcal{L}(V^{(\geq t)})$ is not a Bockstein sub-dgL of $\mathcal{FL}(V)_{[t]}$, but it is a Bockstein sub-dgL of $\mathcal{F}^0\mathcal{L}(V)_{[t]}$. In order to be able to write z_j^0 as an element of $\mathcal{L}^{\geq 2}(V^{(\geq t)})$, which is part of what is assumed possible in Theorem 7.15, we need y_j of (11CH) to satisfy the condition $y_j \in \mathcal{F}^0\mathcal{L}(V)_{[t]}$. This condition together with two other conditions suffices, not only to make $\mathcal{F}^0\mathcal{L}(V)_{[t]}$ a module over $\mathcal{F}(\mathcal{U}M)_{[t]}$, but also to assert that the $\{x_i\}$ and $\{y_j\}$ and VR's constitute a virtual structure for $\mathcal{F}^0\mathcal{L}(V)$. This is the gist of the Virtual Structure Lemma.

Proposition 11.C8. *(Virtual Structure Lemma.) Given the short exact sequence (11CE) of $\mathbf{Z}_{(p)}$-free dgL's, suppose that $d(N) \subseteq pN$ and (N,d) has a linear differential. Fix an indecomposable differential submodule (V,d) as above and identify (N,d) with $\mathcal{L}(V,d)$. For each $t \geq 1$, we find the following sets. Let $I_{[t]}$ denote the set (it is a differential ideal) of all VR's in $\mathcal{F}(\mathcal{U}M)_{[t]}$ for the module $\mathcal{FL}(V)_{[t]}$, and let $\overline{P}_{[t]} = \mathcal{F}(\mathcal{U}M)_{[t]}/I_{[t]}$. Let $\{\bar{x}_i\}_{i\in I}$ be a minimal set of generators for $\overline{V}_{[t]} = \mathcal{F}(V)_{[t]}/(\operatorname{im}(\zeta_t^{t-1}) + \operatorname{im}(d_{[t]}\zeta_t^{t-1}))$ as a left $\overline{P}_{[t]}$-module, and let $\{\bar{y}_j\}_{j\in J}$ be a minimal set of syzygies, $\bar{y}_j = \sum_i \bar{b}_{ij}(\bar{x}_i)$. Lift the $\{\bar{x}_i\}$*

to arbitrary $x_i \in \mathcal{FL}(V)_{[t]}$ *and the* $\bar{b}_{ij}$ *to arbitrary* $b_{ij} \in \mathcal{F}(\mathcal{U}M)_{[t]}$. *Suppose the following three conditions are satisfied for all* t :

(i) *Each virtual generator* x_i *belongs to* $\mathcal{F}^0\mathcal{L}(V)$;

(ii) *For each index* $j \in J$, *the "syzygy element"*

$$y_j = \sum_{i \in I} \psi_{[t]}(b_{ij} \otimes x_i)$$

belongs to $\mathcal{F}^0\mathcal{L}(V)_{[t]}$;

(iii) *For some chain complex decomposition (11CF) of* (V, d), *we have*

$$V^{(\geq t)} \subseteq \text{Span}\{x_i\} + \psi_{[t]}(\mathcal{F}(\mathcal{U}M_+)_{[t]} \otimes \mathcal{F}^0\mathcal{L}(V)_{[t]}) + \overline{\mathcal{F}^0\mathcal{L}(V)}_{[t]}, \tag{11CJ}$$

where

$$\overline{\mathcal{F}^0\mathcal{L}(V)_{[t]}} = \zeta_t^{t-1}(\mathcal{F}^0\mathcal{L}(V)_{[t-1]}) + d_{[t]}\zeta_t^{t-1}(\mathcal{F}^0\mathcal{L}(V)_{[t-1]}) + \mathcal{L}^{\geq 2}(V^{(\geq t)}). \tag{11CK}$$

Then $\mathcal{F}^0\mathcal{L}(V)$ *is a csL over the Dcsa* $\mathcal{F}(\mathcal{U}M)$. *For each* t, $I_{[t]}$ *is an ideal of VR's in* $\mathcal{F}(\mathcal{U}M)_{[t]}$ *for* $\mathcal{F}^0\mathcal{L}(V)_{[t]}$; *the set* $\{x_i\}$ *is a minimal set of virtual generators for* $\mathcal{F}^0\mathcal{L}(V)_{[t]}$, *and the set* J *indexes a minimal set of virtual syzygies. For each index* $j \in J$, *a relation of the form(11CI) holds, with* $z_j^1, z_j^2 \in \mathcal{F}^0\mathcal{L}(V)_{[t]}$ *and* $z_j^0 \in \mathcal{L}^{\geq 2}(V^{(\geq t)}) \subseteq \mathcal{F}^0\mathcal{L}(V)_{[t]}$.

Proof. The last sentence is an immediate consequence of (ii) and 11.B10(c), so we simply assume it throughout.

Let us begin with an observation about VR's. Let r_α be a VR in $\mathcal{F}(\mathcal{U}M)_{[t]}$ for $\mathcal{FL}(V)_{[t]}$, and let $F_{r_\alpha}(x)$ be the corresponding formal linear expression. By definition, $F_{r_\alpha}(x)$ is a linear combination of terms of the form (7BF) and (7BG). A priori, the elements y_t in (7BF) and y_m in (7BG) can be any elements of $\mathcal{U}(\mathcal{FL}(V)_{[t]})$. The observation is that we can assume, by replacing $F_{r_\alpha}(x)$ by an equivalent fle if necessary, that the y_t's (resp. y_m's) all belong to $\mathcal{UL}(V^{(\geq t)})$ (resp. $\mathcal{UL}(V^{\geq m)})$). To see this, use (11BV) and 11.B5(c) to show that each y_t and y_m can be assumed to belong to $\mathcal{F}^0\mathcal{L}(V)$, and then use 11.B10(c) and the csL axioms to eliminate any contribution to a y_t (resp. y_m) other than the contribution form $\mathcal{UL}(V^{(\geq t)})$ (resp. $\mathcal{UL}(V^{(\geq m)})$).

What we are to prove is that

$$\psi_{[t]}(\mathcal{F}(\mathcal{U}M)_{[t]} \otimes \mathcal{F}^0\mathcal{L}(V)_{[t]}) \subseteq \mathcal{F}^0\mathcal{L}(V)_{[t]}. \tag{11CL}$$

Rewrite (11CL) in the form

$$\psi_{[t]}(w \otimes v) \in \mathcal{F}^0\mathcal{L}(V)_{[t]} \text{ for } w \in \mathcal{F}(\mathcal{U}M)_{[t]}, v \in \mathcal{F}^0\mathcal{L}(V)_{[t]}. \tag{11CM}$$

The relation (11CM) can be proved by induction on t and $|v|$, if we assume (11CJ) and the relation

$$\psi_{[t]}(w \otimes x_h) \in \mathcal{F}^0\mathcal{L}(V)_{[t]} \text{ for } w \in \mathcal{F}(\mathcal{U}M)_{[t]}, h \in I. \tag{11CN}$$

It therefore suffices to prove (11CN).

To prove (11CN) we use a double induction on t and on the dimension $q = |w| + |x_h|$ in which (11CN) occurs. When $t = 1$ the assumption $d(V) \subseteq pV$ menas $\mathcal{F}^0\mathcal{L}(V) \subseteq \mathcal{F}\mathcal{L}(V)$ and there is nothing to prove, and for $t > 1$ we are assuming that

(11CO) $$\psi_{[s]}(\mathcal{F}(\mathcal{U}M)_{[s]} \otimes \mathcal{F}^0\mathcal{L}(V)_{[s]}) \subseteq \mathcal{F}^0\mathcal{L}(V)_{[s]} \text{ if } s < t.$$

We are also assuming that

(11CP) $$\psi_{[t]}(u \otimes x_i) \in \mathcal{F}^0\mathcal{L}(V)_{[t]} \text{ if } |u| + |x_i| < q = |w| + |x_h|.$$

One consequence of (11CO) and (11CP) is that $\psi_{[t]}(u \otimes z)$ lies in $\mathcal{F}^0\mathcal{L}(V)$ if $|u| + |z| = q$ and

$$z \in \mathcal{L}^{\geq 2}(V^{(\geq t)}) + \zeta_t^{t-1}(\mathcal{F}^0\mathcal{L}(V)_{[t-1]}) + d_{[t]}\zeta_t^{t-1}(\mathcal{F}^0\mathcal{L}(V)_{[t-1]}).$$

In particular,

(11CQ) $$\sum_{i \in I} \psi_{[t]}(vb_{ij} \otimes x_i) \in \mathcal{F}^0\mathcal{L}(V)_{[t]}$$

for any $v \in \mathcal{F}(\mathcal{U}M)_{[t]}$ and any index $j \in J$ satisfying $|v| + |y_j| = q$. In the same spirit we have

(11CR) $$\psi_{[t]}(u' \otimes x_i) \in \mathcal{F}^0\mathcal{L}(V)_{[t]}$$

whenever $|u'| + |x_i| \leq q$ and $u' \in I_{[t]}$, i.e., when u' is a VR. This follows from our earlier observation about VR's, along with (i) and (11CO) and (11CP).

Returning to (11CN), let $\bar{w}$ be the image of w in $\overline{P}_{[t]}$, and consider $\bar{\delta}^{(0)}(\bar{w}\bar{x}_h) \in \overline{V}_{[t]}$, where $\bar{\delta}^{(0)}$ and $\bar{\delta}^{(1)}$ appear in (11CG). Let $z \in V^{(\geq t)}$ denote any lift of $\bar{\delta}^{(0)}(\bar{w}\bar{x}_h)$ to $V^{(\geq t)} \subseteq \mathcal{F}^0\mathcal{L}(V)_{[t]}$. Using (iii) write $z = \sum_i \psi_{[t]}(u_i \otimes x_i) + z''$, where $u_i \in \mathcal{F}(\mathcal{U}M)_{[t]}$ and $z'' \in \overline{\mathcal{F}^0\mathcal{L}(V)}_{[t]}$. Then $-\bar{w}\bar{x}_h + \sum_i \bar{u}_i\bar{x}_i \in \ker(\bar{\delta}^{(0)}) = \operatorname{im}(\bar{\delta}^{(1)})$, and we can write

(11CS) $$-\bar{w}\bar{x}_h + \sum_i \bar{u}_i\bar{x}_i = \sum_j \bar{v}_j\bar{y}_j = \sum_i (\sum_j \bar{v}_j\bar{b}_{ij})\bar{x}_i,$$

for some $\bar{v}_j \in \overline{P}_{[t]}$. Lift the $\bar{v}_j$ to arbitrary $v_j \in \mathcal{F}(\mathcal{U}M)_{[t]}$.

Think about what equation (11CS) is really saying. It is an equation in $\overline{P}_{[t]} \otimes S$, where S is the $\mathbb{Z}_p$-vector space with basis $\{\bar{x}_i\}$. The equation tells us, for each index i, that $\bar{u}_i = \sum_j \bar{v}_j\bar{b}_{ij}$ if $i \neq h$, and that $\bar{u}_h - \bar{w} = \sum_j \bar{v}_j\bar{b}_{hj}$. Lifting these equations to $\mathcal{F}(\mathcal{U}M)_{[t]}$, we have

$$u_i = \sum_j v_jb_{ij} + u_i' \text{ if } i \neq h;$$

$$u_h - w = \sum_j v_jb_{hj} + u_h',$$

where $u_i' \in I_{[t]}$. Now consider

$$\begin{aligned}\psi_{[t]}(w \otimes x_h) &= \psi_{[t]}(u_h \otimes x_h) - \sum_j \psi_{[t]}(v_jb_{hj} \otimes x_h) - \psi_{[t]}(u_h' \otimes x_h) \\ &= \sum_i \psi_{[t]}(u_i \otimes x_i) - \sum_i \sum_j \psi_{[t]}(v_jb_{ij} \otimes x_i) - \sum_i (u_i' \otimes x_i) \\ &= z - z'' + (\text{ something in } \mathcal{F}^0\mathcal{L}(V)_{[t]}),\end{aligned}$$

by (11CQ) and (11CR). Because $z - z'' \in \mathcal{F}^0\mathcal{L}(V)_{[t]}$, deduce $\psi_{[t]}(w \otimes x_h) \in \mathcal{F}^0\mathcal{L}(V)_{[t]}$, completing the inductive step.

The remaining claims in Proposition 11.C8 are now straightforward to verify, and we omit their proofs.

Proposition 11.C9. *Let $k \geq 0, l \geq 0$, and suppose $p \geq 3$. Recall the chain map g_{kl} : $\mathcal{U}M_+^k \otimes \mathcal{U}M_+^l \longrightarrow N^{kl}$ of Theorem 5.12, and put*

$$V = V^{kl} = g_{kl}(\mathcal{U}M_+^k \otimes \mathcal{U}M_+^l) \subseteq N^{kl}.$$

Then the sub-csL $\mathcal{F}^0\mathcal{L}(V)$ of $\mathcal{F}\mathcal{L}(V) = \mathcal{F}(N^{kl})$ is a csL over $\mathcal{F}(\mathcal{U}M^k \amalg \mathcal{U}M^l)$, and the virtual structure described in Example 7.17 is a virtual structure for $\mathcal{F}^0\mathcal{L}(V)$.

Proof. We are clearly going to apply the Virtual Structure Lemma, to the virtual structure described in Example 7.17. We need to check that the three hypotheses in Proposition 11.C8 are fulfilled by this example. All hypotheses are trivial when $t \leq r$, because $V^{(<t)} = 0$ and consequently $\mathcal{F}^0\mathcal{L}(V)_{[t]} = \mathcal{F}\mathcal{L}(V)_{[t]}$, and all are vacuous for $t > r + \min(k, l)$. So we may assume $r < t \leq r + \min(k, l)$. In particular, we may assume $k \geq 1$ and $l \geq 1$.

By induction on k and l, we may assume that Proposition 11.C9 is true for $N^{k-1,l}$ and for $N^{k,l-1}$. Recall the #-gradings on N^{kl} (defined just prior to Lemma 5.11). Let $\bar{i}$ denote any of the inclusions $M^{k-1} \longrightarrow M^k$, $\mathcal{U}M^{k-1} \longrightarrow \mathcal{U}M^k$, or $N^{k-1,l} \longrightarrow N^{kl}$. We have

$$g_{kl}(\bar{i} \otimes 1) = \bar{i} g_{k-1,l} : \mathcal{U}M_+^{k-1} \otimes \mathcal{U}M_+^l \longrightarrow N^{kl}$$

in $\#^{(1)}$-grade $< p^k$, since it is not until #-grade p^k that the obstruction set for $\mathcal{U}M^{(\ )}$ changes, and hence the definition of $g_{(\ )l}$ in (5V) changes, as we go from $k-1$ to k. As a result,

$$\bar{i}(\mathcal{F}^0\mathcal{L}(V^{k-1,l})) \subseteq \mathcal{F}^0\mathcal{L}(V^{kl})$$

in $\#^{(1)}$-grade $< p^k$, where we write $V^{k-1,l}$ for $g_{k-1,l}(\mathcal{U}M^{k-1}+ \otimes \mathcal{U}M_+^l)$ and V^{kl} for V. This implies, by the inductive hypothesis, that VG's (virtual generators) x_i having $\#^{(1)}$-grade below p^k automatically lie in $\mathcal{F}^0\mathcal{L}(V^{kl})$. Similarly, y_j's of $\#^{(1)}$-grade $< p^k$ lie in $\mathcal{F}^0\mathcal{L}(V)$. The same is true for x_i's and y_j's in $\#^{(2)}$-grade below p^l. So we need only to attend to x_i's and y_j's in #-grade (p^k, p^l).

Consider (i). Refer to the list of VG's in Example 7.17, i.e., (7DC) and (7DD). When $k = l$, the only VG's of #-grade (p^k, p^l) are the basic VG's for $t = r + k$, which are automatically in $V^{(r+k)}$ by definition of V, and the VG $[a_k^{(1)}, a_k^{(2)}]$ for $t < r + k$. It suffices to show that $[a_k^{(1)}, a_k^{(2)}] \in \mathcal{F}^0\mathcal{L}(V)_{[r+k-1]}$, since for smaller t we may simply apply θ_t^{r+k-1}. When $k > l$, the only VG's we need to be concerned with (for #-grade reasons) are $\mathrm{ad}^{(2)}(c_l)(a_k^{(1)})$ and $\mathrm{ad}^{(2)}(b_l)(a_k^{(1)})$ for $t = r + l$, and $[a_k^{(1)}, a_l^{(2)}]$ for $t = r + l - 1$. When $k < l$, we need only be concerned with $\mathrm{ad}^{(1)}(c_l)(a_k^{(2)})$ and $\mathrm{ad}^{(1)}(b_l)(a_k^{(2)})$ for $t = r + k$, and with $[a_k^{(1)}, a_l^{(2)}]$ for $t = r + k - 1$. This classification reduces the problem to manageable proportions.

View $\mathcal{F}^0\mathcal{L}(V)_{[t]}$ as the subset of $\mathcal{L}(V)$ described in Lemma 11.B10(b). For now consider only the case $k \geq l$. We are going to have to retrace part of the proof of Theorem 5.12

in order to get our VG's to lie in $\mathcal{F}^0\mathcal{L}(V)$. In the proof of Theorem 5.12, the element $g(w_3)$ is a cycle, which we now view as an element of $\mathcal{FL}(V)_{[r+l-1]}$. Since $\mathcal{F}^0\mathcal{L}(V)$ and $\mathcal{FL}(V)$ coincide in $*$-degree 3 ($\widetilde{W}^{(t)}$ is zero except in $*$-degrees $\equiv 1$ or 2 mod $(2p)$), we have $g(w_3) \in \mathcal{F}^0\mathcal{L}(V)_{[r+l-1]}$. The element w_2 can be chosen so that $g(w_2)$ belongs to $\mathcal{F}^0\mathcal{L}(V)_{[r+l-1]}$ also (to see this, use the decomposition (11BZ)). The element $g(w_2)$ is then used in Equation (5Y) to define y_2. Since y_2 is later taken as our definition of $g_{kl}(c_k \otimes c_l)$, we know that $y_2 \in V^{(r+l)}$.

Equation (5Y) has two nice consequences. First, viewed as an equation in $\mathcal{FL}(V)_{[r+l]}$ when $k > l$, we can solve for $\mathrm{ad}^{(2)}(c_l)(a_k^{(1)})$, to obtain directly that $\mathrm{ad}^{(2)}(c_l)(a_k^{(1)}) \in \mathcal{F}^0\mathcal{L}(V)_{[r+l]}$. Second, view (5Y) as an equation in $\mathcal{FL}(V)_{[r+l-1]}$ when $k \geq l$. We may expand $\mathrm{ad}^{(2)}(c_l)$ as $\mathrm{ad}^{(2)}(a_l) + \mathrm{ad}^{(2)}(e_l)$. The term $\mathrm{ad}^{(2)}(e_l)(a_k^{(1)})$ lies in $\mathcal{F}^0\mathcal{L}(V)_{[r+l-1]}$ by an argument, like that above, that is based upon the fact that $\mathrm{ad}^{(2)}(e_l)(a_k^{(1)})$ lies in $\bar{i}(\mathcal{F}^0\mathcal{L}(V^{k-1,l})_{[r+l-1]})$. When we solve for the term $\mathrm{ad}^{(2)}(a_l)(a_k^{(1)}) = [a_k^{(1)}, a_l^{(2)}]$, we discover that it equals a sum of terms all belonging to $\mathcal{F}^0\mathcal{L}(V)_{[r+l-1]}$, i.e.

$$[a_k^{(1)}, a_l^{(2)}] \in \mathcal{F}^0\mathcal{L}(V)_{[r+l-1]} \text{ for } k \geq l.$$

The proof that $\mathrm{ad}^{(2)}(b_l)(a_k^{(1)}) \in \mathcal{F}^0\mathcal{L}(V)_{[r+l-1]}$ employs similar techniques. The subtlety is to get $\mathrm{ad}^{(2)}(b_l)(a_k^{(1)})$ to belong to $\mathcal{F}^0\mathcal{L}(V)_{[r+l]}$, which requires showing that it belongs to $\bar{i}(\mathcal{F}^0\mathcal{L}(V^{k-1,l})_{[r+l]})$. When $k < l$, all the same arguments work, but with the roles of (1) and (2) exchanged.

Let us move on to verifying hypothesis (ii). The inductive (on t) nature of the proof of Proposition 11.C8, together with Lemma 11.B12, shows that hypothesis (ii) can in general be replaced by the seemingly weaker condition,

(11CT)
$$\begin{aligned} y_j \in [\mathcal{FL}(V)_{[t]}, \mathcal{FL}(V)_{[t]}] &+ \zeta_t^{t-1}\psi_{[t-1]}(\mathcal{F}(\mathcal{U}M)_{[t-1]} \otimes \mathcal{F}^0\mathcal{L}(V)_{[t-1]}) \\ &+ d_{[t]}\zeta_t^{t-1}\psi_{[t-1]}(\mathcal{F}(\mathcal{U}M)_{[t-1]} \otimes \mathcal{F}^0\mathcal{L}(V)_{[t-1]}). \end{aligned}$$

In the specific application at hand, we will find that each of our y_j's actually satisfies

(11CU)
$$y_j \in \zeta_t^{t-1}(\Psi^0_{[t-1]}) + d_{[t]}\zeta_t^{t-1}(\Psi^0_{[t-1]}),$$

where $\Psi^0_{[t-1]} = \psi_{[t-1]}(\mathcal{F}(\mathcal{U}M^k \amalg \mathcal{U}M^l)_{[t-1]} \otimes \mathcal{F}^0\mathcal{L}(V^{kl})_{[t-1]})$. Since we are working over $\mathcal{F}(\mathcal{U}M^k \amalg \mathcal{U}M^l)$, by combining (11BV) with Lemma 11.C3 and (for the a_m's) Lemma 11.B5(c), we see that

(11CV)
$$\Psi_{[t-1]+} \subseteq \Psi^0_{[t-1]},$$

where

$$\Psi_{[t-1]+} = \psi_{[t-1]}(\mathcal{F}((\mathcal{U}M^k \amalg \mathcal{U}M^l)_+)_{[t-1]} \otimes \mathcal{FL}(V^{kl})_{[t-1]}).$$

Hence (11CU) follows from showing that

(11CW)
$$y_j \in \zeta_t^{t-1}(\Psi_{[t-1]+}) + d_{[t]}\zeta_t^{t-1}(\Psi_{[t-1]+}).$$

It turns out that (11CW) is what we can actually prove, for each y_j.

Example 7.17 contains a procedure to obtain each minimal VS (virtual syzygy) from certain basic equations. As noted above, we need only be concerned with those VS's that arise through applying the procedure starting from VG's of #-grade (p^k, p^l). The "simplification" procedure outlined in Example 7.17 is explicit enough that we can recognize each resulting y_j as being a sum of terms of the form (11CW).

The reader who attempts to carry out this calculation will discover that one additional fact is needed in order to reduce all terms in y_j to the form (11CS). Recall from 7.17 that the images of the VG's in $\mathcal{F}(\mathcal{U}M^k \amalg \mathcal{U}M^l)$ serve as a minimal set of VR's, for $\mathcal{F}(N^{kl})$. The additional neded fact is that the fle corresponding to each such VR is positive, in the sense of Corollary 11.B14. Corollary 11.B14 is then used to help get (11CN). The positivity of the fle's is deducible from the fact that the image of N^{kl} in $\mathcal{U}M^k \amalg \mathcal{U}M^l$ lies entirely in the ideal generated by $[e^{(1)}(\mathcal{U}M^k_+), e^{(2)}(\mathcal{U}M^l_+)]$.

Finally, consider (iii). Let $\widetilde{F}^0_{[t]}$ denote the right-hand side of (11CJ). By (5V), (11CJ) will be satisfied if and only if the set $g_{kl}(S^k_{t-r} \otimes S^l_{t-r})$ belongs to $\widetilde{F}^0_{[t]}$ (the notations S^k_s and S^l_s are defined just before Lemma 5.9). Once again, we can reduce to the case of #-grade (p^k, p^l). This means that we need only to show that the four-element set

(11CX) $$\{g(b_k \otimes b_l), g(b_k \otimes c_l), g(c_k \otimes b_l), g(c_k \otimes c_l)\}$$

belongs to $\widetilde{F}^0_{[t]}$, where we now write g for g_{kl}. However, we must show this for every t in the range $r < t \leq r + \min(k, l)$, since the set $\{x_i\}$ changes with t.

In some cases, an element w of (11CX) served or could have served as a virtual generator. In these cases, the argument showing that (i) holds for x_i can be reversed to obtain that (iii) holds for the element w. The cases where this happens are: for $w = g(c_k \otimes c_l)$ (with $x_i = [a^{(1)}_k, a^{(2)}_l]$) when $t < r + \min(k, l)$; for all four basic VG's when $t - r = k = l$; for $g(c_k \otimes c_l)$ and $g(c_k \otimes b_l)$ when $t - r = l < k$; and for $g(c_k \otimes c_l)$ and $g(b_k \otimes c_l)$ when $t - r = k < l$.

Put $s = t - r$. To get $g(b_k \otimes c_l) \in \widetilde{F}^0_{[t]}$ when $s < l$ in general, compare the definition of $g(b_k \otimes c_l)$ in Theorem 5.12 with $\mathrm{ad}^{(1)}((b_s)^{p^{k-s}-1})(\mathrm{ad}^{(1)}(b_s)(a^{(2)}_l))$. Do likewise for $g(c_k \otimes b_l)$ whenever $s < k$. For $g(b_k \otimes b_l)$ when $k \neq l$ or $k = l > s$, use the fact that the difference between $g(b_k \otimes b_l)$ and

$$\mathrm{ad}^{(1)}(b_s^{p^{k-s}-1})\,\mathrm{ad}^{(2)}(b_s^{p^{l-s}-1})g(b_s \otimes b_s)$$

is a decomposable element. We are in $*$-degree zero, and $\mathcal{F}^0\mathcal{L}(V)^{*=0} = \mathcal{F}\mathcal{L}(V)^{*=0}$. Deduce that $g(b_k \otimes b_l) \in \widetilde{F}^0_{[t]}$. The only remaining cases are $g(b_k \otimes c_l)$ when $s = l < k$, and $g(c_k \otimes b_l)$ when $s = k < l$. Apply $d_{[t]}$ to the relation showing $g(b_k \otimes b_l) \in \widetilde{F}^0_{[t]}$, to get these elements to belong to $\widetilde{F}^0_{[t]}$. This completes the verification of (iii), and with it, the proof of Proposition 11.C9.

Notation. The csL $\mathcal{F}^0\mathcal{L}(V^{kl})$ of Proposition 11.C9 will also be denoted as $\mathcal{F}^0(N^{kl})$. It is a sub-csL-over-$\mathcal{F}(\mathcal{U}M^k \amalg \mathcal{U}M^l)$ of $\mathcal{F}(N^{kl})$, and for each $t \geq 1$ $\mathcal{L}(V^{(\geq t)})$ is a Bockstein sub-dgL of $\mathcal{F}^0(N^{kl})_{[t]}$.

Just as the virtual structure described in Example 7.17 is a virtual structure for $\mathcal{F}^0(N^{kl})$, so too the virtual structure of Example 7.19 is a virtual structure for a csL we shall call

$\mathcal{F}^0(N^{klm})$. This is best proved by embedding it in an inductive framework together with a number of related facts. We list these related facts in the next theorem, but we omit the proof entirely. To simplify the statement of the theorem we consider only the case $k = l = m$.

Theorem 11.C10. *Let $k \geq 0$ and $p \geq 3$. Let V^{kkk} denote the differential submodule $fg(U) \subseteq N^{kkk}$ of Corollary 6.21. Let $1 \leq t \leq r + k$, and put $s = \max(0, t - r)$.*

(a) *If $k > 0$ and $t < r + k$, the inclusion $\bar{i} : N^{k-1,k-1} \longrightarrow N^{kk}$ satisfies*

$$\mathcal{F}(\bar{i})(\mathcal{F}^0(N^{k-1,k-1})_{[t]}) \subseteq \mathcal{F}^0(N^{kk})_{[t]},$$

i.e., the csL/csm homomorphism $\mathcal{F}^0(\bar{i}) : \mathcal{F}^0(N^{k-1,k-1}) \longrightarrow \mathcal{F}^0(N^{kk})$ is well-defined.

(b) *Equations (7DK)–(7DN) are true for $\mathcal{F}^0(N^{kk})$, i.e., both sides of each equation lie in $\mathcal{F}^0(N^{kk})$.*

(c)

$$\mathrm{ad}^{(2)}(\mathcal{F}(\mathcal{U}M^k_+)_{[t]})(e^{(1)}\mathcal{F}(\overline{L}^k)_{[t]}) \subseteq \mathcal{F}^0(N^{kk});$$
$$\mathrm{ad}^{(1)}(\mathcal{F}(\mathcal{U}M^k_+)_{[t]})(e^{(2)}\mathcal{F}(\overline{L}^k)_{[t]}) \subseteq \mathcal{F}^0(N^{kk}).$$

(d) *If $k > 0$ and $t < r + k$, the dgL inclusion $\bar{i} : N^{k-1,k-1,k-1} \longrightarrow N^{kkk}$ satisfies*

$$\mathcal{F}(\bar{i})(\mathcal{F}^0\mathcal{L}(V^{k-1,k-1,k-1})_{[t]}) \subseteq \mathcal{F}^0\mathcal{L}(V^{kkk})_{[t]},$$

i.e. the csL/csm homomorphism

$$\mathcal{F}^0(\bar{i}) : \mathcal{F}^0\mathcal{L}(V^{k-1,k-1,k-1}) \longrightarrow \mathcal{F}^0\mathcal{L}(V^{kkk})$$

is well-defined.

(e) *The eight Jacobi identities described in Example 7.19 are true in $\mathcal{F}^0\mathcal{L}(V^{kkk})_{[t]}$, i.e., (7EB), (7ED), (7EE), etc. are true as congruences modulo*

$$\zeta_t^{t-1}\mathcal{F}^0\mathcal{L}(V^{kkk})_{[t-1]} + d_{[t]}\zeta_t^{t-1}\mathcal{F}^0\mathcal{L}(V^{kkk})_{[t-1]}.$$

(f)

$$\mathrm{ad}^{(1)}(\mathcal{F}(\mathcal{U}M^k_+)_{[t]})e^{(23)}\mathcal{F}(N^{kk})_{[t]} \subseteq \mathcal{F}^0\mathcal{L}(V^{kkk});$$
$$\mathrm{ad}^{(2)}(\mathcal{F}(\mathcal{U}M^k_+)_{[t]})e^{(13)}\mathcal{F}(N^{kk})_{[t]} \subseteq \mathcal{F}^0\mathcal{L}(V^{kkk});$$
$$\mathrm{ad}^{(3)}(\mathcal{F}(\mathcal{U}M^k_+)_{[t]})e^{(12)}\mathcal{F}(N^{kk})_{[t]} \subseteq \mathcal{F}^0\mathcal{L}(V^{kkk}).$$

(g) *The structure described in Example 7.19 is a virtual structure for $\mathcal{F}^0\mathcal{L}(V^{kkk})_{[\leq t]}$ as a csL over $\mathcal{F}(\mathcal{U}M^k \amalg \mathcal{U}M^k \amalg \mathcal{U}M^k)_{[\leq t]}$*

(h) *The folding homomorphism $\nabla^{12} : N^{kkk} \longrightarrow N^{kk}$ defined by (7EH) satisfies*

$$\mathcal{F}(\nabla^{(12)})(\mathcal{F}^0\mathcal{L}(V^{kkk})_{[t]} \subseteq \mathcal{F}^0(N^{kk})_{[t]}.$$

Consequently $\mathcal{F}^0(\nabla^{(12)})$ is a well-defined csL/csm homomorphism over the Dcsa $\mathcal{F}(\mathcal{U}M^k \amalg \mathcal{U}M^k \amalg \mathcal{U}M^k)$. The same is true of $\mathcal{F}^0(\nabla^{(13)})$ and $\mathcal{F}^0(\nabla^{(23)})$.

Notation. In view of 11.C10(g) we will write $\mathcal{F}^0(N^{kkk})$ for $\mathcal{F}^0\mathcal{L}(V^{kkk})$. It is a sub-csL-over-$\mathcal{F}(\mathcal{U}M^k \amalg \mathcal{U}M^k \amalg \mathcal{U}M^k)$ of $\mathcal{F}(N^{kkk})$.

Question 11.C11. In the setting where the dgL kernel N in (11CE) has a linear differential and $d(N) \subseteq pN$, does there always exist an indecomposable module $V \subseteq N$ such that $(N,d) = \mathcal{L}(V,d)$ and $\mathcal{F}^0\mathcal{L}(V,d)$ is a sub-csL-over-$\mathcal{F}(\mathcal{U}M)$ of $\mathcal{F}(N)$?

The Universal CSL for a Chain Complex

There is a "universal" csL for a $\mathbb{Z}_{(p)}$-free chain complex (V,d), denoted $\mathcal{E}\mathcal{L}(V,d)$. It has this property: for any csL $\{L_{[s]}\}$ and any coherent sequence of chain maps $f_s : V^{(\geq s)} \longrightarrow L_{[s]}$, there is a unique extension of $\{f_s\}$ to a csL homomorphism $\{f_{[s]}\} : \mathcal{E}\mathcal{L}(V,d) \longrightarrow \{L_{[s]}\}$. We construct $\mathcal{E}\mathcal{L}(V,d)$ and explore its properties in this unit. We also specialize to consider $\mathcal{E}(N^{kk})$ and $\mathcal{E}(N^{kkk})$.

Let us put this unit in the context of our overall goals. Having removed the troublesome $\tilde{\tau}_j$'s and $\tilde{\sigma}_j$'s from $\mathcal{F}\mathcal{L}(V)$ we obtained $\mathcal{F}^0\mathcal{L}(V)$. The csL $\mathcal{F}^0\mathcal{L}(V)$ has free Bockstein sub-dgL's and (often) has a reasonable virtual structure, which is just what we want when we are interested in applying Proposition 7.5 or Theorem 7.15. But we discover to our chagrin that 7.5 and 7.15 do not work, because one of their hypotheses is that the source csL must be proper, and $\mathcal{F}^0\mathcal{L}(V)$ is not proper (Axiom 7.1(h) fails). We threw out the "baby" (properness) along with the "bath water" ($\tilde{\tau}_j$'s and $\tilde{\sigma}_j$'s). Our recourse is to patch $\mathcal{F}^0\mathcal{L}(V)$ with a new piece that looks remarkably like the piece we removed (the new piece will be the desuspension of the removed piece). When we are finished our new csL, denoted $\mathcal{E}\mathcal{L}(V)$, turns out to have everything: free Bockstein sub-dgL's; the universal property described above, and, at least for the examples we care about, the right virtual structure as well.

We begin with an obvious definition.

Definition. A *short exact sequence of csL's* is a sequence of csL homomorphisms

$$(11DA) \qquad 0 \longrightarrow \{K_{[t]}\} \longrightarrow \{M_{[t]}\} \longrightarrow \{L_{[t]}\} \longrightarrow 0,$$

such that for each $t \geq 1$

$$0 \longrightarrow K_{[t]} \longrightarrow M_{[t]} \longrightarrow L_{[t]} \longrightarrow 0$$

is a short exact sequence of dgL's.

By Lemma 11.B6, any csL surjection $\{M_{[t]}\} \longrightarrow \{L_{[t]}\}$ can be embedded in a short exact sequence (11DA).

Throughout this unit, we continue the convention that (V,d) always denotes a $\mathbb{Z}_{(p)}$-free positively graded chain complex having $d(V) \subseteq pV$. As usual we denote a chain complex decomposition a la (5M) for (V,d) by

$$V = V^{(1)} \oplus V^{(2)} \oplus \cdots \oplus V^{(\infty)}.$$

Recall from Lemma 5.6 the notation

$$(11DB) \qquad W^+(V^{(\geq m)}) = \bigoplus_{\alpha \in I^{(m)}} \mathbb{Z}_{(p)}(y_\alpha, p^{-m}d(y_\alpha)),$$

which determines the indexing set $I^{(m)}$. We will also refer to $\widetilde{W}^m(V^{(\geq m)})$, which for the reader's convenience we recall here as

$$\text{(11DC)} \qquad \widetilde{W}^m(V^{(\geq m)}) = \bigoplus_{\alpha \in I^{(m)}} \bigoplus_{j=1}^{\infty} \mathbb{Z}_{(p)}(\tilde{\tau}_j(y_\alpha), \tilde{\sigma}_j(y_\alpha))$$

We shall also use the module $\widetilde{W}^{(t)}$ from Lemma 11.B11, which we now prefer to write in the form

$$\text{(11DD)} \qquad \widetilde{W}^{(t)} = \bigoplus_{s=1}^{t-1} \bigoplus_{\alpha \in I^{(s)}} \bigoplus_{j=1}^{\infty} \mathbb{Z}_{(p)}(\zeta_t^{s+1}\tilde{\tau}_j(y_\alpha), d_{[t]}\zeta_t^{s+1}\tilde{\tau}_j(y_\alpha)).$$

Lemma 11.B11 tells us that we have a short exact sequence of csL's,

$$\text{(11DE)} \qquad 0 \longrightarrow \mathcal{F}^0\mathcal{L}(V) \longrightarrow \mathcal{FL}(V) \longrightarrow \{\widetilde{W}^{(t)} \otimes \mathbb{Z}_p\}_{t \geq 1} \longrightarrow 0.$$

Let us spend a little while being explicit about the csL structure for the quotient, $\{\widetilde{W}^{(t)} \otimes \mathbb{Z}_p\}_{t\geq 1}$. To begin with, as a Lie algebra over $\mathbb{Z}_p$ each $\widetilde{W}^{(t)} \otimes \mathbb{Z}_p$ is abelian (i.e. has a trivial bracket), and the behavior of $d_{[t]}$ is implicit in the notation (11DD) for our choice of $\mathbb{Z}_p$-basis. We have $\zeta_{t+1}^t(\zeta_t^{s+1}\tilde{\tau}_j(y_\alpha)) = \zeta_{t+1}^{s+1}\tilde{\tau}_j(y_\alpha)$, and $\zeta_{t+1}^t(d_{[t]}\zeta_t^{s+1}\tilde{\tau}_j(y_\alpha)) = 0$ since we are now working over $\mathbb{Z}_p$. In particular,

$$\text{(11DF)} \qquad \ker(d_{[t+1]}\zeta_{t+1}^t) = \ker(\zeta_{t+1}^t) = \bigoplus_{s=1}^{t-1} \bigoplus_{\alpha \in I^{(s)}} \bigoplus_{j=1}^{\infty} \mathbb{Z}_p(d_{[t]}\zeta_t^{s+1}\tilde{\tau}_j(y_\alpha)).$$

Also,

$$\text{(11DG)} \qquad (\widetilde{W}^{(t+1)} \otimes \mathbb{Z}_p)/(\mathrm{im}(\zeta_t^{t-1}) + \mathrm{im}(d_{[t]}\zeta_t^{t-1})) \approx \widetilde{W}^t(V^{(\geq t)}) \otimes \mathbb{Z}_p.$$

Moving on to the behavior of θ_t^{t+1} on this csL, we find $\theta_t^{t+1}\zeta_{t+1}^{s+1}\tilde{\tau}_j(y_\alpha) = 0$ if $s < t$, and $\theta_t^{t+1}(\tilde{\tau}_j(y_\alpha)) = 0$ when $s = t$. Also, $\theta_t^{t+1}d_{[t+1]}\zeta_{t+1}^{s+1}\tilde{\tau}_j(y_\alpha) = d_{[t]}\zeta_t^{s+1}\tilde{\tau}_j(y_\alpha)$ if $s < t$, and $\theta_t^{t+1}d_{[t+1]}\tilde{\tau}_j(y_\alpha) = 0$ when $s = t$. This means that

$$\text{(11DH)} \qquad \ker(\theta_t^{t+1}) \cap \ker(\theta_t^{t+1}d_{[t+1]}) = \widetilde{W}^t(V^{(\geq t)}) \otimes \mathbb{Z}_p,$$

for the csL $\{\widetilde{W}^{(t)} \otimes \mathbb{Z}_p\}_{t\geq 1}$

Lemma 11.D1. *The csL $\{W^{(t)} \otimes \mathbb{Z}_p\}_{t\geq 1}$ satisfies axiom 7.1(h).*

Proof. Since $\mathrm{im}(\zeta_{t+1}^t) \cap \mathrm{im}(d_{[t+1]}\zeta_{t+1}^t) = 0$ for this csL, the only way to get $\zeta_{t+1}^t(x) = d_{[t+1]}\zeta_{[t+1]}^t(y)$ is to have $x \in \ker(\zeta_{t+1}^t)$ and $y \in \ker(d_{[t+1]}\zeta_{t+1}^t)$. By (11DF) this means that there are coefficients $c_{s\alpha j} \in \mathbb{Z}_p$ and $c'_{s\alpha j} \in \mathbb{Z}_p$ such that $x = \sum c_{s\alpha j}d_{[t]}\zeta_t^{s+1}(y_\alpha)$ and $y = \sum c'_{s\alpha j}d_{[t]}\zeta_t^{s+1}(y_\alpha)$. Then

$$z = \sum c_{s\alpha j}\zeta_{t+1}^{s+1}(y_\alpha) + \sum c'_{s\alpha j}d_{[t+1]}\zeta_{t+1}^{s+1}(y_\alpha)$$

has the property that $\theta_t^{t+1}(z) = y$ while $\theta_t^{t+1}d_{[t+1]}(z) = x$.

Axiom 7.1(h) has a small but essential role in the proof of Proposition 7.5 and Theorem 7.15. We use it in order to show that the homomorphism g_1 with domain $\operatorname{im}(\zeta^t_{t+1}) + \operatorname{im}(d_{[t+1]}\zeta^t_{t+1})$ is well-defined. The csL $\mathcal{F}^0\mathcal{L}(V)$ does not satisfy the axiom. Let us take a closer look at 7.1(h), and let us measure a csL's failure to satisfy it.

Definition/Notation. Let $s(\ )$ (resp. $s^{-1}(\ )$) denote the suspension (resp. desuspension) of a graded module M, i.e., $(sM)_m = M_{m-1}$ (resp. $(s^{-1}M)_m = M_{m+1}$). Let $\{L_{[q]}, d_{[q]}\}$ be a csL. Define the degree zero homomorphisms Γ and Ξ in the non-exact sequence

(11DI)
$$0 \longrightarrow L_{[t+1]} \xrightarrow{\Gamma} L_{[t]} \oplus sL_{[t]} \xrightarrow{\Xi} sL_{[t+1]} \longrightarrow 0$$

by

$$\Xi(y,x) = d_{[t+1]}\zeta^t_{t+1}(y) - \zeta^t_{t+1}(x)$$

and

$$\Gamma(z) = (\theta^{t+1}_t(z), \theta^{t+1}_t d_{[t+1]}(z)).$$

It is clear that (11DI) is a chain complex for any csL, i.e. $\Xi\Gamma = 0$. Define the tth ***properness obstruction group*** $\mathcal{B}_{[t]}(\{L_{[q]}\})$ to be the homology of (11DI) at the middle spot, i.e., $\mathcal{B}_{[t]}(\{L_{[q]}\}) = \ker(\Xi)/\operatorname{im}(\Gamma)$.

What Axiom 7.1(h) really says is that $\mathcal{B}_{[*]}(\{A_{[s]}\}) = 0$, i.e., $\mathcal{B}_{[t]}(\{A_{[s]}\}) = 0$ for all t. What Proposition 7.5 actually uses is that $\mathcal{B}_{[t-1]}(\{A_{[s]}\}) = 0$. Lemma 11.D1 is saying that $\mathcal{B}_{[*]}(\{\widetilde{W}^{(t)} \otimes \mathbf{Z}_p\}) = 0$. We know from Lemma 7.2(b) that $\mathcal{B}_{[*]}\mathcal{F}(L,d) = 0$ for any $\mathbf{Z}_{(p)}$-free dgL (L,d).

Lemma 11.D2. *There is an isomorphism* $\mathcal{B}_{[t]}\mathcal{F}^0\mathcal{L}(V) \approx \widetilde{W}^t(V^{(\geq t)}) \otimes \mathbf{Z}_p$, *with the set of pairs* $\{(\tau_j(y_\alpha), \sigma_j(y_\alpha))\}_{j\geq 1, \alpha\in I^{(t)}} \cup \{(\sigma_j(y_\alpha), 0)\}_{j\geq 1,\alpha\in I^{(t)}}$ *serving as a* $\mathbf{Z}_p$*-basis for* $\mathcal{B}_{[t]}\mathcal{F}^0\mathcal{L}(V)$.

Proof. Fix $t \geq 1$, and from (11DI) and (11DE) form the following diagram, in which the rows are short exact sequences of $\mathbf{Z}_{(p)}$-modules and the columns are chain complexes. (Space limitations forced us to omit the zero groups at the beginning and end of each row in (11DJ), but the reader should either draw them in or understand that they are implicitly present.)

(11DJ)
$$\begin{array}{ccccc}
0 & & 0 & & 0 \\
\downarrow & & \downarrow & & \downarrow \\
\mathcal{F}^0\mathcal{L}(V)_{[t+1]} & \longrightarrow & \mathcal{F}\mathcal{L}(V)_{[t+1]} & \longrightarrow & \widetilde{W}^{(t+1)} \otimes \mathbf{Z}_p \\
\downarrow\Gamma & & \downarrow\Gamma & & \downarrow\Gamma \\
\mathcal{F}^0\mathcal{L}(V)_{[t]} \oplus s\mathcal{F}^0\mathcal{L}(V)_{[t]} & \longrightarrow & \mathcal{F}\mathcal{L}(V)_{[t]} \oplus s\mathcal{F}\mathcal{L}(V)_{[t]} & \longrightarrow & (\widetilde{W}^{(t)}\mathbf{Z}_p \oplus s\widetilde{W}^{(t)}) \otimes \mathbf{Z}_p \\
\downarrow\Xi & & \downarrow\Xi & & \downarrow\Xi \\
s\mathcal{F}^0\mathcal{L}(V)_{[t+1]} & \longrightarrow & s\mathcal{F}\mathcal{L}(V)_{[t+1]} & \longrightarrow & s\widetilde{W}^{(t+1)} \otimes \mathbf{Z}_p \ .
\end{array}$$

Take the long exact sequence of homology for (11DJ). Because θ_t^{t+1} is one-to-one on $\mathcal{FL}(V)_{[t+1]}$, Γ is one-to-one on $\mathcal{FL}(V)_{[t+1]}$; since $\mathcal{FL}(V)$ is proper, $\ker(\Xi) = \operatorname{im}(\Gamma)$. The homology along the middle column of (11DJ) vanishes, so the connecting homomorphism

$$(\ker(\Gamma) \text{ for } \widetilde{W}^{(t+1)} \otimes \mathbb{Z}_p) \longrightarrow \mathcal{B}_{[t]}\mathcal{F}^0\mathcal{L}(V)$$

is an isomorphism. The lemma now follows from (11DH).

In view of (11DE), we may think of $\mathcal{FL}(V)$ as being composed of the two "pieces" $\mathcal{F}^0\mathcal{L}(V)$ and $\{\widetilde{W}^{(t)} \otimes \mathbb{Z}_p\}$. We took away the second piece in order to get $\mathcal{F}^0\mathcal{L}(V)$, but in the process we created some obstructions to properness. In order to remove those obstructions, we are going to put the two pieces back together, but differently.

The piece we are going to reinsert differs from $\widetilde{W}^{(t)} \otimes \mathbb{Z}_p$ only in that it is a desuspension. Define

$$\{\widetilde{X}^{(t)}\} = \{s^{-1}(\widetilde{W}^{(t)} \otimes \mathbb{Z}_p)\}$$

as a csL. (Normally we cannot desuspend a dgL and still expect to get a dgL. The one exception to this rule is that you can suspend or desuspend an abelian dgL.) While this is adequate as a definition, we prefer a different notation for the elements of $\widetilde{X}^{(t)}$. Specifically, we replace $\tilde{\tau}_j$ by the symbol $\bar{\xi}_j$. In other words,

(11DK)
$$\widetilde{X}^{(t)} = \bigoplus_{s=1}^{t-1} \bigoplus_{\alpha \in I^{(s)}} \bigoplus_{j=1}^{\infty} \mathbb{Z}_p(\zeta_t^{s+1}\bar{\xi}_j(y_\alpha), d_{[t]}\zeta_t^{s+1}\bar{\xi}(y_\alpha)),$$

where for now $\bar{\xi}_j(y_\alpha)$ is merely a name for a basis element. The calculations we did earlier for $\widetilde{W}^{(t)} \otimes \mathbb{Z}_p$ apply equally to its desuspension $\widetilde{X}^{(t)}$. In particular, (11DF) and (11DG) and Lemma 11.D1 tell us

Lemma 11.D3. *For the csL* $\{X^{(s)}\}_{s \geq 1}$ *we have*

(a) $\ker(\zeta_{t+1}^t) = \ker(d_{[t+1]}\zeta_{t+1}^t) = \bigoplus_{s,\alpha,j} \mathbb{Z}_p(d_{[t]}\zeta_t^{s+1}\bar{\xi}_j(y_\alpha))$.

(b) $\widetilde{X}^{(t+1)}/(\operatorname{im}(\zeta_{t+1}^t) + \operatorname{im}(d_{[t+1]}\zeta_{t+1}^t)) \approx s^{-1}\widetilde{W}^t(V^{(\geq t)}) \otimes \mathbb{Z}_p$.

(c) $\mathcal{B}_{[*]}(\{\widetilde{X}^{(t)}\}) = 0$.

Construction 11.D4. We construct a new csL, denoted $\mathcal{EL}(V,d)$ or simply $\mathcal{EL}(V)$. For each $t \geq 1$, as a differential graded $\mathbb{Z}_{(p)}$-module $\mathcal{EL}(V)_{[t]}$ is the direct sum of $\mathcal{F}^0\mathcal{L}(V)_{[t]}$ and $\widetilde{X}^{(t)}$,

(11DL)
$$\mathcal{EL}(V)_{[t]} = \mathcal{F}^0\mathcal{L}(V)_{[t]} \oplus \widetilde{X}^{(t)}.$$

For $t > 1$ define θ_{t-1}^t on $\mathcal{EL}(V)_{[t]}$ component-wise, i.e. on the piece $\mathcal{F}^0\mathcal{L}(V)_{[t]}$ it is the familiar θ_{t-1}^t for $\mathcal{F}^0\mathcal{L}(V)$ and on $\widetilde{X}^{(t)}$ it is the θ_{t-1}^t for the csL $\{\widetilde{X}^{(s)}\}$. On the component $\widetilde{X}^{(t)}$ we define ζ_{t+1}^t to be what it is for $\{\widetilde{X}^{(s)}\}$, but on the component $\mathcal{F}^0\mathcal{L}(V)_{[t]}$ we make a slight alteration. Let us first review what ζ_{t+1}^t does on the csL $\mathcal{F}^0\mathcal{L}(V)$.

According to (11BZ) we may write $\mathcal{F}^0\mathcal{L}(V)_{[t]}$ as a direct sum,

$$\mathcal{F}^0\mathcal{L}(V)_{[t]} = p\widetilde{W}^{(t)} \oplus [\widetilde{W}^t(V^{(\geq t)}) \oplus \widetilde{W}_{t+1}^{\infty}] \oplus E_{[t]},$$

where $\widetilde{W}^{\infty}_{t+1}$ is a shorthand for $\bigoplus_{m=t+1}^{\infty} \widetilde{W}^m(V^{(\geq m)})$ and $E_{[t]}$ stands for all the other summands in (11BZ). We have similarly

$$\mathcal{F}^0\mathcal{L}(V)_{[t+1]} = [p\widetilde{W}^{(t)} \oplus p\widetilde{W}^t(V^{(\geq t)})] \oplus \widetilde{W}^{\infty}_{t+1} \oplus E_{[t+1]}.$$

In the csL structure for $\mathcal{F}^0\mathcal{L}(V)$, ζ^t_{t+1} respects this decomposition, and it is an isomorphism of $\widetilde{W}^t(V^{(\geq t)})$ with $p\widetilde{W}^t(V^{(\geq t)})$. For this point of view it is best to think of $\widetilde{W}^t(V^{(\geq t)})$ as in (11DC) but to think of $p\widetilde{W}^{(t)} \subseteq \mathcal{F}^0\mathcal{L}(V)_{[t]}$ as (cf. (11DD))

$$\text{(11DM)} \qquad p\widetilde{W}^{(t)} = \bigoplus_{s=1}^{t-1} \bigoplus_{\alpha \in I^{(s)}} \bigoplus_{j=1}^{\infty} \mathbb{Z}_{(p)}(\zeta^s_t \tau_j(y_\alpha), d_{[t]}\zeta^s_t \tau_j(y_\alpha)).$$

In the csL $\mathcal{F}^0\mathcal{L}(V)$, we have $\zeta^t_{t+1}(\tilde{\tau}_j(y_\alpha)) = \zeta^t_{t+1}\tau_j(y_\alpha) \in p\widetilde{W}^{(t+1)}$ and $\zeta^t_{t+1}(\tilde{\sigma}_j(y_\alpha)) = d_{[t+1]}\zeta^t_{t+1}\tau_j(y_\alpha) \in p\widetilde{W}^{(t+1)}$.

To define ζ^t_{t+1} for $\mathcal{EL}(V)_{[t]}$, we let ζ^t_{t+1} act as just described, on all components of $\mathcal{F}^0\mathcal{L}(V)_{[t]}$ except on $\widetilde{W}^t(V^{(\geq t)})$. On that component, we put $\zeta^t_{t+1}(\tilde{\tau}_j(y_\alpha)) = \zeta^t_{t+1}\tau_j(y_\alpha)$ (same as for $\mathcal{F}^0\mathcal{L}(V)$) but on the basis element $\tilde{\sigma}_j(y_\alpha)$ we put

$$\zeta^t_{t+1}(\tilde{\sigma}_j(y_\alpha)) = d_{[t+1]}\zeta^t_{t+1}\tau_j(y_\alpha) - \bar{\xi}_j(y_\alpha).$$

To summarize, on the basis elements $\{\sigma_j(y_\alpha) | j \geq 1, \alpha \in I^{(t)}\}$ of $\mathcal{EL}(V)_{[t]}$ we alter ζ^t_{t+1} by subtracting the perturbation terms $\bar{\xi}_j(y_\alpha)$, but on all other basis elements of $\mathcal{EL}(V)_{[t]}$, ζ^t_{t+1} has the same meaning that it has for the individual components $\mathcal{F}^0\mathcal{L}(V)_{[t]}$ and $\widetilde{X}^{(t)}$.

Lastly, we describe the bracket on $\mathcal{EL}(V)$. Again referring to (11BZ), another way to regroup the summands of (11BZ) is to recognize via Lemma 5.6 that the first, second, and fourth components of (11BZ) add up to $\mathcal{L}(V^{(\geq t)})$ (cf. (5Q)). Consequently

$$\mathcal{EL}(V)_{[t]} = \mathcal{L}(V^{(\geq t)}) \oplus (\bigoplus_{s=1}^{t-1} \tilde{\zeta}^s_t(L^{(s,s)})) \oplus (p\widetilde{W}^{(t)}) \oplus \widetilde{X}^{(t)}.$$

Let $J^{(s)}$ denote an indexing set for which

$$L^{(s,s)} = \bigoplus_{\beta \in J^{(s)}} \mathbb{Z}_{(p)}(z_\beta, p^{-s}d(z_\beta)).$$

Put

$$Y^{(s)} = \mathbb{Z}_{(p)}(\{z_\beta | \beta \in J^{(s)}\} \cup \{\tau_j(y_\alpha) | j \geq 1, \alpha \in I^{(s)}\}).$$

Then ζ^s_t and $d_{[t]}\zeta^s_t$ are one-to-one on $Y^{(s)}$ for $s < t$, and

$$\text{(11DN)} \qquad \mathcal{EL}(V)_{[t]} = \mathcal{L}(V^{(\geq t)}) \oplus \left(\bigoplus_{s=1}^{t-1} \zeta^s_t(Y^{(s)})\right) \oplus \left(\bigoplus_{s=1}^{t-1} d_{[t]}\zeta^s_t(Y^{(s)})\right) \oplus \widetilde{X}^{(t)}.$$

The definition of [,] on $\mathcal{EL}(V)_{[t]}$ is recursive on t. When $t = 1$, we have $\mathcal{EL}(V)_{[1]} = \mathcal{L}(V, d)$, and the bracket is the bracket for $\mathcal{L}(V, d)$. For $t > 1$, assume [,] has been defined on $\mathcal{EL}(V)_{[s]}$ for $s < t$ and refer to (11DN). First, $\mathcal{L}(V^{(\geq t)})$ is a sub-dgL, i.e. on the component $\mathcal{L}(V^{(\geq t)})$ the bracket agrees with the bracket for this free Lie algebra. Second, $\widetilde{X}^{(t)}$ is in the center of $\mathcal{EL}(V)_{[t]}$, i.e. $[\widetilde{X}^{(t)}, \mathcal{EL}(V)_{[t]}] = 0$. Third, for $y \in Y^{(s)}$ and arbitrary $x \in \mathcal{EL}(V)_{[t]}$, define $[\zeta^s_t(y), x]$ to be $\zeta^s_t[y, \theta^t_s(x)]$ and put

$$[d_{[t]}\zeta^s_t(y), x] = d_{[t]}\zeta^s_t[y, \theta^t_s(x)] - (-1)^{|y|}\zeta^s_t[y, \theta^t_s d_{[t]}(x)].$$

Fourth, make [,] antisymmetric, $[x, y] = -(-1)^{|x||y|}[y, x]$.

Lemma 11.D5.

(a) *$\mathcal{EL}(V,d)$ is a csL.*

(b) *When the inclusion and projection implicit in (11DL) are used to rewrite (11DL) in the form*

$$0 \longrightarrow \{\widetilde{X}^{(t)}\} \longrightarrow \mathcal{EL}(V) \longrightarrow \mathcal{F}^0\mathcal{L}(V) \longrightarrow 0, \tag{11DO}$$

then (11DO) is a short exact sequence of csL's.

(c) *In $\mathcal{EL}(V)$ we have $\xi_j(y_\alpha) = \bar{\xi}_j(y_\alpha)$, where ξ_j is the homomorphism given by (11BK).*

Proof. Part (a) is a straightforward but tedious verification of axioms. First show that $\mathcal{EL}(V)$ is a csc. The rest of the check is embedded in an induction on t. Show that $[\ ,\ ]$ is well-defined and satisfies 7.1(f') and 7.1(e'), then that d is a derivation for $[\ ,\ ]$, and lastly that $[\ ,\ ]$ satisfies the Jacobi identity. Parts (b) and (c) are straightforward.

Proposition 11.D6. *(a) $\mathcal{B}_{[*]}\mathcal{EL}(V,d) = 0$, i.e. $\mathcal{EL}(V)$ satisfies Axiom 7.1(h).*

(b) $\mathcal{EL}(V)_{[t+1]}/(\mathrm{im}(\zeta^t_{t+1}) + \mathrm{im}(d_{[t]}\zeta^t_{t+1})) \approx \mathcal{L}(V^{(\geq t+1)}) \otimes \mathbb{Z}_p$.

(c) *$\mathcal{L}(V^{(\geq t)})$ is a Bockstein sub-dgL of $\mathcal{EL}(V)_{[t]}$.*

Proof. Fix $t \geq 1$, and form a diagram like (11DJ) using (11DI) and (11DO). (Once again, we have omitted both zero groups from each row, for space management reasons. Each row of (11DP) should be understood to be a short exact sequence.)

$$\begin{array}{ccccc}
\widetilde{X}^{(t+1)} & \longrightarrow & \mathcal{EL}(V)_{[t+1]} & \longrightarrow & \mathcal{F}^0\mathcal{L}(V)_{[t+1]} \\
\downarrow \Gamma & & \downarrow \Gamma & & \downarrow \Gamma \\
\widetilde{X}^{(t)} \oplus s\widetilde{X}^{(t)} & \longrightarrow & \mathcal{EL}(V)_{[t]} \oplus s\mathcal{EL}(V)_{[t]} & \longrightarrow & \mathcal{F}^0\mathcal{L}(V)_{[t]} \oplus s\mathcal{F}^0\mathcal{L}(V)_{[t]} \\
\downarrow \Xi & & \downarrow \Xi & & \downarrow \Xi \\
s\widetilde{X}^{(t+1)} & \longrightarrow & s\mathcal{EL}(V)_{[t+1]} & \longrightarrow & s\mathcal{F}^0\mathcal{L}(V)_{[t+1]} \\
\downarrow & & \downarrow & & \downarrow \\
0 & & 0 & & 0
\end{array} \tag{11DP}$$

Take the long exact sequence of homology for (11DP). We obtain the long exact sequence

$$\begin{aligned}
&\mathcal{B}_{[t]}(\{\widetilde{X}^{(s)}\}) \longrightarrow \mathcal{B}_{[t]}\mathcal{EL}(V) \longrightarrow \mathcal{B}_{[t]}\mathcal{F}^0\mathcal{L}(V) \xrightarrow{\ \delta\ } sQ^t(\{\widetilde{X}^{(s)}\}) \\
&\longrightarrow sQ^t\mathcal{EL}(V) \longrightarrow sQ^t\mathcal{F}^0\mathcal{L}(V) \longrightarrow 0,
\end{aligned} \tag{11DQ}$$

where the notation $Q^t(\{L_{[t]}\})$ for a csL $\{L_{[t]}\}$ means

$$L_{[t+1]}/(\mathrm{im}(\zeta^t_{t+1}) + \mathrm{im}(d_{[t+1]}\zeta^t_{t+1})).$$

The first group in (11DQ) is zero, by Lemma 11.D3(c). By Lemmas 11.D2 and 11.D3(b) the third and fourth groups in (11DQ) are isomorphic. It is easy to check that the connecting homomorphism δ is this isomorphism. (The pair $(\tau_j(y_\alpha), \sigma_j(y_\alpha))$ pulls back to

$(\tau_j(y_\alpha), \sigma_j(y_\alpha))$, which projects via Ξ to $d_{[t+1]}\zeta^t_{t+1}\tau_j(y_\alpha) - \zeta^t_{t+1}\sigma_j(y_\alpha) = \xi_j(y_\alpha) = \bar{\xi}_j(y_\alpha)$ by 11.D5(c), which pulls back to $\bar{\xi}_j(y_\alpha)$. Similarly we compute that $\delta((\sigma_j(y_\alpha), 0)) = d_{[t+1]}\bar{\xi}_j(y_\alpha)$.) It follows that the second group in (11DQ) is zero, completing part (a).

The fact that δ is an isomorphism in (11DQ) also tells us that

$$Q^t\mathcal{EL}(V) \xrightarrow{\approx} Q^t\mathcal{F}^0\mathcal{L}(V).$$

By Lemma 11.B10 we have

$$Q^t\mathcal{F}^0\mathcal{L}(V) = \mathcal{L}(V^{(\geq t+1)})/(p)\mathcal{L}(V^{(\geq t+1)}).$$

Part (b) follows at once. Part (c) is trivial for $t = 1$ and for $t \geq 2$ it follows from part (b) as soon as we recall that $\mathcal{L}(V^{(\geq t+1)})$ is indeed a sub-dgL of $\mathcal{EL}(V)_{[t+1]}$, by the construction of the bracket on $\mathcal{EL}(V)_{[t+1]}$.

Notice that Proposition 11.D6 does not assert that $\mathcal{EL}(V)$ is proper. Axiom 7.1(g) fails for $\mathcal{EL}(V)$. Fortunately the failure of 7.1(g) is immaterial in this case. In any csL $\{L_{[q]}\}$ we always have

$$\text{(11DR)} \qquad d_{[t]}\zeta^{s+1}_t\xi_j(L_{[s]}) \subseteq \ker(\zeta^t_{t+1})$$

because $\zeta^t_{t+1}d_{[t]}\zeta^{s+1}_t\xi_j = 0$ by (11BL) and 11.B3(a), so true properness makes the unreasonable demand that $d_{[t]}\zeta^{s+1}_t\xi_j$ vanish identically. Define a csL to be ***semi-proper*** if Axiom 7.1(h) holds and

$$\text{(11DS)} \qquad \ker(\zeta^t_{t+1}) = \sum_{j=1}^{\infty}\sum_{s=1}^{t-1} \operatorname{im}(d_{[t]}\zeta^{s+1}_t\xi_j)$$

for each $t \geq 1$. In view of (11DR), semi-properness says that $\ker(\zeta^t_{t+1})$ is only as large as the csL axioms force it to be. Reviewing the proofs of 7.5 and 7.15, Axiom 7.1(g) is used only once, to show that $\zeta^t_{t+1}(x) = \zeta^t_{t+1}(y)$ implies $\zeta^t_{t+1}f_{[t]}(x) = \zeta^t_{t+1}f_{[t]}(y)$ if $\{f_{[s]}\}_{s\leq t}$ is a csL homomorphism. But if $x - y \in \sum_{s,j}\operatorname{im}(d_{[t]}\zeta^{s+1}_t\xi_j)$ then $f_{[t]}(x-y) \in \operatorname{im}(d_{[t]}\zeta^{s+1}_t\xi_j f_{[s]}) \subseteq \ker(\zeta^t_{t+1})$ because (11DR) is true for the target csL. This argument shows

Lemma 11.D7. *Proposition 7.5, Corollary 7.6, and Theorem 7.15 are still true, if the source csL is assumed to be semi-proper instead of proper.*

Lemma 11.D8. *$\mathcal{EL}(V)$ is semi-proper.*

Proof. By Proposition 11.D6(a) we need only to compute $\ker(\zeta^t_{t+1})$. From Lemma 11.D5(b) we have the commuting diagram

$$\text{(11DT)} \qquad \begin{array}{ccccccccc} 0 & \longrightarrow & \widetilde{X}^{(t)} & \longrightarrow & \mathcal{EL}(V)_{[t]} & \longrightarrow & \mathcal{F}^0\mathcal{L}(V)_{[t]} & \longrightarrow & 0 \\ & & \downarrow{\scriptstyle \zeta^t_{t+1}} & & \downarrow{\scriptstyle \zeta^t_{t+1}} & & \downarrow{\scriptstyle \zeta^t_{t+1}} & & \\ 0 & \longrightarrow & \widetilde{X}^{(t+1)} & \longrightarrow & \mathcal{EL}(V)_{[t+1]} & \longrightarrow & \mathcal{F}^0\mathcal{L}(V)_{[t+1]} & \longrightarrow & 0, \end{array}$$

in which the rows are exact and the right-most vertical arrow is a monomorphism. Then

$$\ker(\zeta^t_{t+1} \text{ for } \mathcal{EL}(V)) = \operatorname{im}(\ker(\zeta^t_{t+1} \text{ for } \{\widetilde{X}^{(s)}\})).$$

From Lemmas 11.D3(a) and 11.D5(c) we get (11DS) for $\mathcal{EL}(V)$.

Proposition 11.D9. *$\mathcal{EL}(V)$ is the free csL generated by $\mathcal{F}(V)$. I.e., given any csL $\{L_{[t]}, e_{[t]}\}$ and any csc homomorphism $\{f'_{[t]}\} : \mathcal{F}(V) \longrightarrow \{L_{[t]}\}$ there is a unique extension $\{f_{[t]}\} : \mathcal{EL}(V) \longrightarrow \{L_{[t]}, e_{[t]}\}$ that is a csL homomorphism. Equivalently, given any collection of chain maps*

$$f_s : (V^{(\geq s)}, p^{-s}d) \longrightarrow (L_{[s]}, e_{[s]})$$

that satisfy $f_s i = \theta_s^{s+1} f_{s+1} : V^{(\geq s+1)} \longrightarrow L_{[s]}$, where $i : V^{(\geq s+1)} \longrightarrow V^{(\geq s)}$ is the inclusion, there is a unique csL homomorphism

$$\{f_{[t]}\} : \mathcal{EL}(V) \longrightarrow \{L_{[t]}, e_{[t]}\}$$

for which $f_{[t]}\big|_{V^{(\geq t)}} = f_t$.

Proof. We have seen that $\mathcal{EL}(V)$ is semi-proper and that $\mathcal{L}(V^{(\geq t)})$ is a Bockstein sub-dgL of $\mathcal{EL}(V)_{[t]}$, for every t. Assuming $f_{[s]}$ has been constructed for $s < t$, take the homomorphism g in Proposition 7.5 to be $g = f_t$ (or $g = f'_{[t]}\big|_{V^{(\geq t)}}$). This g extends uniquely to $g : \mathcal{L}(V^{(\geq t)}) \longrightarrow L_{[t]}$, and this extension satisfies (7AG) and (7AH) on the generating submodule $V^{(\geq t)}$. Apply Proposition 7.5, with an assist from Lemma 11.D7.

Remark 11.D10. Although $\mathcal{EL}(V)$ is indeed the free csL on a $\mathbb{Z}_{(p)}$-free chain complex, it is not itself $\mathbb{Z}_{(p)}$-free! By (11DL) all of $\widetilde{X}^{(t)}$ is a p-torsion component of $\mathcal{EL}(V)_{[t]}$. If $V^{(m)} \neq 0$ for some $m \geq 1$, then $\mathcal{EL}(V)_{[t]}$ has p-torsion for all $t \geq m+1$. This torsion is apparently built into the csL axioms. If we considered only those csL's which satisfy the additional axiom that $\xi_j \equiv 0$ for all j, then $\mathcal{F}^0\mathcal{L}(V)$ would be the universal object for this smaller category of csL's.

Proposition 11.D11. *$\mathcal{EL}$ is a functor, on the category of $\mathbb{Z}_{(p)}$-free positively graded chain complexes (V, d) satisfying $d(V) \subseteq pV$, and (11DO) consists of natural transformations between functors from this category to the category CSL of csL's.*

Proof. A chain map $f : V \longrightarrow U$ induces the csc homomorphism $\mathcal{F}(f) : \mathcal{F}(V) \longrightarrow \mathcal{F}(U)$. By Proposition 11.D9, $\mathcal{F}(f)$ induces a canonical csL homomorphism that we may denote as $\mathcal{EL}(f) : \mathcal{EL}(V) \longrightarrow \mathcal{EL}(U)$. It is clearly functorial, since $\mathcal{F}$ is. We have already seen that $\mathcal{F}^0\mathcal{L}$ is a functor, and the quotient homomorphism $\mathcal{EL}(V) \longrightarrow \mathcal{F}^0\mathcal{L}(V)$ consists of dividing out by the ideal $\langle \xi_*(S) \rangle$ (notation from (11BS)), where S consists of all conceivable triples. This quotient homomorphism is clearly natural, hence its csL kernel is functorial also.

Notation. Let

(11DU) $$\nu = \nu(V) : \mathcal{EL}(V) \longrightarrow \mathcal{F}^0\mathcal{L}(V)$$

denote the natural csL surjection of (11DO) and Proposition 11.D11.

We now focus on the two csL's of particular interest for our intended geometric application.

Notation 11.D12. Let $V^{kl} = g_{kl}(\mathcal{U}M_+^k \otimes \mathcal{U}M_+^l)$ as in Proposition 11.C9 and let $V^{kkk} = fg(U)$ be as in Theorem 11.C10. Since $N^{kl} = \mathcal{L}(V^{kl}, d)$ and $N^{kkk} = \mathcal{L}(V^{kkk}, d)$, the notations

$$\mathcal{E}(N^{kl}) = \mathcal{E}\mathcal{L}(V^{kl}, d),\ \mathcal{E}(N^{kkk}) = \mathcal{E}\mathcal{L}(V^{kkk}, d) \tag{11DV}$$

are reasonable.

When $\mathcal{F}^0\mathcal{L}(V)$ is a csL over a Dcsa, an important question is whether a virtual structure for $\mathcal{F}^0\mathcal{L}(V)$ can be lifted through $\nu(V)$ to a virtual structure for $\mathcal{E}\mathcal{L}(V)$. In general the hard part seems to be to get $\mathcal{E}\mathcal{L}(V)$ to be a csL over the Dcsa. If this csL-over structure lifts, then it is nearly automatic that the virtual structure lifts as well. It is very plausible that a csL/csm structure would lift through ν, since the kernel of ν is the very small central csL $\langle \xi_*(S) \rangle$, as noted in the proof of Proposition 11.D11.

The next theorem will list a number of properties of $\mathcal{E}(N^{kk})$ and $\mathcal{E}(N^{kkk})$ as csL's over $\mathcal{F}(\mathcal{U}M^k \amalg \mathcal{U}M^k)$ and $\mathcal{F}(\mathcal{U}M^k \amalg \mathcal{U}M^k \amalg \mathcal{U}M^k)$ respectively. They are essentially the lifted versions of the properties listed in Proposition 11.C9 and Theorem 11.C10. We omit the proofs, except to give very occasional hints. As usual, the proof requires us to verify all the properties together in a double induction framework.

Theorem 11.D13.

(a) *$\mathcal{E}(N^{kk})$ is a csL over $\mathcal{F}(\mathcal{U}M^k \amalg \mathcal{U}M^k)$, and the virtual structure described in 7.17 and summarized in 7.18 is a virtual structure for it. The natural homomorphism ν of (11DU) is a homomorphism of csm's. (Proof uses Remark 11.C6 and 7.9.) Since we have lifted V^{kk}, the notation*

$$g_{kk} : (R_s^k)_+ \otimes (R_s^k)_+ \longrightarrow \mathcal{E}(N^{kk})_{[t]} \tag{11DW}$$

makes sense, where $s = \max(0, t - r)$.

(b) *The inclusion $\bar{i} : N^{k-1,k-1} \longrightarrow N^{kk}$ for $k \geq 1$ induces a csL/csm homomorphism*

$$\mathcal{E}(\bar{i}) : \mathcal{E}(N^{k-1,k-1}) \longrightarrow \mathcal{E}(N^{kk}) \tag{11DX}$$

which commutes with ν, i.e.,

$$\nu\mathcal{E}(\bar{i}) = \mathcal{F}^0(\bar{i})\nu,$$

where $\mathcal{F}^0(\bar{i})$ is described in 11.C10(a). (Proof uses 11.D9 and 7.14).

(c) *Theorem 11.C10(c) can easily be extended to the slightly stronger assertion that there are homomorphisms of csm's over $\mathcal{F}(\mathcal{U}M^k \amalg \mathcal{U}M^k)$,*

$$\mathrm{ad}^{(2)}(\ )e^{(1)}(\ ) : \mathcal{F}(\mathcal{U}M_+^k \otimes \overline{L}^k) \longrightarrow \mathcal{F}^0(N^{kk});$$
$$\mathrm{ad}^{(1)}(\ )e^{(2)}(\ ) : \mathcal{F}(\mathcal{U}M_+^k \otimes \overline{L}^k) \longrightarrow \mathcal{F}^0(N^{kk}).$$

Both of these lift through ν to csm homomorphisms over $\mathcal{F}(\mathcal{U}M^k \amalg \mathcal{U}M^k)$, for which we employ the same notation:

$$\mathrm{ad}^{(2)}(\ )e^{(1)}(\ ) : \mathcal{F}(\mathcal{U}M_+^k \otimes \overline{L}^k) \longrightarrow \mathcal{E}(N^{kk});$$
$$\mathrm{ad}^{(1)}(\ )e^{(2)}(\ ) : \mathcal{F}(\mathcal{U}M_+^k \otimes \overline{L}^k) \longrightarrow \mathcal{E}(N^{kk}).$$

(d) *In particular, there exist lifts to $\mathcal{E}(N^{kk})_{[t]}$ of the elements $w_s^{12},\ldots,z_s^{12},w_s^{21},\ldots,z_s^{21}$ which make the equations(7DK)–(7DN) true as equations in $\mathcal{E}(N^{kk})_{[t]}$, where $s=\max(0,t-r)$ and $t\le r+k$. We keep the same notations $w_s^{12},\ldots,z_s^{21}$ for the lifts of these elements. The virtual syzygies for $\mathcal{E}(N^{kk})$ are derivable from these equations and from(7DG)–(7DH) and the virtual relator equations, just as they are for $\mathcal{F}^0(N^{kk})$. (To prove this at $t=r+k$ derive it from parts (h) and (i) at $t=r+k-1$.)*

(e) *$\mathcal{E}(N^{kkk})$ is a csL over $\mathcal{F}(\mathcal{U}M^k\amalg\mathcal{U}M^k\amalg\mathcal{U}M^k)$, and the virtual structure described in 7.19 and summarized in 7.20 is a virtual structure for it. The homomorphism*

$$\nu:\ \mathcal{E}(N^{kkk})\longrightarrow\mathcal{F}^0(N^{kkk})$$

of (11DU) is a homomorphism of csm's as well as csL's.

(f) *For $k\ge 1,\bar{i}:\ N^{k-1,k-1,k-1}\longrightarrow N^{kkk}$ induces a csL/csm homomorphism*

$$\mathcal{E}(\bar{i}):\ \mathcal{E}(N^{k-1,k-1,k-1})\longrightarrow\mathcal{E}(N^{kkk}) \tag{11DY}$$

which commutes with ν, i.e.,

$$\nu\mathcal{E}(\bar{i})=\mathcal{F}^0(\bar{i})\nu,$$

where $\mathcal{F}^0(\bar{i})$ is described in 11.C10(d).

(g) *Theorem 11.C10(f) can be extended to the slightly stronger assertion that there are three homomorphisms of csm's over $\mathcal{F}(\mathcal{U}M^k\amalg\mathcal{U}M^k\amalg\mathcal{U}M^k)$, from $\mathcal{F}(\mathcal{U}M_+^k\otimes N^{kk})$ to $\mathcal{F}^0(N^{kkk})$ (cf. part (c)). Each of these lifts to a csm homomorphism, for which we keep the same notations:*

$$\mathrm{ad}^{(1)}(\ )e^{(23)}(\ ):\ \mathcal{F}(\mathcal{U}M_+^k\otimes N^{kkk})\longrightarrow\mathcal{E}(N^{kkk});$$
$$\mathrm{ad}^{(2)}(\ )e^{(13)}(\ ):\ \mathcal{F}(\mathcal{U}M_+^k\otimes N^{kk})\longrightarrow\mathcal{E}(N^{kkk});$$
$$\mathrm{ad}^{(3)}(\ )e^{(12)}(\ ):\ \mathcal{F}(\mathcal{U}M_+^k\otimes N^{kk})\longrightarrow\mathcal{E}(N^{kkk}).$$

(h) *There exist lifts to $\mathcal{E}(N^{kkk})$ of the elements $w_s^0,w_s^1,x_s^1,y_s^{13},z_s^{13}$, etc. which make (7EC), (7EF), (7EG), and the other five Jacobi identities true, as equations in $\mathcal{E}(N^{kkk})_{[t]}$, where $t\le r+k$ and $s=\max(0,t-r)$. The virtual syzygies for $\mathcal{E}(N^{kkk})$ are deducible from these equations and from the virtual relator equations and from(7DG), (7DH), (7DK)–(7DN), just as they are for $\mathcal{F}^0(N^{kkk})$.*

(i) *The folding homomorphism $\nabla^{(12)}:\ N^{kkk}\longrightarrow N^{kk}$ defined by(7EH) induces a homomorphism of csL/csm's over $\mathcal{F}(\mathcal{U}M^k\amalg\mathcal{U}M^k\amalg\mathcal{U}M^k)$,*

$$\mathcal{E}(\nabla^{(12)}):\ \mathcal{E}(N^{kkk})\longrightarrow\mathcal{E}(N^{kk}). \tag{11DZ}$$

$\mathcal{E}(\nabla^{(12)})$ commutes with ν, i.e.,

$$\nu\mathcal{E}(\nabla^{(12)})=\mathcal{F}^0(\nabla^{(12)})\nu,$$

where $\mathcal{F}^0(\nabla^{(12)})$ is described in 11.C10(h). The same is true of $\nabla^{(13)}$ and $\nabla^{(23)}$.

Discussion of the Homotopy CSL

In this unit we deal with the last snag mentioned in Discussion 11.A1, namely, the occurrence of a non-zero kernel for $(\Omega f)_*h$ in (6H). We first consider some substitutions that may be made for some of the hypotheses in Theorem 6.4. We look at the possibility of replacing $\pi_*(\Omega X;\mathbb{Z}_{p^{t+1}})$ by a suitable quotient of $\pi_*(\Omega X;\mathbb{Z}_{p^{t+1}})$, in certain applications. After some technical computations for two wedges of Moore spaces, we finish with some discussion and open questions regarding the homotopy csL.

Lemma 11.E1. *Let q, t, and $X \xrightarrow{f} Y \xrightarrow{g} Z$ be as in Theorem 6.4. The hypotheses of Theorem 6.4 may be modified in the following ways.*

(a) *The hypotheses that (6H) be exact at $p^t\pi_{q-2}(\Omega X)$ and that $\pi\delta^t(\hat{z}) = 0$ can be replaced by the pair of hypotheses*

(11EA) $$p^{t+1}\pi_j(X) = 0 \text{ for } j = q-1 \text{ and } j = q$$

and

(11EB) $$\hat{z}\omega^t\rho^{t+1} = 0.$$

Then the conclusion of Theorem 6.4 remains true. Condition (11EB) is implied by: there exists $\hat{z}' \in \pi_q(X; \mathbb{Z}_{p^{t+1}})$ such that ${}_\pi\delta^{t+1}(z') = 0$ and $\hat{z}'\omega_t^{t+1} = \hat{z}$.

(b) *The hypotheses that (6H) be exact at $p^t\pi_{q-1}(\Omega X)$ and that ${}_\pi\delta^t(\hat{y}) = p\hat{z}$ can be replaced by (11EA) together with*

(11EC) $$\hat{y}\omega^t\rho^{t+1} = \hat{z}\rho^t_{t+1},$$

but in the conclusion of Theorem 6.4 we may no longer have uniqueness.

Proof. In csL language (11EB) says

(11ED) $$d_{[t+1]}\zeta^t_{t+1}(\hat{z}) = 0$$

and (11EC) says

(11EE) $$d_{[t+1]}\zeta^t_{t+1}(\hat{y}) = \zeta^t_{t+1}(\hat{z}).$$

For (a), exactness of (6H) at $p^t\pi_{q-2}(\Omega X)$ is used only in the first paragraph of the proof of 6.4, in order to establish that $\hat{z}\omega^t = 0$. We have instead

$$\hat{z}\omega^t \in \ker(\rho^{t+1})^\# = p^{t+1}\pi_{q-1}(X) = 0.$$

If $\hat{z} = (\omega_t^{t+1})^\#(\hat{z}')$ then $\hat{z}\omega^t\rho^{t+1} = \hat{z}'\omega_t^{t+1}\omega^t\rho^{t+1} = (p)\hat{z}'\omega^{t+1}\rho^{t+1} = (p)_\pi\delta^{t+1}(\hat{z}') = 0$. For (b), in the existence portion of the proof of 6.4, exactness of (6H) at $p^t\pi_{q-1}(\Omega X)$ is used only during the third paragraph, to get $\hat{x} = 0$. We have instead

$$\hat{x}\rho^{t+1} = \hat{y}\omega^t\rho^{t+1} - (p)\hat{z}_2\rho^{t+1} = \hat{z}\rho^t_{t+1} - \hat{z}_2\rho^t\rho^t_{t+1} = 0,$$

so $\hat{x} \in \ker(\rho^{t+1})^\# = p^{t+1}\pi_q(X) = 0$.

Lemma 11.E2. *Suppose that all the hypotheses of Theorem 6.4 are satisfied except that*

$$K_* = \ker((\Omega f)_* h) : \; p^t\pi_*(\Omega X) \longrightarrow p^t H_*(\Omega Y))$$

is not assumed to be zero in any dimensions. Let $J \subseteq \pi_*(\Omega X; \mathbb{Z}_{p^{t+1}})$ be a submodule satisfying these conditions:

(i) *J is closed under ${}_\pi\delta^{t+1}$;*

(ii) *$(\omega_t^{t+1})^\#(J) = 0$;*

(iii) *$K_* \subseteq (\omega^{t+1})^\#(J) \subseteq \ker((\Omega f)_* h)$, in dimensions $q-1$ and q. Then the uniqueness argument of Theorem 6.4 is valid, modulo J. I.e., if both $\hat{y}_2$ and $\hat{y}_2'$ satisfy the conditions in Theorem 6.4, then $\hat{y}_2 - \hat{y}_2' \in J$.*

Proof. Let $\hat{\pi}_*(\Omega X;\mathbb{Z}_{p^t}) = \pi_*(\Omega X;\mathbb{Z}_{p^t})$, but let the notation $\hat{\pi}_*(\Omega X;\mathbb{Z}_{p^{t+1}})$ denote $\pi_*(\Omega X;\mathbb{Z}_{p^{t+1}})/J$ and let $\hat{\pi}_*(\Omega X)$ signify $\pi_*(\Omega X)/(\omega^{t+1})^{\#}(J)$. The idea is, that the uniqueness proof remains valid if π_* is everywhere replaced by $\hat{\pi}_*$. By (iii), $\underline{a}_f$ is still well-defined on $\hat{\pi}_*(\Omega X)$, and (6H) becomes exact in dimensions $q-1$ and q when $\hat{\pi}_*$ replaces π_*. The exactness of

$$\hat{\pi}_j(\Omega X) \xrightarrow{p^t} \hat{\pi}_j(\Omega X) \xrightarrow{(\rho^t)^{\#}} \hat{\pi}_j(\Omega X;\mathbb{Z}_{p^t}), \quad j = q-1 \text{ or } q,$$

and of

$$\hat{\pi}_q(\Omega X) \xrightarrow{(\rho^{t+1})^{\#}} \hat{\pi}_q(\Omega X;\mathbb{Z}_{p^{t+1}}) \xrightarrow{(\omega^{t+1})^{\#}} \hat{\pi}_{q-1}(\Omega X)$$

are left as easy exercises. These are all of the facts about homotopy groups that are used in the uniqueness proof, so the proof goes through for $\hat{\pi}_*$.

Lemma 11.E3. *For any space X, the homotopy csc $\{\pi_*(X;\mathbb{Z}_{p^t})\}$ satisfies Axiom 7.1(h).*

Proof. Consider the homotopy push-out diagram,

$$\text{(11EF)}\qquad \begin{array}{ccccc} P^m(p^{t+1}) & \xrightarrow{\rho^{t+1}} & S^m & \xrightarrow{\omega^t} & P^{m+1}(p^t) \\ \downarrow{\scriptstyle \rho^t_{t+1}} & & \downarrow{\scriptstyle p} & & \downarrow{\scriptstyle \omega^{t+1}_t} \\ P^m(p^t) & \xrightarrow{\rho^t} & S^m & \xrightarrow{\omega^{t+1}} & P^{m+1}(p^{t+1}). \end{array}$$

Let $y \in \pi_{m+1}(X;\mathbb{Z}_{p^t})$ be viewed as a map from the upper right corner of (11EF) to X, and think of $x \in \pi_m(X;\mathbb{Z}_{p^t})$ as a map from the lower left corner of (11EF) to X. The condition

$$d_{[t+1]}\zeta^t_{t+1}(y) = \zeta^t_{t+1}(x)$$

is precisely the condition that the diagram (11EF) together with the maps x and y commutes. The homotopy push-out property gives us a map $z \in \pi_{m+1}(X;\mathbb{Z}_{p^{t+1}})$. It satisfies $y = z\omega^{t+1}_t = \theta^{t+1}_t(z)$ and $x = \rho^t\omega^{t+1}_t(z) = \theta^{t+1}_t d_{[t+1]}(z)$.

Remark. The homotopy csc $\{\pi_*(X;\mathbb{Z}_{p^s})\}$ is not proper, nor, when X is a loop space, is it semi-proper as a csL. Instead, $\ker(\zeta^t_{t+1})$ equals in $\operatorname{im}(d_{[t]}\zeta^1_t)$, which is as small as the kernel of ζ^t_{t+1} can be for a csc $\{L_{[s]}\}$ that satisfies $(p)L_{[1]} = 0$.

In Chapter 12 we will encounter a space X (it is denoted D_{kk} in Chapter 12) having the following properties. It is a wedge of Moore spaces, $X = D_{kk} \in \mathcal{W}^{r+k}_r$, so we may write $X = X^{(r)} \vee X^{(r+1)} \vee \ldots \vee X^{(r+k)}$, where $X^{(j)}$ is a finite type bouquet of mod p^j Moore spaces of various dimensions. For each s, $0 \le s \le k$,

$$\text{(11EG)}\qquad X^{(r+s)} = P^{4np^s}(p^{r+s}) \ \vee\ P^{4np^s+1}(p^{r+s}) \ \vee \text{(more highly connected Moore spaces)}$$

Let $\hat{v}$ denote the inclusion of the unique summand $P^{4np^{k-1}+1}(p^{r+k-1})$ into X. Looking in dimensions near $m = 4np^k$, we find (cf. (6K)) that

$$\text{(11EH)}\qquad K_j = \ker(h:\ p^{r+k-1}\pi_j(\Omega X) \longrightarrow p^{r+k-1}H_j(\Omega X))$$

is given by

$$\text{(11EI)}\qquad K_m = 0;\ K_{m-1} = 0;\ K_{m-3} = 0;\ K_{m-2} \text{ is the } \mathbf{Z}_p\text{-vector space spanned by } p^{r+k-2}\tau_1(\hat{v}^a)\omega^{r+k-1}.$$

The information in (11EI) is an easy computation based upon the same principles as Lemma 10.2. The tiny non-zero kernel K_{m-2}, a single copy of $\mathbf{Z}_p$, is difficult to work around, and it has caused for the author no small amount of exasperation.

Later in Chapter 12 we shall encounter another wedge X of Moore spaces, denoted D_{kkk}. We have $X = D_{kkk} \in \mathcal{W}_r^{r+k}$ as well, and its components $X^{(j)}$ satisfy

$$\text{(11EJ)}\qquad X^{(r+s)} = P^{6np^s-1}(p^{r+s})^{(\vee 2)} \vee P^{6np^s}(p^{r+s})^{(\vee 4)} \vee P^{6np^s+1}(p^{r+s})^{(\vee 2)} \vee \text{(more highly connected Moore spaces)},$$

for $0 \le s \le k$. Here $Y^{(\vee i)}$ means a bouquet of i copies of Y. Let $\hat{v}_1, \hat{v}_2$ denote the inclusions into X of the two copies of $P^{6np^{k-1}+1}(p^{r+k-1})$. Letting K_j be the kernel (11EH) for this X, and setting $m = 6np^k$, we find

$$\text{(11EK)}\qquad K_{m-i} = 0 \text{ for } i = 0,1,3,4;\ K_{m-2} \text{ is the } \mathbf{Z}_p\text{-vector space of rank two spanned by } p^{r+k-2}\tau_1(\hat{v}_1^a)\omega^{r+k-1} \text{ and } p^{r+k-2}\tau_1(\hat{v}_2^a)\omega^{r+k-1}$$

(Note: the summands $P^{6np^{k-1}-1}(p^{r+k-1})$ do contribute to K_{m-2p-2} but not to K_{m-i} for $0 \le i \le 4$.) This non-zero kernel K_{m-2} causes much less trouble than the previous one, because we are only interested in the uniqueness aspect of Theorem 6.4 for $X = D_{kkk}$, whereas for $X = D_{kk}$ we want the existence aspect too. For $X = D_{kkk}$, the module

$$\text{(11EL)}\qquad J = p^{r+k-2}\zeta_{r+k}^{r+k-1}\,\mathrm{Span}\{\tau_1(\hat{v}_1^a), \sigma_1(\hat{v}_1^a), \tau_1(\hat{v}_2^a), \sigma_1(\hat{v}_2^a)\}$$

satisfies conditions (i)–(iii) of Lemma 11.E2. This is a special case of a very general result.

Lemma 11.E4. *Let $\{y_\alpha\} \subseteq \pi_*(\Omega X; \mathbf{Z}_{p^t})$ be any set of even-dimensional homotopy classes. The module*

$$J = \mathrm{Span}\{p^{t-1}\zeta_{t+1}^t\tau_1(y_\alpha), p^{t-1}\zeta_{t+1}^t\sigma_1(y_\alpha)\}$$

satisfies ${}_\pi\delta^{t+1}(J) \subseteq J$ and $\theta_t^{t+1}(J) = 0$ and $(\omega^{t+1})^\#(J) \subseteq \ker(h)$.

Proof. We reduce immediately to the universal case, which is $X = P^{2q+1}(p^t)$ and $y_\alpha = 1^a \in \pi_{2q}(\Omega P^{2q+1}(p^t); \mathbf{Z}_{p^t})$. We have $\theta_t^{t+1}(J) \subseteq p^t\pi_*(\Omega P^{2q+1}(p^t); \mathbf{Z}_{p^t}) = 0$, and

$$h(\omega^{t+1})^\#(J) = p^{t-1}h(\mathrm{Span}(\tau_1(1^a)\omega^t, \sigma_1(1^a)\omega^t)).$$

Since ${}_\pi\delta^t\tau_1(1^a)$ and ${}_\pi\delta^t\sigma_1(1^a)$ are both divisible by p, $\tau_1(1^a)\omega^t$ and $\sigma_1(1^a)\omega^t$ are divisible by p, and then

$$h(\omega^{t+1})^\#(J) \subseteq p^tH_*(\Omega P^{2q+1}(p^t)) = 0.$$

When $t > 1$, $(p)\xi_1(1^a) = 0$ by Lemma 11.B3(a), and $(p)d_{[t+1]}\xi_1(1^a) = 0$, so multiply by p^{t-2} to get $p^{t-1}d_{[t+1]}\zeta_{t+1}^t\tau_1(1^a) = p^{t-1}\zeta_{t+1}^t\sigma_1(1^a)$ and $p^{t-1}d_{[t+1]}\zeta_{t+1}^t\sigma_1(1^a) = 0$. When $t = 1$ we require the additional fact that

$$\text{(11EM)}\qquad \xi_1 = 0,$$

i.e., ξ_1 is identically zero as a homomorphism, in the homotopy csL for any loop space. The identity (11EM) will be proved in Chapter 13. We begin next a discussion of this identity and some related issues.

Discussion 11.E5. What can we say about $\tau_j(y), \sigma_j(y)$, and $\xi_j(y)$, for $j \geq 1$? Let $1^a \in \pi_{2q}(\Omega P^{2q+1}(p^t); \mathbf{Z}_{p^t})$ be the universal example. The fact that ${}_\pi\delta^t\tau_j(1^a) = (p)\sigma_j(1^a)$ means that $\tau_j(1^a)\omega^t\rho^1 = (p)\sigma_j(1^a)\omega_1^t \in (p)\pi_*(\Omega P^{2q+1}(p^t); \mathbf{Z}_p) = 0$. This says that $\tau_j(1^a) \in \ker(\omega^t\rho^1)^\#$, and $\ker(\omega^t\rho^1)^\# = \operatorname{im}(\omega_t^{t+1})^\#$ by (6C), i.e., $\tau_j(1^a) = \theta_t^{t+1}\tilde{\tau}_j(1^a)$ for some $\tilde{\tau}_j(1^a) \in \pi_{2qp^j-1}(\Omega P^{2q+1}(p^t); \mathbf{Z}_{p^{t+1}})$. Similarly, $\sigma_j(1^a) = \theta_t^{t+1}(\tilde{\sigma}_j(1^a))$ for some $\tilde{\sigma}_j(1^a)$. We may use $\tilde{\tau}_j(1^a)$ and $\tilde{\sigma}_j(1^a)$ to define $\tilde{\tau}_j(y)$ and $\tilde{\sigma}_j(y)$ for an arbitrary even-dimensional homotopy class $y \in \pi_{2q}(\Omega X; \mathbf{Z}_{p^t})$ by $\tilde{\tau}_j(y) = (\Omega y^{\bar{a}})_\# \tilde{\tau}_j(1^a)$ and $\tilde{\sigma}_j(y) = (\Omega y^{\bar{a}})_\# \tilde{\sigma}_j(1^a)$. Then the $\tilde{\tau}_j$ and $\tilde{\sigma}_j$ defined in this way are natural operations from $\pi_{\text{even}}(\Omega-; \mathbf{Z}_{p^t})$ to $\pi_*(\Omega-; \mathbf{Z}_{p^{t+1}})$, just as we have operations $\tilde{\tau}_j$ and $\tilde{\sigma}_j$ on any csL of the form $\mathcal{F}(L, d)$ for a $\mathbf{Z}_{(p)}$-free dgL (L, d). Summarizing, we have:

(11EN) $$\tilde{\tau}_j\omega_t^{t+1} = \tau_j, \quad \tilde{\sigma}_j\omega_t^{t+1} = \sigma_j, \quad (p)\tilde{\tau}_j = \tau_j\rho_{t+1}^t, \quad (p)\tilde{\sigma}_j = \sigma_j\rho_{t+1}^t.$$

How far does the similarity to $\mathcal{F}(L, d)$ extend? For instance, is it true in general that (cf. (11BP))

(11EO) $${}_\pi\delta^{t+1}\tilde{\tau}_j(y) = \tilde{\sigma}_j(y) \quad ?$$

One difficulty here is that we did not define $\tilde{\tau}_j$ and $\tilde{\sigma}_j$ canonically, since there is indeterminacy in the choice of $\tilde{\tau}_j(1^a)$ and $\tilde{\sigma}_j(1^a)$. The indeterminacy equals $\ker(\omega_t^{t+1})^\# = \operatorname{im}(\rho_{t+1}^1)^\#$ (by (6C)). The right question is: is there any way to choose $\tilde{\tau}_j(1^a)$ and $\tilde{\sigma}_j(1^a)$ so that (11EO) is true? Interestingly, the answer is "yes" when $j = 1$, as we shall see in Chapter 13, but for $j > 1$ it is open.

Pursuing the chain of implications (11BP) $\Longrightarrow$ (11BQ) $\Longrightarrow$ (11BR), a consequence of (11EO) would be the relation

(11EP) $$\theta_t^{t+1} d_{[t+1]}\tilde{\tau}_j(y) = \tilde{\tau}_j(y)\omega^{t+1}\rho^t = \sigma_j(y) \quad (?)$$

In a sense (11EP) is equivalent to (11EO). If (11EP) is true, then we could define $\tilde{\sigma}_j(y)$ to be ${}_\pi\delta^{t+1}\tilde{\tau}_j(y)$, and this definition would be consistent with the requirements (11EN).

Applying ζ_{t+1}^t to (11EP) gives

(11EQ) $$d_{[t+1]}\zeta_{t+1}^t\tau_j(y) = {}_\pi\delta^{t+1}(p)\tilde{\tau}_j(y) = \zeta_{t+1}^t\sigma_j(y), \quad (?)$$

i.e.,

(11ER) $$\xi_j(y) = 0 \; (?)$$

which is true for $j = 1$ but open for $j \geq 2$. Again, (11EQ) and (11ER) are equivalent to (11EP). If (11EQ) holds then Lemma 11.E3 constructs $z = \tilde{\tau}_j(y)$ satisfying $\theta_t^{t+1}\tilde{\tau}_j(y) = \tau_j(y)$ while $\theta_t^{t+1}d_{[t+1]}\tilde{\tau}_j(y) = \sigma_j(y)$. The author thinks of (11ER) as the best way to pose these equivalent questions, since no indeterminacy is involved. A consequence of (11ER) that is not clearly equivalent to it is the open problem: does

(11ES) $$d_{[t+1]}\xi_j(y) = 0 \; (?)$$

for any y, for $j \geq 2$?

The elements $\xi_j(y)$, if they are non-zero, are quite interesting. By Lemma 11.B3(c) we see that $\xi_j(y)$ lies in the center of $\pi_*(\Omega X; \mathbb{Z}_{p^{t+1}})$. We apply this fact to the universal example for the stable homotopy operation $\Phi(\xi_j(y))$ of Lemma 8.9, which is the fibration

$$\text{(11ET)} \qquad X'(m, p^{t+1}) \xrightarrow{f} X \vee P^m(p^{t+1}) \longrightarrow X.$$

Then

$$\text{(11EU)} \qquad \xi_j(y) \in \ker(\Phi).$$

Let us spell out the proof of (11EU), since the reader has come a long distance from Chapter 8. Let $\hat{v}_m \in \pi_m(X'(m, p^{t+1}); \mathbb{Z}_{p^{t+1}})$ be the canonical homotopy class of Chapter 8. When we view $\xi_j(y)^{\bar{a}}$ as an element of $\pi_*(X \vee P^m(p^{t+1}); \mathbb{Z}_{p^{t+1}})$, the Whitehead product $[\xi_j(y)^{\bar{a}}, f_\#(\hat{v}_m)]$ vanishes. Since $f_\#$ is one-to-one, the Samelson product $[\xi_j(y), \hat{v}_m^a]$ vanishes in $\pi_*(\Omega X'(m, p^{t+1}); \mathbb{Z}_{p^{t+1}})$. Since $\hat{v}_m^a$ is the universal example for stable homotopy operations under X, we have shown (11EU).

Consider some consequences of (11EU). Automatically $d_{[t+1]}\xi_j(y) \in \ker(\Phi)$, and $\xi_j \in \ker(h)$, and any composition $(\xi_j(y)^{\bar{a}} \circ \alpha)^a$ for $\alpha \in \pi_*(P^{|y|+1}(p^{t+1}); \mathbb{Z}_{p^{t+1}})$ belongs to $\ker(\Phi)$. The author likes to call $\xi_j(y)$ a "mirage" element of the homotopy of X. You think you can see it, but if you try to do anything with it — like bracket with something, apply θ_t^{t+1}, multiply by p, suspend it, or measure its Hurewicz image — it vanishes.

There exist non-zero mirages. The only examples the author knows of for sure in $\pi_*(\Omega X; \mathbb{Z}_{p^t})$ all have the form $p^{s-1}\zeta_t^s\tau_j(z)$ or $p^{s-1}d_{[t]}\zeta_t^s\tau_j(z)$ for some $z \in \pi_{\text{even}}(\Omega X; \mathbb{Z}_{p^s})$. That these are mirages follows from Lemma 11.B8 (the other mirage properties are straightforward), and in the universal case of $z = 1^a$ for $X = P^{2q+1}(p^{t-1})$ these elements are non-zero.

As with phantom maps, there are many things we would like to know about mirages. Is the set of mirages (it is a submodule of $\pi_*(\Omega X; \mathbb{Z}_{p^t})$) closed under composition on the right? Is a mirage always a co-H map? That $\xi_j(y)$ and $p^{s-1}\rho_t^s\tau_j(z)$ are co-H follows from their additivity (cf. Lemmas 8.4, 11.B3(b), 11.B8(a)). What happens to the homotopy of X if we use a mirage as the attaching map to attach a mod p^t Moore space to X? Do there exist non-zero mirages at $t = 1$? Does every finite 1-connected X have mirages in its mod p^t loop space homotopy, for every $t \geq 2$?

Discussion 11.E6. Returning to the ξ_j, suppose it turns out that (11ER) holds for all j, i.e. $\xi_j \equiv 0$ is an identity that holds for homotopy csL's. This could greatly alter the face of csL theory. Basically, the two "problems" with $\mathcal{F}(L, d)$ that are mentioned in Discussion 11.B4, and which nearly all of Chapter 11 has been devoted to resolving, would no longer be problems.

Consider the first "problem" with $\mathcal{F}(L, d)$, namely the presence of $\tilde{\tau}_j$'s and $\tilde{\sigma}_j$'s. We discuss only the case of $(L, d) = \mathcal{L}(V, d)$, since that is the only situation we have really looked at. If (11ER) holds, then any homotopy csL probably also has "reasonable" $\tilde{\tau}_j$'s and $\tilde{\sigma}_j$'s for all j, where by "reasonable" we mean that they would satisfy all the same axioms that $\tilde{\tau}_j$ and $\tilde{\sigma}_j$ satisfy for $\mathcal{F}(L, d)$ (e.g. Lemma 11.B5), Then there would be no need to remove the $\tilde{\tau}_j$'s and $\tilde{\sigma}_j$'s from $\mathcal{FL}(V)$ because we have somewhere to send them in the target.

We have formerly presented the "problem" of the $\tilde{\tau}_j$'s and $\tilde{\sigma}_j$'s differently. We accused them of standing in the way of $\mathcal{L}(V^{(\geq t)})$ being a Bockstein subalgebra of $\mathcal{FL}(V)_{[t]}$. However, there is an alternative approach, which is to realize that we may have the wrong definition of "Bockstein subalgebra."

The alternative idea is to define a new object, called an "extended coherent sequence of Lie algebras" or "ecsL", which is a csL together with (non-homomorphic) operations $\tilde{\tau}_j : (L_{[t]})_{2q} \longrightarrow (L_{[t+1]})_{2qp^j-1}$. The $\tilde{\tau}_j$'s of an ecsL are required to satisfy a set of axioms, and the axioms are designed so that $\mathcal{F}(L,d)$ will be an ecsL. (This procedure for choosing axioms is modeled after the procedure for obtaining the axioms for a restricted Lie algebra or for a divided power algebra.) Given an ecsL $\{L_{[t]}, \tilde{\tau}_j\}$, define $N_{[t]}$ to be a "Bockstein subalgebra" of $L_{[t]}$ if $N_{[t]} + \bar{L}_{[t]} = L_{[t]}$ and $N_{[t]} \cap \overline{L}_{[t]} = pN_{[t]}$, where

$$\overline{L}_{[t]} = \zeta_t^{t-1}(L_{[t-1]}) + d_{[t]}\zeta_t^{t-1}(L_{[t-1]}) + \sum_j \tilde{\tau}_j(L_{[t-1]}) + \sum_j d_{[t]}\tilde{\tau}_j(L_{[t-1]}).$$

With this new definition, $\mathcal{L}(V^{(\geq t)})$ becomes a Bockstein subalgebra of $\mathcal{FL}(V)$.

To get a reasonable theory of ecsL's we need versions of Proposition 7.5 and Theorem 7.15 that work for ecsL's. This will involve a fair amount of work, but can be done. The new versions will construct homomorphisms of ecsL's once the images of the generators are specified, and once a few hypotheses are checked. The rub is, that the target has to be an ecsL. In order to apply all this new theory to homotopy, $\{\pi_*(\Omega X; \mathbb{Z}_{p^t})\}$ must satisfy all the ecsL axioms, and foremost among the axioms will be (11ER) (cf. (11BR) for $\mathcal{F}(L,d)$).

If the answer to (11ER) turns out to be "no", it may be possible to salvage the above approach by building a universal central extension $\mathcal{E}^+\mathcal{L}(V,d)$ of $\mathcal{FL}(V,d)$. In other words, $\mathcal{E}^+$ would do for $\mathcal{FL}(V)$ what $\mathcal{E}$ does for $\mathcal{F}^0\mathcal{L}(V)$. This approach would enjoy the virtue that the Virtual Syzygy Lemma would become unnecessary, since we would not be throwing anything away from $\mathcal{FL}(V)$. However, the resulting machine $\mathcal{E}^+$ would be universal for a very large set of axioms, and we might need to examine very carefully how the $\tilde{\tau}_j$'s and $\tilde{\sigma}_j$'s behave in homotopy, in order to discern what the correct axioms are. On the other hand, once this initial investment was made, the functor $\mathcal{E}^+$ would greatly streamline the process of building ecsL homomorphisms, from $\mathcal{E}^+(L,d)$ to $\{\pi_*(\Omega X; \mathbb{Z}_{p^t})\}$. A lot of theorems about mod p^t homotopy that are now only dreamed of might fall out as simple consequences. Conceivably, $\mathcal{E}^+$ could eventually play the role for p-primary homotopy (especially when p is large relative to $\dim(X)$) that the Sullivan and Quillen functors play for rational homotopy theory.

CHAPTER 12

Construction of β_t^k and α_t^k

Let D_{kk} denote the homotopy–theoretic fiber of the inclusion $D_k \vee D_k \to D_k \times D_k$ of the wedge into the product of two copies of D_k. In Chapter 12 we continue the proof of Theorem 9.1 by constructing a csL/csm homomorphism $\{\beta_t^k\}$ that partially splits the Hurewicz homomorphism $\pi_*(\Omega D_{kk}; \mathbb{Z}_{p^t}) \to H_*(\Omega D_{kk}; \mathbb{Z}_{p^t})$. To begin, we construct stable homotopy operations under $D_k \vee D_k$ and under $D_k \vee D_k \vee D_k$. Then we present Theorem 12.4, a kind of continuation of Theorem 9.1, that states various facts about β_t^k. We prove all of Theorem 12.4 for $\beta_{<r+k}^k$ and we prove all but three parts of it for β_{r+k}^k. We state and essentially prove another theorem that is the three–factor analog of Theorem 12.4, concerning a csL/csm homomorphism α_t^k for $D_k \vee D_k \vee D_k$. Embedded as part of this latter theorem is a group of eight formulas we call the "third Jacobi identity for D_k". The third Jacobi identity will be the key to the virtual syzygy needed in order to construct γ_{r+k}^k in Theorem 9.1. It is really our principal motivation for both this chapter and for Chapter 11.

All of Chapter 12 is part of the grand inductive scheme initiated in Chapter 9. We think of n, r, and $p \geq 5$ as fixed, and all our results are understood to be embedded in an induction on k.

Our first results for Chapter 12 are a two–factor and a three–factor analog of Theorem 9.1(vi)–(viii). Recall the notations $\mathcal{M}_A$ and K_g from Chapter 8.

Theorem 12.1. *Let $k_1 \geq 0$, $k_2 \geq 0$. (To keep this within the grand inductive scheme suppose $\max(k_1, k_2) \leq k$.) On the category $\mathcal{M}_{D_{k_1} \vee D_{k_2}}$, there is a natural structure for $\{\pi_*(\Omega K_g; \mathbb{Z}_{p^t})\}$ as a csL over $\mathcal{F}(\mathcal{U}M^{k_1} \amalg \mathcal{U}M^{k_2})$. We denote the module structure homomorphism by*

$$\psi_t^{k_1 k_2} : \mathcal{F}(\mathcal{U}M^{k_1} \amalg \mathcal{U}M^{k_2})_{[t]} \otimes \pi_*(\Omega K_g; \mathbb{Z}_{p^t}) \to \pi_*(\Omega K_g; \mathbb{Z}_{p^t}).$$

$\psi_t^{k_1 k_2}$ factors through a structure homomorphism $\bar{\psi}_t^{k_1 k_2}$ for a csm structure over the csa $\{H_*(\mathcal{U}M^{k_1} \amalg \mathcal{U}M^{k_2}; \mathbb{Z}_{p^t})\} = \{H_*(\Omega(D_{k_1} \vee D_{k_2}); \mathbb{Z}_{p^t})\}$. Dualizing $\psi_t^{k_1 k_2}$ gives us $\Psi_t^{k_1 k_2}$, and dualizing $\bar{\psi}_t^{k_1 k_2}$ gives us

$$\bar{\Psi}_t^{k_1 k_2} : H_*(\Omega(D_{k_1} \vee D_{k_2}); \mathbb{Z}_{p^t}) \to \mathcal{O}_{D_{k_1} \vee D_{d_2}}[t],$$

which satisfies $\mathcal{H}\bar{\Psi}_t^{k_1k_2} = (1,0)$ *for each* $t \geq 1$. $\bar{\Psi}_t^{k_1k_2}$ *splits* $\mathcal{H}_0|_{\ker(\mathcal{H}_1)}$ *as dga's, and Corollary 8.17 is true for* $D_{k_1} \vee D_{k_2}$. *Lastly,* $\Psi_t^{k_1k_2}$ *is compatible both with* $\hat{\imath} \vee 1 : D_{k_1-1} \vee D_{k_2} \to D_{k_1} \vee D_{k_2}$ *when* $k_1 \geq 1$ *and with* $1 \vee \hat{\imath} : D_{k_1} \vee D_{k_2-1} \to D_{k_1} \vee D_{k_2}$ *when* $k_2 \geq 1$.

Proof. By Remark 11.C6 each $\pi_*(\Omega K_g; \mathbb{Z}_{p^t})$ for $(f,g) \in \mathcal{M}_{D_{k_1} \vee D_{k_2}}$ is a csL over $\mathcal{F}(\mathcal{U}M^{k_1} \amalg \mathcal{U}M^{k_2})$, since its csL structure is simultaneously a csL over $\mathcal{F}(\mathcal{U}M^{k_1})$ and a csL over $\mathcal{F}(\mathcal{U}M^{k_2})$. The proofs of the other properties are routine, using 9.1(vi)–(viii).

Likewise, the proof of the three–factor version is routine.

Theorem 12.2. *Let* $k_1 \geq 0$, $k_2 \geq 0$, $k_3 \geq 0$. *(To keep this within the grand inductive scheme suppose* $\max(k_1, k_2, k_3) \leq k$.*) On the category* $\mathcal{M}_{D_{k_1} \vee D_{k_2} \vee D_{k_3}}$, *there is a natural structure for* $\{\pi_*(\Omega K_g; \mathbb{Z}_{p^t})\}$ *as a csL over* $\mathcal{F}(\mathcal{U}M^{k_1} \amalg \mathcal{U}M^{k_2} \amalg \mathcal{U}M^{k_3})$. *Denote the module structure homomorphism by*

$$\psi_t^{k_1k_2k_3} : \mathcal{F}(\mathcal{U}M^{k_1} \amalg \mathcal{U}M^{k_2} \amalg \mathcal{U}M^{k_3})_{[t]} \otimes \pi_*(\Omega K_g; \mathbb{Z}_{p^t}) \to \pi_*(\Omega K_g; \mathbb{Z}_{p^t}).$$

$\psi_t^{k_1k_2k_3}$ *factors through a structure homomorphism* $\bar{\psi}_t^{k_1k_2k_3}$ *for a csm structure over the csa*

$$\{H_*(\mathcal{U}M^{k_1} \amalg \mathcal{U}M^{k_2} \amalg \mathcal{U}M^{k_3}; \mathbb{Z}_{p^t})\} = \{H_*(\Omega(D_{k_1} \vee D_{k_2} \vee D_{k_3}); \mathbb{Z}_{p^t})\}.$$

Dualizing $\psi_t^{k_1k_2k_3}$ *gives us* $\Psi_t^{k_1k_2k_3}$, *and dualizing* $\bar{\psi}_t^{k_1k_2k_3}$ *gives us*

$$\bar{\Psi}_t^{k_1k_2k_3} : H_*(\Omega(D_{k_1} \vee D_{k_2} \vee D_{k_3}); \mathbb{Z}_{p^t}) \to \mathcal{O}_{D_{k_1} \vee D_{k_2} \vee D_{k_3}}[t],$$

which satisfies $\mathcal{H}\bar{\Psi}_t^{k_1k_2k_3} = (1,0)$ *for each* $t \geq 1$. $\bar{\Psi}_t^{k_1k_2k_3}$ *splits* $\mathcal{H}_0|_{\ker(\mathcal{H}_1)}$ *as dga's, and Corollary 8.17 is true for* $D_{k_1} \vee D_{k_2} \vee D_{k_3}$. *Lastly,* $\Psi_t^{k_1k_2k_3}$ *is compatible (in the sense of 9.1(viii)) with* $\hat{\imath} \vee 1 \vee 1$ *and with* $1 \vee \hat{\imath} \vee 1$ *and with* $1 \vee 1 \vee \hat{\imath}$.

We now turn our attention to the space D_{kk} and the csL/csm homomorphism $\{\beta_t^k\}$. Define D_{kk} to be the homotopy–theoretic fiber of the inclusion map $(\bar{\pi}_1, \bar{\pi}_2)$ from $D_k \vee D_k$ into $D_k \times D_k$, and denote the fiber inclusion map by η_{kk}. We have a fibration up to homotopy

$$D_{kk} \xrightarrow{\eta_{kk}} D_k \vee D_k \xrightarrow{(\bar{\pi}_1, \bar{\pi}_2)} D_k \times D_k. \tag{12A}$$

As discussed in Chapter 6, we know that $D_{kk} \approx \Sigma(\Omega D_k) \wedge (\Omega D_k)$ and that (12A) splits when looped once. The inclusion $\hat{\imath} : D_{k-1} \to D_k$ induces a map on fibers, also denoted $\hat{\imath}$, that makes this diagram commute:

$$\begin{array}{ccccc} D_{k-1,k-1} & \xrightarrow{\eta_{k-1,k-1}} & D_{k-1} \vee D_{k-1} & \xrightarrow{(\bar{\pi}_1, \bar{\pi}_2)} & D_{k-1} \times D_{k-1} \\ \downarrow{\scriptstyle \hat{\imath}} & & \downarrow{\scriptstyle \hat{\imath} \vee \hat{\imath}} & & \downarrow{\scriptstyle \hat{\imath} \times \hat{\imath}} \\ D_{kk} & \xrightarrow{\eta_{kk}} & D_k \vee D_k & \xrightarrow{(\bar{\pi}_1, \bar{\pi}_2)} & D_k \times D_k. \end{array} \tag{12B}$$

The algebraic analog of $\hat{\imath}$ is the induced homomorphism $\bar{\imath} : N^{k-1,k-1} \to N^{kk}$ on dgL kernels in the diagram (the rows are the short exact sequences of dgL's (5R))

$$\begin{array}{ccccccccc} 0 & \longrightarrow & N^{k-1,k-1} & \longrightarrow & M^{k-1} \amalg M^{k-1} & \longrightarrow & M^{k-1} \oplus M^{k-1} & \longrightarrow & 0 \\ & & \downarrow{\scriptstyle \bar{\imath}} & & \downarrow & & \downarrow & & \\ 0 & \longrightarrow & N^{kk} & \longrightarrow & M^k \amalg M^k & \longrightarrow & M^k \oplus M^k & \longrightarrow & 0. \end{array} \tag{12C}$$

Notation 12.3. One thing we will be very interested in doing is applying the homotopy operations associated with the first component of $D_k \vee D_k$ to the homotopy classes associated with the second component, and vice versa. By analogy with the notations $e^{(1)}$ and $e^{(2)}$ for the dgL inclusions of Chapter 5, we let $\hat{e}^{(1)} : D_k \to D_k \vee D_k$ and $\hat{e}^{(2)} : D_k \to D_k \vee D_k$ denote the two geometric inclusions. Where no confusion can result, $\Psi_t^{kk}(w^{(1)})$ denotes $\Psi_t^{kk}(e^{(1)}(w))$ for $w \in \mathcal{F}(\mathcal{U}M^k)_{[t]}$, and likewise for $\Psi_t^{kk}(e^{(2)}(w))$. When $y \in \mathcal{F}(\bar{L}^k)_{[t]}^{*\leq 2p}$, let $\gamma_t^k(y^{(1)})$ denote the image of y under the composition

$$\mathcal{F}(\bar{L}^k)_{[t]}^{*\leq 2p} \xrightarrow{\gamma_t^k} \pi_*(\Omega E_k; \mathbf{Z}_{p^t}) \xrightarrow{(\Omega\lambda_k\eta_k)_\#} \pi_*(\Omega D_k; \mathbf{Z}_{p^t})$$
$$\xrightarrow{(\Omega\hat{e}^{(1)})_\#} \pi_*(\Omega(D_k \vee D_k); \mathbf{Z}_{p^t}),$$

where γ_t^k is given by (9J). Because $\mathcal{O}_{(\bar{\pi}_1)}\Psi_t^{kk}(w^{(2)}) = 0$ whenever $w \in \mathcal{F}(\mathcal{U}M_+^k)_{[t]}$, and $(\Omega\bar{\pi}_2)_\#\gamma_t^k(y^{(1)}) = 0$ for any $y \in \mathcal{F}(\bar{L}^k)_{[t]}^{*\leq 2p}$, we have

$$\psi_t^{kk}(w^{(2)} \otimes \gamma_t^k(y^{(1)})) \in \ker(\Omega(\bar{\pi}_1, \bar{\pi}_2)_\#) = \operatorname{im}(\eta_{kk})_\#.$$

Since $(\eta_{kk})_\#$ is one–to–one (recall that (12A) splits when looped once), $(\eta_{kk})_\#^{-1}$ is defined on $\operatorname{im}(\eta_{kk})_\#$. The notation

$$\Psi_t^{kk}(w^{(2)})\gamma_t^k(y^{(1)}),$$

which technically does not make sense for the map pair $D_k \vee D_k \xrightarrow{=} D_k \vee D_k \xrightarrow{(\bar{\pi}_1,\bar{\pi}_2)} D_k \times D_k$, will nevertheless be used to denote the element

$$(\Omega\eta_{kk})_\#^{-1}\psi_t^{kk}(w^{(2)} \otimes \gamma_t^k(y^{(1)})), \tag{12D}$$

where in (12D) ψ_t^{kk} is evaluated for the map pair $D_k \vee D_k \xrightarrow{=} D_k \vee D_k \to (*)$. The notation (12D) makes sense whenever $w \in \mathcal{F}(\mathcal{U}M_+^k)_{[t]}$.

We are about to state Theorem 12.4. We encourage the reader to think of Theorems 12.4 and 12.6 as continuations of Theorem 9.1. In other words, they are all part of one big theorem with many parts that is broken up for convenience only. For this reason we begin the numbering of the parts of Theorems 12.4 and 12.6 with (xiii), since 9.1 ended with part (xii). When we refer to these parts we will drop the 12.4 or 12.6, e.g., (xv) means Theorem 12.4 part (xv).

Theorem 12.4. *There is a homomorphism of csL's over $\mathcal{F}(\mathcal{U}M^k \amalg \mathcal{U}M^k)$, denoted*

$$\{\beta_t^k\}_{1\leq t\leq r+k} : \mathcal{E}(N^{kk}) \to \{\hat{\pi}_*(\Omega D_{kk}; \mathbf{Z}_{p^t})\}, \tag{12E}$$

where $\{\hat{\pi}_*(\Omega D_{kk}; \mathbf{Z}_{p^t})\}$ is a quotient csL of $\hat{\pi}_*(\Omega D_{kk}; \mathbf{Z}_{p^t})$ to be described in Note 12.5 below, and $\mathcal{E}(N^{kk})$ is introduced in 11.D12. It has the following properties.

(xiii) *Let $\underline{q} = \{\underline{q}_{[t]}\}$ denote the quotient csL homomorphism of (7AB), and let $\bar{e}$ be the dgL inclusion*

$$\bar{e} : N^{kk} \to M^k \amalg M^k \to \mathcal{U}M^k \amalg \mathcal{U}M^k. \tag{12F}$$

This square commutes:

$$\begin{array}{ccc} \mathcal{E}(N^{kk})_{[t]} & \xrightarrow{\beta_t^k} & \hat{\pi}_*(\Omega D_{kk}; \mathbb{Z}_{p^t}) \\ \Big\downarrow {\scriptstyle \underline{q}_{[t]}} & & \Big\downarrow {\scriptstyle (\Omega\eta_{kk})_* h} \\ H_*(N^{kk}; \mathbb{Z}_{p^t}) & \xrightarrow{(\bar{e})_*} & H_*(\mathcal{U}M^k \amalg \mathcal{U}M^k; \mathbb{Z}_{p^t}), \end{array} \tag{12G}$$

where h is the Hurewicz homomorphism. Equivalently,

$$\underline{a}_{\eta_{kk}}(\beta_t^k(w)) = \bar{e}(w) \text{ for } w \in \mathcal{E}(N^{kk})_{[t]}. \tag{12H}$$

(xiv) *For $k \geq 1$, β_t^k is compatible with $\hat{i}$, i.e., this diagram commutes if $t \leq r + k - 1$:*

$$\begin{array}{ccc} \mathcal{E}(N^{k-1,k-1})_{[t]} & \xrightarrow{\beta_t^{k-1}} & \hat{\pi}_*(\Omega D_{k-1,k-1}; \mathbb{Z}_{p^t}) \\ \Big\downarrow {\scriptstyle \mathcal{E}(\bar{i})_{[t]}} & & \Big\downarrow {\scriptstyle (\Omega\hat{i})_\#} \\ \mathcal{E}(N^{kk})_{[t]} & \xrightarrow{\beta_t^k} & \hat{\pi}_*(\Omega D_{kk}; \mathbb{Z}_{p^t}), \end{array} \tag{12I}$$

where $\mathcal{E}(\bar{i})$ is described in Theorem 11.D13(b).

(xv) *("The Second Jacobi Identity") For $w \in \mathcal{E}(N^{kk})_{[t]}$, $1 \leq t \leq r + k$,*

$$\Phi((\Omega\eta_{kk})_\# \beta_t^k(w)) = \Psi_t^{kk}(\mathcal{F}(\bar{e})(w)) \text{ in } \mathcal{O}_{D_k \vee D_k}[t]. \tag{12J}$$

(xvi) *$D_{kk} \in \mathcal{W}_r^{r+k}$, i.e., D_{kk} has the homotopy type of a bouquet of Moore spaces whose orders are p^m for various m in the range $r \leq m \leq r + k$.*

(xvii) *Let $g = g_{kk} : \mathcal{U}M_+^k \otimes \mathcal{U}M_+^k \to N^{kk}$ be the chain map defined in Theorem 5.12. In $\hat{\pi}_*(\Omega D_{kk}; \mathbb{Z}_{p^{r+k}})$, we have the equation*

$$\begin{aligned} \psi_{r+k}^{kk}(c_k^{(1)} \otimes \beta_{r+k}^k g(b_k \otimes b_k)) + \psi_{r+k}^{kk}(b_k^{(1)} \otimes \beta_{r+k}^k g(c_k \otimes b_k)) \; + \; \Psi_{r+k}^{kk}(b_k^{(2)})\gamma_{r+k}^k(y_{2p^k}^{(1)}) \\ = \zeta_{r+k}^{r+k-1} \beta_{r+k-1}^k(x_k^{12}) + d_{[r+k]} \zeta_{r+k}^{r+k-1} \beta_{r+k}^k(w_k^{12}), \end{aligned} \tag{12K}$$

where x_k^{12} and w_k^{12} are the elements of $\mathcal{E}(N^{kk})_{[r+k]}$ defined in Theorem 11.D13(d). We think of Equation (12K) as the topological counterpart of (7DK). The equation (7DM) likewise has its counterpart in $\hat{\pi}_*(\Omega D_{kk}; \mathbb{Z}_{p^{r+k}})$ hold true.

(xviii) *In $\hat{\pi}_*(\Omega D_{kk}; \mathbb{Z}_{p^{r+k}})$, we have the equation*

$$\begin{aligned} \psi_{r+k}^{kk}(c_k^{(1)} \otimes \beta_{r+k}^k g(b_k \otimes c_k)) - \psi_{r+k}^{kk}(b_k^{(1)} \otimes \beta_{r+k}^k g(c_k \otimes c_k)) \; + \; \Psi_{r+k}^{kk}(c_k^{(2)})\gamma_{r+k}^k(y_{2p^k}^{(1)}) \\ = \zeta_{r+k}^{r+k-1} \beta_{r+k-1}^k(z_k^{12}) + d_{[r+k]} \zeta_{r+k}^{r+k-1} \beta_{r+k-1}^k(y_k^{12}), \end{aligned} \tag{12L}$$

where z_k^{12} and y_k^{12} are the elements defined in Theorem 11.D13(d). We think of (12L) as the topological counterpart of Equation (7DL). The equation (7DN) likewise has its counterpart in $\hat{\pi}_*(\Omega D_{kk}; \mathbb{Z}_{p^{r+k}})$ hold true.

(xix) For $1 \le t \le r+k$, $w \in \mathcal{F}(\mathcal{U}M_+^k)_{[t]}$, $y \in \mathcal{F}(\bar{L}^k)_{[t]}^{*\le 2p}$,

$$\beta_t^k(\mathrm{ad}^{(2)}(w)e^{(1)}(y)) = \Psi_t^{kk}(w^{(2)})\gamma_t^k(y^{(1)})\ ; \tag{12M}$$

$$\beta_t^k(\mathrm{ad}^{(1)}(w)e^{(2)}(y)) = \Psi_t^{kk}(w^{(1)})\gamma_t^k(y^{(2)})\ . \tag{12N}$$

Note 12.5. We define $\hat{\pi}_*(\Omega D_{kk};\mathbb{Z}_{p^t})$ to be $\pi_*(\Omega D_{kk};\mathbb{Z}_{p^t})/J'_{[t]}$ for a certain csL ideal J' of $\pi_*(\Omega D_{kk};\mathbb{Z}_{p^t})$. We have segregated the description of $J' = \{J'_{[t]}\}$ into its own note so as not to interrupt the flow of Theorem 12.4. Let S be the set of triples

$$S = \{(r+s, 1, (\Omega\hat{i})_\#\beta_{r+s}^{k-1}(\mathrm{ad}^{(i)}(b_s)g_{ss}(b_s\otimes b_s))) \mid 0 \le s < k, i = 1 \text{ or } 2\}.$$

We put

$$J' = \langle p^*\tau_*(S)\rangle,$$

where this notation is defined right after Lemma 11.C5.

The csL ideal J' is one of those "very small" csL ideals described in Chapter 11. It consists entirely of mirages (see Discussion 11.E5). It is annihilated by $\mathcal{F}((\mathcal{U}M^k \amalg \mathcal{U}M^k)_+)$ (Lemma 11.C5 and Corollary 11.C7) and it lies in the center of $\{\pi_*(\Omega D_{kk};\mathbb{Z}_{p^t})\}$. Because $J' \subseteq \ker(\Phi)$, (12J) still makes sense. Dividing out by J' hardly affects the csL $\{\pi_*(\Omega D_{kk};\mathbb{Z}_{p^t})\}$ at all, and in reading the properties or proofs the reader should pretend that $\hat{\pi}_*$ is π_*, except when the issue of indeterminacy in J' is directly relevant. Also, $J'_{[t]} \subseteq (p)\pi_*(\Omega D_{kk};\mathbb{Z}_{p^t})$, because of the general fact that $\zeta_{s+1}^s\tau_1(y) = (p)\tau_1(y)$ for any $y \in \pi_*(\Omega -;\mathbb{Z}_{p^s})$.

Overall Plan for the Proof

The overall plan for the proof of Theorem 12.4 imitates closely the pattern used in Chapter 10 for the construction of ψ_t^k. We begin by doing the case $t \le r$. Because t stays in the range $1 \le t \le r+k$ in Theorem 12.4, the case of $t \le r$ subsumes the case of $k = 0$. The construction of β_t^k for $t > r$ follows the framework suggested by Theorem 7.15, so we are doing a dimension–wise induction for each fixed t, within a larger induction on t (all within the grand inductive scheme on k).

We separate the cases $r < t < r+k$ and $t = r+k$ because they are very different. For $t < r+k$, a "pure" application of Theorem 7.15 would require some difficult technical work, especially when it comes to verifying Conditions 7.15 (i,ii). We can make the proof much easier if we modify Theorem 7.15 by carrying along properties (xiv) and (xix) as part of the dimension–wise induction process. Other than this change, the proof for $r < t < r+k$ is largely straightforward, and it provides a good example of the ability of Theorem 7.15 to handle a fairly intricate virtual structure.

The wedge decomposition (xvi) can be proved using $\beta_{<r+k}^k$ only, so we squeeze this in between the construction of $\beta_{<r+k}^k$ and the construction of β_{r+k}^k.

When $t = r+k$, there are no virtual syzygies, so lifting the virtual generators and verifying the virtual relators becomes our focus. To lift the virtual generators we must construct some mod p^{r+k} homotopy classes where none (other than ζ_{r+k}^{r+k-1}–images) are previously known, so we try to invoke Theorem 6.4. Here we run up against the fourth snag mentioned in Discussion 11.A1, and we need several new ideas in order to push the construction through. The virtual relators rely upon Corollary 6.7. The fibration to which we apply Corollary 6.7 is (6MM) for the thin product $\mathcal{R}(D_k, D_k, P^{q+1}(p^{r+k}))$, so we think of the resulting formula (12J) as a "second Jacobi identity". (Recall that the first Jacobi identity involved a thin product of one D_k and two Moore spaces.)

Equations (12M) and (12N) at $t = r + k$ and the equations in (xvii) and (xviii) involve γ^k_{r+k}. We will not construct γ^k_{r+k} until Chapter 13, so we postpone the proof of these equations.

Construction of β^k_t for $1 \le t \le r$

When $1 \le t \le r$, the construction of β^k_t is merely a matter of assembling some components that we already have. One reason the case $t \le r$ is so simple is that

$$\mathcal{E}(N^{kk})_{[t]} = \mathcal{F}^0(N^{kk})_{[t]} = \mathcal{F}(N^{kk})_{[t]} = (N^{kk}, p^{-t}d)$$

when $t \le r$. We may use ψ^k_t and γ^k_t and all their properties, since either $k = 0$ and all of Theorem 9.1 has been established at $k = 0$ (with the possible exception of (xii)), or else $k \ge 1$ in which case $t < r + k$ and (ix)–(xi) have been established at t.

Let $\hat{i}^k_m$ denote the inclusion map from D_m to D_k, for $0 \le m \le k$. Define $\tilde{b} \in \pi_{2n}(\Omega D_k; \mathbb{Z}_{p^t})$ and $\tilde{a}_0 \in \pi_{2n-1}(\Omega D_k; \mathbb{Z}_{p^t})$ by $\tilde{b} = (\hat{i}^k_0 \omega^r_t)^a$ and $\tilde{a}_0 = -(\hat{i}^k_0 \omega^r \rho^t)^a$. Thus $\tilde{b}$ and $\tilde{a}_0$ depend upon t (as well as n, p, r, k) but we suppress the t from the notation. We include the minus sign in the definition of $\tilde{a}_0$ because we want the relation

(12O1) $$_\pi\delta^t(\tilde{b}) = -(p^{r-t})\tilde{a}_0$$

to hold in $\pi_*(\Omega D_k; \mathbb{Z}_{p^t})$, just as the relation

$$\delta^t(b) = -(p^{r-t})a_0$$

is true in the dgL $\mathcal{F}(M^k)_{[t]}$.

Recall from Chapter 10 the homotopy class $\hat{a}_m \in \pi_{2np^m-1}(\Omega D_m; \mathbb{Z}_{p^{r+m-1}})$, defined by (10J), for $1 \le m \le k$. For $1 \le m \le k$ put

(12O2) $$\tilde{a}_m = (\Omega\hat{i}^k_m)_\#(\omega^{r+m-1}_t)^\#(\hat{a}_m) \in \pi_{2np^m-1}(\Omega D_k; \mathbb{Z}_{p^t}).$$

Recalling the definition of $\gamma^k_t(y^{p^k})$, we see that

(12O3) $$\tilde{a}_m = (\Omega\lambda_k\eta_k)_\#\gamma^k_t(y_{p^m}).$$

For $0 \le m < k$, the equation

(12O4) $$\mathrm{ad}^{p^{m+1}-p^m}(\tilde{b})(\tilde{a}_m) = (p)\tilde{a}_{m+1}$$

is true. To prove (12O4), consider the cases $k > m > 0$ and $k > m = 0$ separately. When $k > m > 0$, start with the equation in $\pi_*(\Omega E_k; \mathbb{Z}_{p^t})$,

(12O5) $$\Phi((\tilde{b})^{p^{m+1}-p^m})\gamma^k_t(y_{p^m}) = (p)\gamma^k_t(y_{p^{m+1}}),$$

which holds because γ_t^k is a csm homomorphism and the equation

$$\mathrm{ad}(b^{p^{m+1}-p^m})(y_{p^m}) = (p)y_{p^{m+1}}$$

holds in $\mathcal{F}(\bar{L}^k)_{[t]}$. Apply $(\Omega\lambda_k\eta_k)_\#$ to both sides of (12O5) to get (12O4). When $k > m > 0$, we need the additional fact, implicit in the definition of γ_t^0, that $\gamma_t^k(y_2)$ is the unique element of $\pi_{4n-1}(\Omega E_k; \mathbb{Z}_{p^t})$ satisfying

$$(\Omega\lambda_k\eta_k)_\#\gamma_t^k(y_2) = [\tilde{b}, \tilde{a}_0].$$

Because

$$\mathrm{ad}(b^{p-2})(y_2) = (p)y_p$$

in $\mathcal{F}(\bar{L}^k)_{[t]}$, we have

$$\Phi((\tilde{b})^{p-2})\gamma_t^k(y_2) = (p)\gamma_t^k(y_p).$$

Again, apply $(\Omega\lambda_k\eta_k)_\#$ to both sides.

Having established (12O4), it is clear from the presentation of M^k in Chapter 3 that $b \mapsto \tilde{b}$, $a_m \mapsto \tilde{a}_m$ defines a Lie algebra homomorphism

$$\widetilde{\gamma}_t^k : M^k \longrightarrow \pi_*(\Omega D_k; \mathbb{Z}_{p^t}).$$

This $\widetilde{\gamma}_t^k$ satisfies

(12O6) $$_\pi\delta^t\widetilde{\gamma}_t^k = \widetilde{\gamma}_t^k(p^{-t}d).$$

We need only to check (12O6) on generators. Formula (12O1) gives us (12O6) on $\tilde{b}$, and we leave the check for $\tilde{a}_m$ as an exercise. Thus

$$\widetilde{\gamma}_t^k : (M^k, p^{-t}d) \longrightarrow \pi_*(\Omega D_k; \mathbb{Z}_{p^t})$$

is a homomorphism of dgL's. It should come as no surprise, using Proposition 7.5, that

$$\{\widetilde{\gamma}_t^k\}_{1\le t\le r} : \mathcal{F}(M^k) \longrightarrow \{\pi_*(\Omega D_k; \mathbb{Z}_{p^t})\}$$

is a homomorphism of csL's. The diagram of csL homomorphisms

(12O7) $$\begin{array}{ccc} \mathcal{F}(\bar{L}^k)_{[t]} & \xrightarrow{\mathcal{F}(\bar{\epsilon})} & \mathcal{F}(M^k)_{[t]} \\ {\scriptstyle \gamma_t^k}\big\downarrow & & {\scriptstyle \widetilde{\gamma}_t^k}\big\downarrow \\ \pi_*(\Omega E_k; \mathbb{Z}_{p^t}) & \xrightarrow{(\Omega\lambda_k\eta_k)_\#} & \pi_*(\Omega D_k; \mathbb{Z}_{p^t}) \end{array}$$

commutes, for $1 \le t \le r$. Furthermore, it is straightforward that $\{\widetilde{\gamma}_t^k\}_{1\le t\le r}$ is a homomorphism of csL's over $\mathcal{F}(\mathcal{U}M^k)$, and that (12O7) is a commuting diagram of csL/csm homomorphisms.

Consider next the csL $\mathcal{F}(M^k \amalg M^k)$. For $1 \le t \le r$ we have

$$\mathcal{F}(M^k \amalg M^k)_{[t]} = (M^k \amalg M^k, p^{-t}d).$$

A set of generators for $M^k \amalg M^k$ can be denoted as

$$\{b^{(1)}, a_0^{(1)}, \ldots, a_k^{(1)}, b^{(2)}, a_0^{(2)}, \ldots, a_k^{(2)}\}.$$

Define a csL homomorphism

$$\text{(12O8)} \qquad \{\widetilde{\beta}_t^k\}_{1\le t\le r} : \mathcal{F}(M^k \amalg M^k) \to \{\pi_*(\Omega(D_k \vee D_k); \mathbb{Z})_{p^t})\}$$

by

$$\widetilde{\beta}_t^k(b^{(i)}) = (\Omega\hat{e}^{(i)})_\# \widetilde{\gamma}_t^k(b) \ , \ \widetilde{\beta}_t^k(a_m^{(i)}) = (\Omega\hat{e}^{(i)})_\# \tilde{\gamma}_t^k(a_m^{(i)}) \ , \ i = 1, 2 \ .$$

Let $q = (\bar{\pi}_1, \bar{\pi}_2)$ denote the inclusion $D_k \vee D_k \longrightarrow D_k \times D_k$. The $(\Omega q)_\#$–image of $\hat{w} \in \pi_*(\Omega(D_k \vee D_k); \mathbb{Z}_{p^t})$ is zero, if $\hat{w}$ is an iterated Samelson product having at least one factor in $\operatorname{im}(\Omega\hat{e}^{(1)})_\#$ and at least one factor in $\operatorname{im}(\Omega\hat{e}^{(2)})_\#$. In particular, $(\Omega q)_\# \widetilde{\beta}_t^k(w) = 0$ if w lies in the image of

$$\mathcal{F}(N^{kk}) \xrightarrow{\mathcal{F}(\bar{e})} \mathcal{F}(M^k \amalg M^k)$$

for $t \le r$, $\bar{e}$ now denoting the dgL inclusion of N^{kk} into $M^k \amalg M^k$. Referring to (12A), each such $\widetilde{\beta}_t^k(w)$ lifts through $\pi_*(\Omega D_{kk}; \mathbb{Z}_{p^t})$. Because $(\Omega\eta_{kk})_\#$ is a monomorphic (since (12A) splits) csL homomorphism, there is a well–defined csL homomorphism

$$\{\beta_t^k\}_{1\le t\le r} : \mathcal{F}(N^{kk}) \to \{\pi_*(\Omega D_{kk}; \mathbb{Z}_{p^t})\}$$

for which

$$\text{(12O9)} \qquad (\Omega\eta_{kk})_\# \beta_t^k = \widetilde{\beta}_t^k \mathcal{F}(\bar{e}) \ .$$

It is straightforward that $(\Omega\eta_{kk})_\#$ and (12O8) are homomorphisms of csL's over the Dcsa $\mathcal{F}(\mathcal{U}M^k \amalg \mathcal{U}M^k)$, so $\{\beta_t^k\}_{1\le t\le r}$ is also a csL/csm homomorphism. This completes the construction of $\{\beta_t^k\}_{1\le t\le r}$, for any $k \ge 0$.

We leave it to the reader to check that $\{\beta_t^k\}_{1\le t\le r}$ satisfies all the relevant parts of Theorem 12.4. For the case $k = 0$ we add two remarks. First, (xvi) at $k = 0$ is the well–known fact that $D_{00} = \Sigma(\Omega P^{2n+1}(p^r)) \wedge (\Omega P^{2n+1}(p^r))$ has the homotopy type of a bouquet of mod p^r Moore spaces. Second, (xvii) and (xviii) are just complicated–looking ways of writing down some Jacobi identities, when $k = 0$.

Construction of β_t^k for $r < t < r + k$

To construct β_t^k for $r < t < r + k$, we assume that β_{t-1}^k exists having all the stated properties, and we follow the dimension–by–dimension inductive procedure outlined in Theorem 7.15. We have in Example 7.17, and summarized in 7.18, a list of virtual generators (VG's), virtual syzygies (VS's), and virtual relators (VR's) for the source csL $\mathcal{E}(N^{kk})$ (see Theorem 11.D13(a)). Assuming β_t^k has been constructed below some dimension m, we must choose images for any VG's arising in dimension $\le m+1$, and we must verify that any VS's in dimension m and VR's in dimension $m-1$ are also VS's and VR's for the target csL. As it turns out, we prefer to prove (xix) and (xiv) as part of the dimension–wise induction. When constructing β_t^k in dimension m, we can assume that (xix) and (xiv) hold for β_{t-1}^k in all dimensions, and that they hold for β_t^k below dimension m. (To be precise, (12M) and (12N) are assumed to be true if $|w| + |y| < m$.)

Property (xiv) compels our choice of where to send each basic VG in (7DC). Since $r < t < r+k$, we let $s = t - r < k$, and we define $\beta_t^k(w) = (\Omega\hat{i})_{\#}\beta_t^{k-1}(w)$ when w is one of the four elements listed in (7DC). They have the right θ_{t-1}^t–images and the right boundaries, as required by 7.15(i,ii), because these properties are true of β_t^{k-1}. Similarly, VG's in the list (7DD) other than those involving $a_k^{(1)}$ or $a_k^{(2)}$ have their β_t^k–images determined by (12I).

Now suppose our VG is $\mathrm{ad}^{(1)}(c_s)(a_k^{(2)})$. We define

$$\beta_t^k(\mathrm{ad}^{(1)}(c_s)(a_k^{(2)})) = \Psi_t^{kk}(c_s^{(1)})\gamma_t^k(y_{p^k}^{(2)}).$$

For compatibility with θ_{t-1}^t (i.e. 7.15(i)), we have

$$\begin{aligned}\theta_{t-1}^t\beta_t^k(\mathrm{ad}^{(1)}(c_s)(a_k^{(2)})) &= \Psi_{t-1}^{kk}(c_s^{(1)})\gamma_{t-1}^k(y_{p^k}^{(2)})\\ &= \beta_{t-1}^k\mathrm{ad}^{(1)}(c_s)(a_k^{(2)}) = \beta_{t-1}^k\theta_{t-1}^t(\mathrm{ad}^{(1)}(c_s)(a_k^{(2)})),\end{aligned}$$

via our inductive assumption that (xix) holds for β_{t-1}^k. The verification of 7.15(i) for the other VG's that involve $a_k^{(1)}$ or $a_k^{(2)}$ is quite similar. Likewise, the check (7.15(ii)) that

$$d_{[t]}\beta_t^k(\mathrm{ad}^{(1)}(c_s)(a_k^{(2)})) = \beta_t^k d_{[t]}(\mathrm{ad}^{(1)}(c_s)(a_k^{(2)}))$$

uses the inductive assumption that (xix) holds below dimension m, and it works similarly for the other VG's. Condition (7BK) is also easily checked.

When we encounter a VS at dimension m, we must say why the same VS holds for $\hat{\pi}_*(\Omega D_{kk}; \mathbf{Z}_{p^t})$. We saw in 7.17 that the VS's are all consequences of two types of relations. The first type are the relations (7DK)–(7DN), whose counterparts hold in $\hat{\pi}_*(\Omega D_{kk}; \mathbf{Z}_{p^t})$ by the grand inductive assumption that (xvii) and (xviii) are valid at $s < k$. The second type arise by applying $\mathrm{ad}^{(1)}(w)$ or $\mathrm{ad}^{(2)}(w)$ for suitable w to the relations (7DG) or (7DH). Because γ_t^k is a csL homomorphism, these second type equations are easily translated into the corresponding equations in $\hat{\pi}_*(\Omega D_{kk}; \mathbf{Z}_{p^t})$, so we can deduce the appropriate VS in $\hat{\pi}_*(\Omega D_{kk}; \mathbf{Z}_{p^t})$ as well.

Now consider Condition (7.15(iv)) for a VR at dimension $m-1$. It suffices to check the condition for those VR's that are the $\mathcal{F}(\bar{e})$–images of the VG's (cf. 7.18). For those VG's that do not involve $a_k^{(1)}$ or $a_k^{(2)}$, the corresponding VR is deducible from the grand inductive scheme and (xv) for β_t^{k-1}. For a VG of the form $w = \mathrm{ad}^{(2)}(u)(a_k^{(1)})$, use the first Jacobi identity to evaluate the right–hand side of (12J) applied to an arbitrary $\hat{x} \in \hat{\pi}_*(\Omega D_{kk}; \mathbf{Z}_{p^t})$. The answer comes out to the left–hand side of (12J).

This completes verification of the four hypotheses of Theorem 7.15, so the conclusion of the theorem holds, i.e., β_t^k exists as needed in dimension $\leq m$. But we strengthened these hypotheses by adjoining the dimensionwise versions of (xiv) and (xix), so we check these now at dimension m. Property (xiv) is a straightforward application of the uniqueness of a csL/csm homomorphism, i.e., (xiv) commutes because (by construction) it commutes on the virtual generators of $\mathcal{E}(N^{k-1,k-1})$.

For (xix), we perform the usual routine of building up to the full equation in small steps. Let $Q_y = \{w \mid \text{(12M) holds for } w \text{ and } y\}$. We suppress the implicit dimension restrictions from our notation in this and the next paragraph, e.g., "$Q_y = (R_s^k)_+$" means

"$Q_y = ((R_s^k)_+)_{\leq m-|y|}$". Observe first that $Q_y \supseteq \mathrm{im}(\zeta_t^{t-1})$ for any y, and consequently $Q_y \supseteq \mathrm{im}(d_{[t]}\zeta_t^{t-1})$ for any y. Next, $Q_y \supseteq (R_s^k)_+$ if and only if Q_y contains b_s and c_s and a_m for $s < m \leq k$. Recalling how β_t^k is defined on the VG's, it follows that $Q_{a_q} \supseteq (R_s^k)_+$, hence $Q_{a_q} = \mathcal{F}(\mathcal{U}M_+^k)_{[t]}$, for $s < q \leq k$. Since the same is true about a_q if the roles of (1) and (2) are switched, we have by (9N) that $a_q \in Q_y$ for any y. As a result, we need only to check that $b_s \in Q_y$ and $c_s \in Q_y$, in order to get $Q_y \supseteq (R_s^k)_+$, and then (since $Q_y \supseteq \mathrm{im}(\zeta_t^{t-1}) + \mathrm{im}(d_{[t]}\zeta_t^{t-1})$) we will have $Q_y = \mathcal{F}(\mathcal{U}M_+^k)_{[t]}$. One consequence of this characterization is that the set $Y = \{y \in \mathcal{F}(\bar{L}^k)_{[t]}^{*\leq 2p} \mid Q_y = \mathcal{F}(\mathcal{U}M_+^k)_{[t]}\}$ is closed under $d_{[t]}$.

Consider $u = y_{2p^s} \in \mathcal{F}(\bar{L}^k)_{[t]}$. Because β_t^k is a csL/csm homomorphism at or below any dimension we are currently interested in, (12L) actually shows that $b_s \in Q_u$. Similarly, (12M) says that $c_s \in Q_u$. So $Q_u = \mathcal{F}(\mathcal{U}M_+^k)_{[t]}$, i.e., $u \in Y$. Next, by using the VR's, see that Y is closed under $\mathrm{ad}(b_s)$ and $\mathrm{ad}(c_s)$. Since $Y \subseteq \{y_{2p^s}, a_{s+1}, \ldots, a_k)$, it follows that Y contains all the virtual generators $\{y_j, z_j' | j \in I_s\}$ of $\mathcal{F}(\bar{L}^k)_{[t]}$. Just two more steps remain. Y contains $\mathrm{im}(\zeta_t^{t-1})$ by the usual manipulations, so $Y \supseteq \mathrm{im}(\zeta_t^{t-1}) + \mathrm{im}(d_{[t]}\zeta_t^{t-1})$. Lastly, Y is closed under brackets: to see this, take apart both sides of (12M) when $y = [y', y'']$ using the csL–over–$\mathcal{F}(\mathcal{U}M^k)$ axiom. The result on each side is a sum of decomposables and ζ_t^{t-1}–images and $d_{[t]}\zeta_t^{t-1}$–images, so by the inductive assumption that (12M) holds below m, get these two sums to be equal. Conclude, finally, that $Y = \mathcal{F}(\bar{L}^k)_{[t]}^{*\leq 2p}$, i.e., (12M) holds for all w and y satisfying $|w| + |y| \leq m$.

This completes the construction of β_t^k for $t < r + k$, and it proves (xiv) and (xix) for $t < r+k$. To prove (xiii), it suffices to see that (12G) commutes on the VG's. For the basic VG's this is true because (xiii) holds at $k-1$; for the other VG's it is a trivial consequence of the definition of β_t^k.

Let us prove (xv) for $t < r + k$. We have already established that (xv) holds on the VG's. Put $W = \{w \in \mathcal{E}(N^{kk})_{[t]} \mid \text{(12J) is true for } w\}$. Now W is closed under $d_{[t]}$, and it contains $\zeta_t^{t-1}(\mathcal{E}(N^{kk})_{[t-1]})$, so we need only show that W is closed under $\mathrm{ad}^{(1)}(b_s)$ and $\mathrm{ad}^{(2)}(b_s)$. Once again, the first Jacobi identity enables us to confirm this. This completes the proof of all of Theorem 12.4, for $t < r + k$.

Equipped with $\{\beta_t^k\}_{t<r+k}$ we can decompose D_{kk} as a bouquet of Moore spaces. Recall the identification $g = g_{kk}$ of $\mathcal{L}(\mathcal{U}M_+^k \otimes \mathcal{U}M_+^k)$ with N^{kk} as a dgL (Theorem 5.12). By Lemma 5.4 we may write $\mathcal{U}M_+^k \otimes \mathcal{U}M_+^k = \mathrm{Span}\{w_\alpha, p^{-r_\alpha}d(w_\alpha)\}$, as α runs through an indexing set, and where each r_α is in the range $r \leq r_\alpha \leq r + k$.

For each index α having $r_\alpha < r+k$, let $\tilde{w}_\alpha : P^{|w_\alpha|}(p^{r_\alpha}) \to \Omega D_{kk}$ be any homotopy class in $\pi_*(\Omega D_{kk}; \mathbb{Z}_{p^{r_\alpha}})$ whose image when projected to the quotient Lie algebra $\hat{\pi}_*(\Omega D_{kk}; \mathbb{Z}_{p^{r_\alpha}})$ is $\beta_{r+k-1}^k g(w_\alpha)$. For an indexing α having $r_\alpha = r + k$, observe that $d_{[r+k-1]}\beta_{r+k-1}^k g(w_\alpha)$ is divisible by p. Let $\tilde{w}_\alpha' : P^{|w_\alpha|}(p^{r+k-1}) \to \Omega D_{kk}$ be a mod p^{r+k-1} homotopy class whose image in $\hat{\pi}_*(\Omega D_{kk}; \mathbb{Z}_{p^{r+k-1}})$ is $\beta_{r+k-1}^k g(w_\alpha)$. Since the kernel J' of $(\pi_* \to \hat{\pi}_*)$ is contained in $(p)\pi_*$, we see that ${}_\pi\delta^{r+k-1}(\tilde{w}_\alpha')$ is also divisible by p. Extend $\tilde{w}_\alpha'$ to a mod p^{r+k} homotopy class $\tilde{w}_\alpha : P^{|w_\alpha|}(p^{r+k}) \to \Omega D_{kk}$, i.e., $\tilde{w}_\alpha \omega_{r+k-1}^{r+k} = \tilde{w}_\alpha'$. Put $W = \bigvee_\alpha P^{|w_\alpha|}(p^{r_\alpha})$ and $\tilde{w} = \bigvee_\alpha \tilde{w}_\alpha : W \to \Omega D_{kk}$. Checking Hurewicz images, we find that the mod p Hurewicz image of W is exactly a submodule of indecomposable for the free associative $\mathbb{Z}_p$–algebra $H_*(\Omega D_{kk}; \mathbb{Z}_p) = H_*(\mathcal{U}N^{kk}; \mathbb{Z}_p)$, hence $\Omega\tilde{w}^{\bar{a}} : \Omega\Sigma W \to \Omega D_{kk}$ is a homotopy equivalence between 0–connected H–spaces. Deduce that $\tilde{w}^{\bar{a}} : \Sigma W \to D_{kk}$ provides the wedge decomposition of D_{kk} called for in (xvi).

Construction of β^k_{r+k} when $k>0$

Our next step is to construct β^k_{r+k}. Part (xiv) is vacuous at $t=r+k$. We will prove (xiii) and (xv) here, (xvi) is done, and (xvii)–(xix) will have to wait until after we define γ^k_{r+k} in Chapter 13. To construct β^k_{r+k}, again we follow the dimensionwise inductive framework suggested by Theorem 7.15. The hardest part by far is lifting the four VG's to $\pi_*(\Omega D_{kk};\mathbf{Z}_{p^{r+k}})$, i.e., in terms of 7.15, the hardest part is constructing $\tilde{x}_i$'s that satisfy 7.15(i) and 7.15(ii).

According to Example 7.17, there are only four VG's for $\mathcal{E}(N^{kk})_{[r+k]}$, namely the basic VG's, and there are only the four corresponding VR's. There are no VS's, which simplifies matters considerably.

We would like to apply Theorem 6.4 in order to obtain suitable β^k_{r+k}–images of the basic VG's. In Lemma 6.5 take $t=r+k-1$ and $q=4np^k-1$ or $q=4np^k$ and (in 6.5(iv)) $m=3$. All of conditions 6.5(i)–(iv) are satisfied, except – and this exception turns out to be a major snag – (6K) has a non–zero kernel in dimension $j=4np^k-2$. Let us identify this kernel precisely. Because of (xvi), an easy homology calculation for $H_*(D_{kk})$ shows that $X=D_{kk}$ has the wedge decomposition described in (11EG). In dimensions $\leq 4np^k$ the kernel of (6K), which is also (11EH), is the single copy of $\mathbf{Z}_p$ described in (11EI). According to (11EI), a generator of this kernel is

(12P1) $$\hat{w}_0 = p^{r+k-2}\tau_1(\hat{v}^a)\omega^{r+k-1} \in \pi_{4np^k-2}(\Omega D_{kk}),$$

where

(12P2) $$\hat{v}^a = \beta^k_{r+k-1}g(b_{k-1}\otimes b_{k-1}) \in \pi_{4np^k-1}(\Omega D_{kk};\mathbf{Z}_{p^{r+k-1}}).$$

Let us temporarily set aside the problem of the non–zero kernel of (6K), and let us check the other hypotheses of Theorem 6.4. We will apply Theorem 6.4 twice, since there are two pairs of VG's to be lifted. In the first application we take $\hat{y}=(\beta^k_{r+k-1}g(c_k\otimes b_k))^{\bar{a}}$ and $\hat{z}=(\beta^k_{r+k-1}g(c_k\otimes c_k))^{\bar{a}}$, so $q=4np^k-1$. (Note: By Note 12.5 $\hat{\pi}_m=\pi_m$ whenever m is not of the form $6np^s-2$ or $6np^s-1$, so there is no indeterminacy in the homotopy classes $\hat{y}$ and $\hat{z}$.) Since $\beta^k_{r+k-1}g$ is a chain map we certainly have ${}_\pi\delta^{r+k-1}(\hat{y})=p\hat{z}$ and ${}_\pi\delta^{r+k-1}(\hat{z})=0$. By (12H) we know $y=\underline{a}_{\eta_{kk}}(\hat{y})=\bar{e}(g(c_k\otimes b_k))\in \mathcal{U}M^k \amalg \mathcal{U}M^k$ and $z=\underline{a}_{\eta_{kk}}(\hat{z})=\bar{e}(g(c_k\otimes c_k))$, so the condition $d(y)=p^{r+k}z$ is fulfilled.

The hypothesis in Theorem 6.4 that p^{r+k} divides $\mathcal{A}(\bar{\pi}_1,\bar{\pi}_2)(y)$ is not immediately obvious. Corollary 6.6 permits us to replace the homomorphism $\mathcal{A}(\bar{\pi}_1,\bar{\pi}_2)$ by the projection of $\mathcal{U}M^k \amalg \mathcal{U}M^k$ onto $\mathcal{U}M^k \otimes \mathcal{U}M^k$, because this diagram commutes by (iv) and [AH, Sec. 4]:

$$\begin{array}{ccc}
\mathcal{A}(D_k\vee D_k) & \xrightarrow{\mathcal{A}(\bar{\pi}_1,\bar{\pi}_2)} & \mathcal{A}(D_k\times D_k)\\
{\scriptstyle =}\downarrow & & \downarrow{\scriptstyle \simeq}\\
\mathcal{A}(D_k)\amalg\mathcal{A}(D_k) & \longrightarrow & \mathcal{A}(D_k)\otimes\mathcal{A}(D_k)\\
{\scriptstyle \simeq}\downarrow & & \downarrow{\scriptstyle \simeq}\\
\mathcal{U}M^k\amalg\mathcal{U}M^k & \xrightarrow{Q_{(\bar{\pi}_1,\bar{\pi}_2)}} & \mathcal{U}M^k\otimes\mathcal{U}M^k
\end{array}$$

Since the composite $N^{kk}\xrightarrow{\bar{e}}\mathcal{U}M^k\amalg\mathcal{U}M^k\xrightarrow{Q_{(\bar{\pi}_1,\bar{\pi}_2)}}\mathcal{U}M^k\amalg\mathcal{U}M^k$ is the zero homomorphism, we have $Q_{(\bar{\pi}_1,\bar{\pi}_2)}(y)=0$. Now all the hypotheses of Theorem 6.4 are seen to be satisfied, except that (6K) has the indicated kernel.

It took the author a long time to work out the details of how to circumvent the snag of the non–zero kernel. We will provide only an outline since in the context of the entire Theorem 12.4 this is a minor technical point. We know by (xvi) and Lemma 10.2 that $p^{r+k}\pi_{\leq 4np^k}(\Omega D_{kk}) = 0$, so by Lemma 11.E1(b) we can drop the requirement that (6K) be one–to–one if we can verify instead that (11EC) holds, i.e., that

(12P3) $$d_{[r+k]}\zeta_{r+k}^{r+k-1}(\hat{y}) = \zeta_{r+k}^{r+k-1}(\hat{z}).$$

It may be possible to prove (12P3) directly from the definitions of $\hat{y}$ and $\hat{z}$ (i.e. from the definition of β_{r+k-1}^k), but after finding several incorrect proofs that for subtle reasons involved circular logic the author abandoned this approach. (E.g. it does not work to say that the equation $d_{[r+k]}\zeta_{r+k}^{r+k-1}g(c_k \otimes b_k) = \zeta_{r+k}^{r+k-1}g(c_k \otimes c_k)$ holds in $\mathcal{E}(N^{kk})_{[r+k]}$ so we merely apply β_{r+k-1}^k to get an equation in $\pi_*(\Omega D_{kk}; \mathbf{Z}_{p^{r+k}})$).

Instead, we employ an indirect strategy to prove (12P3). We have verified all the hypotheses of 6.4 except the injectivity of (6K). Let the module J be

$$J = p^{r+k-2}\ \mathrm{Span}\{\zeta_{r+k}^{r+k-1}\tau_1(\hat{v}^a), d_{[r+k]}\zeta_{r+k}^{r+k-1}\tau_1(\hat{v}^a)\}.$$

By Lemmas 11.E4 and 11.E2 we find easily that (12P3) is true modulo J. We are in dimension $4np^k - 2$, so

(12P4) $$d_{[r+k]}\zeta_{r+k}^{r+k-1}(\hat{y}) - \zeta_{r+k}^{r+k-1}(\hat{z}) = (c)p^{r+k-2}d_{[r+k]}\zeta_{r+k}^{r+k-1}\tau_1(\hat{v}^a)$$

for some $c \in \mathbf{Z}_p$. What is required of us is to show that $c = 0$.

The key to getting $c = 0$ in (12P4) is a trick that we call "detecting" c. The trick is precisely the geometric analog of the "detection" argument at the end of the proof of Theorem 5.12. The idea is to map the equation into a simpler space where c becomes more plainly visible. What works is to use the map $\tilde{\rho}\rho_k : D_k \to S^{2n+1}$ of (9C) to induce a map ρ' between homotopy–theoretic fibers in the diagram of fibrations

(12P5) $$\begin{array}{ccccc} D_{kk} & \xrightarrow{\eta_{kk}} & D_k \vee D_k & \longrightarrow & D_k \times D_k \\ {\scriptstyle\rho'}\downarrow & & \downarrow{\scriptstyle\tilde{\rho}\rho_k \vee 1} & & \downarrow{\scriptstyle\tilde{\rho}\rho_k \times 1} \\ S' & \xrightarrow{s'} & S^{2n+1} \vee D_k & \longrightarrow & S^{2n+1} \times D_k. \end{array}$$

Here $S' = \Sigma(\Omega S^{2n+1}) \wedge (\Omega D_k)$ is a wedge of Moore spaces, $S' \in \mathcal{W}_r^{r+k}$, and $(s')_\#$ is monomorphic because the bottom row of (12P5) splits when looped once.

We claim that both $(\beta_{r+k-1}^k g(c_k \otimes b_k))^{\bar{a}}$ and $(\beta_{r+k-1}^k g(c_k \otimes c_k))^{\bar{a}}$ lie in the kernel of $(\rho')^\#$. Since $(s')_\#$ is one–to–one, the way to see this is to look at their $(\tilde{\rho}\rho_k \vee 1)_\#(\eta_{kk})_\#$–images in $\pi_*(S^{2n+1} \vee D_k; \mathbf{Z}_{p^{r+k-1}})$. Omitting many details, after unraveling the definitions it boils down to the fact that $\mathcal{O}_{(\tilde{\rho}\rho_k)}[r+k-1]$ takes $\Psi_{r+k-1}^k(c_k) \in \mathcal{O}_{D_k}[r+k-1]$ to the zero operation in the ring $\mathcal{O}_{S^{2n+1}}[r+k-1]$.

So the left–hand side of (12P4) becomes zero when $(\rho')_\#$ is applied to it. However, the Moore space summand $P^{4np^{k-1}+1}(p^{r+k-1})$ of D_{kk} on which $\hat{v}$ lives ($\hat{v}$ is given by (12P2)) is sent by $(\rho')_\#$ to a Moore space summand of the same order in S', so the composite

(12P6) $$P^{4np^{k-1}+1}(p^{r+k-1}) \xrightarrow{\hat{v}} D_{kk} \xrightarrow{\rho'} S'$$

has a left inverse. (Obviously, we are again omitting many details.) Consequently $(\rho')_\#$ is one–to–one on $\mathrm{im}(\hat{v}_\#)$, and $\mathrm{im}(\hat{v}_\#)$ contains the element

$$(c)p^{r+k-2}(d_{[r+k]}\zeta_{r+k}^{r+k-1}\tau_1(\hat{v}^a))^{\bar{a}}. \tag{12P7}$$

Since the left–hand side of (12P4) is in $\ker(\rho')_\#$, so is (12P7). But $(\rho')_\#$ is one–to–one on a submodule containing (12P7). The only way this can happen is if $c = 0$.

Where are we in the proof? We finally got c to be zero in (12P4), meaning that (12P3) is true. Now Lemma 11.E1(b) applies, and Theorem 6.4 gives us a homotopy class denoted $\hat{y}_2 \in \pi_{4np^k}(D_{kk}; \mathbb{Z}_{p^{r+k}})$ for which $\theta_{r+k-1}^{r+k}(\hat{y}_2) = \hat{y}$ and $\theta_{r+k-1}^{r+k}d_{[r+k]}(\hat{y}_2) = \hat{z}$ and $\underline{a}_{\eta_{kk}}(\hat{y}_2) = y = \bar{e}(g(c_k \otimes b_k))$. Define $\beta_{r+k}^k g(c_k \otimes b_k)$ to be $\hat{y}_2^a$ and $\beta_{r+k}^k g(c_k \otimes c_k)$ to be ${}_\pi\delta^{r+k}(\hat{y}_2^a)$. Notice that (12G) commutes on these virtual generators, i.e., (12H) is true for $w = g(c_k \otimes b_k)$ and for $w = g(c_k \otimes c_k)$.

This completes the lifting for the first pair of VG's. Next, we attempt to use Theorem 6.4 to construct suitable β_{r+k}^k–images for the second pair of VG's, which consists of $g(b_k \otimes b_k)$ and $d_{[r+k]}g(b_k \otimes b_k) = g(-c_k \otimes b_k - b_k \otimes c_k)$. We put $q = 4np^k$ and $t = r + k - 1$ and $\hat{y} = (\beta_{r+k-1}^k g(b_k \otimes b_k))^{\bar{a}}$ and $\hat{z} = (\beta_{r+k-1}^k g(-c_k \otimes b_k - b_k \otimes c_k))^{\bar{a}}$. All the hypotheses of Theorem 6.4 are satisfied with the exception that (6K) still has $\hat{w}_0$ in its kernel. This time, since $|\hat{w}_0| = 4np^k - 2$ equals $q - 2$ instead of $q - 1$, we will need Lemma 11.E1(a) instead of 11.E1(b) in order to get around the problem. According to (11EB) what needs to be demonstrated is

$$d_{[r+k]}\zeta_{r+k}^{r+k-1}\beta_{r+k-1}^k g(b_k \otimes c_k + c_k \otimes b_k) = 0. \tag{12P8}$$

Since we already showed (this is what (12P3) says) that

$$d_{[r+k]}\zeta_{r+k}^{r+k-1}\beta_{r+k-1}^k g(c_k \otimes b_k) = \zeta_{r+k}^{r+k-1} g(c_k \otimes c_k),$$

what remains to prove is

$$d_{[r+k]}\zeta_{r+k}^{r+k-1}\beta_{r+k-1}^k g(b_k \otimes c_k) = -\zeta_{r+k}^{r+k-1} g(c_k \otimes c_k). \tag{12P9}$$

Equation (12P9) is also proved using a detection argument. For (12P9) we "simplify" on the second coordinate, i.e. in place of (12P5) we utilize the diagram

$$\begin{array}{ccccc}
D_{kk} & \xrightarrow{\eta_{kk}} & D_k \vee D_k & \longrightarrow & D_k \times D_k \\
{\scriptstyle \rho''}\downarrow & & \downarrow{\scriptstyle 1 \vee \tilde{\rho}\rho_k} & & \downarrow{\scriptstyle 1 \times \tilde{\rho}\rho_k} \\
S'' & \xrightarrow{s''} & D_k \vee S^{2n+1} & \longrightarrow & D_k \times S^{2n+1}
\end{array}$$

Again we omit the proof itself, but (12P9) follows. Then (12P8) follows, and Lemma 11.E1(a) permits us to apply Theorem 6.4.

Theorem 6.4 supplies us with a homotopy class denoted $\hat{y}_2 \in \pi_{4np^k+1}(D_{kk}; \mathbb{Z}_{p^{r+k}})$ for which $\theta_{r+k-1}^{r+k}(\hat{y}_2) = \hat{y}$ and $\theta_{r+k-1}^{r+k}d_{[r+k]}(\hat{y}_2) = \hat{z}$ and $\underline{a}_{\eta_{kk}}(\hat{y}_2) = \bar{e}(g(b_k \otimes b_k))$. Put $\beta_{r+k}^k g(b_k \otimes b_k) = \hat{y}_2^a$ and $\beta_{r+k}^k g(b_k \otimes c_k) = -{}_\pi\delta^{r+k}(\hat{y}_2^a) - \beta_{r+k}^k(c_k \otimes b_k)$. Notice that (12H) is true, when $w = g(b_k \otimes b_k)$ or $w = g(b_k \otimes c_k)$.

This completes the job of choosing lifts in $\pi_*(\Omega D_{kk}; \mathbb{Z}_{p^{r+k}})$ (which coincides with $\hat{\pi}_*(\Omega D_{kk}; \mathbb{Z}_{p^{r+k}})$ in these dimensions) for all the VG's. As the reader can readily check, we did this in such a way that criteria 7.15(i,ii) are satisfied. Since there are no VS's, 7.15(iii) is vacuously satisfied. Let us turn to 7.15(iv).

Why is each of the basic VG's a VR when it is carried by $\bar{e}$ into $\mathcal{U}M^k \amalg \mathcal{U}M^k$? It suffices to prove that (12J) holds, for w a basic VG. The universal example for (12J) is the homotopy–theoretic fiber in the fibration

$$(D_k \vee D_k)'(q+1, p^{r+k}) \xrightarrow{d'_\vee} (D_k \vee D_k) \vee P^{q+1}(p^{r+k})) \xrightarrow{\bar{\pi}_1} D_k \vee D_k,$$

but it will come as no surprise (cf. Remark 10.3) that this fibration is not universal enough. We look instead at the top row in the diagram of fibration sequences,

(12Q)

$$\begin{array}{ccccc}
\mathcal{G}(D_k, D_k, P^{q+1}) & \longrightarrow & D_k \vee D_k \vee P^{q+1} & \xrightarrow{\tilde{\lambda}} & \mathcal{R}(D_k, D_k, P^{q+1}) \\
\downarrow & & \downarrow = & & \downarrow \mathcal{R}(1,1,*) \\
(D_k \vee D_k)'(q+1, p^{r+k}) & \xrightarrow{d'_\vee} & D_k \vee D_k \vee P^{q+1} & \longrightarrow & D_k \vee D_k = \mathcal{R}(D_k, D_k, *),
\end{array}$$

where we have abbreviated $P^{q+1}(p^{r+k})$ as P^{q+1} in order to fit the diagram on the page.

Denote the fiber $\mathcal{G}(D_k, D_k, P^{q+1}(p^{r+k}))$ in (12Q) by $D_k^\#$. We have that $D_k^\#$ is a wedge of Moore spaces by Proposition 6.25(e), since each summand $\Sigma(\Omega D_k)^{\wedge a_1} \wedge (\Omega D_k)^{\wedge a_2} \wedge (\Omega P^{q+1}(p^{r+k}))^{\wedge a_3}$ is a wedge of Moore spaces. The top row of (12Q) splits when looped once, by Lemma 6.24(b). Let $\hat{v}$ denote the homotopy class of the inclusion of $P^{q+1}(p^{r+k})$ into $D_k \vee D_k \vee P^{q+1}(p^{r+k})$, and suppose w is any one of the four basic VG's. We leave it to the reader to check that all the hypotheses of Corollary 6.7 are fulfilled, for

$$\text{(12R)} \qquad \begin{aligned} \hat{y}_0 &= [(\Omega\eta_{kk})_\# \beta^k_{r+k}(w), \hat{v}^a]^{\bar{a}} - \Psi^{kk}_{r+k}(\mathcal{F}(\bar{e})(w))(\hat{v}) \\ &\in \pi_*(\Omega(D_k \vee D_k \vee P^{q+1}(p^{r+k})); \mathbb{Z}_{p^{r+k}}). \end{aligned}$$

Deduce from Corollary 6.7 that $\hat{y}_0 = 0$. Since the map $d'_\vee$ in (12Q) induces a monomorphism on homotopy, (12J) is true for basic VG's.

We have now completed the verification of the four hypotheses in Theorem 7.15, so there exists β^k_{r+k} making $\{\beta^k_t\}_{1 \le t \le r+k}$ into a csL/csm homomorphism over $\mathcal{F}(\mathcal{U}M^k \amalg \mathcal{U}M^k)$. Now (xiii) holds in general at $t = r+k$ because it holds for the virtual generators. The proof we gave before that $\{w \in \mathcal{E}(N^{kk})_{[t]} \mid \text{(12J) is true for } w\}$ equals all of $\mathcal{E}(N^{kk})_{[t]}$ works equally well for $t = r+k$ as for $t < r+k$. This completes the proof of Theorem 12.4, with the exception of the three parts that we are postponing until after we can define γ^k_{r+k}.

Remark. Referring to Note 12.5, nowhere in the proof given above did we need the fact that we were working with $\hat{\pi}_*$ rather than π_*. We express Theorem 12.4 in terms of $\hat{\pi}_*$ because we are only going to prove (12K) and (12L) modulo $J'_{[r+k]}$. The need to divide out by J' will become plausible with Theorem 12.6, and apparent only with Theorem 13.15.

The Third Jacobi Identity

We have one more theorem to consider in Chapter 12. It establishes the absolutely critical "third Jacobi identity". Theorem 12.6 is a "three–factor" analog of Theorem 12.4, i.e., it deals with $D_k \vee D_k \vee D_k$ where Theorem 12.4 dealt with $D_k \vee D_k$. Since we are thinking of (6MM) as the three–factor analog of (6NN), the fibration we are interested in is

$$\text{(12S)} \qquad D_{kkk} = \mathcal{G}(D_k, D_k, D_k) \xrightarrow{\eta_{kkk}} D_k \vee D_k \vee D_k \xrightarrow{\bar\lambda} R = \mathcal{R}(D_k, D_k, D_k),$$

which serves to define the space D_{kkk} and the map η_{kkk}. The inclusion map $\hat{i}: D_{k-1} \to D_k$ induces a map on fibers, also denoted $\hat{i}$:

$$\text{(12T)} \qquad \begin{array}{ccccc} D_{k-1,k-1,k-1} & \xrightarrow{\eta_{k-1,k-1,k-1}} & D_{k-1} \vee D_{k-1} \vee D_{k-1} & \xrightarrow{\bar\lambda} & \mathcal{R}(D_{k-1}, D_{k-1}, D_{k-1}) \\ \downarrow{\scriptstyle \hat{i}} & & \downarrow{\scriptstyle \hat{i}\vee\hat{i}\vee\hat{i}} & & \downarrow{\scriptstyle \mathcal{R}(\hat{i},\hat{i},\hat{i})} \\ D_{kkk} & \xrightarrow{\eta_{kkk}} & D_k \vee D_k \vee D_k & \xrightarrow{\bar\lambda} & \mathcal{R}(D_k, D_k, D_k). \end{array}$$

By Lemma 6.24(b), (12S) splits when looped once.

The space D_{kkk} has the homotopy type of a wedge of Moore spaces; specifically, $D_{kkk} \in \mathcal{W}_r^{r+k}$. To see this, use Proposition 6.25(e) to split D_{kkk} into pieces of the form $\Sigma(\Omega D_k)^{\wedge m}$, $m \geq 3$. Use Theorem 12.4(xvi) to write the pieces as a wedge of smash products of (ΩD_k)'s and $P^{q+1}(p^t)$'s, for $r \leq t \leq r+k$. Then use the fact (proved in Chapter 10) that $\Sigma^2 \Omega D_k$ and $P^{q+1}(p^t) \wedge (\Omega D_k)$ belong to $\mathcal{W}_r^{r+k}$, if $r \leq t \leq r+k$.

The homology of D_{kkk} can be computed from Proposition 6.25 as well. The information we need from this homology computation is that $X = D_{kkk}$ satisfies the description (11EJ).

Notation. By analogy with the situation described in Notation 12.3, we are particularly interested in the result of applying homotopy operations associated with one component in $D_k \vee D_k \vee D_k$ to homotopy classes associated with the other two components. Let $\hat{e}^{(i)}: D_k \to D_k \vee D_k \vee D_k$ denote the three geometric inclusion maps, $i = 1, 2, 3$, and let

$$\hat{e}^{(12)}, \hat{e}^{(13)}, \hat{e}^{(23)}: D_k \vee D_k \to D_k \vee D_k \vee D_k$$

denote the inclusion maps for pairs of summands. Similarly, let

$$e^{(12)}, e^{(13)}, e^{(23)}: N^{kk} \to M^k \amalg M^k \to M^k \amalg M^k \amalg M^k$$

denote the three dgL inclusions. It is clear that $\mathrm{ad}^{(1)}(w)e^{(23)}(y)$ belongs to N^{kkk} whenever $w \in \mathcal{U}M_+^k$ and $y \in N^{kk}$, and likewise if we permute (1), (2), (3). When $w \in \mathcal{F}(\mathcal{U}M_+^k)_{[t]}$ and $y \in \mathcal{E}(N^{kk})_{[t]}$, then $\mathrm{ad}^{(1)}(w)e^{(23)}(y)$ belongs to $\mathcal{E}(N^{kkk})_{[t]}$, by Theorem 11.D13(g). Accordingly, as with (12D), we allow the notation

$$\Psi_t^{kkk}(w^{(1)}\beta_t^k(y^{(23)}))$$

to denote the element

$$\text{(12U)} \qquad (\Omega\eta_{kkk})_{\#}^{-1}\psi_t^{kkk}(e^{(1)}(w) \otimes (\Omega\hat{e}^{(23)}\eta_{kk})_{\#}\beta_t^k(y)) \in \hat\pi_*(\Omega D_{kkk}; \mathbb{Z}_{p^t}),$$

and likewise with (1), (2), (3) permuted. When y is denoted as $g(u)$, where $g = g_{kk} : \mathcal{L}(\mathcal{U}M_+^k \otimes \mathcal{U}M_+^k) \to N^{kk}$ is the dgL isomorphism of Theorem 5.12, it is sometimes notationally inconvenient to impose a superscript on y. In such a case we write $y^{(23)}$ as $g^{(23)}(u)$. For example, if $y = g(b_k \otimes c_k)$, our notation for (12U) would be

$$\Psi_t^{kkk}(w^{(1)})\beta_t^k g^{(23)}(b_k \otimes c_k).$$

We label the parts of Theorem 12.6 starting with (xx), to emphasize their continuity with Theorems 9.1 and 12.4. The theorem concerns the dgL N^{kkk}, which is discussed in 5.15, 6.21, 7.19, 7.20, and 11.D13. It also mentions the csL/csm homomorphism $\mathcal{E}(\bar{i})$ of Theorem 11.D13(f).

Theorem 12.6. *For each $k \geq 0$ there exists a csL/csm homomorphism over $\mathcal{F}(\mathcal{U}M^k \amalg \mathcal{U}M^k \amalg \mathcal{U}M^k)$,*

$$\{\alpha_t^k\}_{1\leq t\leq r+k} : \mathcal{E}(N^{kkk}) \to \{\hat{\pi}_*(\Omega D_{kkk}; \mathbf{Z}_{p^t})\},$$

where $\{\hat{\pi}_*(\Omega D_{kkk}; \mathbf{Z}_{p^t})\}$ denotes a certain quotient csL of $\{\pi_*(\Omega D_{kkk}; \mathbf{Z}_{p^t})\}$ which is described in Note 12.7 below and $\mathcal{E}(N^{kkk})$ is defined in 11.D12. It has the following properties.

(xx)

$$\underline{a}_{\eta_{kkk}}(\alpha_t^k(w)) = \bar{e}(w) \text{ for } w \in \mathcal{E}(N^{kkk})_{[t]}, \tag{12V}$$

where $\bar{e} : N^{kkk} \to M^k \amalg M^k \amalg M^k \to \mathcal{U}M^k \amalg \mathcal{U}M^k \amalg \mathcal{U}M^k$ is the inclusion homomorphism.

(xxi) *For $k \geq 1$ and $t \leq r+k-1$, this diagram commutes:*

$$\begin{array}{ccc} \mathcal{E}(N^{k-1,k-1,k-1})_{[t]} & \xrightarrow{\alpha_t^{k-1}} & \hat{\pi}_*(\Omega D_{k-1,k-1,k-1}; \mathbf{Z}_{p^t}) \\ \mathcal{E}(\bar{i})_{[t]} \downarrow & & \downarrow (\Omega\bar{i})_\# \\ \mathcal{E}(N^{kkk})_{[t]} & \xrightarrow{\alpha_t^k} & \hat{\pi}_*(\Omega D_{kkk}; \mathbf{Z}_{p^t}), \end{array} \tag{12W}$$

where $\mathcal{E}(\bar{i})$ is described in Theorem 11.D13(f).

(xxii) *("The third Jacobi identity") In $\hat{\pi}_*(\Omega D_{kkk}; \mathbf{Z}_{p^{r+k}})$,*

$$\begin{aligned} &\Psi_{r+k}^{kkk}(b_{r+k}^{(1)} g^{(23)}(b_k \otimes b_k) + \Psi_{r+k}^{kkk}(b_k^{(3)})\beta_{r+k}^k g^{(12)}(b_k \otimes b_k) \\ &= \Psi_{r+k}^{kkk}(b_k^{(2)})\beta_{r+k}^k g^{(13)}(b_k \otimes b_k) + \zeta_{r+k}^{r+k-1}\alpha_{r+k-1}^k(w_k^0), \end{aligned} \tag{12X}$$

where w_k^0 denotes the element of $\mathcal{E}(N^{kkk})$ defined in Theorem 11.D13(h). We think of (12X) as the topological counterpart of (7EC). Seven similar formulas (containing both ζ_{r+k}^{r+k-1} and $d_{[r+k]}\zeta_{r+k}^{r+k-1}$ terms in general) also hold in $\hat{\pi}_*(\Omega D_{kkk}; \mathbf{Z}_{p^{r+k}})$, corresponding to the seven other Jacobi identities in 7.19.

(xxiii) *For $w \in \mathcal{F}(\mathcal{U}M_+^k)_{[t]}$ and $y \in \mathcal{E}(N^{kk})_{[t]}$, $1 \leq t \leq r+k$,*

$$\alpha_t^k(\mathrm{ad}^{(1)}(w)e^{(23)}(y)) = \Psi_t^{kkk}(w^{(1)})\beta_t^k(y^{(23)}) \tag{12Y}$$

in $\hat{\pi}_*(\Omega D_{kkk};\mathbf{Z}_{p^t})$. *The other two formulas obtained by permuting* (1), (2), (3) *in* (12Y) *are also true in* $\hat{\pi}_*(\Omega D_{kkk};\mathbf{Z}_{p^t})$.

Note 12.7. Let S denote the set of triples

$$S = \{(r+s, 1, (\Omega\hat{i})_\# \alpha_{r+s}^{k-1}(\mathrm{ad}^{(2)}(b_s) g^{(13)}(b_s \otimes b_s))) \mid 0 \le s < k\}$$
$$\cup \{(r+s, 1, (\Omega\hat{i})_\# \alpha_{r+s}^{k-1}(\mathrm{ad}^{(1)}(b_s) g^{(23)}(b_s \otimes b_s))) \mid 0 \le s < k\}.$$

Put

$$J'' = \langle p^* \tau_*(S) \rangle \subseteq \{\pi_*(\Omega D_{kkk};\mathbf{Z}_{p^t})\},$$

where the notation is described in connection with Lemma 11.C5. It is part of the grand inductive scheme that the set we have written down is well-defined. (When $k = 0$ let $J'' = 0$.) Like the csL ideal J' of Note 12.5, J'' consists entirely of mirages, it is contained in $\ker(\Phi)$, and it is zero except in dimensions of the form $6np^s - 1$ and $6np^s - 2$, for $0 < s \le k$.

Proof assuming (xvii) *and* (xviii). We omit even major details, since we are following a pattern that we have seen before. As with Theorem 12.4, we employ the dimension-wise inductive framework of Theorem 7.15, embedded in a larger induction on t. Also mimicking the proof of Theorem 12.4, we carry along the dimension-wise versions of (xxi) and (xxiii) as part of the induction.

We have enough information about the virtual structure of $\mathcal{E}(N^{kkk})$, from 7.20. Property (xxiii) dictates where we must send each VG. That the $\bar{e}$-image of each VG is a VR for $\hat{\pi}_*(\Omega D_{kkk};\mathbf{Z}_{p^t})$ can be proved in a straightforward manner, modeled after the proof for β_t^k when $t < r+k$. (For α_t^k this technique also suffices at $t = r+k$.) Both (xv) and the first Jacobi identity are needed in order to carry out the proof. The VS's will hold in $\hat{\pi}_*(\Omega D_{kkk};\mathbf{Z}_{p^t})$ as long as the eight Jacobi identities of (xxii) and the four formulas of (xvii) and (xviii) are true. So the principal point of interest is the formulas (12X).

The formula (12X) and its seven sisters are proved using Corollary 6.7, modified according to Lemma 11.E2. As the fibration (6G) we take (12S). Conditions 6.5(i)–(iv) are satisfied, with the exception that (6K) has a non-zero kernel in dimension $6np^k - 2$. See (11EK) for a precise description of this kernel. The two homotopy classes $\hat{v}_1$ and $\hat{v}_2$ that occur in connection with (11EK) can be chosen to be

$$\hat{v}_1 = \hat{i}_\#(\alpha_{r+k-1}^{k-1}(\mathrm{ad}^{(1)}(b_{k-1}) g^{(23)}(b_{k-1} \otimes b_{k-1})))^{\bar{a}}$$

and

$$\hat{v}_2 = \hat{i}_\#(\alpha_{r+k-1}^{k-1}(\mathrm{ad}^{(2)}(b_{k-1}) g^{(13)}(b_{k-1} \otimes b_{k-1})))^{\bar{a}}.$$

Let J be the module (11EL). What Lemma 11.E2 tells us is that the eight Jacobi identities are true, modulo J. (If $\hat{y}_0$ is the difference between the two sides of (12X) or of another of the eight Jacobi identities then the "four formulas" of Corollary 6.7 are easily checked, for $\hat{y}_0$). It seems likely that a detection type of argument could eliminate this indeterminacy, but we are content instead to keep the indeterminacy as part of the formula. Since $J \subseteq J''_{[r+k]}$, the eight Jacobi identities are true in $\hat{\pi}_*(\Omega D_{kkk};\mathbf{Z}_{p^{r+k}})$.

Remark 12.8. Since we have not yet proved (xvii) and (xviii) at $r+k$, yet the above proof assumes them, what have we really proved? The above proof certainly constructs α_t^k for $t < r+k$, since this construction does not involve (xvii) or (xviii) at $r+k$. Actually, (xvii) and (xviii) at $r+k$ are only used to help demonstrate condition 7.15(iii), for VS's at $t = r+k$. By 7.20 there are no VS's below dimension $12np^k - 6$. So we have definitely constructed α_t^k for $t < r+k$, and we have constructed α_{r+k}^k and proved its properties up to some moderately high cutoff that exceeds $6np^k$, say, $8np^k$. In Chapter 13 we will see how (xvii) and (xviii) follow from (xxii); and we have proved (xxii).

Remark 12.9. Although we whizzed right through it, Theorem 12.6 and its proof are in some sense the high point of our journey through the world of the D_k's. We have been climbing steadily upwards, and the remainder of the construction will be mostly a downhill coast. By obtaining some results at $t = r+k$ above dimension $6np^k$, we have crossed a certain psychological and mathematical threshhold. We took the liberty of omitting so much of the proof of Theorem 12.6 only because the groundwork for it – including the theory of thin products, the virtual structure for $\mathcal{E}(N^{kkk})$, the properties of $\{\hat{\pi}_*(\Omega D_{kkk}; \mathbb{Z}_{p^t})\}$ as a csL over $\mathcal{F}(\mathcal{U}M^k \amalg \mathcal{U}M^k \amalg \mathcal{U}M^k)$, and Corollary 6.7 – had been laid so carefully. It was the search for this proof that ultimately led to most of the ideas, definitions, and approaches in Chapters 5, 6, 7, and 11.

CHAPTER 13

Completion of the Proof

In Chapter 13 we complete the inductive step of the grand inductive scheme begun in Chapter 9. This finishes the proof of Theorems 9.1, 12.4, 12.6, and firmly establishes the existence and properties of the infinite family $\{D_k\}_{k\geq 0}$ of spaces.

Chapter 13 moves along at a rapid clip. We are clearing up all of the unfinished business from Chapters 9 and 12. There are a lot of different things to prove, and the results fall one after another like dominoes.

We divide Chapter 13 into four units. The first unit consists of results that hold below dimension $8np^k$, the somewhat arbitrary cutoff we saw in connection with Theorem 12.6 and Remark 12.8. When this part is complete we will have proved the long–awaited virtual syzygy for γ^k_{r+k}, and the construction of γ^k_{r+k} in all dimensions can go through. The second very short unit then establishes our various equations and properties in all dimensions. The third unit focuses on the existence and properties of $\hat{y}_{p^{k+1}}$. In this third part we prove 9.1(xii), the last step of the grand inductive scheme. As a coda, in the fourth unit we take a closer look at $\hat{y}_p$ (i.e. the case $k = 0$) since this serves as our definition of the natural homotopy operation $\tilde{\tau}_1$.

We continue to refer to the parts of Theorems 9.1, 12.4, and 12.6 by unadorned lower case Roman numerals, e.g., (xii) means Theorem 9.1(xii) and (xxii) means Theorem 12.6(xxii).

Except for the very last material on $\tilde{\tau}_1$, all of Chapter 13 is part of the grand inductive scheme. Once again, n, $p \geq 5$, and r are viewed as fixed. In Chapter 13, k is allowed to be zero as well as any positive integer, so our proofs cover some aspects of the initial step of the grand induction, as well as the inductive step.

Results in low dimensions

Recalling Summary 10.4 and Discussion 11.A1, our intermediate range goal at this point is still to construct γ^k_{r+k}. We plan to use Theorem 7.15 to do this, and the relevant virtual structure was described in Example 7.16. This first unit of Chapter 13 establishes that all the hypotheses of Theorem 7.15 are satisfied, for this particular application of that theorem.

Showing that the hypotheses of Theorem 7.15 are fulfilled is neither as dull nor as trivial as it sounds. The first step is to choose virtual generators in $\pi_*(\Omega E_k; \mathbf{Z}_{p^{r+k}})$. Since these virtual generators are the first mod p^{r+k} homotopy classes for E_k that we encounter, we will use Theorem 6.4 to construct them. However, unlike all the other fibrations to which we have so far applied Theorem 6.4, this time the fiber, which is E_k, is not a wedge of Moore spaces. We need to do something special in order to verify hypothesis (6H) for ΩE_k. This "something special" is to factor ΩF_k, and use the factorization to help us write ΩE_k as a product of two factors. Each of the two factors of ΩE_k has a homotopy exponent, hence E_k does too. (Note: the author claims no credit for this idea; it merely reproduces for E_k what Cohen–Moore–Neisendorder did for E_0 in [CMN2]).

Once the virtual generators, named $\hat{y}_{2p^k}$ and $\hat{z}'_{2p^k}$, are constructed, one might expect that the logical next step is to prove that $w = b_s c_s - c_s b_s$ is a virtual relator as in (7CA). However, that is not the correct next step, because the virtual relators are actually consequences of the second Jacobi identity (i.e. (xv)). So we begin the process of relating the csL homomorphism β^k_{r+k} of Theorem 12.4 to $\hat{y}_{2p^k}$ and $\hat{z}'_{2p^k}$. This process entails the folding map ∇_k hinted at in (11AA), which we now construct for real. Making frequent use of the uniqueness aspect of Theorem 6.4, we eventually deduce the virtual relators from (xv).

The last task of this unit is to prove the long-awaited virtual syzygies in $\pi_*(\Omega E_k, \mathbf{Z}_{p^{r+k}})$. By combining everything we have already done, these are remarkably easy. Using the folding map we first deduce (xvii) and (xviii) from (xxii), and then the virtual syzygies in $\pi_*(\Omega E_k; \mathbf{Z}_{p^{r+k}})$ are straightforward corollaries.

To begin, recall that $H_*(\Omega F_k) = H_*(\mathcal{U}L^k)$, because $\mathcal{U}L^k$ is an AH model for F_k by (iii). Among other things, this tells us that ΩF_k is a $(2n-2)$–connected H–space and that $\pi_{2n-1}(\Omega F_k) \otimes \mathbf{Z}_{(p)} = H_{2n-1}(\Omega F_k) = \mathbf{Z}_{(p)}$. (Recall that the default coefficient ring for homology is not $\mathbf{Z}$ but $\mathbf{Z}_{(p)}$.) Let $f_k : S^{2n-1} \to \Omega F_k$ denote a generator satisfying $h(f_k) = x_1 \in H_*(\mathcal{U}L^k)$.

Notation. Let X be an H–space, with H–space multiplication map denoted $\mu : X \times X \to X$. If $g_1 : Y_1 \to X$ and $g_2 : Y_2 \to X$ are maps, let $g_1 \cdot g_2$ denote the composition

$$Y_1 \times Y_2 \xrightarrow{g_1 \times g_2} X \times X \xrightarrow{\mu} X,$$

and likewise for products in X of three or more maps.

Lemma 13.1. *There exists a wedge of Moore spaces $\widehat{W}_k \in \mathcal{W}_r^{r+k}$ together with a map*

$$\widehat{\chi}_k : \widehat{W}_k \to E_k$$

having the following property. With $f_k : S^{2n-1} \to \Omega F_k$ as above, and with η_k as in (9A), the map

$$f_k \cdot \Omega(\eta_k \widehat{\chi}_k) : S^{2n-1} \times \Omega \widehat{W}_k \to \Omega F_k$$

is a homotopy equivalence, below dimension $2np^{k+1} - 3$. When $\widehat{W}_k$ is written as

$$\widehat{W}_k = \widehat{W}_k^{(r)} \vee \ldots \vee \widehat{W}_k^{(r+k)},$$

where $\widehat{W}_k^{(j)}$ is a bouquet of mod p^j Moore spaces, then

$$\widehat{W}_k^{(r+k-1)} = P^{4np^{k-1}}(p^{r+k-1}) \vee P^{6np^{k-1}}(p^{r+k-1}) \vee (\text{higher}), \tag{13AA}$$

while

$$\widehat{W}_k^{(r+k)} = P^{4np^k}(p^{r+k}) \vee P^{6np^k}(p^{r+k}) \vee (\text{higher}). \tag{13AB}$$

Proof. Recall the notation $I_s = \{jp^s | j \geq 2\}$. Put

$$\widehat{W}_k = \bigvee_{s=0}^{k} \bigvee_{j \in I_s - I_{s+1}} P^{2nj}(p^{r+s}). \tag{13AC}$$

Define $\widehat{\chi}_k : \widehat{W}_k \to E_k$ by defining it on the jth Moore space summand as follows. If $j \in I_s - I_{s+1}$, $0 \leq s < k$, let $\widehat{\chi}_k$ on $P^{2nj}(p^{r+s})$ be the homotopy class $(\gamma^k_{r+s}(y_j))^{\bar{a}}$. If $j \in I_k$ then $\gamma^k_{r+k-1}(y_j)$ has the property that

$$_{\pi}\delta^{r+k-1}\gamma^k_{r+k-1}(y_j) = \gamma^k_{r+k-1}(d_{[r+k-1]}(y_j)) = (p)\gamma^k_{r+k-1}(p^{-(r+k)}d(y_j))$$

is divisible by p, so $\gamma^k_{r+k-1}(y_j) : \ P^{2nj-1}(p^{r+k-1}) \to \Omega E_k$ may be extended to a map $g_j : \ P^{2nj-1}(p^{r+k}) \to \Omega E_k$. Define $\widehat{\chi}_k$ on the summand $P^{2nj}(p^{r+k})$ to be $(g_j)^{\bar{a}}$. We leave it to the reader to check that $f_k \cdot \Omega(\eta_k \widehat{\chi}_k)$ is a mod p homology equivalence, hence a (p–local) homotopy equivalence, below dimension $2np^{k+1} - 3$. The relevant formulas are (9K) and (9L), along with Proposition 2.7.

Notation. The (p–local) homotopy–theoretic fiber of the double suspension map on S^{2n-1} is denoted $C(n)$, i.e.,

$$C(n) \to S^{2n-1} \xrightarrow{\Sigma^2} \Omega^2 S^{2n+1} \tag{13AD}$$

is a fibration up to homotopy.

Lemma 13.2. *$C(n)$ has homotopy exponent p, i.e., $(p)\pi_*(C(n)) = 0$*

Proof. See [CMN2].

Lemma 13.3. *There is a map, denoted*

$$f'_k : \ C(n) \to \Omega E_k,$$

for which

$$f'_k \cdot (\Omega \widehat{\chi}_k) : \ C(n) \times \Omega \widehat{W}_k \to \Omega E_k$$

is a homotopy equivalence, below dimension $2np^{k+1} - 4$.

Proof. Consider the diagram of fibration sequences

$$\begin{array}{ccccc} C(n) & \longrightarrow & S^{2n-1} & \xrightarrow{\Sigma^2} & \Omega^2 S^{2n+1} \\ {\scriptstyle f'_k} \downarrow & & {\scriptstyle f_k} \downarrow & & \downarrow = \\ \Omega E_k & \xrightarrow{\Omega \eta_k} & \Omega F_k & \xrightarrow{\Omega \mu_k} & \Omega^2 S^{2n+1}, \end{array} \tag{13AE}$$

where the bottom row comes from looping the third–from–right column of (9C). The right-hand square of (13AE) commutes by an easy homology calculation (e.g. put $j = 1$ in (9D)), so f'_k exists.

Now use f'_k to make the diagram of fibration sequences

(13AF)
$$\begin{array}{ccccc} C(n)\times\Omega\widehat{W}_k & \longrightarrow & S^{2n-1}\times\Omega\widehat{W}_k & \xrightarrow{\Sigma^2\bar{\pi}_1} & \Omega^2S^{2n+1} \\ \downarrow{\scriptstyle f'_k\cdot(\Omega\chi_k)} & & \downarrow{\scriptstyle f_k\cdot(\Omega\eta_k\widehat{\chi}_k)} & & \downarrow{=} \\ \Omega E_k & \xrightarrow{\Omega\eta_k} & \Omega F_k & \xrightarrow{\Omega\mu_k} & \Omega^2S^{2n+1} \end{array}$$

which clearly commutes up to homotopy. Since two out of three of the vertical arrows in (13AF) are homotopy equivalences below dimension $2np^{k+1}-3$, the ladder of long exact homotopy sequences plus the five–lemma makes the third vertical arrow a homotopy equivalence too, below dimension $2np^{k+1}-4$.

Corollary 13.4. *$(\Omega\eta_k)_*h : p^{r+k-1}\pi_j(\Omega E_k) \to p^{r+k-1}H_j(\Omega F_k)$ is a monomorphism for $j < 4np^k + 2np - 5$, and an epimorphism for $2n-1 < j < 4np+2n-3$.*

Proof. The epimorphism comes straight from Lemma 13.1 and the Kunneth formula. For the monomorphism, we have by Lemma 13.3

$$\pi_j(\Omega E_k) = \pi_j(C(n)) \oplus \pi_j(\Omega\widehat{W}_k)$$

for relevant dimensions j. Since $(p)\pi_j(C(n)) = 0$, we need only to see that

$$(\Omega\chi_k)_*h : p^{r+k-1}\pi_j(\Omega\widehat{W}_k) \to p^{r+k-1}H_j(\Omega F_k)$$

is one–to–one. This is easy using (13AA) and (13AB) and Lemma 10.2(d), since the even-dimensional Moore space summands in (13AA) make no contribution to $p^{r+k-1}\pi_j(\Omega\widehat{W}_k)$ for $j < 8np^k - 2p - 2$.

Proposition 13.5. *There exist unique elements $\hat{y}_{2p^k} \in \pi_{4np^k-1}(\Omega E_k; \mathbb{Z}_{p^{r+k}})$ and $\hat{z}'_{2p^k} \in \pi_{4np^k-2}(\Omega E_k; \mathbb{Z}_{p^{r+k}})$ having the following properties.*

(a) $\hat{y}_{2p^k}\omega^{r+k}_{r+k-1} = \gamma^k_{r+k-1}(y_{2p^k})$

(b) $\hat{z}'_{2p^k}\omega^{r+k}_{r+k-1} = \gamma^k_{r+k-1}(z'_{2p^k})$

(c) ${}_\pi\delta^{r+k}(\hat{y}_{2p^k}) = (-2)\hat{z}'_{2p^k}$

(d) ${}_\pi\delta^{r+k}(\hat{z}'_{2p^k}) = 0$

(e) $\underline{a}_{\eta_k}(\hat{y}_{2p^k}) = x_{2p^k}$

(f) $\underline{a}_{\eta_k}(\hat{z}'_{2p^k}) = \bar{e}(z'_{2p^k})$, *where $\bar{e} : \bar{L}^k \to \mathcal{U}L^k = \mathcal{A}(F_k)$ is the inclusion.*

Proof. We apply Theorem 6.4 using as (6G) the fibration

(13AG)
$$E_k \xrightarrow{\eta_k} F_k \xrightarrow{\mu_k} \Omega S^{2n+1},$$

with $q = 4np^k - 1$ and $t = r + k - 1$. Checking the hypotheses of Theorem 6.4, we see that (6I) is clear from Proposition 2.7 (since $(p)H_j(\Omega^2 S^{2n+1}) = 0$ for $j \geq 2n$) and the exactness of (6H) at $p^t\pi_j(\Omega X)$ follows from Corollary 13.4. Since $p^t H_j(\Omega Z) = 0$ for $Z = \Omega S^{2n+1}$ for these j, the exactness of (6H) at $p^t H_j(\Omega Y)$ also follows from Corollary 13.4 as long as $n \geq 2$. We will return shortly to the case $n = 1$.

Put $\hat{y} = (\gamma^k_{r+k-1}(y_{2p^k}))^{\bar{a}}$ and $\hat{z} = (-2)(\gamma^k_{r+k-1}(z'_{2p^k})^{\bar{a}}$; then $y = \underline{a}_{\eta_k}(\hat{y}) = x_{2p^k}$ and $z = \underline{a}_{\eta_k}(\hat{z}) = -2z'_{2p^k}$ by (ix). Also, p^{r+k} divides $\mathcal{A}(\mu_k)(y)$; this detail is the principal purpose for including (9D) in Theorem 9.1. All the hypotheses of Theorem 6.4 are satisfied, so the theorem provides us with a unique $\hat{y}^a_2 \in \pi_q(\Omega E_k; \mathbb{Z}_{p^{r+k}})$. Set $\hat{y}_{2p^k} = \hat{y}^a_2$ and $\hat{z}'_{2p^k} = -(\frac{1}{2})_\pi \delta^{r+k}(\hat{y}_{2p^k})$. Then (a)–(f) are all true.

When $n = 1$ the exactness of (6H) at $p^t H_j(\Omega Y)$ fails for $j = 4np^k - 1$. We escape this problem by resorting to the 2–connected cover $F_k\langle 2\rangle$ of F_k. Since $E_k \xrightarrow{\eta_k} F_k \to \mathbb{CP}^\infty$ is null–homotopic (because E_k is 2–connected), η_k lifts through $F_k\langle 2\rangle$, say $E_k \xrightarrow{\eta'_k} F_k\langle 2\rangle \xrightarrow{f'} F_k$ is the factorization of η_k. It is straightforward to see that $\eta'_k\widehat{\chi}_k : \widehat{W}_k \to F_k\langle 2\rangle$ is a homotopy equivalence below dimension $2np^{k+1} - 4$, and that $\mathcal{U}\bar{L}^k$ is an Adams–Hilton model for $F_k\langle 2\rangle$ with $\mathcal{A}(f')$ being the inclusion $\mathcal{U}\bar{L}^k \to \mathcal{U}L^k$. All of the hypotheses of Theorem 6.4 are satisfied if we use

$$E_k \xrightarrow{\eta'_k} F_k\langle 2\rangle \xrightarrow{\mu_k f'} \Omega S^3 \tag{13AH}$$

instead of (13AG), so Lemma 13.5 is valid for $n = 1$ as well. (Note: the sequence (13AH) will be the only sequence we shall use as (6G) that is not a fibration. Of course, $\mu_k f'\eta'_k \simeq \mu_k\eta_k$ is null–homotopic.)

We now see immediately

Corollary/Definition 13.6. Define $\gamma^k_{r+k}(y_{2p^k})$ to be $\hat{y}_{2p^k}$ and $\gamma^k_{r+k}(z'_{2p^k})$ to be $\hat{z}'_{2p^k}$. Then these satisfy criteria 7.15(i,ii) for the images in $\pi_*(\Omega E_k; \mathbb{Z}_{p^{r+k}})$ of the virtual generators y_{2p^k} and z'_{2p^k} of $\mathcal{F}(\bar{L}^k)_{[r+k]}$.

We have completed the work described in the second paragraph of the introduction to this unit, and we now move on to study the folding map.

Stasheff showed in [St] that any H–space (X, μ) can be replaced up to homotopy by a strict H–space. We call (X, μ) *strict* if the diagram

$$\begin{array}{ccc} X \vee X & \xrightarrow{\nabla} & X \\ \downarrow & & \downarrow = \\ X \times X & \xrightarrow{\mu} & X \end{array}$$

commutes on the nose, where ∇ is the folding map. We also know from [N3] that $S^{2n+1}\{p^r\}$ is an H–space, so we can give it an H–space structure that is strict. Then the right–hand square in

$$\begin{array}{ccccc} D_{kk} & \xrightarrow{\eta_{kk}} & D_k \vee D_k & \xrightarrow{(\bar{\pi}_1, \bar{\pi}_2)} & D_k \times D_k \\ \nabla_k \vdots & & \downarrow \nabla & & \downarrow \rho_k \cdot \rho_k \\ E_k & \xrightarrow{\lambda_k \eta_k} & D_k & \xrightarrow{\rho_k} & S^{2n+1}\{p^r\} \end{array} \tag{13AI}$$

commutes on the nose, so there is a canonical induced map, dubbed ∇_k, between the homotopy–theoretic fibers. The map ∇_k is going to permit us to transfer information from D_{kk} to E_k and from D_{kkk} to D_{kk}.

Lemma 13.7. *For $t < r+k$, if $w \in \mathcal{F}(\mathcal{U}M_+^k)_{[t]}$ and $y \in \mathcal{F}(\mathcal{U}\bar{L}^k)_{[t]}$ satisfy $|w|_* + |y|_* \leq 2p$,*

$$(\Omega\nabla_k)_\# \Psi_t^{kk}(w^{(2)})\gamma_t^k(y^{(1)}) = \gamma_t^k(\psi_{[t]}(w \otimes y)) \tag{13AJ}$$

in $\pi_*(\Omega E_k; \mathbf{Z}_{p^t})$. Likewise,

$$(\Omega\nabla_k)_\# \Psi_t^{kk}(w^{(1)})\gamma_t^k(y^{(2)}) = \gamma_t^k(\psi_{[t]}(w \otimes y)). \tag{13AK}$$

Proof. We only prove (13AJ); the proof of (13AK) is identical. Consider these two diagrams, in which the rows but not the columns are fibrations. The diagrams determine maps that we denote as f, g, f', g', g''.

$$\begin{array}{ccccc}
\Sigma\Omega E_k \wedge \Omega D_k & \xrightarrow{f} & E_k \vee D_k & \longrightarrow & E_k \times D_k \\
{\scriptstyle g}\downarrow & & \downarrow{\scriptstyle \lambda_k\eta_k \vee 1} & & \downarrow{\scriptstyle \lambda_k\eta_k \times 1} \\
D_{kk} & \xrightarrow{\eta_{kk}} & D_k \vee D_k & \longrightarrow & D_k \times D_k \\
{\scriptstyle \nabla_k}\downarrow & & \downarrow{\scriptstyle \nabla} & & \downarrow{\scriptstyle \rho_k \cdot \rho_k} \\
E_k & \xrightarrow{\lambda_k\eta_k} & D_k & \xrightarrow{\rho_k} & S^{2n+1}\{p^r\}\ ;
\end{array} \tag{13AL}$$

$$\begin{array}{ccccc}
\Sigma\Omega E_k \wedge \Omega D_k & \xrightarrow{f} & E_k \vee D_k & \longrightarrow & E_k \times D_k \\
{\scriptstyle g'}\downarrow & & \downarrow{\scriptstyle =} & & \downarrow{\scriptstyle i_0 \times 1} \\
E_k \rtimes (\Omega D_k) & \xrightarrow{f'} & E_k \vee D_k & \longrightarrow & CE_k \times D_k \\
{\scriptstyle g''}\downarrow & & \downarrow{\scriptstyle \nabla(\lambda_k\eta_k \vee 1)} & & \downarrow{\scriptstyle F \cdot \rho_k} \\
E_k & \xrightarrow{\lambda_k\eta_k} & D_k & \xrightarrow{\rho_k} & S^{2n+1}\{p^r\}\ .
\end{array} \tag{13AM}$$

In (13AM), CE_k is the (reduced) cone on E_k and i_0 is the inclusion of E_k into it. F is the canonical null–homotopy of $\rho_k\lambda_k\eta_k$ (see (9C)). Both right–hand squares in both diagrams commute on the nose. Since the composites $(\rho_k \cdot \rho_k)(\lambda_k\eta_k \times 1)$ and $(F \cdot \rho_k)(i_0 \times 1)$ are homotopic, deduce that

$$\nabla_k g \simeq g''g' : \ \Sigma\Omega E_k \wedge \Omega D_k \to E_k. \tag{13AN}$$

Every row in both (13AL) and (13AM) except the middle row of (13AL) has an obvious inclusion from D_k into the total space, so ψ_t^k is defined on all their fibers. For the middle row of (13AL), ψ_t^{kk} is defined on D_{kk}.

Allowing $\hat{e}^{(1)}$ to denote the inclusion of E_k into $E_k \vee D_k$, we have by Definition (12D),

$$\begin{aligned}
\Psi_t^{kk}(w^{(2)})\gamma_t^k(y^{(1)}) &= (\Omega\eta_{kk})_\#^{-1}\psi_t^{kk}(w^{(2)} \otimes (\Omega\hat{e}^{(1)}\lambda_k\eta_k)_\#\gamma_t^k(y)) \\
&= (\Omega\eta_{kk})_\#^{-1}\Omega(\lambda_k\eta_k \vee 1)_\#(z),
\end{aligned}$$

where $z = \psi_t^k(w \otimes (\Omega\hat{e}^{(1)})_\# \gamma_t^k(y))$. Since $f_\#$ is one–to–one, and because $w \in \mathcal{F}(\mathcal{U}M_+^k)_{[t]}$ (i.e. $|w| > 0$), $(f_\#)^{-1}(z)$ is well-defined. Being sloppy about Ω's, we have

$$\begin{aligned}\Psi_t^{kk}(w^{(2)})\gamma_t^k(y^{(1)}) &= (\eta_{kk})_\#^{-1}(\lambda_k\eta_k \vee 1)_\# f_\#(f_\#)^{-1}(z)\\ &= g_\#(f_\#)^{-1}(z).\end{aligned}$$

Using (13AN), we get (again omitting most Ω's for the calculation)

$$\begin{aligned}(\Omega\nabla_k)_\#\Psi_t^{kk}(w^{(2)})\gamma_t^k(y^{(1)}) &= (\nabla_k)_\# g_\#(f_\#)^{-1}(z)\\ &= (g'')_\#(g')_\#(f_\#)^{-1}(z) = (g'')_\#(f')_\#^{-1}(z)\\ &= (g'')_\#\psi_t^k(w \otimes (\Omega v)_\#\gamma_t^k(y)) = \psi_t^k(w \otimes \gamma_t^k(y)),\end{aligned}$$

where $v : E_k \to E_k \rtimes (\Omega D_k)$ is the canonical section for g''. To get from $\psi_t^k(w \otimes \gamma_t^k(y))$ to $\gamma_t^k(\psi_{[t]}(w \otimes y))$ is merely the fact (ix) that γ_t^k is a csm homomorphism over $\mathcal{F}(\mathcal{U}M^k)$.

Remark. The proof given above would actually be valid for all $t \le r + k$, except that γ_{r+k}^k has not yet been defined. What the argument actually shows is that

$$(\Omega\nabla_k)_\#\psi_t^{kk}(w^{(2)} \otimes (\Omega e^{(1)}\lambda_k\eta_k)_\#(\hat{y})) = \psi_t^k(w \otimes \hat{y}) \qquad \text{(13AO)}$$

for any $t \le r + k$ and for any $\hat{y} \in \pi_*(\Omega E_k; \mathbb{Z}_{p^t})$. Also, as soon as γ_{r+k}^k is defined and shown to be a homomorphism of csm's over $\mathcal{F}(\mathcal{U}M^k)$, Lemma 13.7 will be valid at $t = r + k$ as well.

The next result is one that we think of as part of our list of the properties of $\{D_k\}$, so we distinguish it by calling it a theorem. Recall from Definition 5.13 the dgL homomorphism $\nabla : N^{kk} \to \bar{L}^k$. It clearly induces $\mathcal{F}(\nabla) : \mathcal{F}(N^{kk}) \to \mathcal{F}(\bar{L}^k)$. Let $\mathcal{E}(\nabla)$ denote the composition

$$\mathcal{E}(\nabla) : \mathcal{E}(N^{kk}) \to \mathcal{F}^0(N^{kk}) \to \mathcal{F}(N^{kk}) \xrightarrow{\mathcal{F}(\nabla)} \mathcal{F}(\bar{L}^k).$$

Notice that $\mathcal{E}(\nabla)$ is a csL/csm homomorphism over $\mathcal{F}(\mathcal{U}M^k \amalg \mathcal{U}M^k)$. The action of $\mathcal{F}(\mathcal{U}M^k \amalg \mathcal{U}M^k)$ on $\mathcal{F}(\bar{L}^k)$ comes from the usual action of $\mathcal{F}(\mathcal{U}M^k)$ on $\mathcal{F}(\bar{L}^k)$ and the Dcsa homomorphism $\mathcal{F}(\nabla) : \mathcal{F}(\mathcal{U}M^k \amalg \mathcal{U}M^k) \to \mathcal{F}(\mathcal{U}M^k)$. In other words, the action of $w \in \mathcal{F}(\mathcal{U}M^k \amalg \mathcal{U}M^k)_{[t]}$ on $\mathcal{F}(\bar{L}^k)_{[t]}$ is describable as $\psi_{[t]}(\mathcal{F}(\nabla)(w) \otimes -)$.

Theorem 13.8. *This diagram commutes:*

$$\begin{array}{ccc}\mathcal{E}(N^{kk})_{[t]}^{*\le 2p} & \xrightarrow{\beta_t^k} & \hat{\pi}_*(\Omega D_{kk}; \mathbb{Z}_{p^t})\\ \Big\downarrow{\scriptstyle \mathcal{E}(\nabla)} & & \Big\downarrow{\scriptstyle (\Omega\nabla_k)_\#}\\ \mathcal{F}(\bar{L}^k)_{[t]}^{*\le 2p} & \xrightarrow{\gamma_t^k} & \pi_*(\Omega E_k; \mathbb{Z}_{p^t}),\end{array}$$

and the induced homomorphism $(\Omega\nabla_k)_\#$ here is well-defined. Equivalently, for $w \in \mathcal{E}(N^{kk})_{[t]}^{*\le 2p}$,

$$\gamma_t^k\mathcal{E}(\nabla)(w) = (\Omega\nabla_k)_\#\beta_t^k(w). \qquad \text{(13AP)}$$

Proof for $t < r+k$. Because of the grand inductive framework, we may assume that (13AP) holds for $k-1$. Since (by Note 12.5) $\hat{\pi}_*(\Omega D_{kk}; \mathbf{Z}_{p^t})$ is obtained from $\pi_*(\Omega_{kk}; \mathbf{Z}_{p^t})$ by dividing out by only a few elements and these elements are in the image of $\pi_*(\Omega D_{k-1,k-1}; \mathbf{Z}_{p^t})$, it is part of the grand inductive assumption that $(\Omega\nabla_k)_\#$ here is well-defined for $t < r+k$. To prove (13AP), it suffices to check it on the virtual generators of $\mathcal{E}(N^{kk})_{[t]}$, since all four homomorphisms are csL/csm homomorphisms over $\mathcal{F}(\mathcal{U}M^k \amalg \mathcal{U}M^k)$. For the basic virtual generators it follows from our assuming the theorem at $k-1$, along with (x) and (xiv). For the non-basic virtual generators it follows from Lemma 13.7 and (xix).

Lemma 13.9. *(a)* $(\Omega\nabla_k)_\#\beta^k_{r+k}g(b_k \otimes b_k) = 0.$

(b) $(\Omega\nabla_k)_\#\beta^k_{r+k}g(b_k \otimes c_k) = \hat{y}_{2p^k}.$

(c) $(\Omega\nabla_k)_\#\beta^k_{r+k}g(c_k \otimes b_k) = -\hat{y}_{2p^k}.$

(d) $(\Omega\nabla_k)_\#\beta^k_{r+k}g(c_k \otimes c_k) = 2\hat{z}'_{2p^k}.$

Proof. We omit details, as they are straightforward. To prove (c) and (d), check that $-(\Omega\nabla_k)_\#\beta^k_{r+k}g(c_k \otimes b_k)$ and $(\frac{1}{2})(\Omega\nabla_k)_\#\beta^k_{r+k}g(c_k \otimes c_k)$ satisfy conditions (a)–(f) of 13.5, and then use the uniqueness aspect of Proposition 13.5 to show that these elements coincide respectively with $\hat{y}_{2p^k}$ and $\hat{z}'_{2p^k}$. Properties (a)–(f) for these elements follow from Theorem 13.8 at $r+k-1$, Lemma 5.14, and (xiii). The check uses the fact that $\mathcal{A}(\nabla)\underline{a}_{\eta_{kk}} = \mathcal{A}(\lambda_k)\underline{a}_{\eta_k\nabla_k}$, which is implicit in the commutativity of (13AI) and Lemma 6.3. It also uses the fact that $H_*(\mathcal{A}(\lambda_k); \mathbf{Z}_{p^{r+k}})$, which is the homomorphism $H_*(\mathcal{U}L^k; \mathbf{Z}_{p^k}) \to H_*(\mathcal{U}M^k; \mathbf{Z}_{p^{r+k}})$ by (iv), is one-to-one on primitives, in dimensions between $2n$ and $2np^{k+1}-2$.

The proof of (a) and (b) is similar, but we employ the uniqueness aspect of Theorem 6.4 directly. Use the fibration (13AG) if $n \geq 2$ and use (13AH) if $n = 1$. Put

$$\hat{y} = (\Omega\nabla_k)_\#\beta^k_{r+k}g(b_k \otimes b_k)$$

and

$$\hat{z} = (\Omega\nabla_k)_\#\beta^k_{r+k}g(-b_k \otimes c_k - c_k \otimes b_k).$$

After computing the analog of properties 13.5(a)–(f) for this $\hat{y}$ and $\hat{z}$, uniqueness in 6.4 shows that $\hat{y} = 0$ and $\hat{z} = 0$.

Corollary 13.10. *(a) The homomorphism $(\Omega\nabla_k)_\#$ in Theorem 13.8 is well-defined for $t = r+k$.*

(b) Formula (13AP) is true for

$$w \in Span\{g(b_k \otimes b_k), g(b_k \otimes c_k), g(c_k \otimes b_k), g(c_k \otimes c_k)\}.$$

Proof. (a) Since Lemma 13.9 is part of the grand inductive scheme, it is also true at $k-1$. Since $\beta^k_{r+k-1}g(b_{k-1} \otimes b_{k-1})$ is in the kernel of $(\Omega\nabla_k)_\#$, and since the ideal we divide out by in going from π_* to $\hat{\pi}_*$ is built upon $\beta^k_{r+k-1}g(b_{k-1} \otimes b_{k-1})$, this ideal lies in $\ker(\Omega\nabla_k)_\#$, and $(\Omega\nabla_k)_\#$ is well-defined on the quotient. (b) Lemmas 13.9 and 5.14, and Definition 13.6.

Lemma 13.11.

$$\mathcal{O}_{(\nabla)}\Psi_t^{kk} = \Psi_t^k\mathcal{F}(\nabla) : \mathcal{F}(\mathcal{U}M^k \amalg \mathcal{U}M^k)_{[t]} \to \mathcal{O}_{D_k}[t].$$

for all t.

Proof. This is implicit in thc definition of Ψ_t^{kk}, in Theorem 12.1. For any map pair $(f,g) \in \mathcal{M}_{D_k}$, $\mathcal{O}_{(\nabla)}\Psi_t^{kk}$ and $\Psi_t^k\mathcal{F}(\nabla)_{[t]}$ are two structures for $\{\tilde{\pi}_*(K_g;\mathbb{Z}_{p^t})\}$ as a csm over $\mathcal{F}(\mathcal{U}M^k \amalg \mathcal{U}M^k)$. By Theorem 12.1 they coincide on the generators of the Bockstein sub–dga $R_{t-r}^k \amalg R_{t-r}^k$ of $\mathcal{F}(\mathcal{U}M^k \amalg \mathcal{U}M^k)_{[t]}$, for every t. By Proposition 7.9, they coincide.

Lemma 13.12. *Let $\bar{e} : \bar{L}^k \to \mathcal{U}M^k$ denote the inclusion. Then*

$$\Psi_{r+k}^k\mathcal{F}(\bar{e})(y_{2p^k}) = \Phi((\Omega\lambda_k\eta_k)_\#(\hat{y}_{2p^k})) \in \mathcal{O}_{D_k}[r+k]$$

and

$$\Psi_{r+k}^k\mathcal{F}(\bar{e})(z'_{2p^k}) = \Phi((\Omega\lambda_k\eta_k)_\#(\hat{z}'_{2p^k})) \in \mathcal{O}_{D_k}[r+k].$$

In other words, (9N) *is true at $t = r+k$, for y_{2p^k} and for z'_{2p^k}.*

Proof. Let $\bar{e}' : N^{kk} \to \mathcal{U}M^k \amalg \mathcal{U}M^k$ denote the inclusion of (12F). By (xv) we have

$$\Phi((\Omega\eta_{kk})_\#\beta_{r+k}^k g(b_k \otimes c_k)) = \Psi_{r+k}^{kk}\mathcal{F}(\bar{e}')g(b_k \otimes c_k) \in \mathcal{O}_{D_k\vee D_k}[r+k].$$

Apply $\mathcal{O}_{(\nabla)}$ to both sides of this equation and use Lemma 13.11 to obtain

$$\Phi((\Omega\nabla\eta_{kk})_\#\beta_{r+k}^k g(b_k \otimes c_k)) = \Psi_{r+k}^k\mathcal{F}(\bar{e}')g(b_k \otimes c_k).$$

By (13AI) and (5JJ) this becomes

$$\Phi((\Omega\lambda_k\eta_k\nabla_k)_\#\beta_{r+k}^k g(b_k \otimes c_k)) = \Psi_{r+k}^k\mathcal{F}(\bar{e})\mathcal{E}(\nabla)g(b_k \otimes c_k).$$

Simplifying with Lemmas 13.9 and 5.14 we get

$$\Phi((\Omega\lambda_k\eta_k)_\#(\hat{y}_{2p^k})) = \Psi_{r+k}^k\mathcal{F}(\bar{e})(y_{2p^k}),$$

as promised. The other formula is proved similarly, with $(\frac{1}{2})g(c_k\otimes c_k)$ in place of $g(b_k\otimes c_k)$; or one can simply differentiate both sides of this formula.

Proposition 13.13. *(cf.* (7CA)*)* $w = b_kc_k - c_kb_k \in \mathcal{F}(\mathcal{U}M^k)_{[r+k]}$ *is a virtual relator for* $\pi_*(\Omega E_k;\mathbb{Z}_{p^{r+k}})$. *It satisfies*
(13AQ)

$$\psi_{[r+k]}(w \otimes -) = [\hat{y}_{2p^k}, -] + \sum_{i=2}^{2p^k-1} (\mu_i p^{m-k})\zeta_{r+k}^{r+m}[\gamma_{r+m}^k(y_i), \psi_{[r+m]}(b^{2p^k-i} \otimes \theta_{r+m}^{r+k}(-))],$$

where in the summation m is the least integer for which $i \in I_m$, and the $\{\mu_i\}$ are the coefficients appearing in Lemma 4.4(c). Likewise, $(c_k)^2$ is a virtual relator for $\pi_*(\Omega E_k;\mathbb{Z}_{p^{r+k}})$. Consequently condition 7.15(iv) is satisfied, for both of the minimal virtual relators for $\mathcal{F}(\bar{L}^k)_{[r+k]}$ as a $\mathcal{F}(\mathcal{U}M^k)_{[r+k]}$-module (see Example 7.16).

Proof. We proved in Chapter 10 that Formula (9N) for $w = y_{2p^s}$ implies that $b_sc_s - c_sb_s$ is a virtual relator, when $s < k$. The same proof works for $s = k$. For $(c_k)^2$, apply $d_{[r+k]}$ to both sides of (13AQ), and solve for $\psi_{[r+k]}((c_k)^2 \otimes -)$.

We have now completed the material promised in the third paragraph of the introduction to this unit. At last, we are ready to tackle the virtual syzygy equation.

Lemma 13.14. *For all* $t \geq 1$,

(13AR) $$\mathcal{O}_{(\nabla\vee 1)}\Psi_t^{kkk} = \Psi_t^{kk}\mathcal{F}(\nabla \amalg 1): \mathcal{F}(\mathcal{U}M^k \amalg \mathcal{U}M^k \amalg \mathcal{U}M^k) \to \mathcal{O}_{D_k \vee D_k}[t],$$

and likewise for $1 \vee \nabla$.

Proof. Like Lemma 13.11.

Notation/Definition. Observe that the thin product $\mathcal{R}(X,Y,*)$ can be identified as a space with $X \vee Y$ when one of the three inputs is a point. The composite

$$X \vee Y \vee Z \xrightarrow{\tilde{\lambda}} \mathcal{R}(X,Y,Z) \xrightarrow{\mathcal{R}(1,1,*)} \mathcal{R}(X,Y,*) = X \vee Y$$

is simply the map that collapses the summand Z. Similarly, the composite

$$X \vee Y \vee Z \xrightarrow{\tilde{\lambda}} \mathcal{R}(X,Y,Z) \xrightarrow{\mathcal{R}(*,*,1)} \mathcal{R}(*,*,Z) = Z$$

is simply the projection onto the last summand. Define

$$\tilde{r}^{(12)}: \mathcal{R}(D_k, D_k, D_k) \to D_k \times D_k$$

by

$$\tilde{r}^{(12)} = (\nabla \circ \mathcal{R}(1,1,*), \mathcal{R}(*,*,1)).$$

Then the right-hand square in

(13AS) $$\begin{array}{ccccc} D_{kkk} & \xrightarrow{\eta_{kkk}} & D_k \vee D_k \vee D_k & \xrightarrow{\tilde{\lambda}} & R(D_k, D_k, D_k) \\ \Big\downarrow{\scriptstyle \nabla_k^{(12)}} & & \Big\downarrow{\scriptstyle \nabla\vee 1} & & \Big\downarrow{\scriptstyle \tilde{r}^{(12)}} \\ D_{kk} & \xrightarrow{\eta_{kk}} & D_k \vee D_k & \longrightarrow & D_k \times D_k \end{array}$$

commutes on the nose and induces a map, called $\nabla_k^{(12)}$, between the homotopy-theoretic fibers. The diagram (13AS) commutes. Likewise there are maps $\nabla_k^{(13)}$ and $\nabla_k^{(23)}$.

Theorem 13.15. *For* $t \leq r + k$, *this diagram commutes:*

$$\begin{array}{ccc} \mathcal{E}(N^{kkk}) & \xrightarrow{\alpha_t^k} & \hat{\pi}_*(\Omega D_{kkk}; \mathbb{Z}_{p^t}) \\ \Big\downarrow{\scriptstyle \mathcal{E}(\nabla^{(12)})} & & \Big\downarrow{\scriptstyle (\Omega\nabla_k^{(12)})_\#} \\ \mathcal{E}(N^{kk}) & \xrightarrow{\beta_t^k} & \hat{\pi}_*(\Omega D_{kk}; \mathbb{Z}_{p^t}). \end{array}$$

Equivalently,

(13AT) $$(\Omega\nabla_k^{(12)})_\# \alpha_t^k(w) = \beta_t^k \mathcal{E}(\nabla^{(12)})(w)$$

for $w \in \mathcal{E}(N^{kkk})_{[t]}$. *Also,*

(13AU) $$(\Omega\nabla_k^{(13)})_\# \alpha_t^k(w) = \beta_t^k \mathcal{E}(\nabla^{(13)})(w);$$

$$(\Omega\nabla_k^{(23)})_{\#}\alpha_t^k(w) = \beta_t^k\mathcal{E}(\nabla^{(23)})(w). \tag{13AV}$$

Proof. As usual it suffices to prove these equations on the virtual generators (VG's) of $\mathcal{E}(N^{kkk})_{[t]}$. By 7.20 every VG has the form $\mathrm{ad}^{(1)}(w)(u^{(23)})$ or $\mathrm{ad}^{(2)}(w)(u^{(13)})$, as indicated in (7EA). There are six cases (two types of VG's in each of three equations); since they are all so similar we work through only the first case. Using (12U) and (xxiii), we have

$$\begin{aligned}(\Omega\nabla_k^{(12)})_{\#}\alpha_t^k(\mathrm{ad}^{(1)}(w)(u^{(23)})) &= (\Omega\nabla_k^{(12)})_{\#}\Psi_t^{kkk}(w^{(1)})\beta_t^k(u^{(23)})\\ &= (\Omega\nabla_k^{(12)})_{\#}(\Omega\eta_{kkk})_{\#}^{-1}\psi_t^{kkk}(z),\end{aligned}$$

where $z = e^{(1)}(w)\otimes(\Omega\hat{e}^{(23)}\eta_{kk})_{\#}\beta_t^k(u)$. Using (13AR) and (13AS), we get

$$\begin{aligned}(\Omega\nabla_k^{(12)})_{\#}\alpha_t^k(\mathrm{ad}^{(1)}(w)(u^{(23)})) &= (\Omega\eta_{kk})_{\#}^{-1}\Omega(\nabla\vee 1)_{\#}\psi_t^{kkk}(z)\\ &= (\Omega\eta_{kk})_{\#}^{-1}\psi_t^{kk}(\mathcal{F}(\nabla\amalg 1)\otimes\Omega(\nabla\vee 1)_{\#})(z)\\ &= (\Omega\eta_{kk})_{\#}^{-1}\psi_t^{kk}(e^{(1)}(w)\otimes(\Omega\eta_{kk})_{\#}\beta_t^k(u))\\ &= \psi_t^{kk}(e^{(1)}(w)\otimes\beta_t^k(u)) = \beta_t^k(\psi_{[t]}(w^{(1)}\otimes u))\\ &= \beta_t^k\mathcal{E}(\nabla^{(12)})(\mathrm{ad}^{(1)}(w)(u^{(23)})).\end{aligned}$$

Remark. Since α_{r+k}^k has not really been defined yet above dimension $8np^k$ (see Remark 12.8), the above proof is not valid in quite the generality stated. What the proof actually shows is that Theorem 13.15 is a formal consequence of (xxiii). In particular, (13AT,AU,AV) are true for $t < r+k$, and this is the only fact we need in order to prove the following.

Corollary 13.16. *(xvii) and (xviii) are true at $t = r+k$.*

Proof. There are four equations to be proved, namely (12K) and (12L) and the counterparts with (1) and (2) exchanged. Suppose we apply $(\Omega\nabla_k^{(12)})_{\#}$ to both sides of (12X) and simplify each term. The simplification process was illustrated in the proof of Theorem 13.15. The outcome, using 13.9(a), boils down to the trivial equation

$$0 = 0 + \beta_{r+k}^k\zeta_{r+k}^{r+k-1}\mathcal{E}(\nabla^{(12)})(w_k^0),$$

and the right–hand side is actually zero because the equation $\zeta_{r+k}^{r+k-1}\mathcal{E}(\nabla^{(12)})(w_k^0) = 0$ is deducible by applying $\nabla^{(12)}$ to both sides of (7EC). However, if we start with another of the eight Jacobi identities, such as the equation corresponding to (7EF) instead of (7EC), the net result (using (7EJ)) is precisely (12K). Likewise (12L) and the remaining two equations are derivable from (xxii) by applying a suitable $(\nabla_k^{(ij)})_{\#}$.

We are finally ready for the virtual syzygies in $\pi_*(\Omega E_k;\mathbb{Z}_{p^{r+k}})$. What do these virtual syzygy equations really say? We can determine one equation by applying ∇ to both sides of (7DL). In (7DL) let $s = k$. The result, using Lemma 5.14, is

$$\begin{aligned}\mathrm{ad}(c_k)(y_{2p^k}) - (2)\,\mathrm{ad}(b_k)(z'_{2p^k}) + \mathrm{ad}(c_k)(y_{2p^k})&\\ = \zeta_{r+k}^{r+k-1}\nabla(z_k^{12}) + d_{[r+k]}\zeta_{r+k}^{r+k-1}\nabla(y_k^{12}),&\end{aligned}$$

or, combining like terms,

$$\text{(13AW)}\quad \mathrm{ad}(c_k)(y_{2p^k}) = \mathrm{ad}(b_k)(z'_{2p^k}) + (\tfrac{1}{2})\zeta_{r+k}^{r+k-1}\nabla(z_k^{12}) + (\tfrac{1}{2})d_{[r+k]}\zeta_{r+k}^{r+k-1}\nabla(y_k^{12}).$$

We knew in advance from 7.16 that the virtual syzygy would have this form, except that perhaps we did not predict that the decomposable term, i.e. the term denoted z_j^0 in (7BJ), would be zero. The absence of a decomposable term from (13AW) is not surprising at all, once we realize something. Both virtual generators are in dimensions $\geq 4np^k - 2$, so all non-zero decomposables have dimension $\geq 8np^k - 4$, but (13AW) lives in dimension $6np^k - 2$. So automatically z_j^0 is zero here.

According to Example 7.16 there is a second virtual syzygy. We can obtain it by applying $d_{[r+k]}$ to both sides of (13AW):

$$(2)\mathrm{ad}(c_k)(z'_{2p^k}) = -\mathrm{ad}(c_k)(z'_{2p^k}) + (\tfrac{1}{2})d_{[r+k]}\zeta_{r+k}^{r+k-1}\nabla(z_k^{12}),$$

or after rearranging,

$$\text{(13AX)}\qquad \mathrm{ad}(c_k)(z'_{2p^k}) = 0 + (\tfrac{1}{6})d_{[r+k]}\zeta_{r+k}^{r+k-1}\nabla(z_k^{12}).$$

What criterion 7.15(iii) requires is that the corresponding two syzygy equations should be true in $\pi_*(\Omega E_k; \mathbf{Z}_{p^{r+k}})$, i.e.,

$$\begin{aligned}\psi_{r+k}^k(c_k \otimes \hat{y}_{2p^k}) &= \psi_{r+k}^k(b_k \otimes \hat{z}'_{2p^k})\\ \text{(13AY)}\qquad &+ (\tfrac{1}{2})\zeta_{r+k}^{r+k-1}\gamma_{r+k-1}^k(\nabla(z_k^{12})) + (\tfrac{1}{2})d_{[r+k]}\zeta_{r+k}^{r+k-1}\gamma_{r+k-1}^k(\nabla(y_k^{12}))\ ;\end{aligned}$$

$$\text{(13AZ)}\qquad \psi_{r+k}^k(c_k \otimes \hat{z}'_{2p^k}) = (\tfrac{1}{6})d_{[r+k]}\zeta_{r+k}^{r+k-1}(\nabla(z_k^{12}))\ .$$

Proposition 13.17. *Equations* (13AY) *and* (13AZ) *are true in* $\pi_*(\Omega E_k; \mathbf{Z}_{p^{r+k}})$.

Proof. Now that (xviii) is known to be true, we merely apply $(\Omega\nabla_k)_\#$ to both sides of (12L). To simplify the equation that results, we use 13.10(b), 13.8, and (13AO). The result is precisely (13AY). For (13AZ), apply $_\pi\delta^{r+k}$ to both sides of (13AY), and solve for $\psi_{r+k}^k(c_k \otimes \hat{z}'_{2p^k})$.

Construction of the csL/csm homomorphisms

In this very short second unit of Chapter 13, we complete the construction and properties of our three csL/csm homomorphisms at the $(r+k)^{\text{th}}$ stage. All the needed arguments have actually been given previously, so we need only to refer to them.

Construction and properties of γ_{r+k}^k. Propositions 13.5, 13.13, and 13.17 establish that all the hypotheses of Theorem 7.15 are satisfied, in any dimension where they might be non-vacuous. As a result we have $\gamma_{r+k}^k : \mathcal{F}(\bar{L}^k)_{[r+k]}^{*\leq 2p} \to \pi_*(\Omega E_k; \mathbf{Z}_{p^{r+k}})$ making

$$\{\gamma_t^k\}_{1\leq t\leq r+k} : \mathcal{F}(\bar{L}^k)_{[r+k]}^{*\leq 2p} \to \{\pi_*(\Omega E_k; \mathbf{Z}_{p^t})\}$$

into a csL/csm homomorphism over $\mathcal{F}(\mathcal{U}M^k)$.

For (ix), we use the same method of proof that gave us equations (9K) and (9L) for $t < r + k$. Like for $t < r + k$, both (9K) and (9L) are special cases of the commutativity of (10Z). Once again, it suffices that (10Z) commute when restricted to the virtual generators of $\mathcal{F}(\bar{L}^k)_{[r+k]}$. But this is precisely what parts (e) and (f) of Proposition 13.5 tells us.

The proof of (xi) at $t = r + k$ also follows the same pattern as the proof for $t < r + k$ at the end of Chapter 10. Put

$$Q = \{w \in \mathcal{F}(\bar{L}^k)^{*\leq 2p}_{[r+k]} \mid (9N) \text{ is true for } w\}.$$

Then Q is closed under brackets and under $d_{[r+k]}$, and $Q \supseteq \zeta^{r+k-1}_{r+k}(\mathcal{F}(\bar{L}^k)^{*\leq 2p}_{[r+k-1]})$. Using the first Jacobi identity and a little combinatorics we can show that Q is closed under $\psi_{[r+k]}(b_k \otimes -)$. Because Q contains the virtual generators y_{2p^k} and z'_{2p^k} by Lemma 13.12, Q is all of $\mathcal{F}(\bar{L}^k)^{*\leq 2p}_{[r+k]}$.

Since (x) is vacuous at $t = r + k$, this completes the proof of all parts of Theorem 9.1, except (xii). As a corollary, Lemma 13.7 is true for $t = r + k$ (see the Remark following the proof).

Properties of β^k_{r+k}. Consider Theorem 13.8 when $t = r + k$. The same proof works as for $t < r + k$. Formula (13AP) is true when w is a virtual generator of $\mathcal{E}(N^{kk})_{[r+k]}$, by Corollary 13.10. This suffices to establish (13AP) on all of $\mathcal{E}(N^{kk})_{[r+k]}$.

Because Corollary 13.16 took care of (xvii) and (xviii), the only part of Theorem 12.4 that remains to be proved is (xix). The proof in Chapter 12 for $t < r + k$ works equally well for $t = r + k$. In fact, the proof at $t = r + k$ is easier, since there are no a_q's to be concerned with, and we do not need to embed the proof in a dimension-wise induction. This completes the proof of Theorem 12.4.

Construction of α^k_{r+k}. As noted in connection with Remark 12.8 in Chapter 12, the only obstacle to proving the existence and properties of α^k_{r+k} in all dimensions was the possible need for (xvii) and (xviii). Now that this obstacle has been removed, the proof of Theorem 12.6 is also complete. As a corollary, Theorem 13.15 is also true for α^k_{r+k}, in all dimensions (see the Remark after the proof of 13.15).

The decompositions for ΩF_k and ΩE_k

In this third unit of Chapter 13 we prove (xii), thereby completing the inductive step of the grand inductive scheme. In the process we find factorizations for the spaces ΩF_k and ΩE_k.

We begin with some infrastructure regarding the space $S^{2q+1}\{p^t\}$, because $S^{2q+1}\{p^t\}$'s will be some of the factors in our factorizations. We have already noted that $S^{2q+1}\{p^t\}$ is an H-space. From the fibration

(13CA) $$\Omega S^{2q+1} \to S^{2q+1}\{p^t\} \xrightarrow{\tilde{\rho}} S^{2q+1} \xrightarrow{p^t} S^{2q+1}$$

we see that the mod p^s homology of $S^{2q+1}\{p^t\}$ is a free $H_*(\Omega S^{2q+1}; \mathbb{Z}_{p^s})$-module of rank two, for any $s \leq t$. For the calculations we shall be doing, we shall focus most of our attention on homology with coefficients in $\mathbb{Z}_p$.

Recall the notation $\mathcal{U}\Xi(2q+1,t)$ from Chapter 2, i.e.,

$$\mathcal{U}\Xi(2q+1,t) = \mathbb{Z}_{(p)}[u] \otimes \Lambda(v),$$

with $|u| = 2q$ and $|v| = 2q+1$. We make $\mathcal{U}\Xi(2q+1,t)$ into a dga by setting $d(u) = 0$, $d(v) = -p^t u$. For $s \leq t$ we have

$$H_*(S^{2q+1}\{p^t\}; \mathbb{Z}_{p^s}) = \mathcal{U}\Xi(2q+1,t) \otimes \mathbb{Z}_{p^s}$$

as dga's. We have no need for this, but $\{H_*(S^{2q+1}\{p^t\}; \mathbb{Z}_{p^s})\}_{s\geq 1}$ and $\{H_*(\mathcal{U}\Xi(2q+1,t); \mathbb{Z}_{p^s}\}_{s\geq 1}$ are actually equal as csa's. We shall use the names u and v as well for the images of u and v in $H_*(S^{2q+1}\{p^t\}; \mathbb{Z}_{p^s})$, for $s \leq t$. Thus we may write

$$H_*(S^{2q+1}\{p^t\}; \mathbb{Z}_{p^s}) = \mathbb{Z}_{p^s}(u^m, u^m v \mid m \geq 0)$$

as free $\mathbb{Z}_{p^s}$–modules.

We denote the inclusion of the $(2q+1)$–skeleton into $S^{2q+1}\{p^t\}$ by

$$\tilde{\rho}_0 : \; P^{2q+1}(p^t) \to S^{2q+1}\{p^t\}.$$

We may assume that our homology classes u and v satisfy $h(\tilde{\rho}_0 \omega^t_s) = -v$ and $h(\tilde{\rho}_0 \omega^t \rho_s) = u$.

Lemma 13.18. *Let X be a 0–connected homotopy–associative H–space, and let $g : P^{2q+1}(p^t) \to X$ be any map. Looking in $H_*(\;;\mathbb{Z}_p)$, let $x = h(g\omega^t_1)$ and $z = h(g\omega^t \rho^1)$. There exists an extension (through $\tilde{\rho}_0$) of g,*

$$\tilde{g} : \; S^{2q+1}\{p^t\} \to X,$$

such that $\tilde{g}_* : \; H_*(S^{2q+1}\{p^t\}; \mathbb{Z}_p) \to H_*(X; \mathbb{Z}_p)$ satisfies

(13CB) $$\tilde{g}_*(u^m) = z^m \text{ and } \tilde{g}_*(u^m v) = -z^m x.$$

Proof. There exists a unique extension (through $(\Sigma 1)^a$) of g to an H–map g' from the James construction on $P^{2q+1}(p^t)$ to X, i.e., $g' : \; \Omega\Sigma P^{2q+1}(p^t) \to X$. In [CMN1] it is shown that $S^{2q+1}\{p^t\}$ is a retract of $\Omega P^{2q+2}(p^t)$. It is possible to choose the inclusion $f : \; S^{2q+1}\{p^t\} \to \Omega P^{2q+2}(p^t)$ such that

$$f_* : \; H_*(S^{2q+1}\{p^t\}; \mathbb{Z}_p) \to H_*(\Omega P^{2q+2}(p^t); \mathbb{Z}_p) = \mathbb{Z}_p\langle v_{2q+1}, u_{2q}\rangle$$

satisfies $f_*(u^m) = (u_{2q})^m$ and $f_*(u^m v) = -(u_{2q})^m (v_{2q+1})$. Another property of f is that $f\tilde{\rho}_0 = (\Sigma 1)^a$. Define $\tilde{g}$ to be the composition, $\tilde{g} = g'f : \; S^{2q+1}\{p^t\} \to X$. Then $\tilde{g}\tilde{\rho}_0 = g' f \tilde{\rho}_0 = g'(\Sigma 1)^a = g$, and $\tilde{g}_* = g'_* f_*$ is given by (13CB) because g' is an H–map.

Remark. From here until the proof of (xii), we will adhere closely to the path laid out by Cohen–Moore–Neisendorfer in [CMN2]. They found product decompositions for ΩF_0 and ΩE_0, and we are generalizing their techniques to obtain decompositions of ΩF_k and ΩE_k.

Notation/Definition. Recall the map $f_k : \; S^{2n-1} \to \Omega F_k$ constructed for Lemma 13.1, satisfying $h(f_k) = x_1 \in H_{2n-1}(\mathcal{U}L^k) = H_{2n-1}(\Omega F_k)$. Combining Lemma 2.4(g) and (2O), observe that for any $j \geq 1$, $d(x_{p^{k+j}})$ is divisible by p^{r+k+1} in L^k. Consequently $d(y_{p^{k+j}})$ is divisible by p^{r+k+1} in $\bar{L}^k$, and $d_{[r+k]}(y_{p^{k+j}})$ is divisible by p when $y_{p^{k+j}}$ is viewed as an element of $\mathcal{F}(\bar{L}^k)_{[r+k]}$. Taking it one more step, ${}_\pi\delta^{r+k}\gamma^k_{r+k}(y_{p^{k+j}})$ is divisible by p, hence $\gamma^k_{r+k}(y_{p^{k+j}})$ extends through ω^{r+k+1}_{r+k}. Choose such an extension and call it

$$\delta_{(j)} : P^{2np^{k+j}-1}(p^{r+k+1}) \to \Omega E_k.$$

By (9K) we have

$$\overline{\underline{a}_{\eta_k}(\delta_{(j)})} = \bar{x}_{p^{k+j}} \in H_*(\mathcal{U}L^k; \mathbf{Z}_p) = H_*(\Omega F_k; \mathbf{Z}_p),$$

the overbar denoting reduction to mod p homology. Let $\bar{z}_{(j)}$ denote the mod p reduction of $p^{-(r+k+1)}d(x_{p^{k+j}})$, in $H_*(\mathcal{U}L^k; \mathbf{Z}_p)$.

By Lemma 13.18, we may for each $j \geq 1$ choose an extension of $\delta_{(j)}$,

$$\widetilde{\delta}_{(j)} : S^{2np^{k+j}-1}\{p^{r+k+1}\} \to \Omega E_k.$$

If we write $H_*(S^{2np^{k+j}-1}\{p^{r+k+1}\}; \mathbf{Z}_p)$ as $\mathbf{Z}_p[u_{(j)}] \otimes \Lambda(v_{(j)})$, then we have by (13CB)

(13CC) $$(\widetilde{\delta}_{(j)})_*(u^m_{(j)}) = (\bar{z}_{(j)})^m, \quad (\widetilde{\delta}_{(j)})_*(u^m_{(j)}v_{(j)}) = -(\bar{z}_{(j)})^m \bar{x}_{p^{k+j}} .$$

Let $\delta_k : V_k \to \Omega E_k$ denote the weak infinite sequential product of all the $\widetilde{\delta}_{(j)}$'s. I.e., V_k is the weak infinite product

(13CD) $$V_k = \prod_{j=1}^{\infty} S^{2np^{k+j}-1}\{p^{r+k+1}\} = \lim_{m\to\infty} \prod_{j=1}^{m} S^{2np^{k+j}-1}\{p^{r+k+1}\},$$

and

(13CE) $$\delta_k = \lim_{m\to\infty} (\widetilde{\delta}_{(1)} \cdot \widetilde{\delta}_{(2)} \cdot \ldots \cdot \widetilde{\delta}_{(m)}).$$

Lemma 13.19. *There exists a wedge of Moore spaces, $W_k \in \mathcal{W}^{r+k}_r$, together with a map*

$$\chi_k : W_k \to E_k,$$

having the following properties.

(a) *When W_k is written as*

$$W_k = W^{(r)} \vee \ldots \vee W^{(r+k)},$$

where $W^{(m)}$ is a bouquet of mod p^m Moore spaces, then

(13CF) $$W^{(r+s)} = P^{4np^s}(p^{r+s}) \vee P^{6np^s}(p^{r+s}) \vee (\text{higher}),$$

for $0 \leq s \leq k$.

(b) The map

$$f_k^+ = f_k \cdot (\Omega\eta_k)\delta_k \cdot (\Omega\eta_k\chi_k) : S^{2n-1} \times V_k \times \Omega W_k \to \Omega F_k \tag{13CG}$$

is a homotopy equivalence.

(c) If $n = 1$, let $\eta'_k : E_k \to F_k\langle 2\rangle$ be any lifting of η_k through the 2–connected cover of F_k (cf. (13AH)). Then

$$f_k^- = (\Omega\eta'_k)(\delta_k \cdot \Omega\chi_k) : V_k \times \Omega W_k \to \Omega F_k\langle 2\rangle \tag{13CH}$$

is a homotopy equivalence.

Proof. Theorem 2.8 defines a free dgL with linear differential, denoted $C(\mathcal{P}(\lambda))$, that embeds as a sub–dgL in $\bar{L}^k$. A procedure is outlined whereby $C(\mathcal{P}(\lambda))$ can be written as $\coprod_{m=r}^{r+k} C^{(m)}$, where $C^{(m)} = \mathcal{L}(V^{(m)})$ for a certain $\mathbb{Z}_{(p)}$–free chain complex $(V^{(m)}, d)$ which is acyclic with respect to $p^{-m}d$. To define the bouquet W_k, let $I^{(m)}$ be an indexing set such that

$$V^{(m)} = \bigoplus_{\alpha\in I^{(m)}} \mathbb{Z}_{(p)}(w_\alpha, p^{-m}d(w_\alpha)),$$

and for $r \le m \le r+k$ let

$$W^{(m)} = \bigvee_{\alpha\in I^{(m)}} P^{|w_\alpha|+1}(p^m).$$

Put $W_k = W^{(r)} \vee \ldots \vee W^{(r+k)}$. Part (a) follows from computing the first few generators of each $V^{(r+s)}$. The list begins

$$V^{(r+s)} = \mathrm{Span}\{y_{2p^s}, z'_{2p^s}, y_{3p^s}, z'_{3p^s}, \ldots\},$$

as needed for (13CF).

To define the map χ_k, we define it individually on the summand $P^{|w_\alpha|+1}(p^m)$, for each $\alpha \in \bigcup_{m=r}^{r+k} I^{(m)}$. If $\alpha \in I^{(m)}$, we may view $w_\alpha \in V^{(m)} \subseteq C(P(\lambda)) \subseteq \bar{L}^k$ as an element of $\mathcal{F}(\bar{L}^k)_{[m]}$. Then $p^{-m}d(w_\alpha)$ corresponds to $d_{[m]}(w_\alpha)$. Define $\chi_k(\alpha) : P^{|w_\alpha|+1}(p^m) \to E_k$ to be the map $\gamma_m^k(w_\alpha)^{\bar{a}}$, and put

$$\chi_k = \bigvee_{m=r}^{r+k} \bigvee_{\alpha\in I^{(m)}} \chi_k(\alpha) : W_k \to E_k. \tag{13CI}$$

Let $\bar{e} : \mathcal{U}\bar{L}^k \to \mathcal{U}L^k$ be the dga inclusion. Then the image of $H_*(\Omega\eta_k\chi_k; \mathbb{Z}_p)$ is $\bar{e}(\mathcal{U}C(\mathcal{P}(\lambda))) \otimes \mathbb{Z}_p \subseteq H_*(\Omega F_k; \mathbb{Z}_p)$. This follows from (ix), and even more clearly, from (10Z).

The map f_k^+ is given by (13CG). The image of $H_*(f_k^+; \mathbb{Z}_p)$ equals the sequential product in $H_*(\Omega F_k; \mathbb{Z}_p)$ of the three submodules: $\Lambda(x_1) \otimes \mathbb{Z}_p$ for f_k; $(\bigotimes_{j=1}^{\infty} \mathcal{U}\Xi(2np^{k+j}-1, r+k+1)) \otimes \mathbb{Z}_p$ for $(\Omega\eta_k)\delta_k$; and $\bar{e}(\mathcal{U}C(\mathcal{P}(\lambda)) \otimes \mathbb{Z}_p$ for $\Omega\eta_k\chi_k$. Comparing this with $H_*(\Omega F_k; \mathbb{Z}_p) = \mathcal{U}L^k \otimes \mathbb{Z}_p$ as described in Theorem 2.8, we find that $H_*(f_k^+; \mathbb{Z}_p)$ is an isomorphism. Since both source and target in (13CG) are 0–connected H–spaces of finite type, f_k^+ is a (p–local) homotopy equivalence.

For part (c), observe that the looped covering map $\Omega F_k\langle 2\rangle \xrightarrow{f'} \Omega F_k$ is monomorphic on $H_*(\ ;\mathbf{Z}_p)$, because the fibration

$$\Omega F_k\langle 2\rangle \xrightarrow{f'} \Omega F_k \longrightarrow S^1$$

splits. Since $(f')(\Omega\eta_k')(\delta_k\cdot\Omega\chi_k) = (\Omega\eta_k)(\delta_k\cdot\Omega\chi_k)$ is monomorphic on $H_*(\ ;\mathbf{Z}_p)$ by part (b), we see that $f_k^- = (\Omega\eta_k')(\delta_k\cdot\Omega\chi_k)$ is monomorphic on $H_*(\ ;\mathbf{Z}_p)$. For f_k^- to be an equivalence, it suffices that $H_*(f_k^-;\mathbf{Z}_p)$ be an isomorphism. Since $H_*(f_k^-;\mathbf{Z}_p)$ is one-to-one, it suffices to check that the vector spaces $H_q(\Omega F_k\langle 2\rangle;\mathbf{Z}_p)$ and $H_q(V_k\times\Omega W_k;\mathbf{Z}_p)$ have the same rank for every dimension q. This is an easy Poincaré series argument based upon the two homotopy equivalences

$$S^1\times\Omega F_k\langle 2\rangle \approx F_k \approx S^1\times V_k\times\Omega W_k.$$

Remark. Technically there is a gap in the construction of $\chi_k(\alpha)$ in the above proof. It is true, as we claim, that all generators w_α of $C(\mathcal{P}(\lambda))$ may be viewed as elements of $\mathcal{F}(\bar{L}^k)_{[m]}$, but a few of them will have $*$-degree greater than $2p$, so officially $\gamma_m^k(w_\alpha)$ is undefined. This problem is surmountable. Since γ_m^k is defined and commutes with differentials on all the generators $\{y_i, z_i' \mid i\in I_{m-r}\}$ of the free dgL $\bar{L}_{m-r}^k$, for each fixed m we see that γ_m^k extends to a dgL homomorphism

$$\gamma_m^k:\ \bar{L}_{m-r}^k \longrightarrow \pi_*(\Omega E_k;\mathbf{Z}_{p^m})\ ,$$

with no $*$-degree restrictions. These extended γ_m^k's are compatible with θ_{m-1}^m's but not with ζ_m^{m-1}'s. Indeed, the potential failure of $(\rho_{m+1}^m)^\#\gamma_m^k$ to equal $\gamma_{m+1}^k\zeta_{m+1}^m$ in all cases, is the only thing preventing $\{\gamma_i^k\}$ from being a csL homomorphism in all $*$-degrees. For 13.19, all the w_α's can be chosen to belong to $\bar{L}_{m-r}^k$ (one needs to review the construction of $C(\mathcal{P}(\lambda))$ to see this), and therefore each $\gamma_m^k(w_\alpha)$ can be presumed to be well-defined.

Lemma 13.20. *Let $f:\ Y\to X$ be any map, and denote the connecting map in the fibration sequence for f by q, i.e.,*

$$\cdots\longrightarrow\Omega Y \xrightarrow{\Omega f} \Omega X \xrightarrow{q} F\longrightarrow Y\longrightarrow X,$$

where F is the homotopy-theoretic fiber of f. Suppose $g:\ Z\to\Omega X$ is a map with the property that

$$g\cdot\Omega f:\ Z\times\Omega Y\to\Omega X$$

is a homotopy equivalence. Then the composite $qg:\ Z\to F$ is a homotopy equivalence, i.e., the homotopy-theoretic fiber of f has the homotopy type of Z, and the fiber inclusion map $F\to Y$ is null-homotopic.

Proof. We have $\pi_*(\Omega X) = g_\#(\pi_*(Z))\oplus(\Omega f)_\#(\pi_*(\Omega Y))$, and both $g_\#$ and $(\Omega f)_\#$ are one-to-one. The homomorphisms $g_\#$ and $q_\#$ induce isomorphisms

$$\pi_*(Z)\underset{\approx}{\xrightarrow{g_\#}}\pi_*(\Omega X)/(\Omega f)_\#(\pi_*(\Omega Y))\underset{\approx}{\xrightarrow{q_\#}}\pi_*(F),$$

hence qg induces an isomorphism on homotopy groups, i.e., it is a homotopy equivalence. Now it is clear that q has a section, namely $g(qg)^{-1}$, so the fiber inclusion map is null-homotopic and the fibration $\Omega Y\xrightarrow{\Omega f}\Omega X\xrightarrow{q}F$ splits.

Applying Lemma 13.20 to (13CG) gives

Corollary 13.21. *The homotopy–theoretic fiber of $\eta_k\chi_k : W_k \to F_k$ has the homotopy type of $S^{2n-1} \times V_k$, and the fiber inclusion map is null–homotopic.*

Lemma 13.22. *Let $f_k' : C(n) \to \Omega E_k$ be the map defined in Lemma 13.3.*

(a) The map

$$f_k' \cdot \delta_k \cdot (\Omega\chi_k) : C(n) \times V_k \times \Omega W_k \to \Omega E_k \tag{13CJ}$$

is a homotopy equivalence.

(b) The homotopy–theoretic fiber of $\chi_k : W_k \to E_k$ has the homotopy type of $C(n)\times V_k$, and the fiber inclusion map from $C(n) \times V_k$ to W_k is null–homotopic.

(c) The homomorphism

$$p^{r+k}\pi_j(\Omega E_j) \xrightarrow{(\Omega\eta_k)_* h} p^{r+k}H_j(\Omega F_k)$$

is monomorphic for $j < 2np^{k+1} + 2p - 5$, and it is epimorphic for $2n \le j < 2np^{k+1} + 2n - 3$.

(d) When $n = 1$, the homomorphism

$$p^{r+k}\pi_j(\Omega E_k) \xrightarrow{(\Omega\eta_k')_* h} p^{r+k}H_j(\Omega F_k\langle 2\rangle)$$

is an isomorphism, for $2n \le j \le 2np^{k+1}$.

Proof. (a) Repeat the proof of Lemma 13.3, using $V_k \times \Omega W_k$ in place of $\Omega\widehat{W}_k$. This time there is no dimension restriction.

(b) Apply Lemma 13.20 to (13CJ).

(c) By Lemma 13.2 and (13CF) and Lemma 10.2(d), we have

$$p^{r+k}\pi_*(C(n) \times \Omega W_k) = 0$$

in this range, so

$$p^{r+k}\pi_*(\Omega E_k) = p^{r+k}\pi_*(V_k)$$

is just a single copy of $\mathbb{Z}_p$ in this range, in dimension $2np^{k+1} - 2$. The same is true of $p^{r+k}H_*(\Omega F_k)$ in dimensions between $2n - 1$ and $2np^{k+1} + 2n - 3$, and $\tilde{\delta}_{(1)}$ is designed to induce this isomorphism.

(d) Same argument as in part (c). Once we have removed the factor S^1 from ΩF_k, the Kunneth formula applied to (13CH) tells us that $p^{r+k}H_j(\Omega F_k\langle 2\rangle) = 0$ for $2np^{k+1} - 2 < j < 4np^{k+1} - 4$.

Lemma 13.23. *There is a CMN diagram*

$$\begin{array}{ccccc}
\Omega E_k & \xrightarrow{\Omega\eta_k} & \Omega F_k & \xrightarrow{\Omega\mu_k} & \Omega^2 S^{2n+1} \\
\downarrow & & \downarrow{\scriptstyle \epsilon_k} & & \downarrow{\scriptstyle =} \\
C(n) \times V_k & \longrightarrow & S^{2n-1} \times V_k & \xrightarrow{\Sigma^2\bar{\pi}_1} & \Omega^2 S^{2n+1} \\
{\scriptstyle 0}\downarrow & & {\scriptstyle 0}\downarrow & & \downarrow \\
W_k & \xrightarrow{=} & W_k & \longrightarrow & * \\
{\scriptstyle \chi_k}\downarrow & & \downarrow{\scriptstyle \eta_k\chi_k} & & \downarrow \\
E_k & \xrightarrow{\eta_k} & F_k & \xrightarrow{\mu_k} & \Omega S^{2n+1},
\end{array} \tag{13CK}$$

in which a label of 0 *indicates a null–homotopic map. The bottom row of* (13CK) *comes from the third–from–right column of* (9C), *and the diagram serves as a definition for the map* ε_k.

Proof. Clearly $\mu_k\eta_k \simeq *$ so the lower right square commutes up to homotopy, and the diagram exists. We have already seen, in 13.21 and 13.22(b), that the arrows labeled 0 are null–homotopic. The only thing to check is that the arrow labeled $\Sigma^2\bar{\pi}_1$ really is the projection from $S^{2n-1} \times V_k$ to S^{2n-1} followed by the double suspension.

When we write $S^{2n-1} \times V_k$ in (13CK) for the homotopy–theoretic fiber of $\eta_k\chi_k$, we are implicitly suppressing the homotopy equivalence between these two spaces. The arrow labeled $\Sigma^2\bar{\pi}_1$ is really the composite

$$S^{2n-1} \times V_k \xrightarrow{f_k\cdot(\Omega\eta_k)\delta_k} \Omega F_k \to (\text{fiber of } \eta_k\chi_k) \to \Omega^2 S^{2n+1}$$

which is (by diagram (13CK)) homotopic to

$$S^{2n-1} \times V_k \xrightarrow{f_k\times(\Omega\eta_k)\delta_k} \Omega F_k \times \Omega F_k \xrightarrow{\mu} \Omega F_k \xrightarrow{\Omega\mu_k} \Omega^2 S^{2n+1},$$

where an unsubscripted μ is an H–space multiplication map. This composition occurs in the homotopy–commutative diagram

$$\begin{array}{ccccc} S^{2n-1} \times V_k & \xrightarrow{f_k\times(\Omega\eta_k)\delta_k} & \Omega F_k \times \Omega F_k & \xrightarrow{\mu} & \Omega F_k \\ \downarrow{\scriptstyle \bar{\pi}_1} & & \downarrow{\scriptstyle \Omega\mu_k\times\Omega\mu_k} & & \downarrow{\scriptstyle \Omega\mu_k} \\ S^{2n-1} \times (*) & \xrightarrow{\Sigma^2\times(*)} & \Omega^2 S^{2n+1} \times \Omega^2 S^{2n+1} & \xrightarrow{\mu} & \Omega^2 S^{2n+1}, \end{array}$$

which makes it clear that it is $\Sigma^2\bar{\pi}_1$.

The long–awaited moment has arrived: we are finally about to complete the inductive step of the grand inductive scheme.

Proof of (xii). We apply Theorem 6.4. If $n \geq 2$ use the fibration (13AG) as (6G), and if $n = 1$ use (13AH). Let $t = r + k$ and $q = 2np^{k+1} - 1$. The exactness of (6H) is proved in Lemma 13.22(c,d), and (6I) holds as it did for the proof of Proposition 13.5.

Let $\hat{y} = \gamma^k_{r+k}(y_{p^{k+1}})^{\bar{a}}$ and $\hat{z} = \gamma^k_{r+k}(p^{-(r+k+1)}d(y_{p^{k+1}}))^{\bar{a}}$. Then $y = \underline{a}_{\eta_k}(\hat{y}) = x_{p^{k+1}}$ by (9K), and p^{r+k+1} divides $\mathcal{A}(\mu_k)(y)$ by (9D). We leave it to the reader to check the remaining hypothesis. The conclusion is the existence and uniqueness of the mod p^{r+k+1} homotopy class $\hat{y}_2$. We take $\hat{y}_2^a$ as our definition of $\hat{y}_{p^{k+1}}$. It is characterized by the three properties:

$$\hat{y}_{p^{k+1}}\omega^{r+k+1}_{r+k} = \hat{y}^a = \gamma^k_{r+k}(y_{p^{k+1}})\ ;$$
$$\hat{y}_{p^{k+1}}\omega^{r+k+1}_{r+k} = \hat{z}^a = \gamma^k_{r+k}(p^{-(r+k+1)}d(y_{p^{k+1}}))\ ;$$
$$\text{and} \qquad \underline{a}_{\eta_k}(\hat{y}_{p^{k+1}}) = x_{p^{k+1}}\ .$$

These are precisely the properties of $\hat{y}_{p^{k+1}}$ called for in (xii).

Properties of $\tilde{\tau}_1$

There is just one "loose end": we claimed in (11EM) that ξ_1 vanishes identically in any homotopy csL. We will now prove this, in the equivalent form (11EP). It depends upon the case $k = 0$ of (xii), so it is not part of the grand inductive scheme.

When $k = 0$, we have by (xii) the special element $\hat{y}_p \in \pi_{2np-1}(\Omega E_0; \mathbb{Z}_{p^{r+1}})$. Let $\hat{b} = 1^a \in \pi_{2n}(\Omega P^{2n+1}(p^r); \mathbb{Z}_{p^r})$ and let $\hat{a} = {}_\pi\delta^r(\hat{b}) \in \pi_{2n-1}(\Omega P^{2n+1}(p^r); \mathbb{Z}_{p^r})$. Define $\tilde{\tau}_1(\hat{b})$ to be $-(\Omega\lambda_0\eta_0)_\#(\hat{y}_p) \in \pi_{2np-1}(\Omega P^{2n+1}(p^r); \mathbb{Z}_{p^{r+1}})$. Given an arbitrary $2n$–dimensional homotopy class $\hat{x} \in \pi_{2n}(\Omega X; \mathbb{Z}_{p^r})$, define $\tilde{\tau}_1(\hat{x})$ to be $(\Omega\hat{x}^{\bar{a}})_\#\tilde{\tau}_1(\hat{b})$. Then

$$\tilde{\tau}_1 : \pi_{2n}(\Omega -; \mathbb{Z}_{p^r}) \to \pi_{2np-1}(\Omega -; \mathbb{Z}_{p^{r+1}})$$

is a (non–homomorphic) natural homotopy operation.

Proposition 13.24. *(cf. Lemma 11.B5) $\tilde{\tau}_1$ has the following properties, for $\hat{x}, \hat{y} \in \pi_{2n}(\Omega X; \mathbb{Z}_{p^r})$, $\hat{z} \in \pi_m(\Omega X; \mathbb{Z}_{p^{r+1}})$:*

(a) $\tilde{\tau}_1(\hat{x})\omega_r^{r+1} = \mathrm{ad}^{p-1}(\hat{x})({}_\pi\delta^r(\hat{x})) = \tau_1(\hat{x})$

(b) (cf. (11EP)) $\tilde{\tau}_1(\hat{x})\omega^{r+1}\rho^r = \sigma_1(\hat{x})$

(c) $\tilde{\tau}_1(c\hat{x}) = c^p\tilde{\tau}_1(\hat{x})$

(d) $\tilde{\tau}_1(\hat{x} + \hat{y}) = \tilde{\tau}_1(\hat{x}) + \tilde{\tau}(\hat{y}) + \alpha_{01}(\hat{x}, \hat{y})\omega^r\rho^{r+1} + \alpha_{11}(\hat{x}, \hat{y})\rho^r_{r+1}$*, where* α_{01} *and* α_{11} *are the p–fold binary brackets that appear in* (11BG).

(e)

$$\begin{aligned}[\tilde{\tau}_1(\hat{x}), \hat{z}] = &(\mathrm{ad}^p(\hat{x})(\hat{z}\omega_r^{r+1}))\omega^r\rho^{r+1} - (\mathrm{ad}^p(\hat{x})(\hat{z}\omega^{r+1}\rho^r))\rho^r_{r+1} \\ &- (\mathrm{ad}(e_1(\hat{x}))(\hat{z}\omega_r^{r+1}))\rho^r_{r+1},\end{aligned} \tag{13DA}$$

where $e_1(\hat{x})$ *is described in* (11BD).

Proof. For (a)–(c) it suffices to prove the equation in the universal case, namely $\hat{x} = \hat{b} \in \pi_{2n}(\Omega P^{2n+1}(p^r); \mathbb{Z}_{p^r})$.

(a) This is mainly a matter of unraveling the definition. When $k = 0$, $\gamma_r^0(y_2)$ and $\gamma_r^0(z_2)$ are the unique elements of $\pi_*(\Omega E_0; \mathbb{Z}_{p^r})$ whose $\Omega(\lambda_0\eta_0)_\#$–images are $-[\hat{b}, \hat{a}]$ and $(\frac{1}{2})[\hat{a}, \hat{a}]$ respectively. For any $m \geq 2$,

$$\gamma_r^0(y_m) = \gamma_r^0(\mathrm{ad}(b^{m-2})(y_2)) = \psi_r^0(b^{m-2} \otimes \gamma_r^0(y_2)) = \mathrm{ad}^{m-2}(\hat{b})\gamma_r^0(y_2),$$

since $\Psi_r^0(b)$ equals $\Phi(\hat{b}) = [\hat{b}, -]$. As a result, for any $m \geq 2$,

$$\begin{aligned}(\Omega\lambda_0\eta_0)_\#\gamma_r^0(y_m) &= (\Omega\lambda_0\eta_0)_\#(\mathrm{ad}^{m-2}(\hat{b})\gamma_r^0(y_2)) \\ &= -\mathrm{ad}^{m-2}(\hat{b})[\hat{b}, \hat{a}] = -\mathrm{ad}^{m-1}(\hat{b})(\hat{a})\end{aligned} \tag{13DB}$$

Thus

$$\begin{aligned}\tilde{\tau}_1(\hat{b})\omega_r^{r+1} &= -(\Omega\lambda_0\eta_0)_\#(\hat{y}_p)\omega_r^{r+1} = -(\Omega\lambda_0\eta_0)_\#\gamma_r^0(y_p) \\ &= \mathrm{ad}^{p-1}(\hat{b})(\hat{a}) = \tau_1(\hat{b}).\end{aligned}$$

(b) $\tilde{\tau}_1(\hat{b})\omega^{r+1}\rho^r = -(\Omega\lambda_0\eta_0)_\#(\hat{y}_p)\omega^{r+1}\rho^r = -(\Omega\lambda_0\eta_0)_\#\gamma_r^0(p^{-(r+1)}d(y_p))$. By (2Q) we have

$$d(y_p) = -p^{r+1}z_p - (\tfrac{1}{2})p^r\sum_{j=2}^{p-2}(p|j)[y_j, y_{p-j}],$$

so

$$-p^{-(r+1)}d(y_p) = \mathrm{ad}(a_0)(y_{p-1}) + (\tfrac{1}{2})\sum_{j=2}^{p-2}(p^{-1}(p|j))[y_j, y_{p-j}].$$

Applying γ_r^0 to both sides gives

$$-\gamma_r^0(p^{-(r+1)}d(y_p)) = \psi_r^0(a_0 \otimes \gamma_r^0(y_{p-1})) + \sum_{j=2}^{p-2}(2p)^{-1}(p|j)[\gamma_r^0(y_j), \gamma_r^0(y_{p-j})].$$

Apply $(\Omega\lambda_0\eta_0)_\#$ to both sides. To simplify, use (13DB) and the fact that $\Psi_r^0(a_0) = -\Phi(\hat{a})$. We get

$$\tilde{\tau}_1(\hat{b})\omega^{r+1}\rho^r = [\hat{a}, \mathrm{ad}^{p-1}(\hat{b})(\hat{a})] + \sum_{j=2}^{p-2}(2p)^{-1}(p|j)[\mathrm{ad}^{j-1}(\hat{b})(\hat{a}), \mathrm{ad}^{p-j-1}(\hat{b})(\hat{a})] = \sigma_1(\hat{b}).$$

(c) We will use the uniqueness aspect of Theorem 6.4. We apply it to the fibration

$$E_0 \xrightarrow{\eta_k} F_0 \xrightarrow{\mu_k} \Omega S^{2n+1}$$

if $n \geq 2$, or if $n = 1$ to

$$E_0 \to F_0\langle 2\rangle \to \Omega S^3.$$

Put $t = r$ and $q = 2np - 1$. We have already checked that all hypotheses of 6.4 that are independent of $\hat{y}$ and $\hat{z}$ are satisfied.

Let $c :\ P^{2n+1}(p^r) \to P^{2n+1}(p^r)$ be multiplication by c, induced by the co-H-space structure on $P^{2n+1}(p^r)$. Let $c :\ S^{2n+1}\{p^r\} \to S^{2n+1}\{p^r\}$ also be multiplication by c, defined using the H-space multiplication. Then

$$c\rho_0 \simeq \rho_0 c:\ P^{2n+1}(p^r) \to S^{2n+1}\{p^r\}.$$

Notice that $c_\#(\hat{b}) = c\hat{b}$ and $c_\#(\hat{a}) = c\hat{a}$. There is an induced map, denoted $\tilde{c}$, between homotopy-theoretic fibers,

(13DC)
$$\begin{array}{ccccc} E_0 & \xrightarrow{\lambda_0\eta_0} & P^{2n+1}(p^r) & \xrightarrow{\rho_0} & S^{2n+1}\{p^r\} \\ \downarrow{\scriptstyle \tilde{c}} & & \downarrow{\scriptstyle c} & & \downarrow{\scriptstyle c} \\ E_0 & \xrightarrow{\lambda_0\eta_0} & P^{2n+1}(p^r) & \xrightarrow{\rho_0} & S^{2n+1}\{p^r\}. \end{array}$$

We claim that

(13DD) $$(\Omega\tilde{c})_\#(\hat{y}_p) = c^p\hat{y}_p.$$

To see why (13DD) proves 13.24(c), apply $-(\Omega\lambda_0\eta_0)_\#$ to both sides of (13DD) and use the commutativity of the left square of (13DC):

$$\begin{aligned}\tilde{\tau}_1(c\hat{b}) = (\Omega c)_\#\tilde{\tau}_1(\hat{b}) &= -(\Omega c)_\#(\Omega\lambda_0\eta_0)_\#(\hat{y}_p) = -(\Omega\lambda_0\eta_0)_\#(\Omega\tilde{c})_\#(\hat{y}_p)\\ &= -(c^p)(\Omega\lambda_0\eta_0)_\#(\hat{y}_p) = c^p\tilde{\tau}(\hat{b}).\end{aligned}$$

So we need only to prove (13DD). Since we are now talking about mod p^{r+1} homotopy classes in E_0, this is where Theorem 6.4 comes in. The uniqueness aspect of Theorem 6.4 will yield (13DD), but we need to demonstrate three equations, namely:

$$\begin{aligned}(\Omega\tilde{c})_\#(\hat{y}_p)\omega_r^{r+1} &= (c^p)\hat{y}_p\omega_r^{r+1}\ ;\\ (\Omega\tilde{c})_\#(\hat{y}_p)\omega^{r+1}\rho^r &= (c^p)\hat{y}_p\omega^{r+1}\rho^r\ ;\\ \text{and}\qquad \underline{a}_{\eta_k\tilde{c}}(\hat{y}_p) &= (c^p)\underline{a}_{\eta_k}(\hat{y}_p)\ .\end{aligned}$$

For the first equation, use the fact that

$$\begin{aligned}(\hat{y}_p)\omega_r^{r+1} = \gamma_r^0(y_p) &= -\psi_r^0(b^{p-2}\otimes(\Omega\lambda_0\eta_0)_\#^{-1}[\hat{b},\hat{a}])\\ &= -\Phi(\hat{b})^{p-2}((\Omega\lambda_0\eta_0)_\#^{-1}[\hat{b},\hat{a}])\\ &= -\mathrm{ad}^{p-2}(\hat{b})((\Omega\lambda_0\eta_0)_\#^{-1}[\hat{b},\hat{a}]).\end{aligned}$$

Hence

$$\begin{aligned}(\Omega\tilde{c})_\#(\hat{y}_p)\omega_r^{r+1} &= -(\Omega\tilde{c})_\#\mathrm{ad}^{p-2}(\hat{b})((\Omega\lambda_0\eta_0)_\#^{-1}[\hat{b},\hat{a}])\\ &= \mathrm{ad}^{p-2}(c\hat{b})((\Omega\lambda_0\eta_0)_\#^{-1}[c\hat{b},c\hat{a}])\\ &= -(c^p)\mathrm{ad}^{p-2}(\hat{b})((\Omega\lambda_0\eta_0)_\#^{-1}[\hat{b},\hat{a}]) = (c^p)\hat{y}_p\omega_r^{r+1}.\end{aligned}$$

The second equation is checked in a similar fashion.

For the third equation, we are looking at the induced map, also denoted $\tilde{c}$, from F_0 to F_0 in the commuting diagram

$$\text{(13DE)}\qquad \begin{array}{ccccccc} E_0 & \xrightarrow{\eta_0} & F_0 & \xrightarrow{\lambda_0} & P^{2n+1}(p^r) & \xrightarrow{\tilde{\rho}\rho_0} & S^{2n+1}\\ \downarrow{\scriptstyle\tilde{c}} & & \downarrow{\scriptstyle\tilde{c}} & & \downarrow{\scriptstyle c} & & \downarrow{\scriptstyle c}\\ E_0 & \xrightarrow{\eta_0} & F_0 & \xrightarrow{\lambda_0} & P^{2n+1}(p^r) & \xrightarrow{\tilde{\rho}\rho_0} & S^{2n+1}. \end{array}$$

Thus $\underline{a}_{\eta_k\tilde{c}}(\hat{y}_p) = \mathcal{A}(\tilde{c})\underline{a}_{\eta_k}(\hat{y}_p)$, and the equation boils down to showing that

$$\mathcal{A}(\tilde{c})(x_p) = c^p x_p \text{ in } \mathcal{A}(F_0).$$

At $k=0$, $\mathcal{A}(F_0)$ is the cobar construction on the differential coalgebra $(H_*(F_0;\mathbb{Z}_{(p)}),0)$, which as a graded $\mathbb{Z}_{(p)}$–module may be written as $\mathbb{Z}_{(p)}(1,\tilde{x}_1,\tilde{x}_2,\dots)$, x_i being the generator of $H_{2ni}(F_0;\mathbb{Z}_{(p)}) \approx \mathbb{Z}_{(p)}$. Since $\tilde{c}_*(\tilde{x}_1) = c\tilde{x}_1$, an easy cohomology ring argument gives $\tilde{c}_*(\tilde{x}_i) = c^i\tilde{x}_i$ for all i. In particular, $\tilde{c}_*(\tilde{x}_p) = c^p\tilde{x}_p$, which gives $\mathcal{A}(\tilde{c})(x_p) = c^p x_p$ once we apply the cobar construction.

We leave the case $n = 1$ as an exercise; it is quite similar.

(d) In principle, (d) is a lot like (c), so we will merely set up the context for the proof. We need a universal example for addition. Write P as a shorthand for $P^{2n+1}(p^r)$ and S as a shorthand for $S^{2n+1}\{p^r\}$. The CMN diagram

$$\text{(13DF)}\qquad \begin{array}{ccccc} \Sigma\Omega P \wedge \Omega P & \xrightarrow{=} & \Sigma\Omega P \wedge \Omega P & \longrightarrow & * \\ \downarrow & & \downarrow & & \downarrow \\ E_{00} & \longrightarrow & P \vee P & \longrightarrow & S \times S \\ \downarrow & & \downarrow & & \downarrow \\ E_0 \times E_0 & \longrightarrow & P \times P & \longrightarrow & S \times S \end{array}$$

serves to define the space E_{00}. It also shows that the fibration along the left column of (13DF) splits if looped once, so $\pi_*(E_{00}) = \pi_*(E_0) \oplus \pi_*(E_0) \oplus \pi_*(\Sigma\Omega P \wedge \Omega P)$. For $j < 2np+2p-5$, we see that $p^r\pi_*(E_{00})$ is zero except when $j = 2np-2$, and $p^r\pi_{2np-2}(E_{00})$ consists of two copies of $\mathbb{Z}_p$.

The diagram

$$\text{(13DG)}\qquad \begin{array}{ccccc} E_0 & \xrightarrow{\lambda_0\eta_0} & P & \longrightarrow & S \\ {\scriptstyle\widetilde{+}}\downarrow & & \downarrow & & \downarrow{\scriptstyle\Delta} \\ E_{00} & \longrightarrow & P \vee P & \longrightarrow & S \times S \end{array}$$

plays the role that (13DC) played for part (c). In (13DG), the middle vertical arrow is the comultiplication, and $\widetilde{+}$ is the induced map on fibers. What we are doing in part (d) is comparing $(\widetilde{+})_\#(\hat{y}_p)^{\bar{a}}$ and $e^{(1)}_\#(\hat{y}_p)^{\bar{a}} + e^{(2)}_\#(\hat{y}_p)^{\bar{a}}$, where $e^{(i)} : E_0 \to E_{00}$ are induced by the summand inclusions $e^{(i)} : P \to P \vee P$, for $i = 1, 2$. We also need lifts of α_{01} and α_{11} to $\pi_*(E_{00}; \mathbb{Z}_{p^r})$.

We need two more diagrams. The induced map on fibers $F_0 \to F_{00}$ appears in

$$\text{(13DH)}\qquad \begin{array}{ccccccc} E_0 & \xrightarrow{\eta_0} & F_0 & \xrightarrow{\lambda_0} & P & \longrightarrow & S^{2n+1} \\ {\scriptstyle\widetilde{+}}\downarrow & & \downarrow & & \downarrow & & \downarrow{\scriptstyle\Delta} \\ E_{00} & \longrightarrow & F_{00} & \longrightarrow & P \vee P & \longrightarrow & S^{2n+1} \times S^{2n+1}. \end{array}$$

which plays the role for (d) that (13DE) played for (c). Finally, if $n \geq 2$, the fibration that serves as (6G) in Theorem 6.4 is the top row of the CMN diagram

$$\text{(13DI)}\qquad \begin{array}{ccccc} E_{00} & \longrightarrow & F_{00} & \longrightarrow & \Omega S^{2n+1} \times \Omega S^{2n+1} \\ \downarrow & & \downarrow & & \downarrow \\ P \vee P & \xrightarrow{=} & P \vee P & \longrightarrow & * \\ \downarrow & & \downarrow & & \downarrow \\ S \times S & \xrightarrow{\tilde{\rho}\times\tilde{\rho}} & S^{2n+1} \times S^{2n+1} & \xrightarrow{p^r \times p^r} & S^{2n+1} \times S^{2n+1}. \end{array}$$

If $n = 1$, replace F_{00} by $F_{00}\langle 2\rangle$. We omit the remainder of the proof.

(e) This one is much easier. Use the fibration

$$\Sigma\Omega P^{2n+1}(p^r) \wedge \Omega P^{m+1}(p^{r+1}) \to P^{2n+1}(p^r) \vee P^{m+1}(p^{r+1}) \to P^{2n+1}(p^r) \times P^{m+1}(p^{r+1})$$

in Corollary 6.7. Put $t = r$ and put $q = 2np + m - 1$. Let $\hat{y}_0$ be the difference between the two sides of (13DA). The hypotheses of Corollary 6.7 are easily checked. (When computing $h(\hat{y}_0^a)$ we use $h(\tilde{\tau}_1(\hat{b})) = -\underline{a}_{\lambda_0\eta_0}(\hat{y}_p) = -\mathcal{A}(\lambda_0)(x_p) = -\mathrm{ad}^{p-1}(b)(a_0) = \tau_1(b) \in \mathcal{U}M^0$.) Conclude that $\hat{y}_0 = 0$.

CHAPTER 14

The Space T_k

In their study of the odd-dimensional Moore spaces $P^{2n+1}(p^r)$, Cohen, Moore, and Neisendorfer discovered an intriguing space that they denoted $T^{2n+1}\{p^r\}$. It is (with one exception) the atomic factor of $\Omega P^{2n+1}(p^r)$ that carries the lowest-dimensional non-zero homotopy. It plays an important role in mod p^r homotopy theory and in the theory of loop space decompositions.

Since our family $\{D_k\}$ of spaces generalizes the odd-dimensional Moore space, with D_0 equaling $P^{2n+1}(p^r)$ itself, we can expect that there will be spaces $\{T_k\}$ that generalize $T^{2n+1}\{p^r\}$. In this chapter we define the T_k's and begin the study of their properties. Our space T_0 is the Cohen–Moore–Neisendorfer $T^{2n+1}\{p^r\}$. We compute the Bockstein spectral sequence and cohomology ring structure of T_k, and we give a sufficient condition for T_k to split off of ΩD_k as a factor. We finish with a quick look at some homotopy exponents for the spaces we have constructed.

We continue to work over a prime $p \geq 5$. Whenever we mention any of the spaces D_k, V_k, etc., there are implicitly associated with it, but suppressed from the notation, the connectivity parameter n and the torsion order parameter r. We continue to reserve these meanings for n and r.

Our first lemma does not concern T_k, but it does generalize a very important result of Cohen–Moore–Neisendorfer.

Lemma 14.1. *Let $n \geq 1$, $r \geq 1$, $p \geq 5$. Associated to each $k \geq 0$, there is a map*

(14A) $$\phi_k : \Omega^2 S^{2n+1} \to S^{2n-1}$$

having the two properties $\phi_k \Sigma^2 = p^r : S^{2n-1} \to S^{2n-1}$ and

(14B) $$\Sigma^2 \phi_k = \Omega^2 p^r : \Omega^2 S^{2n+1} \to \Omega^2 S^{2n+1}$$

Proof. From the lower left corner of (9C) we have the equation $\mu_k \kappa_k = \Omega p^r$, which we loop to obtain

$$(\Omega \mu_k)(\Omega \kappa_k) = \Omega^2 p^r.$$

From (13CK) we have $\Omega\mu_k = \Sigma^2\bar{\pi}_1\varepsilon_k$. Substitute this for $\Omega\mu_k$ to obtain

$$\Sigma^2(\bar{\pi}_1\varepsilon_k)(\Omega\kappa_k) = \Omega^2 p^r.$$

Define ϕ_k to be $\bar{\pi}_1\varepsilon_k(\Omega\kappa_k) : \Omega^2 S^{2n+1} \to S^{2n-1}$. Then $\Sigma^2\phi^k = \Omega^2 p^r$. That $\phi_k\Sigma^2 = p^r$ is a simple computation of H_{2n-1}; indeed, (14B) always implies $\phi_k\Sigma^2 = p^r$.

Remark. The existence of a map (14A) satisfying (14B) was first proved by Cohen–Moore–Neisendorder in [CMN2]. Using such a map, they gave a very elegant proof that p^n is an H–space exponent for $\Omega^{2n+1}S^{2n+1}\langle 2n+1\rangle$, if $p \geq 5$, and consequently that p^n is a homotopy exponent for S^{2n+1}. The proof given here is essentially a copy of their proof; when $k = 0$ it is exactly their proof. It is not clear whether or not the ϕ_k's are different from each other, for different values of k.

Definition/Notation 14.2. Because $\rho_k\lambda_k\eta_k$ is null–homotopic by (9C), so is the composite $(\tilde{\rho}\rho_k)(\lambda_k\eta_k\chi_k)$, where χ_k and W_k are defined in Lemma 13.19. As a result there is a CMN diagram

$$\text{(14C)}\qquad \begin{array}{ccccccc}
 & & & & \Omega W_k & & \\
 & & & & \Big\downarrow{\scriptstyle \Omega(\lambda_k\eta_k\chi_k)} & & \\
\Omega^2 S^{2n+1} & \xrightarrow{\Omega\kappa_k} & \Omega F_k & \xrightarrow{\Omega\lambda_k} & \Omega D_k & \xrightarrow[\Omega(\tilde{\rho}\rho_k)]{} & \Omega S^{2n+1} \\
\Big\downarrow{\scriptstyle =} & & \Big\downarrow{\scriptstyle \varepsilon_k} & & \Big\downarrow{\scriptstyle \nu_k} & & \Big\downarrow{\scriptstyle =} \\
\Omega^2 S^{2n+1} & \xrightarrow{(\phi_k,\phi'_k)} & S^{2n-1}\times V_k & \longrightarrow & T_k & \longrightarrow & \Omega S^{2n+1} \\
\Big\downarrow & & \Big\downarrow{\scriptstyle 0} & & \Big\downarrow & & \Big\downarrow \\
* & \longrightarrow & W_k & \xrightarrow{=} & W_k & \longrightarrow & * \\
\Big\downarrow & & \Big\downarrow & & \Big\downarrow{\scriptstyle \lambda_k\eta_k\chi_k} & & \Big\downarrow \\
\Omega S^{2n+1} & \xrightarrow{\kappa_k} & F_k & \xrightarrow{\lambda_k} & D_k & \xrightarrow{\tilde{\rho}\rho_k} & S^{2n+1}.
\end{array}$$

We take (14C) as the definition of the space T_k and the map $\nu_k : \Omega D_k \to T_k$, i.e., T_k is the homotopy–theoretic fiber of the map $\lambda_k\eta_k\chi_k : W_k \to D_k$.

Question 14.3. In diagram (14C), the first component of the map from $\Omega^2 S^{2n+1}$ to $S^{2n-1} \times V_k$ is $\bar{\pi}_1\varepsilon_k(\Omega\kappa_k)$, which was just our definition of ϕ_k in Lemma 14.1. Is the map $\phi'_k : \Omega^2 S^{2n+1} \to V_k$ null–homotopic? This question can be asked component–wise, since by (13CD) V_k is a (weak) infinite product.

Notation/Examples 14.4. Let $\Gamma(x)$ denote the divided power algebra over $\mathbb{Z}_{(p)}$ or $\mathbb{Z}_p$ on an even–dimensional class x. The mth divided power of x is denoted by $x^{(m)}$. It is well–known and easily checked that

$$H^*(\Omega S^{2n+1}; \mathbb{Z}_{(p)}) = \Gamma(y_0)$$

for a $2n$–dimensional cohomology class y_0. Another example is

$$H^*(S^{2n+1}\{p^t\}; \mathbb{Z}_p) = \Gamma(u) \otimes \Lambda(v),$$

$|u| = 2n$, $|v| = 2n+1$, with cohomology Bockstein spectral sequence described by $\beta^t(u^{(m)}) = u^{(m-1)}v$, $\beta^t(u^{(m-1)}v) = 0$, $\beta^j = 0$ for $j \neq t$. Similarly,

$$H^*(\Omega^2 S^{2n+1}; \mathbb{Z}_p) = \Lambda(z_0) \otimes \bigotimes_{i=1}^{\infty} (\Lambda(x_i) \otimes \Gamma(y_i)),$$

where $|z_0| = 2n-1$, $|x_i| = 2np^i - 1$, and $|y_i| = 2np^i - 2$. The cohomology Bockstein spectral sequence is described by

$$\beta^1(z_0) = 0,\ \beta^1((y_i)^{(m)}) = (y_i)^{(m-1)}x_i,\ \beta^j = 0 \text{ for } j > 1.$$

Lemma 14.5. *The fibrations*

$$S^{2n-1} \times V_k \to T_k \to \Omega S^{2n+1} \tag{14D}$$

and

$$\Omega W_k \to \Omega D_k \xrightarrow{\nu_k} T_k \tag{14E}$$

appearing in (14C) are both totally non–cohomologous to zero (henceforth TNCZ), i.e., the cohomology Serre spectral sequence over $\mathbb{Z}_p$ degenerates. The cohomology ring $H^*(T_k; \mathbb{Z}_p)$, as a ring, equals the tensor product

$$\begin{aligned} &H^*(S^{2n-1}; \mathbb{Z}_p) \otimes H^*(V_k; \mathbb{Z}_p) \otimes H^*(\Omega S^{2n+1}; \mathbb{Z}_p) \\ &= (\Lambda(z_0) \otimes \Gamma(y_0)) \otimes \Big(\bigotimes_{j=k+1}^{\infty} (\Gamma(u_j) \otimes \Lambda(v_j))\Big), \end{aligned} \tag{14F}$$

where $|u_j| = 2np^j - 2$, $|v_j| = 2np^j - 1$, $|z_0| = 2n-1$, $|y_0| = 2n$.

Proof. We use a Poincaré series argument. Poincaré series are computed over $\mathbb{Z}_p$, and the series for the space X is denoted $X(t)$. We have

$$(S^{2n-1} \times V_k)(t)\Omega S^{2n+1}(t) \ \geq\ T_k(t),$$

with equality if and only if (14D) is TNCZ. Similarly, we have

$$\Omega W_k(t) T_k(t) \ \geq\ \Omega D_k(t),$$

with equality if and only if (14E) is TNCZ. Putting them together,

$$S^{2n-1}(t) V_k(t) \Omega W_k(t) \Omega S^{2n+1}(t) \ \geq\ \Omega D_k(t)\ , \tag{14G}$$

with equality if and only if both fibrations are TNCZ. We know that

$$\Omega D_k(t) \ =\ \mathcal{U}M^k(t) \ =\ \mathcal{U}L^k(t)(\mathcal{U}\mathbb{L}^{ab}[b])(t) \ =\ \Omega F_k(t)\Omega S^{2n+1}(t)$$

by (3B), and by (13CG)

$$\Omega F_k(t) = S^{2n-1}(t)V_k(t)\Omega W_k(t).$$

Conclude that we do have equality in (14G), so both fibrations are TNCZ.

Let us discuss the ring structure of $H^*(T_k;\mathbb{Z}_p)$. By "polynomial ring" we now mean a tensor product of a symmetric algebra on even–dimensional elements with an exterior algebra on odd–dimensional elements; these are the free objects in the category of graded commutative rings. In general, if $F \to X \to B$ is a TNCZ fibration and $H^*(F;\mathbb{Z}_p)$ is a polynomial ring, then automatically $H^*(X;\mathbb{Z}_p)$ equals $H^*(F;\mathbb{Z}_p)\otimes H^*(B;\mathbb{Z}_p)$ as a ring. To see why, arbitrarily lift each generator of $H^*(F;\mathbb{Z}_p)$ to $H^*(X;\mathbb{Z}_p)$. This defines a ring homomorphism from $H^*(F;\mathbb{Z}_p)\otimes H^*(B;\mathbb{Z}_p)$ to $H^*(X;\mathbb{Z}_p)$ which is an isomorphism on vector spaces, hence a ring isomorphism. If $H^*(F;\mathbb{Z}_p)$ fails to be free, i.e. if nontrivial relations hold among the generators, the argument can fail unless we know for some reason that the same relations hold among the lifts of the generators.

For the fibration (14D), $H^*(S^{2n-1}\times V_k;\mathbb{Z}_p)$ is not a polynomial ring, but it is free as a divided power algebra. Viewing it as a quotient of a polynomial ring, all its minimal relations have the form "a pth power is zero." The above argument would still work if we knew that all p^{th} powers vanish in $H^*(T_k;\mathbb{Z}_p)$. Actually, we do know this. We just showed that $H^*(T_k;\mathbb{Z}_p)$ embeds in $H^*(\Omega D_k;\mathbb{Z}_p)$ as algebras, and in $H^*(\Omega D_k;\mathbb{Z}_p)$ all p^{th} powers vanish because $H_*(\Omega D_k;\mathbb{Z}_p)$ is primitively generated as a Hopf algebra (see [MM]).

We can compute the Bockstein spectral sequence (BSS) for T_k in a similar fashion, but we need to make use of the interaction between the Serre and the Bockstein spectral sequences. For this discussion and the next proposition only, let $\{E^m(X),\beta^m\}$ denote the cohomology rather than the homology BSS for X, at the prime p. Suppose $F\to X\to B$ is an orientable fibration. The BSS for X respects the Serre filtration. This means that the chain complex $(E^m(X),\beta^m)$ is a filtered complex, so there is a bigraded Serre spectral sequence whose 0th term is $E^m(X)$ and which converges to $H^*(E^m(X),\beta^m)=E^{m+1}(X)$. We need a new notation, so let $\{F_q^{(m+1)**}, d_q^{(m+1)**}\}_{q\geq 0}$ denote the latter spectral sequence. Notice that $\{F_q^{(1)**}\}_{q\geq 2}$ is the usual cohomology Serre spectral sequence with coefficients in $\mathbb{Z}_p$. Some properties are that $F_2^{(1)}=E^1(B)\otimes E^1(F)$, since this is $H^*(B;H^*(F;\mathbb{Z}_p))$, and that $d_q^{(m)}$ has bidegree $(q,1-q)$ for $q\geq 0$. The link between one spectral sequence and the next is that $F_\infty^{(m)}=E^m(X)=F_0^{(m+1)}$, for $m\geq 1$. If $d_q^{(1)}=0$ for all $q\geq 2$, i.e., if the usual Serre spectral sequence with $\mathbb{Z}_p$ coefficients collapses, then $F_0^{(2)}=F_\infty^{(1)}=E^1(B)\otimes E^1(F)$, with $d_0^{(2)}=1\otimes\beta^1$ and $d_1^{(2)}=\beta^1\otimes 1$, implying $F_2^{(2)}=E^2(B)\otimes E^2(F)$. Generalizing this last fact, if $d_2^{(m)}=0$ for $q\geq 2$ for all m in the range $1\leq m<t$, then $F_2^{(t)}=E^t(B)\otimes E^t(F)$.

Proposition 14.6.

(a) *In the Bockstein–Serre spectral sequence described above, for the fibration*

$$S^{2n-1}\times V_k\to T_k\to\Omega S^{2n+1},$$

we have

(14H) $$F_2^{(r+1)}=F_2^{(1)}=H^*(S^{2n-1}\times V_k;\mathbb{Z}_p)\otimes H^*(\Omega S^{2n+1};\mathbb{Z}_p).$$

For $0\leq s\leq k$ *we have*

(14I) $$\begin{aligned} E^{r+s}(T_k) &= F_\infty^{(r+s)} = F_0^{(r+s+1)} = F_{2n}^{(r+s+1)} \\ &= \Lambda(y_0^{(p^s-1)}z_0)\otimes\Gamma(y_0^{(p^s)})\otimes\Big(\bigotimes_{j=k+1}^{\infty}(\Gamma(u_j)\otimes\Lambda(v_j))\Big)\end{aligned}$$

and

(14J) $$d_{2n}^{(r+s+1)}(y_0^{(p^s-1)}z_0) = y_0^{(p^s)}.$$

If $s < k$ then $d_q^{(r+s+1)} = 0$ for $q \neq 2n$, i.e., $d_{2n}^{(r+s+1)}$ is the only non-zero differential. For $s = k$ however we have for $j \geq k+1$

$$\begin{aligned} d_{2np^j}^{(r+k+1)}(v_j) &= y_0^{(p^j)} \; ; \; d_{2np^{k+1}-1}^{(r+k+1)}(u_{k+1}) = y_0^{(p^{k+1}-1)}z_0 \; ; \\ d_{2np^{j+1}-1}^{(r+k+1)}(u_{j+1}) &= y_0^{(p^{j+1}-p^j)}v_j \; . \end{aligned}$$

Furthermore, $\bar{F}_\infty^{(r+k+1)} = 0$, i.e., the BSS for T_k becomes trivial starting at $E^{r+k+1}(T_k)$, which means that $p^{r+k}\bar{H}^*(T_k) = 0$.

(b) *In the Bockstein–Serre spectral sequence for the fibration*

$$\Omega W_k \to \Omega D_k \xrightarrow{\nu_k} T_k,$$

we have $d_q^{(m)} = 0$ for $q \geq 2$ for all m.

(14K) $$E^m(\Omega D_k) = F_\infty^{(m)} = F_2^{(m)} = E^m(T_k) \otimes E^m(\Omega W_k).$$

Also, $F_\infty^{(r)}(\Omega D_k) = E^1(T_k) \otimes E^1(\Omega W_k)$ and $\bar{E}^{r+k+1}(\Omega D_k) = 0$.

Proof. For $0 \leq s \leq k$, suppose inductively that $F_\infty^{(r+s)}$ is as stated for both fibrations. This is true for $s = 0$. Since $E^r(\Omega D_k) = E^r(\mathcal{U}M^k) = \mathcal{U}M^k \otimes \mathbb{Z}_p$, the Poincaré series argument of Lemma 14.5 works to show that all differentials before $F_\infty^{(r)}$ must vanish, for both (14D) and (14E). It is part of our inductive assumption that $y_0^{(p^s-1)}z_0$ and $y_0^{(p^s)}$ survive to $F_0^{(r+s+1)}$, so (14J) makes sense. To see why $d_q^{(r+s+1)} = 0$ for $q < 2n$ and why (14J) is true, look at the Serre spectral sequence over $\mathbb{Z}_{(p)}$ for (14D) to see that d_{2n} is the first non-zero differential and that $d_{2n}(z_0) = p^r(y_0)$ (e.g. because $H^{2n}(T_k) = H^{2n}(\Omega D_k) = \mathbb{Z}_{p^r}$), and (14J) is a consequence of $d_{2n}(z_0) = p^r(y_0)$. The expression (14I) at $s+1$ is precisely the homology of $F_{2n}^{(r+s+1)}$ with respect to $d_{2n}^{(r+s+1)}$, i.e. it equals $F_{2n+1}^{(r+s+1)}$, so (14I) equals $F_\infty^{(r+s+1)}$ if and only if all subsequent differentials on $F_*^{(r+s+1)}$ vanish. To express this in terms of Poincaré series, let the series for (14I) be denoted $S^{(r+s)}(t)$. We have so far shown that

$$E^{r+s+1}(T_k) = F_\infty^{(r+s+1)}(t) \leq S^{(r+s+1)}(t),$$

with equality if and only if $d_q^{(r+s+1)} = 0$ for all $q > 2n$.

Since our inductive assumption also includes that (14K) holds for $F_\infty^{(r+s)} = F_0^{(r+s+1)}$, for the spectral sequence associated to (14E), we can say that $F_2^{(r+s+1)} = E^{r+s+1}(T_k) \otimes E^{r+s+1}(\Omega W_k)$. Hence

$$E^{r+s+1}(\Omega D_k)(t) = F_\infty^{(r+s+1)}(t) \leq E^{r+s+1}(T_k)(t)E^{r+s+1}(\Omega W_k)(t),$$

with equality if and only if $d_q^{(r+s+1)} = 0$ for (14E), for all $q \geq 2$. Combining this with the inequality $E^{r+s+1}(T_k)(t) \leq S^{(r+s+1)}(t)$, we get

(14L) $$E^{r+s+1}(\Omega D_k)(t) \leq E^{r+s+1}(\Omega W_k)(t)S^{(r+s+1)}(t),$$

with equality if and only if all $d^{(r+s+1)}$'s other than those already indicated are zero.

We know what $E^{r+s+1}(\Omega D_k) = E^{r+s+1}(\mathcal{U}M^k)$ is from Theorem 4.8, and $S^{(r+s+1)}(t)$ looks difficult at first but it is really just $V_k(t)(1+t^{2np^{s+1}-1})(1-t^{2np^{s+1}})^{-1}$. We can determine $E^{r+s+1}(\Omega W_k)$ from Theorem 2.8. We omit this computation, but when $s<k$ the net result is that equality holds in (14L). This completes the inductive step, for $s<k$.

When $s=k$, the first part of the proof, computing $d_{2n}^{(r+k+1)}$ for (14D), remains intact. However, let us put (14D) aside for a moment and consider the fibration (14E). Because $W_k \in \mathcal{W}_r^{r+k}$, $\bar{E}^{r+k+1}(\Omega W_k) = 0$. We had

$$F_0^{(r+k+1)} = F_\infty^{(r+k)} = E^{r+k}(T_k) \otimes E^{r+k}(\Omega W_k) ,$$

so now we have $F_1^{(r+k+1)} = E^{r+k}(T_k) \otimes E^{r+k+1}(\Omega W_k) = E^{r+k}(T_k) \otimes \mathbb{Z}_p$. This bigraded complex lives only in row zero, so $F_\infty^{(r+k+1)} = F_2^{(r+k+1)} = E^{r+k+1}(T_k) \otimes \mathbb{Z}_p \approx E^{r+k+1}(T_k)$. But we know that $F_\infty^{(r+k+1)} = E^{r+k+1}(\Omega D_k) = E^{r+k+1}(\mathcal{U}M^k) = \mathbb{Z}_p$ by Theorem 4.8. We deduce that $E^{r+k+1}(T_k) = \mathbb{Z}_p$.

Now we return to the fibration (14D). Thinking ahead, we have to end up with $F_\infty^{(r+k+1)} = E^{r+k+1}(T_k) = \mathbb{Z}_p$, starting from the known $F_{2n+1}^{(r+k+1)}$. This forces the remaining $d_q^{(r+k+1)}$'s to cause everything to cancel out. It is not hard to see that (except for scalar multiples that we absorb into the generators) $d_q^{(r+k+1)}$ must be stated in 14.6(a).

Corollary 14.7. *The map $\nu_k : \Omega D_k \to T_k$ induces a split surjection of BSS's, and $\Omega(\lambda_k \eta_k \chi_k) : \Omega W_k \to \Omega D_k$ induces a split monomorphism of BSS's.*

The monomorphism of BSS's in Corollary 14.7 permits us to give a sufficient condition for splitting the fibration (14E). We first describe a certain chain complex splitting for $\mathcal{U}M_+^k$, which we then realize geometrically. Recall from Chapter 5 the differential submodule $S_0^k = \mathbb{Z}_{(p)}(b_0, c_0, b_1, c_1, \ldots, b_k, c_k)$ of $\mathcal{U}M_+^k$. Recall also from Lemma 5.9 the set B, which is the set of all words on the alphabet $\{b_0, c_0, \ldots, b_k, c_k\}$ that do not have as a subword any element of the obstruction set (5S). Let

$$U_0^k = \text{Span}\{\text{words in } B \text{ of length} \geq 2\} \subseteq \mathcal{U}M_+^k.$$

An easy consequence of Lemma 5.9 is that $d(U_0^k) \subseteq U_0^k$, hence

$$\mathcal{U}M_+^k = S_0^k \oplus U_0^k \tag{14M}$$

is a splitting of $\mathcal{U}M_+^k$ as chain complexes over $\mathbb{Z}_{(p)}$. Obviously we have a consequent splitting of the BSS,

$$E^m(\mathcal{U}M_+^k) = E^m(S_0^k) \oplus E^m(U_0^k),$$

for every $m \geq 1$. Notice that not only does U_0^k consist of decomposables, i.e. $U_0^k \subseteq (\mathcal{U}M_+^k)^2$, but we can make the stronger statement that $E^m(U_0^k) \subseteq (E^m(\mathcal{U}M_+^k))^2$, for every $m \geq 1$.

We want to take the splitting (14M) a step further, in order to relate it to W_k. In Theorem 2.8 we have a sub–dgL embedding

$$C(\mathcal{P}(\lambda)) \to \bar{L}^k \to M^k.$$

Define C to be the image in $\mathcal{U}M^k$ of

$$\mathcal{U}C(\mathcal{P}(\lambda)) \to \mathcal{U}\bar{L}^k \to \mathcal{U}M^k. \tag{14N}$$

Since it is isomorphic with $\mathcal{U}C(\mathcal{P}(\lambda))$, C is a tensor subalgebra (over $\mathbb{Z}_{(p)}$) of $\mathcal{U}M^k$, and as a dga it has a linear differential. It was shown during the proof of Lemma 13.19 that the image of

$$H_*(\Omega W_k;\mathbb{Z}_p) \xrightarrow{(\Omega\eta_k\chi_k)_*} H_*(\Omega F_k;\mathbb{Z}_p) \;=\; \mathcal{U}L^k \otimes \mathbb{Z}_p$$

is $\bar{e}(\mathcal{U}C(\mathcal{P}(\lambda))) \otimes \mathbb{Z}_p$, hence the image of

$$H_*(\Omega W_k;\mathbb{Z}_p) \xrightarrow{(\Omega\lambda_k\eta_k\chi_k)_*} H_*(\Omega D_k;\mathbb{Z}_p) = \mathcal{U}M^k \otimes \mathbb{Z}_p$$

is precisely $C \otimes \mathbb{Z}_p$.

One consequence of the split monomorphism on BSS's in Corollary 14.7 is that C is a chain complex summand of $\mathcal{U}M^k$. Also, $S_0^k \cap C = 0$. To see this, S_0^k lies only in $*$–degrees 0 and 1 while $(C_+)^{*=0} = 0$, so if there is an intersection it must occur entirely in $*$–degree 1. But $(C)^{*=1} = \mathrm{Span}\{x_i\}_{i\geq 2}$, which $(S_0^k)^{*=1} = \mathrm{Span}\{c_s\}_{0\leq s\leq k}$ does not meet. So $S_0^k \oplus C$ is a summand of $\mathcal{U}M^k$. We may choose the complementary summand, call it K, to be contained in U_0^k. Summarizing, we have shown

Lemma 14.8. *There is a splitting as $\mathbb{Z}_{(p)}$–free chain complexes*

$$\mathcal{U}M_+^k = S_0^k \oplus C_+ \oplus K, \tag{14O}$$

where $K \subseteq U_0^k$, and C is the image of (14N).

Theorem 14.9. *Let ϕ denote the projection coming from (14M),*

$$\begin{aligned}\phi:\ \bar{H}_*(\Sigma\Omega D_k;\mathbb{Z}_p) &\xrightarrow{\cong} H_*(\Omega D_k;\mathbb{Z}_p) = (\mathcal{U}M_+^k)\otimes\mathbb{Z}_p\\ &= (S_0^k \otimes \mathbb{Z}_p) \oplus (U_0^k \otimes \mathbb{Z}_p) \to S_0^k \otimes \mathbb{Z}_p.\end{aligned}$$

Suppose there exists a map $f:\ Z \to \Sigma\Omega D_k$ such that the composition

$$\bar{H}_*(Z;\mathbb{Z}_p) \xrightarrow{f_*} \bar{H}_*(\Sigma\Omega D_k;\mathbb{Z}_p) \xrightarrow{\phi} S_0^k \otimes \mathbb{Z}_p \tag{14P}$$

is an isomorphism. Then

(a) *The fibration $\Omega W_k \to \Omega D_k \xrightarrow{\nu_k} T_k$ splits.*
(b) *ΩD_k has the homotopy type of $\Omega W_k \times T_k$.*
(c) *The space T_k is an H–space.*
(d) *The space Z is a wedge summand of $\Sigma\Omega D_k$.*

Proof. Parts (b) and (c) obviously follows immediately from (a). For (a), by [CMN3, Lemma 1.6] it suffices to construct a left inverse for

$$\chi' = \Sigma\Omega(\lambda_k\eta_k\chi_k):\ \Sigma\Omega W_k \to \Sigma\Omega D_k.$$

We will do this by exhibiting $\Sigma\Omega D_k$ as a bouquet of three summands, one of which is $\Sigma\Omega W_k$, corresponding to the decomposition (14O) for $\mathcal{U}M_+^k$.

Consider the composition

$$g_1 : D_{kk} = \Sigma\Omega D_k \wedge \Omega D_k \to \Sigma(\Omega D_k \times \Omega D_k) \xrightarrow{\Sigma\mu} \Sigma\Omega D_k,$$

where μ is the H–space multiplication. By Theorem 12.4(xvi), D_{kk} has the homotopy type of a wedge of Moore spaces. The image on the reduced homology BSS induced by g_1, denoted $\mathrm{im}(\bar{E}(g_1))$, contains all of $(\bar{E}^m(\Omega D_k))^2 = (E(\mathcal{U}M_+^k))^2$, hence it contains all of $\bar{E}^m(U_0^k)$, hence all of $\bar{E}^m(K)$. If we use Lemma 5.4 to write K as $K = \bigoplus_{\alpha\in I} \mathbf{Z}_{(p)}(y_\alpha, x_\alpha)$ for some indexing set I, with $d(y_\alpha) = (p^{r_\alpha})x_\alpha$, then there must exist for each index α an element $w_\alpha \in H_*(D_{kk};\mathbf{Z}_p)$ that survives to $E^{r_\alpha}(D_{kk})$ and which is sent by $E^{r_\alpha}(g_1)$ to y_α, where $y_\alpha$is now viewed as an element of $E^{r_\alpha}(K)$. Then $E^{r_\alpha}(g_1)$ also carries $\beta^{r_\alpha}(w_\alpha)$ to $x_\alpha \in E^{r_\alpha}(K)$. Because D_{kk} is a wedge of Moore spaces, w_α must be a mod p^{r_α} Hurewicz image, and we may define

$$g_0 : U = \bigvee_{\alpha\in I} P^{|y_\alpha|}(p^{r_\alpha}) \to D_{kk} \tag{14Q}$$

so that $\mathrm{im}(\bar{H}_*(g_1g_0;\mathbf{Z}_p)) = K \otimes \mathbf{Z}_p$ and $H_*(g_1g_0;\mathbf{Z}_p)$ is one–to–one.

Combine the three maps: f given in the hypothesis, $\chi' = \Sigma\Omega\lambda_k\eta_k\chi_k$, and $g_2 = g_1g_0$ just constructed. Let

$$g = f \vee \chi' \vee g_2 : Z \vee \Sigma\Omega W_k \vee U \to \Sigma\Omega D_k.$$

We claim that g_* injects on $H_*(;\mathbf{Z}_p)$. Granting this claim, g_* is a monomorphism between spaces having identical Poincaré series (use (14O) to compare series), hence a homotopy equivalence. Now χ' obviously has a left inverse, and part (d) follows as well.

Let us prove the claimed injectivity of g_*. If we knew that the decompositions (14M) and (14O) were compatible, i.e. that $C_+ \oplus K \subseteq U_0^k$, the claim would be trivial. We have to work around the fact that $C_+ \not\subseteq U_0^k$. Since $(C_+ \otimes \mathbf{Z}_p) \cap (K \otimes \mathbf{Z}_p) = 0$, the only way g_* could fail to inject would be if $\mathrm{im}(f_*) \cap (C_+ + K) \otimes \mathbf{Z}_p$ were non–zero. $\mathrm{Im}(f_*)$ has rank 0 or 1 in each dimension, and we may write the $\mathbf{Z}_p$–vector space $\mathrm{im}(f_*)$ as $\mathbf{Z}_p(b_0', c_0', \ldots, b_k', c_k')$, where $\phi(b_s') = b_s$, $\phi(c_s') = c_s$. It suffices to show that $b_s', c_s' \notin (C_+ + K) \otimes \mathbf{Z}_p$.

In even dimensions C_+ consists only of elements of $*$–degree ≥ 2, so $(C_+)_{\mathrm{even}} \subseteq U_0^k$. Since $K \subseteq U_0^k$, get $(C_+ + K)_{\mathrm{even}} \subseteq (U_0^k)_{\mathrm{even}}$. Then $(C_+ + K)_{\mathrm{even}} \otimes \mathbf{Z}_p \subseteq \ker(\phi)$, so $b_s' \notin (C_+ + K) \otimes \mathbf{Z}_p$.

The elements $\{c_s'\}$ are in odd dimensions, but they are necessarily infinite cycles in the BSS, being the Bockstein images of the $\{b_s'\}$. Let $J = \{\text{infinite cycles}\} \subseteq \mathcal{U}M_+^k \otimes \mathbf{Z}_p$. For $A \in \{S_0^k, U_0^k, C_+, K\}$ let an overbar $\bar{A}$ denote $(A \otimes \mathbf{Z}_p) \cap J$. Then $J = \bar{S}_0^k \oplus \bar{C}_+ \oplus \bar{K}$ because the decomposition (14O) respects the BSS. Since $(C_+)^{*=1} = \mathrm{Span}\{x_j\}_{j\geq 2}$ and none of these are infinite cycles when reduced mod p, we have $\bar{C}_+ = (\bar{C}_+)^{*\geq 2} \subseteq \bar{U}_0^k$. Like before $\bar{K} \subseteq \bar{U}_0^k$ yields $(\bar{K} + \bar{C}_+) \subseteq \bar{U}_0^k \subseteq \ker(\phi)$ and $c_s' \notin (K + C_+) \otimes \mathbf{Z}_p$. This finishes the proof.

Remark. There does exist for each k a map $f : Z \to \Sigma\Omega D_k$ making (14P) an isomorphism, but its construction is beyond the scope of this monograph. When $k = 0$ we may take $f = \Sigma(1^a) : P^{2n+1}(p^r) \to \Sigma\Omega P^{2n+1}(p^r)$. The resulting splitting of $\Omega P^{2n+1}(p^r) = \Omega D_0$ can be found in [CMN3].

We finish the chapter with some easy corollaries regarding homotopy exponents for the spaces we have constructed. Recall that a space X has ***homotopy exponent*** p^e (at the prime p) if p^e annihilates the p–torsion component of $\pi_*(X)$ but p^{e-1} does not. We have already mentioned in Lemma 10.1 the homotopy exponent p^{r+1} for the Moore space $P^q(p^r)$. The odd sphere S^{2n+1} has homotopy exponent p^n [CMN2], and $S^{2q+1}\{p^r\}$ has homotopy exponent p^r [N3]. If $X \in \mathcal{W}_1^t$ then $p^{t+1}\pi_*(X) = 0$ [N2].

Proposition 14.10.

(a) $p^{r+k+1}\pi_*(E_k) = 0$, *and* $p^{r+k}\pi_*(E_k) \neq 0$.

(b) *Let* $e = \max(n-1, r+k+1)$. $\pi_{<2n}(F_k) = 0$ *and* $\pi_{2n}(F_k) = \mathbb{Z}$ *and* $p^e(\pi_{>2n}(F_k) \otimes \mathbb{Z}_{(p)}) = 0$ *but* $p^{e-1}(\pi_{>2n}(F_k) \otimes \mathbb{Z}_{(p)}) \neq 0$.

(c) $p^{2r+k+1}\pi_*(T_k) = 0$.

(d) $p^{2r+k+1}\pi_*(D_k) = 0$.

Proof. (a) and (b) are immediate from (13CJ) and (13CG). For (c) and (d), use the top and bottom rows respectively of the CMN diagram

$$
\text{(14R)}\qquad
\begin{array}{ccccc}
C(n) \times V_k & \longrightarrow & T_k & \longrightarrow & \Omega S^{2n+1}\{p^r\} \\
\downarrow & & \downarrow & & \downarrow \\
W_k & \xrightarrow{=} & W_k & \longrightarrow & * \\
{\scriptstyle \chi_k}\downarrow & & \downarrow{\scriptstyle \lambda_k \eta_k \chi_k} & & \downarrow \\
E_k & \xrightarrow{\lambda_k \eta_k} & D_k & \xrightarrow[\rho_k]{} & S^{2n+1}\{p^r\}.
\end{array}
$$

Conjecture 14.11. $p^{r+k+1}\pi_*(T_k) = 0$ and $p^{r+k+1}\pi_*(D_k) = 0$. This conjecture for D_k is a special case of the following strengthened form of Moore's conjecture, for p–torsion spaces. Let $p \geq 5$ and let $\{X_0, X_1, \ldots, X_m\}$ be a sequence of spaces for which $X_0 = (*)$ and for $0 \leq j < m$ there are cofibrations $P^{t_j}(p) \to X_j \to X_{j+1}$, with $t_j \geq 2$. Then $p^{m+1}\pi_*(X_m) = 0$.

Remark. Conjecture 14.11 is true at $k = 0$. It is shown in [N2] that $p^{r+1}\pi_*(T_0) = 0$.

CHAPTER 15

The Limit Space D_∞

In this final chapter, we construct and explore briefly the direct limit space for our family of spaces, namely $D_\infty = \varinjlim\{D_k\}$. There are also colimit spaces $E_\infty, F_\infty, W_\infty, T_\infty$. We construct these too, and we consider the relationships among them. With minor exceptions, the results of Sections 9, 10, 12, and 13 remain true when we put $k = \infty$. One consequence of particular interest is the construction of a fibration

(15A) $$S^{2n-1} \to X \to \Omega S^{2n+1}$$

in which the transgression is multiplication by p^r.

Let $p \geq 5$, and for now think of the parameters $n \geq 1$ and $r \geq 1$ as fixed. By (9B) each D_k for $k \geq 1$ comes with a map $\hat{\imath} : D_{k-1} \to D_k$. We now attach subscripts to the $\hat{\imath}$'s and we consider the entire sequence at once:

(15B) $$P^{2n+1}(p^r) = D_0 \xrightarrow{\hat{\imath}_0} D_1 \xrightarrow{\hat{\imath}_1} D_2 \xrightarrow{\hat{\imath}_2} \cdots$$

Define $D_\infty = \varinjlim\{D_k\}$. Since $\hat{\imath}_j$ is an equivalence below dimension $2np^{k+1} - 1$ once j is $\geq k$, the same is true of the inclusion $i_k : D_k \to D_\infty$. A CW structure for D_∞ is

$$D_\infty = P^{2n+1}(p^r) \cup (e^{2np} \underset{p}{\cup} e^{2np+1}) \cup (e^{2np^2} \underset{p}{\cup} e^{2np^2+1}) \cup \ldots \cup (e^{2np^k} \underset{p}{\cup} e^{2np^k+1}) \cup \ldots,$$

which gives the same information as the description of the reduced homology of D_k in Theorem 9.1(i) (put $k = \infty$ in that description).

Theorem 9.1 also tells us that there exist compatible maps $\rho_k : D_k \to S^{2n+1}\{p^r\}$ satisfying $\rho_{k+1}\hat{\imath}_k = \rho_k$, for all k. We put

$$\rho_\infty = \varinjlim\{\rho_k\} : D_\infty \to S^{2n+1}\{p^r\}.$$

The CMN diagram (9C) exists because the lower right square commutes up to homotopy. The homotopy is merely ρ_k composed with some particular null–homotopy of $p^r \circ \tilde{\rho}$: $S^{2n+1}\{p^r\} \to S^{2n+1}$. The diagrams (9C) for $k = 0, 1, 2, \ldots$ are therefore compatible with the maps $\{\hat{\imath}_k\}$, and we obtain an infinite commuting diagram of CMN diagrams. (Think of the $\hat{\imath}_k$'s as extending along an axis perpendicular to the plane of (9C).) The direct limits of the spaces and maps in this infinite diagram comprise a CMN diagram for $k = \infty$, namely

$$\text{(15C)}\qquad \begin{array}{ccccccc}
 & & \Omega^2 S^{2n+1} & \longrightarrow & * & \longrightarrow & \Omega S^{2n+1} \\
 & & \downarrow & & \downarrow & & \downarrow \\
 & & E_\infty & \xrightarrow{\lambda_\infty \eta_\infty} & D_\infty & \xrightarrow{\rho_\infty} & S^{2n+1}\{p^r\} \\
 & & {\scriptstyle \eta_\infty}\downarrow & & {\scriptstyle =}\downarrow & & \downarrow{\scriptstyle \bar{\rho}} \\
\Omega S^{2n+1} & \xrightarrow{\kappa_\infty} & F_\infty & \xrightarrow{\lambda_\infty} & D_\infty & \xrightarrow{\bar{\rho}\rho_\infty} & S^{2n+1} \\
{\scriptstyle \Omega p^r}\downarrow & & {\scriptstyle \mu_\infty}\downarrow & & \downarrow & & \downarrow{\scriptstyle p^r} \\
\Omega S^{2n+1} & \xrightarrow{=} & \Omega S^{2n+1} & \longrightarrow & * & \longrightarrow & S^{2n+1}.
\end{array}$$

The results in the algebraic chapters, Chapters 2–5, have their $k = \infty$ versions as well. On the algebraic level, we have dgL inclusions $M^0 \to M^1 \to M^2 \to \dots,\ L^0 \to L^1 \to L^2 \to \dots$, and so on. Put $M^\infty = \varinjlim\{M^k\}$, $L^\infty = \varinjlim\{L^k\}$, and so on. All of the results of Chapters 2–5 remain valid for these limit objects as well. Of course, they have to be suitably interpreted for $k = \infty$.

To illustrate the process of passing to $k = \infty$, we briefly discuss $\mathcal{U}M^\infty$. By Theorem 4.8, the BSS $\{E^t(\mathcal{U}M^k)\}$ has $\bar{E}^{r+k}(\mathcal{U}M^k) \neq 0$ and $\bar{E}^{r+k+1}(\mathcal{U}M^k) = 0$. As $k \to \infty$ more and more terms of the BSS become non-zero (although they also very rapidly become very highly connected), so in fact $\bar{E}^t(\mathcal{U}M^\infty)$ is non-zero for all t. The algebra $\mathcal{U}M^\infty$ is not finitely generated. It has the intriguing presentation (cf. 4.3(e))

$$\text{(15D)}\qquad \mathcal{U}M^\infty = \mathbb{Z}_{(p)}\langle b, c_0, c_1, c_2, \dots\rangle / \langle \sigma_0, \sigma_1, \sigma_2, \dots\rangle,$$

$\sigma_s = pc_{s+1} - \epsilon(b^{p^s}, c_s)$, $|b| = 2n$, $|c_s| = 2np^s - 1$, $d(b) = -p^r c_0$, $d(c_s) = 0$. The differential satisfies $d(b^{p^s}) = -p^{r+s}c_s$ for all $s \geq 0$ (cf. 4.3(d)).

The dgL $\bar{L}^\infty$ exhibits a subtle but important difference from the case of k finite. In expression (2R), as $k \to \infty$, there are fewer and fewer "exceptions" excluded from the summation (since $I_k \subseteq I_{k+1} \subseteq \dots$), and at $k = \infty$ there are no exceptions at all (since $\cap I_k = 0$). This means that (2S) and (2T) serve to define the differential d for $\bar{L}^\infty$ on a complete set of generators. Thus $\bar{L}^\infty$ has a linear differential. An explicit presentation for it is

$$\text{(15E)}\qquad \bar{L}^\infty = \mathbb{L}\langle y_j, z'_j | j \geq 2\rangle,\ d(y_j) = -(p^r \langle j|1\rangle_\infty) z'_j,$$

where $\langle m|i\rangle_\infty$ now denotes $(m|i)p^{\alpha(i)+\alpha(m-i)-\alpha(m)}$, where $\alpha(t)$ is merely $[\log_p(t)]$. Another way to understand $\bar{L}^\infty$ is that the component involving Ξ in (2X) becomes more and more highly connected as $k \to \infty$. When $k = \infty$ that component completely vanishes (i.e. reduces to $\mathbb{Z}_{(p)}$) and we have $\bar{L}^\infty = C(\mathcal{P}(\lambda))$, a free dgL which has a linear differential.

All of Theorems 9.1, 12.4, 12.5, 13.8 and 13.15 remain valid for $k = \infty$. (When $k = \infty$, in place of (viii), (x), (xiv), (xxi) we have results concerning $\hat{\imath} : D_j \to D_\infty$, $\hat{\imath} : E_j \to E_\infty$, $\hat{\imath} : D_{jj} \to D_{\infty\infty}$, $\hat{\imath} : D_{jjj} \to D_{\infty\infty\infty}$, for any $j < \infty$.) We have been careful throughout to keep all our constructions compatible with $\hat{\imath}_k$'s. Indeed, this compatibility has been an essential feature in the proofs of many of the properties of our spaces.

For any $t \geq 1$ the ring of stable homotopy operations under D_∞, i.e. $\mathcal{O}_{D_\infty}[t]$, contains operations $\Psi_t^\infty(b^{p^s})$ and $\Psi_t^\infty(c_s)$, where $s = \max(0, t-r)$. When $s > 0$ these operations do not belong to im(Φ). There is a csL/csm homomorphism

$$\text{(15F)} \qquad \{\gamma_t^\infty\}_{1 \leq t < \infty} : \mathcal{F}(\bar{L}^\infty)^{* \leq 2p} \to \{\pi_*(\Omega E_\infty; \mathbb{Z}_{p^t})\} .$$

In fact, since $\bar{L}^\infty$ has a linear differential, we can probably get rid of the annoying $*$–degree restriction and replace $\mathcal{F}(\bar{L}^\infty)^{*\leq 2p}$ by the much more elegant $\mathcal{E}(\bar{L}^\infty)$, i.e., we would have

$$\{\gamma_t^\infty\}_{1 \leq t < \infty} : \mathcal{E}(\bar{L}^\infty) \to \{\pi_*(\Omega E_\infty; \mathbb{Z}_{p^t})\}$$

in place of (15F).

A subtle point is that there exist induced maps $\hat{i}_k : W_k \to W_{k+1}$ as well, where W_k is defined in Lemma 13.19, so $W_\infty = \varinjlim\{W_k\}$ and $\chi_\infty = \varinjlim\{\chi_k\} : W_\infty \to E_\infty$ exists. To see why, we must inspect the construction of χ_k in Lemma 13.19. We defined χ_k summand–wise. On the summand $P^{|w_\alpha|+1}(p^{m(\alpha)})$ of W_k associated to an index α, we took $\chi_k(\alpha)$ to be $\gamma^k_{m(\alpha)}(w_\alpha)^{\bar{a}}$. Let $\bar{i}_k : \bar{L}^k \to \bar{L}^{k+1}$ denote the dgL inclusion and rename as C^k the sub–dga of $\bar{L}^k$ that we called $C(\mathcal{P}(\lambda))$ in Theorem 2.8. The indexing set through which α runs is chosen so that the set $\bigcup_\alpha \{w_\alpha, p^{-m(\alpha)}d(w_\alpha)\}$ freely generates C^k.

It turns out that $\bar{i}_k(C^k) \subseteq C^{k+1}$. Consequently the $\bar{i}_k$–image $\bar{i}_k(w_\alpha)$ belongs to C^{k+1} for each index α, and there is an associated map $\hat{i}_k(\alpha) : P^{|w_\alpha|+1}(p^{m(\alpha)}) \to W_{k+1}$. Put these together to form $\hat{i}_k = \bigvee_\alpha \hat{i}_k(\alpha) : W_k \to W_{k+1}$. Using Theorem 9.1(x), we have

$$\chi_{k+1}(\hat{i}_k(\alpha)) = \gamma^{k+1}_{m(\alpha)}(\bar{i}_k(w_\alpha))^{\bar{a}} = (\hat{i}_k)\gamma^k_{m(\alpha)}(w_\alpha)^{\bar{a}} = \hat{i}_k \circ \chi_k(\alpha),$$

i.e., $\chi_{k+1}\hat{i}_k \simeq \hat{i}_k\chi_k : W_k \to E_{k+1}$.

The existence of $\chi_\infty : W_\infty \to E_\infty$ alone can be deduced without needing the intermediary maps $W_k \to W_{k+1}$. We may define χ_∞ directly, using the representation (15E) for $\bar{L}^\infty$ as a free dgL with a linear differential, and using the csL/csm homomorphism $\{\gamma_t^\infty\}$ of (15F). Viewed this way, W_α has one Moore space summand for each generator pair (y_j, z_j'), $j \geq 2$. The bouquet W_∞ has the very nice form

$$\text{(15G)} \qquad W_\infty = \bigvee_{j=2}^{\infty} P^{2nj}(p^{r+s(j)}),$$

where $s(j)$ denotes the unique integer such that $j \in I_{s(j)} - I_{s(j)+1}$.

The one space that is an exception to our scheme of compatibility is V_k. The space V_k and the map δ_k had their own ad hoc definitions that are not compatible with $\hat{i}_k$'s. However, the proofs of Lemma 13.19 through 13.23 can be performed afresh at $k = \infty$. The correct space to use for V_∞ is a point! We find $\Omega F_\infty \approx S^{2n-1} \times \Omega W_\infty$ and $\Omega E_\infty \approx$

$C(n) \times W_\infty$. The $k = \infty$ analog of the CMN diagram (13CK) is

$$
\text{(15H)} \qquad
\begin{array}{ccccc}
\Omega E_\infty & \xrightarrow{\Omega\eta_\infty} & \Omega F_\infty & \xrightarrow{\Omega\mu_\infty} & \Omega^2 S^{2n+1} \\
\downarrow & & \downarrow & & \downarrow{=} \\
C(n) & \longrightarrow & S^{2n-1} & \xrightarrow{\Sigma^2} & \Omega^2 S^{2n+1} \\
{\scriptstyle 0}\downarrow & & {\scriptstyle 0}\downarrow & & \downarrow \\
W_\infty & \xrightarrow{=} & W_\infty & \longrightarrow & * \\
{\scriptstyle \chi_\infty}\downarrow & & \downarrow & & \downarrow \\
E_\infty & \xrightarrow{\eta_\infty} & F_\infty & \xrightarrow{\mu_\infty} & \Omega S^{2n+1}
\end{array}
$$

Lastly, let us mention the space T_∞. It is defined as the homotopy–theoretic fiber of the map $\lambda_\infty\eta_\infty\chi_\infty : W_\infty \to D_\infty$. It also has the homotopy type of $\varinjlim\{T_k\}$, and it comes with maps $T_k \to T_\infty$. The diagram that takes the place of (14C) at $k = \infty$ is

$$
\text{(15I)} \qquad
\begin{array}{ccccccc}
\Omega^2 S^{2n+1} & \xrightarrow{\Omega\kappa_\infty} & \Omega F_\infty & \xrightarrow{\Omega\lambda_\infty} & \Omega D_\infty & \xrightarrow{\Omega(\tilde{p}\rho_\infty)} & \Omega S^{2n+1} \\
{=}\downarrow & & {\scriptstyle \epsilon_\infty}\downarrow & & {\scriptstyle \nu_\infty}\downarrow & & {=}\downarrow \\
\Omega^2 S^{2n+1} & \xrightarrow{\phi_\infty} & S^{2n-1} & \longrightarrow & T_\infty & \longrightarrow & \Omega S^{2n+1} \\
\downarrow & & \downarrow & & \downarrow & & \downarrow \\
* & \longrightarrow & W_\infty & \longrightarrow & W_\infty & \longrightarrow & * \\
\downarrow & & \downarrow{\scriptstyle \eta_\infty\chi_\infty} & & \downarrow & & \downarrow \\
\Omega S^{2n+1} & \xrightarrow{\kappa_\infty} & F_\infty & \xrightarrow{\lambda_\infty} & D_\infty & \xrightarrow{\tilde{p}\rho_\infty} & S^{2n+1}.
\end{array}
$$

The map $\phi_\infty : \Omega^2 S^{2n+1} \to S^{2n-1}$ again satisfies

$$
\text{(15J)} \qquad \Sigma^2\phi_\infty \simeq \Omega^2 p^r : \Omega^2 S^{2n+1} \to \Omega^2 S^{2n+1},
$$

since the proof of Lemma 14.1 works equally well at $k = \infty$. Question 14.3 becomes moot at $k = \infty$, since there is no second component ϕ'_∞.

This completes our tour of the limit spaces that occur by letting k go to ∞. We list a few results and conjectures about them next.

The fibration in the second row of (15I) is quite interesting.

Proposition 15.1. *The fibration*

$$
\text{(15K)} \qquad S^{2n-1} \to T_\infty \to \Omega S^{2n+1}
$$

has the property that the transgression in the Serre spectral sequence with coefficients in $\mathbb{Z}_{(p)}$ is multiplication by p^r. The $\mathbb{Z}_{(p)}$–cohomology of T_∞ is given by

$$
\text{(15L)} \qquad H^q(T_\infty) = \begin{cases} \mathbb{Z}_{(p)}/(p^r m)\mathbb{Z}_{(p)} & \text{if } q = 2nm,\ m \geq 0; \\ 0 & \text{if } 2n \nmid q. \end{cases}
$$

As a ring,

$$H^*(T_\infty;\mathbf{Z}_p) = \Lambda(z_{2n-1}) \otimes \Gamma(y_{2n}), \tag{15M}$$

subscript indicating dimension, with $\beta^r(z_{2n-1}) = y_{2n}$, and

$$E^{r+s}(T_\infty) = \Lambda(z_{2n-1}y_{2n}^{(p^s-1)}) \otimes \Gamma(y_{2n}^{(p^s)}),\ s \geq 0. \tag{15N}$$

Proof. Because W_∞ is $(4n-2)$-connected by (15G), we have $H^{2n}(T_\infty) = H^{2n}(\Omega D_\infty) = \mathbf{Z}_{p^r}$, so the transgression is multiplication by p^r. (15L) is a straightforward and rather pretty calculation via the Serre spectral sequence. The cohomology ring structure $H^*(T_\infty;\mathbf{Z}_p)$ has no choice but to be the tensor product (15M). (If a formal argument is desired refer to the proof of Lemma 14.5: the cohomology of the fiber is free.) To compute the BSS for T_∞, the answer as a vector space is implicit in (15L). The structure of $E^{r+s}(T_\infty)$ as a ring is a trivial consequence of (15M).

Lemma 15.2. $p^{r+1}\pi_*(T_\infty) = 0$.

Proof. Consider the top row of this CMN diagram, which is the $k = \infty$ case of (14R), and use Lemma 13.2:

$$\begin{array}{ccccc}
C(n) & \longrightarrow & T_\infty & \longrightarrow & \Omega S^{2n+1}\{p^r\} \\
\downarrow & & \downarrow & & \downarrow \\
W_\infty & \xrightarrow{=} & W_\infty & \longrightarrow & * \\
{\scriptstyle \chi_\infty}\downarrow & & \downarrow & & \downarrow \\
E_\infty & \xrightarrow{\lambda_\infty \eta_\infty} & D_\infty & \xrightarrow{\rho_\infty} & S^{2n+1}\{p^r\}.
\end{array} \tag{15O}$$

Conjecture 15.3. $p^r\pi_*(T_\infty) = 0$.

Remark. Of all the spaces in Proposition 14.10, T_∞ is the only space at $k = \infty$ that has a homotopy exponent. The others all have a factor involving W_∞, which contains Moore space summands of arbitrarily high order.

We can also form CMN diagrams that mix different k's, or which juxtapose spaces defined for k and for ∞. One of the more interesting such diagrams is

$$\begin{array}{ccccccc}
 & & V_k & \xrightarrow{=} & V_k & \longrightarrow & * \\
 & & \downarrow & & \downarrow & & \downarrow \\
\Omega^2 S^{2n+1} & \xrightarrow{(\phi_k,\phi'_k)} & S^{2n-1} \times V_k & \longrightarrow & T_k & \longrightarrow & \Omega S^{2n+1} \\
= \downarrow & & \downarrow & & \downarrow & & \downarrow = \\
\Omega^2 S^{2n+1} & \xrightarrow{\phi_\infty} & S^{2n-1} & \longrightarrow & T_\infty & \longrightarrow & \Omega S^{2n+1}.
\end{array} \tag{15P}$$

To construct (15P), proceed as follows. First, the lower right square commutes up to homotopy by the compatibility arguments given earlier, so the diagram exists (but we have not yet identified the spaces along the top row). Looking in $H_{2n-1}(\ )$, and knowing that both ϕ_k and ϕ_∞ induce multiplication by p^r, the map from $S^{2n-1} \times V_k$ to S^{2n-1} is seen to be an isomorphism on $H_{2n-1}(\ )$. The map consequently has a right inverse, and its homotopy–theoretic fiber has the homotopy type of V_k. Because the upper right corner space is contractible, the top center is also a V_k. We obtained this way the fibration sequence

$$V_k = \prod_{j=k+1}^{\infty} S^{2np^j-1}\{p^{r+k+1}\} \to T_k \to T_\infty . \tag{15Q}$$

Joe Neisendorfer [unpublished] proved that a certain fibration splits. His fibration is very much like the fibration

$$\Omega V_0 \to \Omega T_0 \to \Omega T_\infty \tag{15R}$$

obtained by looping (15Q) once, when $k = 0$ and $r \geq 2$. (The difference is that, in place of ΩT_∞, which is the homotopy–theoretic fiber of ϕ_∞, Neisendorfer has the homotopy–theoretic fiber of ϕ_0.) This motivates the following.

Question 15.4. Does (15Q) split if looped once?

It cannot split as it stands, because the cohomology of T_∞ has torsion of all orders, but the BSS for T_k becomes trivial at E^{r+k+1}.

If we do not wish to suppress the parameters n and p^r from the notation we may put them in, e.g., D_k is $D_k^n(p^r)$, T_k is $T_k^n\{p^r\}$, and so on. For the map ϕ_k we can write $\phi_k^n\{p^r\}$. We are following the Cohen–Moore–Neisendorfer convention, that parentheses as used in notations for spaces defined via cofibrations, and curly brackets are used in notations for spaces defined via fibrations.

The reader who is familiar with $\Omega S^3\langle 3\rangle$, the loops on the 3–connected cover of S^3, will recognize that its cohomology ring looks a lot like (15L). It is true that

$$\Omega S^3\langle 3\rangle_{(p)} \;=\; T_\infty^p\{p\}, \tag{15S}$$

but the proof of this fact goes beyond the scope of this book.

There seem to be some subtle connections among our spaces when n, p^r, and k are allowed to vary together. Two perhaps surprising compatibles are given in the next lemma.

Lemma 15.5. *If $k \geq 1$, there are homotopy equivalences*

(a) $V_k^n\{p^r\} \;\approx\; V_{k-1}^{np}\{p^{r+1}\}$;

(b) $W_k^n(p^r) \;\approx\; W_{k-1}^{np}(p^{r+1}) \vee W'$, *where $W' \in \mathcal{W}_r^r$. In other words, when $W_k^n(p^r)$ is written as $W^{(r)} \vee W^{(r+1)} \vee \ldots \vee W^{(r+k)}$, where $W^{(m)}$ is a bouquet of mod p^m Moore spaces, then $W_{k-1}^{np}(p^{r+1}) \approx W^{(r+1)} \vee \ldots \vee W^{(r+k)}$.*

Proof. (a) In (13CD), substitute $k-1$ for k, $r+1$ for r, and np for n. The expression is unchanged.

(b) This is essentially the trick used in proving Lemma 2.6. Write $\bar{L}^k(n,p^r)$ for the dgL $\bar{L}^k$ constructed for the parameters n and r. As in Lemma 2.6, there is an isomorphism of dgL's,

$$\text{(15T)} \qquad \bar{L}^k(n,p^r) \approx \bar{L}^{k-1}(np,p^{r+1}) \amalg \mathbb{L}\langle y_j, z'_j | j \in I_0 - I_1\rangle .$$

The dga $\mathcal{U}\mathbb{L}\langle y_j, z'_j | j \in I_0 - I_1\rangle$ is acyclic with respect to $p^{-r}d$. Consequently the BSS's for $\mathcal{U}\bar{L}^k(n,p^r)$ and for $\mathcal{U}\bar{L}^{k-1}(np,p^{r+1})$ coincide from their E^{r+1} terms onward.

Write $W^n_k(p^r) = W' \vee W''$, where $W' \in \mathcal{W}^r_r$ and $W'' \in \mathcal{W}^{r+k}_{r+1}$. Since W'' and $W^{np}_{k-1}(p^{r+1})$ are wedges of Moore spaces, it suffices to show that the Poincaré series of the BSS's for their loop spaces coincide. Both BSS's have $E^1 = E^{r+1}$, and both have trivial E^m for $m > r+k$. Write [A] for the Poincaré or Hilbert series of a space or graded module A. Using (2Y) and (13CG), for $r+k \geq m \geq r+1$ we have

$$\begin{aligned}
[E^m(\mathcal{U}\bar{L}^k(n,p^r))](1+t^{2n-1}) &= [E^m(\mathcal{U}L^k(n,p^r))] = [E^m(\Omega F^n_k\{p^r\})] \\
&= [E^m(S^{2n-1})][E^m(V^n_k\{p^r\})][E^m(\Omega W^n_k(p^r))] \\
&= (1+t^{2n-1})[V^n_k\{p^r\}][E^m(\Omega W'')] ,
\end{aligned}$$

hence

$$[E^m(\mathcal{U}\bar{L}^k(n,p^r))] = [V^n_k\{p^r\}][E^m(\Omega W'')] .$$

By similar reasoning,

$$[E^m(\mathcal{U}\bar{L}^{k-1}(np,p^{r+1}))] = [V^{np}_{k-1}\{p^{r+1}\}][E^m(\Omega W^{np}_{k-1}(p^{r+1}))] .$$

We noted in the previous paragraph that the two left-hand sides are equal. By part (a) the series for the two V's are equal. Part (b) follows.

Since ΩE_k consists of $C(n)$ and V_k and ΩW_k, Lemma 15.5 motivates us to wonder whether the isomorphisms have any topological meaning.

Question 15.6. For $k \geq 1$, write $W^n_k(p^r)$ as $W' \vee W''$, where $W' \in \mathcal{W}^r_r$ and $W'' \in \mathcal{W}^{r+k}_{r+1}$. Does there exist a map $f: E^n_k\{p^r\} \to E^{np}_{k-1}\{p^{r+1}\}$ such that the composite

$$\begin{aligned}
&V^n_k\{p^r\} \times \Omega W'' \to C(n) \times V^n_k\{p^r\} \times \Omega W^n_k(p^r) \xrightarrow{\approx} \Omega E^n_k\{p^r\} \xrightarrow{\Omega f} \Omega E^{np}_{k-1}\{p^{r+1}\} \\
&\xrightarrow{\approx} C(np) \times V^{np}_{k-1}\{p^{r+1}\} \times \Omega W^{np}_{k-1}(p^{r+1}) \to V^{np}_{k-1}\{p^{r+1}\} \times \Omega W^{np}_{k-1}(p^{r+1})
\end{aligned}$$

is a homotopy equivalence?

Similar to Question 15.6, but much easier to state, is the following question, motivated by the observation (cf. (15N)) that $E^{r+1}(T^n_\infty\{p^r\})$ and $E^{r+1}(T^{np}_\infty\{p^{r+1}\})$ coincide.

Question 15.7. Does there exist a map

$$\text{(15U)} \qquad f: T^n_\infty\{p^r\} \to T^{np}_\infty\{p^{r+1}\}$$

which induces a monomorphism on $H^*(\,;\mathbb{Z}_p)$?

Such a map, if it exists, would be a p^r–torsion analog of the Hopf map $H : \Omega S^{2n+1} \to \Omega S^{2np+1}$. Indeed, if the answer to 15.7 is "yes", it is conceivable that f could be chosen so that the square

(15V)
$$\begin{array}{ccc} T^n_\infty\{p^r\} & \xrightarrow{f} & T^{np}_\infty\{p^{r+1}\} \\ \downarrow & & \downarrow \\ \Omega S^{2n+1} & \xrightarrow{H} & \Omega S^{2np+1} \end{array}$$

would commute up to homotopy, where the vertical arrows are the maps that appear in the fibration (15K).

We close with a lovely conjecture due to Brayton Gray. The conjecture illustrates how the ideas developed herein may have unexpected relationships to other aspects of homotopy theory.

Conjecture 15.8. Let $H : \Omega S^{2n+1} \to \Omega S^{2np+1}$ denote the Hopf map as above. Then the composite

(15W)
$$\Omega^2 S^{2n+1} \xrightarrow{\Omega H} \Omega^2 S^{2np+1} \xrightarrow{\phi^{np}_\infty\{p\}} S^{2np-1}$$

is null–homotopic.

Remark. Postcomposing (15W) with Σ^2 gives $(\Omega^2 p)(\Omega H)$ by Lemma 14.1, and the composite $(\Omega^2 p)(\Omega H)$ is null–homotopic [G1]. So it is not so far–fetched that (15W) should be null.

If Conjecture 15.8 is true, a beautiful consequence would be the existence of a double classifying space for the fiber of the double suspension.

Consequence 15.9. If (15W) is null–homotopic, then

(15X)
$$C(n) \approx \Omega^2 T^{np}_\infty\{p\}\ .$$

Proof of (15X) *Assuming 15.8.* Because the lower right–hand square is assumed to homotopy–commute, there is a CMN diagram

(15Y)
$$\begin{array}{ccccccc} C(n) & \longrightarrow & S^{2n-1} & \xrightarrow{\Sigma^2} & \Omega^2 S^{2n+1} & \longrightarrow & \Omega T^{np}_\infty\{p\} \\ \downarrow & & \downarrow & & {\scriptstyle =}\downarrow & & \downarrow \\ \Omega^3 S^{2np+1} & \longrightarrow & \Omega J_{p-1}(S^{2n}) & \longrightarrow & \Omega^2 S^{2n+1} & \xrightarrow{\Omega H} & \Omega^2 S^{2np+1} \\ {\scriptstyle \Omega\phi^{np}_\infty\{p\}}\downarrow & & {\scriptstyle T}\downarrow & & \downarrow & & \downarrow{\scriptstyle \phi^{np}_\infty\{p\}} \\ \Omega S^{2np-1} & \xrightarrow{=} & \Omega S^{2np-1} & \longrightarrow & * & \longrightarrow & S^{2np-1}. \end{array}$$

Here $J_{p-1}(S^{2n})$ denotes the $(p-1)^{\text{st}}$ stage of the James construction on S^{2n}. It is clear that the spaces in the lower two rows of (15Y) are as indicated. The map marked "T" induces an isomorphism on $H_{2np-2}(\ )$, as is easily deduced from the fact that $\Omega\phi_\infty^{np}\{p\}$ and the connecting map $\Omega^3 S^{2np+1} \to \Omega J_{p-1}(S^{2n})$ both induce multiplication by p. Consequently T is a Toda map ([G2]), and a cohomology calculation using the Serre spectral sequence shows that the homotopy–theoretic fiber of T must be S^{2n-1}. Computing on $H_{2n-1}(\ )$, the map marked Σ^2 is readily seen to be just that. Its homotopy–theoretic fiber, the upper left corner of (15Y), is $C(n)$. But this corner also coincides with the loop space on the upper right corner (Lemma 1.3). Deduce (15X).

BIBLIOGRAPHY

[Ad] J.F. Adams, On the cobar construction, *Proc. Nat. Acad. Sci. USA* 42 (1956), 409–412.

[AH] J.F. Adams and P.J. Hilton, On the chain algebra of a loop space, *Comm. Math. Helv.* 30 (1955), 305–330.

[A1] D.J. Anick, Non–commutative graded algebras and their Hilbert series, *J. Algebra* 78 No. 1 (1982), 120–140.

[A2] D.J. Anick, On the homology of associative algebras, *Trans. Amer. Math. Soc.* 296 No. 2 (1986), 641–659.

[A3] D.J. Anick, Hopf algebras up to homotopy, *J. Amer. Math. Soc.* 2 No. 3 (1989), 417–453.

[A4] D.J. Anick, The Adams–Hilton model for a fibration over a sphere, *J. Pure and Appl. Alg.* 75 (1991), 1–35.

[A5] D.J. Anick, Single loop space decompositions, *Trans. Amer. Math. Soc.*, to appear.

[BC] G. Bergman and P.M. Cohn, Symmetric elements in free powers of rings, *J. London Math. Soc.* (2) 1 (1969), 525–534.

[Be] G. Bergman, The diamond lemma for ring theory, *Adv. in Math.* 29 (1978), 178–218.

[BK] A.K. Bousfield and D.M. Kan, *Homotopy Limits, Completions and Localizations*, Lecture Notes in Math. 304, Springer–Verlag (1972).

[CE] H. Cartan and S. Eilenberg, *Homological Algebra*, Princeton University Press, Princeton, NJ (1956).

[CMN1] F. Cohen, J.C. Moore, and J.A. Neisendorfer, Torsion in homotopy groups, *Annals of Math.* 109 (1979), 121–168.

[CMN2] F. Cohen, J.C. Moore, and J.A. Neisendorfer, The double suspension and exponents of the homotopy groups of spheres, *Annals of Math.* 110 (1979), 549–565.

[CMN3] F. Cohen, J.C. Moore, and J.A. Neisendorfer, Exponents in homotopy theory, *Algebraic Topology and Algebraic K–theory*, W. Browder, ed., Annals of Math. Study No. 113, Princeton University Press (1987), 3–34.

[D] A. Dold, Zur Homotopietheorie der Kettenkomplexe, *Math. Annalen* 140 (1960), 278–298.

[EM] S. Eilenberg and J.C. Moore, Homology and fibrations I, *Comm. Math. Helv.* 40 (1966), 199–236.

[FT] Y. Felix and J.–C. Thomas, Homotopie rationnelle, dualité et complémentarité des modèles, *Bull. Soc. Math. Belgique* 23 (1981), 7–19.

[G1] B. Gray, On the sphere of origin of infinite families in the homotopy groups of shperes, *Topology* 8 (1969), 219–232.

[G2] B. Gray, On Toda's fibration, *Math. Proc. Camb. Phil. Soc.* 97 (1985), 289–298.

[Hi] P.J. Hilton, On the homotopy groups of the union of spheres, *J. London Math. Soc.* 30 (1955), 154–172.

[HL] S. Halperin and J.–M. Lemaire, Suites inertes dans les algèbres de Lie graduées, *Math. Scand.* 61 (1987), 39–67.

[Ho] M. Hovey, A new definition of cocategory, Thesis, M.I.T. (1989); (updated version:) Lusternik-Schnirelmann cocategory, submitted.

[J] I.M. James, Reduced product spaces, *Annals of Math.* 62 (1955), 170–197.

[Le] J.–M. Lemaire, *Algèbres Connexes et Homologie des Espaces de Lacets*, Lecture Notes in Math. 422 (1974), Springer–Verlag.

[Mc] J. McCleary, *User's Guide to Spectral Sequences*, Mathematics Lecture Series No. 12, Publish or Perish, Wilmington, Del., USA (1985).

[MM] J. Milnor and J.C. Moore, On the structure of Hopf algebras, *Annals of Math.* 81 (1965), 211–264.

[Mu] H. Munkholm, DGA algebras as a Quillen model category, *J. Pure and Appl. Alg.* 13 (1978), 221–232.

[MW] C.A. McGibbon and C. Wilkerson, Loop space of finite complexes at large primes, *Proc. Amer. Math. Soc.* 96 (1986), 698–702.

[N1] J.A. Neisendorfer, *Primary Homotopy Theory*, Memoirs of the Amer. Math. Soc. 25 No. 232 (1980).

[N2] J.A. Neisendorfer, The exponent of a Moore space, *Algebraic Topology and Algebraic K–Theory*, W. Browder, ed., Annals of Math. Study No. 113, Princeton Univ. Press, Princeton, N.J. (1987), 35–71.

[N3] J.A. Neisendorfer, Properties of certain H–spaces, *Quart. J. Math. Oxford* (2) 34 (1983), 201–209.

[N4] J.A. Neisendorfer, 3–Primary exponents, *Math. Proc. Camb. Phil. Soc.* 90 (1981), 63–83.

[NS] J.A. Neisendorfer and P. Selick, Some examples of spaces with or without exponents, *Can. Math. Soc. Conf. Proc.* 2, Part 1 (1982), 343–357.

[Se] J.–P. Serre, Homologie singulière des espaces fibrés, *Annals of Math.* 54 (1951), 425–505.

[S1] P. Selick, Odd primary torsion in $\pi_k(S^3)$, *Topology* 17 (1978), 407–412.

[S2] P. Selick, On conjectures of Moore and Serre in the case of torsion-free suspensions, *Math. Proc. Camb. Phil. Soc.* 94 (1983), 53–60.

[Sm] L. Smith, Split extensions of Hopf algebras and semi–tensor products, *Math. Scand.* 26 (1970), 17–41.

[St] J.D. Stasheff, *H–spaces from a Homotopy Point of View*, Lecture Notes in Math. 161 (1970), Springer–Verlag.

[W] G.W. Whitehead, *Elements of Homotopy Theory*, Graduate Texts in Mathematics 61, Springer–Verlag (1978).

Milton Keynes UK
Ingram Content Group UK Ltd.
UKHW020022071024
449327UK00032B/2895

9 780367 450038